Environmental Biotechnology

Alan Scragg

Reader in Environmental Biotechnology, University of the West of England

UNIVERSITY PRESS

OXFORD
UNIVERSITY PRESS

Great Clarendon Street, Oxford OX2 6DP

Oxford University Press is a department of the University of Oxford.
It furthers the University's objective of excellence in research, scholarship,
and education by publishing worldwide in

Oxford New York

Auckland Bangkok Buenos Aires Cape Town Chennai
Dar es Salaam Delhi Hong Kong Istanbul Karachi Kolkata
Kuala Lumpur Madrid Melbourne Mexico City Mumbai Nairobi
São Paulo Shanghai Taipei Tokyo Toronto

Oxford is a registered trade mark of Oxford University Press
in the UK and in certain other countries

Published in the United States
by Oxford University Press Inc., New York

First published 2004

British Library Cataloguing in Publication Data
Data available

Library of Congress Cataloging in Publication Data
Data available

ISBN 0-19-926867-3

Typeset by Newgen Imaging Systems (P) Ltd., Chennai, India
Printed in Great Britain by
Ashford Colour Press, Gosport, Hampshire

Contents

Preface vii

Acknowledgments viii

1 Introduction 1

2 Microbiology 15

3 Environmental monitoring 73

4 Sewage treatment 112

5 Bioremediation 173

6 Biotechnology and sustainable technology 230

7 Biofuels 266

8 Natural resource recovery 320

9 Agricultural biotechnology 345

10 Biotechnology of the marine environment 378

11 Specific topics 401

References 419

Index 439

Preface

It has been five years since I produced a book on environmental biotechnology which covered the influence and application of biotechnology to many aspects of the environment. Biotechnology has advanced rapidly over the last five years with the development of areas such as genomics and proteomics. These recent developments have been applied to environmental topics and therefore a second book has been produced to incorporate the new information.

The application of the techniques of molecular biology to environmental studies has seen the discovery of many more microorganisms in the environment, especially in conditions thought to be too extreme to sustain life. Large and varied microbial populations have been found in hydrothermal vents, deep mines and ice sheets which were not detected using conventional techniques. The new molecular biology techniques can estimate species that are either non-growing or difficult to grow helping to discover new species and phylogenic relationships.

Industry and the widespread use of persistent chemicals have left a legacy of pollution and biological methods continue to offer an alternative to engineering and chemical methods of treating this pollution. Ideally it would be best not to produce the pollution in the first place and the development of sustainable and renewable sources of raw material and energy should go some way towards this. The development of alternative supplies of chemicals normally obtained from petrochemicals will be required as fossil fuels are depleted.

The development of alternatives to fossil fuels, which are renewable, sustainable and carbon dioxide neutral is off particular interest. Biofuels such as biomass in the form of grasses and coppiced willow, biogas from anaerobic digestion, biodiesel, bioethanol and bio-hydrogen are possible alternative energy sources.

Other subjects covered are the development of biosensors and bioindicators to monitor pollution, the application of genetic manipulation to agriculture, resource recovery all of which are influenced by biotechnology. The book is intended to cover the application of biotechnology to the environment without going into too much depth into the techniques involved. The further reading should lead the reader to further detail if required.

Acknowledgments

First I would like to thank my family for their patience during the preparation of the book and the proofreaders that have converted my stumbling efforts into something readable.

1 Introduction

1.1	Introduction	1
1.2	Environment	1
1.3	Biotechnology	2
1.4	Areas of application for biotechnology	4
1.5	Genetically manipulated organisms	9
1.6	Legislation	11
1.7	Conclusions	13
1.8	Further reading	14

1.1 Introduction

Environmental biotechnology is the application of biotechnology to all aspects of the environment. Biotechnology deals with the use of living organisms or their products and has been around for some time. Biotechnology has been recently invigorated by the development of genetic engineering, which now has applications in all areas of biotechnology. This book is concerned with how biotechnology can be applied to environmental problems and issues. This will cover existing pollution in terms of the detection and bioremediation of the contaminants. In addition, treatment is required for industrial, agricultural, and domestic process wastes, some of which cannot be eliminated. There is also a need to reduce the production of pollutants at source. One of the most important pollutants is carbon dioxide, which is a cause of global warming. The reduction in the production of industrial pollutants is known as **'clean technology'**. The chapters cover these subjects and also included is a chapter on microbiology because it is one of the core disciplines in biotechnology and genetic engineering.

1.2 Environment

It would appear that the public's awareness of the environment was only noticeable after a number of disastrous environmental accidents. Examples are the oil spills from the tankers *Amoco Cadiz*, off the coast of Brittany in 1978, and *Exxon Valdez*, in Alaskan water in 1989, the toxic gas release from a pesticide plant in Bophal, India, in 1984, and the nuclear power station accident at Chernobyl in 1986. These events achieved great publicity and increased the awareness of environmental issues. However, problems with the environment were recognized earlier, with Rachel Carson's book *Silent Spring* in 1963 dealing with the problems caused by pesticides. This book and others lead to a gradual change in the perception of the environment as unchanging and capable of dealing with any amount of pollution and contamination to one of the environment being a delicate balance of many components. Unfortunately not all the problems in the environment develop rapidly; often the changes are slow and can remain unnoticed until a critical stage has been reached. These types of problems are much more difficult to publicize and deal with. Spectacular environmental accidents do sometimes have positive outcomes. One of the most publicized environmental catastrophes was the grounding of the supertanker *Exxon Valdez* on the Bligh Reef in Prince William Sound, Alaska, in 1989. The tanker released 35 500 tonnes (41.6 million litres) of crude oil, polluting 1700 km of coastline. The spill was subjected to intense media coverage because of its visible impact on this pristine and unspoiled area of Alaska. Pictures of oiled seabirds and otters caused such concern that the oil spill was directly responsible for the passing of the US Oil Pollution Act of 1990.

1.3 Biotechnology

Although the term 'biotechnology' has been in existence for a considerable time since it was first used in 1919, and again in 1938 (Kennedy, 1991), the term was only recognized much later. Probably the first recognition by the wider scientific community was with the publication of the *Journal of Biotechnology and Bioengineering* in 1962 and the *Journal of Biotechnology* in 1979. The general public may have first heard of biotechnology with the publication of the Spinks report in 1980 (*Biotechnology, Report of a Joint Working Party*). The working party was set up to study the industrial applications of biological knowledge and were to review the opportunities in biotechnology, to recommend action by the government and to prepare a report. In this report biotechnology was defined as "The application of biological organisms, systems or processes to manufacturing and service industries". More recently the European Federation of Biotechnology defined biotechnology as "The integrated use of biochemistry, microbiology and engineering sciences in order to achieve applications of the capabilities of microorganisms, cultured animal cells or plant cells or parts thereof in industry, agriculture, health care and in environmental processes" (European Federation of Biotechnology, 1988). The Organization for Economic Cooperation and Development (OECD) defined biotechnology as "the application of scientific and engineering principles to the processing of materials by biological agents to provide goods and services".

The term biotechnology may imply a degree of coherence as a separate subject or discipline but, in practice, it embraces a set of disparate but often interrelated technologies relevant to a broad range of industries and requiring a wide range of science and engineering expertise, often to produce a single development. The choice of subjects combined depends on the individual industrial or scientific problem. Thus biotechnology acts as an interface between scientific disciplines and their areas of application. The relationship between environmental biotechnology and biotechnology is shown in Fig. 1.1.

It was thought that biotechnology would create novel industries with low demands on fossil-fuel energy and the technology would affect industries involved with food and feed, animal feed, energy sources, waste recycling, pollution control, and medical and veterinary care (Houwink, 1989). In this report it was stated that biotechnology would "launch an industry as characteristic of the twenty first century as those based on Physics and Chemistry

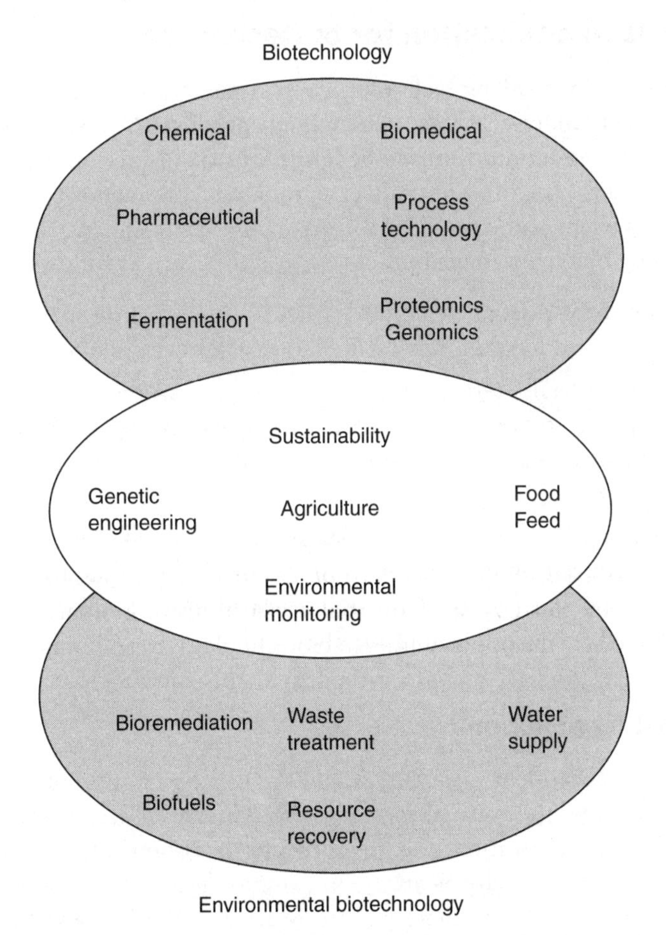

Figure 1.1 The link and crossover between biotechnology and environmental biotechnology.

have been for the twentieth century". Moreover it was to be an industry based in large part on renewable and recyclable materials and thus adapted to the needs of a society in which energy is expensive and scarce. Many of these predictions have come true and genetic manipulation has had a significant impact on almost all industries from food to forensics. Some predictions have not come true to date. For example, the appreciation that the supply of fossil fuels is limited and their use contributes to global warming has only just begun to be a national and international issue. Thus the development of biological energy supplies has only begun recently.

Despite the somewhat later start, environmental biotechnology — the application of biotechnology to environmental issues — has a predicted market in excess of US$300 million worldwide (Golub, 1997) and $84–94 billion dollars in Europe.

1.4 Areas of application for biotechnology

In contrast to medical biotechnology, environmental biotechnology has to deal with high volumes of low-value wastes, products, and services. A list of environmental pollutants that can be released into the environment is given in Table 1.1 and these can be very diverse, ranging from metals to gases. These pollutants present three main problems in the environment, which can be addressed by biotechnological processes, and these are as follows.

- The problem of existing pollution of land, water, and air. The pollution is often a legacy from old industries and industrial processes often long defunct.
- Current industrial, agricultural, and domestic processes still producing pollutants and to this can be added accidents and spills. Some of these pollutants, such as sewage, cannot be reduced or eliminated so that they will have to be treated.
- Industrial, agricultural, and transport sectors create pollutants which could be reduced by introducing a more sustainable, renewable source of raw materials and energy. The application of biotechnology to industrial processes is known as **'clean technology'**.

1.4.1 Existing pollution

Central to the treatment of pollution and the implementation of any environmental act is the ability to determine what chemicals are present or being released, their concentration, and their potential environmental risk. It is estimated that some 60 000 chemicals are in common use and hundreds are introduced each year. Many of these chemicals are released into the environment during processing, use, and accidental spills. These pollutants can be found in

Table 1.1 Environmental pollutants

Type of pollutant	Examples
Inorganic	
Metals	Cd, Hg, Ag, Co, Pb, Cu, Cr, Fe
Radionucleotides	
Nitrates, nitrites, phosphates	
Cyanides	
Asbestos	
Organic	
Biodegradable	Sewage, domestic agricultural, and process waste
Petrochemical	Oil, diesel, BTEX*
Synthetic	Pesticides, organohalogens, PAHs†
Biological	
Pathogens	Bacteria, viruses
Gaseous	
Gases	Sulphur dioxide, carbon dioxide, nitrous oxides, methane
Volatiles	CFCs‡, VOCs§
Particulates	

*BTEX, benzene, toluene, ethylbenzene, xylene.
†PAH, polyaromatic hydrocarbon.
‡CFC, chlorofluorocarbon.
§VOC, volatile organic compound.

air, water, or soil and can be metals or organic compounds not normally found in nature. Chemical techniques are often used to determine the concentration of pollutants in samples but **bioindicators** and **biomarkers** can be used to monitor pollution *in situ*. Bioindicators are whole organisms that are representative of their environment and whose population changes are used to estimate the effects of pollutants. Biomarkers are physiological, biochemical, and molecular characteristics of organisms removed from the environment and the characteristics chosen are those that will be affected by pollutants. Micro-organisms containing a green fluorescent protein under the control of a gene that responds to pollutants *in vivo* have been developed using recombinant gene technology. Genetic manipulation has also produced luminescent micro-organisms that can be used to test for toxicity in addition to tests like the Ames test. Molecular techniques can now be used to follow microbial populations *in situ* and from extreme habitats such as deep-sea hydrothermal vents and volcanic springs. **Biosensor** technology can be used to measure biological oxygen demand (BOD) and follow processes online and in real time (see Chapter 3).

Once the pollution has been detected and quantified, it will need to be removed and degraded, or degraded *in situ*. Micro-organisms provide an alternative to chemical methods in a process known as **bioremediation** (see Chapter 5). A specific problem is the release of heavy metals into the environment. Metals from a number of industrial processes can pollute air, water, and land. Metals are clearly not degraded by biological systems but a number of micro-organisms and plants can accumulate metals in a process known as **bioaccumulation** (see Chapter 5). The ability of biological material to accumulate metals is being put to use in the removal of heavy metals from water and contaminated sites.

Organic industrial waste such as that from food processing can be degraded biologically and treated both aerobically and anaerobically, in a similar way to domestic sewage. However, many of the synthetic chemicals produced by industry are not found in nature and are known as **xenobiotic,** from the Greek *xenos*, meaning new. The Environmental Protection Agency (EPA) and European Union (EU) list the chemicals which can cause problems if released into the environment. These are listed in Chapter 3 (Tables 3.3 and 3.5). The degradation of these synthetic chemicals is dependent on their structure, solubility, and toxicity. Often the lower the solubility of a compound in water the greater its solubility in cellular lipids and thus the greater the accumulation in the fatty tissues of the organism in a process known as **bioaccumulation** or **bioconcentration**. The relationship between the aqueous solubility of organic compounds and the bioconcentration factor is given in Fig. 1.2. As the compound becomes more water soluble the less it is accumulated. Therefore, these types of compounds have a greater toxic potential, as the bioaccumulation may be in an organism that is part of a food chain. In general organic compounds containing halogen atoms are slow to degrade, and the rate of degradation is influenced by the type of halogen atom, its position in the molecule, and the number of atoms present. In some cases it takes up to 15 years or more for these types of compound to be reduced by 50% and therefore, in essence, they are permanent and will accumulate in the environment. Some fungi and bacteria have been isolated that can degrade xenobiotics and although the degradation rate is slow, work is in progress to harness these organisms to treat waste streams and contaminated sites (see Chapter 5).

Another group of industrial wastes are the petrochemicals; oil, petrol, and diesel. Pollution of the environment with petrochemical wastes is often the result of disposal, tank leakage, and the more spectacular major marine spills. There are natural micro-organisms that occur widely in the environment that are capable of degrading petrochemicals. This is not unusual, as petrochemicals were organic material many millions of years ago. Biotechnology is developing ways of enhancing the petrochemical-degrading activity of micro-organisms either *in situ* or *ex situ* (off site).

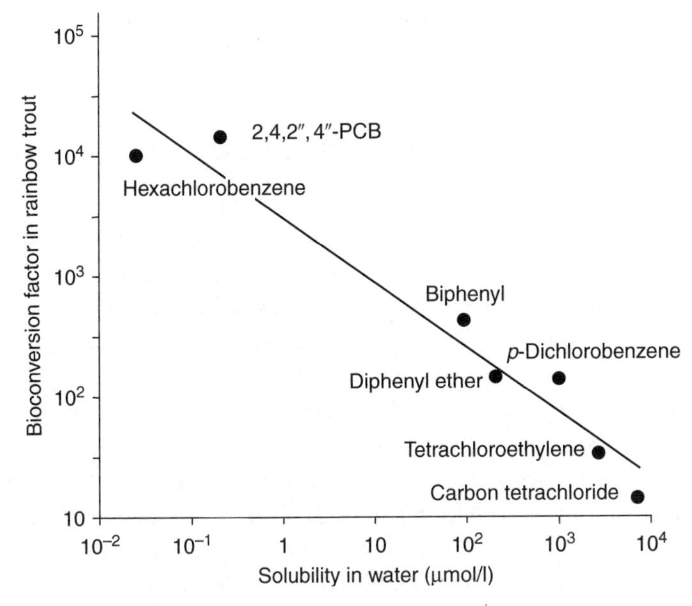

Figure 1.2 The relationship between the water solubility of organic compounds and their bioconcentration factor (Reeve, 1994). PCB, polychlorobiphenyl.

1.4.2 Bioremediation

Bioremediation is the biological treatment and removal of pollution from the environment. A long history of industrialization in developed countries has left a legacy of industrial pollution of land, water, and air. Industries such as petrochemical, smelting, mineral extraction, and gasworks, and other sources such as landfill have left a long list of polluted sites. Legislation will now require local authorities to list contaminated sites and perhaps oversee the cleaning up. It has been estimated that in the UK there are some 100 000 sites, which will take between £10 000 million and £20 000 million to clean up. The use of micro-organisms for the removal of pollution is not a new method, as sewage treatment using activated sludge has been with us since 1914. The principal organisms in bioremediation are bacteria and fungi that have the ability to degrade hydrocarbons such as oil and coal tar, and xenobiotics such as pesticides. Although metals cannot be degraded they can be accumulated by micro-organisms and therefore removed from the environment. Various bioremediation strategies have been developed (see Chapter 5) to treat polluted land and water, although there is only a limited knowledge of the processes involved. As bioremediation uses mixed populations of micro-organisms the dynamics of such populations are complex. In addition, the bioremediation process depends greatly on the quality and quantity of the pollution and is affected by other factors such as the presence of toxic agents, temperature, presence of nutrients, bioavailability of the compounds, and oxygen limitation.

1.4.3 Removal or reduction of pollution from current processes

Bioremediation can be applied to the wastes from various processes in an 'end-of-pipe' system. Biotechnology offers some of the most environmentally friendly solutions. Examples are the removal of metals and radionucleotides from waste streams, reed beds for the degradation of organic pollutants and the accumulation of metals, and the reduction of volatile organic compounds (VOCs) from waste gases by passing through a biofilter (see Chapter 5).

Gaseous pollution consists also of particulates, and gases like carbon dioxide, sulphur dioxide, and nitrous oxide produced by industrial processes. There are physical methods for trapping particulates and removing sulphur dioxide and nitrous oxide but under development are biofiltration units which are capable of removing VOCs, such as dichloromethane (see Chapter 5).

Burning of fossil fuels is responsible for most of the production of sulphur dioxide, nitric oxide, and carbon dioxide and the main source of these pollutants is power stations. Sulphur dioxide and nitric oxide released into the atmosphere will produce acid rain, which has the effect of reducing the pH of waters and soils, killing plants, and corroding buildings. Legislation now seeks to reduce the levels of sulphur and nitrous oxide emissions from power stations. A number of approaches have been made to reduce these:

- burn less fossil fuel, switch to gas or other energy sources;
- use low-sulphur coal or reduce the sulphur content by a biological process of desulphurization (biological; see Chapter 5);
- improve combustion; and
- flue gas desulphurization using an alkali such as limestone.

The land and the seas/oceans exchange carbon dioxide in a balanced manner. Plants fix carbon dioxide during photosynthesis while respiration and decomposition release carbon dioxide back to the atmosphere. The oceans act as a sink, by dissolving carbon dioxide and by algae in the ocean fixing carbon dioxide photosynthetically. The burning of fossil fuel releases large amounts of carbon dioxide which was fixed millions of years ago when the plants, which formed coal, were alive. It is the release of this carbon dioxide that is causing a significant increase in the amount of carbon dioxide in the atmosphere. The carbon dioxide is known as a **greenhouse gas** as it traps heat radiated from the Earth, so that an increase in carbon dioxide will cause **global warming**. An alternative to fossil fuels, other than fission or fusion, would be the use of biologically derived fuels which would result in no net gain in carbon dioxide as the carbon dioxide released was previously removed from the atmosphere during photosynthesis (carbon dioxide-neutral).

1.4.4 Prevention of pollution (clean technology)

Biotechnology can contribute towards the cleaner production of existing products. The drive towards greener processes, which are environmentally sustainable, is a preferred option to the clean up of the pollution once formed, but this approach has not received the attention it deserves. The use of enzymes or bacteria instead of chemical processes has the potential to reduce the use of feedstock and the energy required for the processes, as lower temperatures are used. In addition, the enzymes and bacteria themselves are also biodegradable. One example is the microbial removal of sulphur compounds from coal prior to combustion. Another is the use of fungi to pretreat logs before pulp and paper production, which reduces the use of energy and bleaching materials. The development of biodegradable plastics would be of considerable use in the reduction of solid waste, as much (20%) of the material deposited in landfill sites is plastic. Micro-organisms are capable of producing biopolymers from sustainable resources that can be used as plastics and these are biodegradable. At present the cost of biopolymers has restricted their application (see Chapter 6).

The demand for energy shows no evidence of declining and electricity generation uses mainly coal, oil, and gas in addition to nuclear and hydropower. Coal, oil, and gas supplies are limited, and all produce greenhouse gases on combustion and therefore alternatives are required to replace these fossil fuels and to reduce greenhouse gases. Biological alternatives to fossil fuels are the combustion of rapidly growing plants (willow, *Miscanthus*) for electricity generation, plant oil (rapeseed oil), and ethanol as liquid fuel (**biofuels**) for transport (see Chapter 7). Ethanol can be used in petrol engines with little modification to the engine and ethanol can be produced from sugar and starch by fermentation. Plant oil can be modified so that it can be used in diesel engines. Both fuels represent a sustainable, renewable, and carbon dioxide-neutral alternative to fossil fuels.

To combine the protection of the environment with the operation of an industrial process is known as **integrated pollution control**. The aims of integrated pollution control are to prevent, minimize, or render harmless releases of prescribed substances using the best available technology not entailing excessive cost ('BATNEEC'). The concept was that pollutants could transfer from air to land and to water, using complex routes. Therefore, any civil and industrial development should consider both public and expert advice as to the effect on the environment as a whole.

1.5 Genetically manipulated organisms

The last era recognized by Houwink (1989) was the 'new biotechnologies', which are principally involved with the application of genetic engineering to all areas of biotechnology including environmental biotechnology. Genetic engineering or recombinant DNA technology was developed in the 1970s and the techniques have revolutionized our ability to isolate, manipulate, and express genes and therefore proteins. The ability to isolate a particular gene, to multiply it if required, and to insert it into another organism means that traditional species barriers can be crossed in ways not possible by traditional breeding. An example is the production of human proteins in bacteria, animal cells, and plants. The ability to manipulate and transfer genes has had a dramatic impact on all areas of biotechnology. The transformed organism can be animal, plant, or micro-organism and the genetic engineering, irrespective of the organism, can be of three main types.

- The insertion of a single gene which gives the recipient a new characteristic, such as herbicide resistance in cotton plants, or starch degradation (amylase) in *Saccharomyces cerevisiae.*
- The alteration of the operation of existing genes which may change the characteristics of the recipient, for example the change in fruit ripening by reducing the activity of polygalacturonase by antisense technology in tomato plants (Flavr Savr™) or changes in oil quality in rapeseed.
- The insertion of a gene so that the recipient produces a specific product — usually a protein — which is not designed to alter the characteristics of the organism; instead, the recipient acts as a supply of the product. This is often known as **biopharming** and an example is the production of human insulin in the bacterium *Escherichia coli.* The recipient can be bacterial, yeast, insect, plant, or mammalian cultures depending on the product required and the posttranslational processing.

The first medical transgenic product was produced in 1982 and was human insulin (humulin) where the gene was cloned into the bacterium *E. coli* and the bacterium used to produce insulin for the treatment of diabetes. Human growth hormone is now produced in bacteria and although the market for this hormone is limited the hormone was previously extracted from human pituitary glands and any extraction from human material carries a risk of viral contamination. Thus many of the transgenic products that have been developed are for medicine, providing material which was difficult to extract, carries other risks, or was very expensive. In contrast, agriculture has seen the genetic engineering of common crops such as maize, cotton, and tomato. In these cases the transgenic plants have been given characteristics such as herbicide and insect resistance or changes to fruit quality. The first genetically engineered crop was introduced in 1994 and was the Flavr Savr™ tomato. This tomato has a prolonged shelf life due to inhibition of any enzyme which delays ripening. Since that time the adoption of genetically modified (GM) crops has increased continuously, with the main crops being soybean, maize (corn), rapeseed (canola) cotton, potato, and tomato. The characteristics transferred have been principally herbicide resistance, insect resistance, and delayed ripening. It is clear that there are tremendous advances to be made with transgenic organisms (**genetically manipulated organisms**, GMOs) in all areas of biotechnology including environmental biotechnology.

Despite the successful introduction into the UK of GM tomato puree by Zeneca in 1996–97 there has been as backlash against GM crops and food (Moses, 2002; Lassen et al., 2002). This is despite the safety evaluation of foods for GM crops laid out by organizations like the OECD, which introduced the Principle of Substantial Equivalence (Kok and Kuiper, 2003). There is public concern over GMOs in food production as the public perceives the development of genetically engineered plants is being mainly for profit, and believes that the crops and animals are unnatural and carry unknown risks, and that transgenic organisms or products should be labelled. The introduction of GM food also has occurred at an unfortunate time with the worries about bovine spongiform encephalopathy (BSE), food-contamination scares, and foot and mouth, which means that GM foods are regarded in the same light. It may be that GM crops and GM foods have become the scapegoat for other issues like globalization, the power of multinational companies, US dominance, ethical values, etc. (Braun, 2002).

It has been stated that "genetic engineering is not some minor biotechnology development. It is a radical new technology that violates fundamental laws of nature" (Kareiva and Stark, 1994). Environmental newsletters have had titles such as "Are you ready for frankenfood?" (Kareiva and Stark, 1994). All this publicity has seen sales of genetically engineered tomatoes staying at low levels in the USA (Golub, 1997). This resistance may be overcome if the technology seeks to solve the perceived problems rather than being introduced just for profit. As a consequence some supermarkets in the UK have dropped any GM foods or have labelled products as not containing genetically engineered ingredients. Companies involved in the use of transgenic plants have run a series of explanations of biotechnology in newspapers in the UK, including a web site (www.monsanto.co.uk) with links to the web site for Friends of the Earth (www.foe.co.uk). In contrast, the application of genetic modification to the medical and screening sector remains popular.

1.6 Legislation

There has been legislation concerning the environment in terms of public health in the UK since 1848. The Public Health Act (1848) and many of the subsequent acts produced were in response to continued industrialization and were related mainly to sanitation and housing. The response to present concerns about the environment in the UK initiated the Environmental Protection Act (1990) (Table 1.2), which saw the setting up of the Environment Agency for England and Wales and the Scottish Environment Protection Agency in the 1995 Environmental Act. These two acts define the environment as "consisting of the air, water and land and the medium of air includes the air within buildings and the air within other natural or man-made structures above or below ground" and pollution as "the release into any environmental medium from any process of substances which are capable of causing harm to man or any other living organisms supported by the environment". In the USA acts dealing with clean air, clean water, and hazardous waste were passed between 1977 and 1982 and in 1994 the Maritime Pollution Treaty was signed.

However, pollution is not constrained within individual countries and therefore many attempts at cleaning up the environment have had to be carried out on an international scale. One of the most public international meetings was the United Nations Conference on Environment and Development (UNCED), held in Rio de Janeiro in 1992, which resulted in conventions on biodiversity and climate change; however, these have achieved little in real terms to date.

The Rio Declaration also included a number of principles for future action, which are:

- access to environmental information;
- that the precautionary principle should be widely applied, where national authorities should adopt the policy that the polluter pays.

Conventions are really international treaties under international law, which may also contain subsidiary agreements called protocols. The convention on climate change asked developed countries to submit plans to reduce the production of greenhouse gasses. The Montreal Protocol agreed in 1987 covered the production and use of chlorofluorocarbons (CFCs) which were reducing the ozone layer around the world. There have been a number of meetings since Rio de Janeiro, including at Kyoto in 1997, which have ratified the previous agreements and 'polluter pays' has been adopted in some countries but otherwise little progress has been made in these meetings on greenhouse gas emissions. The World Summit on Sustainable Development in Johannesburg in 2002 made commitments to water and sanitation, energy, health (air pollution), agriculture, and biodiversity. The hope is to reduce natural resource degradation, restore fisheries, introduce ozone friendly chemicals, and to stop marine pollution. There was another UNCED in South Korea in 2004.

In the Treaty of Maastricht (1991) obligations were added to the original Treaty of Rome (1957) which included to preserve, protect, and improve the quality of the environment, contribute to the protection of the health of individuals, the prudent and rational use of natural resources, and to promote at an international level methods to deal with environmental problems. These activities have seen the adoption of a number of acts in various countries and the setting up of agencies to monitor specific environments such as clean beaches.

Table 1.2 Legislation and summits on environmental biotechnology

Date	UK and International
1848	Public Health Act
1957	Treaty of Rome
1969	USA National Environmental Policy Act (NEPA)
1970	Environmental Protection Agency (EPA) established
1970	USA Clean Air Act (CAA)
1972	Stockholm Declaration established United Nations (UN) Environmental Programme
1973	USA Endangered Species Act (ESA)
1974	UK Control of Pollution Act
1977	USA Clean Water Act
1987	Montreal Protocol (UN) elimination of CFC* use
1990	USA Oil Pollution Act
1990	UK Environmental Protection Act
1990	USA Pollution Prevention Act
1991	UK Water Resources Act
1991	Treaty of Maastricht (European Union)
1991	UK Water Industries Act
1992	UN Conference on Environment and Development (UNCED), Rio de Janeiro
1994	UN Maritime Pollution Treaty (MARPOL)
1995	UK Environment Act
1996	USA Food Quality Protection Act
1997	UN Conference on Environment and Development, Kyoto
2002	UN World Summit on Sustainable Development, Johannesberg
2004	UN World Environmental Summit, South Korea

*CFC, chlorofluorocarbon.

1.7 Conclusions

Thus, in conclusion, environmental biotechnology will be applied in the following areas.

- Transgenic organisms will continue to be generated but with more emphasis on improved food quality rather than just agricultural advantages. Acceptance of GM food will depend on whether the public can see benefits to society rather than to multinational corporations. An example of this type of advantage is the golden rice which has vitamin A engineered into the grain.

- Bioremediation will continue to expand as a cheap, environmentally sustainable method for the clean up of contaminants.
- The monitoring of environmental contamination is essential.
- Biofuels and clean technology will become increasingly important as fuel stocks diminish and global warming threatens to change the climate in significant ways.

1.8 Further reading

Creuger, W. and Creuger, A. (1991) *Biotechnology – A Textbook of Industrial Microbiology.* Freeman, New York.

Glazer, A.N. and Nikaido, N. (1994) *Microbial Biotechnology.* Freeman, New York.

McEldowney, S., Hardman, D.J., and Waite, S. (1993) *Pollution: Ecology and Biotreatment.* Longman Scientifc & Technical, Harlow.

McEldowney, J.F. and McEldowney, S. (1996) *Environment and the Law.* Addison Wesley Longman, Harlow.

Moses, A. and Cape, R.A. (1992) *Biotechnology, the Science and Business.* Harwood Academic Press, London.

Moses, A. and Moses, S. (1995) *Exploiting Biotechnology.* Harwood Academic Publishers, Chur, Switzerland.

2 Microbiology

2.1	Introduction	15
2.2	Taxonomy	16
2.3	Microbial structure	20
2.4	Shape and size	22
2.5	Direct methods for the determination of cell number or mass	27
2.6	Methods for the analysis of the microbial population *in situ*	38
2.7	Recombinant DNA technology	40
2.8	Indirect methods for the determination of microbial numbers and activity	48
2.9	Medium and conditions	51
2.10	Metabolism	55
2.11	Growth and growth kinetics	61
2.12	Relationship to the environment	68
2.13	Identification of micro-organisms	69
2.14	Conclusions	71
2.15	Further reading	72

2.1 Introduction

Micro-organisms are those organisms that are less than 0.1 mm (0.004 inch) in size and are not visible to the eye, and include fungi, algae, protozoa, and bacteria. Micro-organisms can be either free-living or grow on some form of surface, and can occur as part of mixed communities. We tend to associate micro-organisms with disease (bugs) but micro-organisms can be found in all types of natural habitat, such as fresh and salt water, soil, on plants (**phytosphere**), and on and within animals. This chapter is an introduction to microbiology and micro-organisms and how they relate to the environment. Considerably more detail of microbiology is given in the microbiology books in the Further reading section at the end of the chapter. The shape, structure, and metabolism of micro-organisms are introduced, and the cultivation, enumeration, and growth kinetics of micro-organisms are discussed along with the formation of biofilms. The taxonomy of micro-organisms is illustrated with particular reference to the Archaea.

2.1.1 The development of microbiology

Microbiology probably started with the development of a microscope powerful enough to observe bacteria in 1675 by Antonie van Leeuwenhoek. Some of

BOX 2.1 **Landmarks in the development of microbiology**

Development	Discoverer	Date
Microscope	Van Leeuwenhoek	1632–1723
Cell structure	Hooke	1635–1703
Taxonomy	Linnaeus	1707–1778
Spontaneous generation refuted	Spallianzani	1729–1799
	Schulze	1832
	Schwann	1837
Micro-organisms in air	Pasteur	1822–1895
	Tyndall	1820–1893
Geochemical changes	Beijerinck	1851–1931
	Winogradsky	1856–1953
Endospore	Cohn	1877
Yeast causes fermentation	Caniard-Latour, Schwann, Kutzing, Pasteur	1837
Growth on solid medium	Koch, Cohn	1875
Petri dish	Petri, Koch, Esmarch, Frankland	1877
Causation of disease	Koch	1876
Activated sludge	–	1914
Production of citric acid	–	1920
Penicillin	Fleming, Chain, Florie	1940
Development of cloning	–	1974
First genetically manipulated product	–	1982
Small-subunit taxonomy	Woese	1990

the historical milestones in the development of microbiology are given in Box 2.1. It started with the realization that micro-organisms existed and that they were responsible for many processes. Techniques for purifying cultures lead to the use of micro-organisms in processes and the synthesis of a number of products. Microbial techniques greatly assisted with the development of molecular biology, which lead to the development of genetic engineering. Molecular biology techniques are now being applied to almost all aspects of environmental biotechnology.

2.2 Taxonomy

Micro-organisms, which by definition are small, include organisms such as bacteria, algae, fungi, and yeasts. It was early recognized that micro-organisms

BOX 2.2 **Classification of micro-organisms**

The sequence of taxonomic hierarchy runs from domain, kingdom, division, class, family, and genus to species. The genus/species levels are used to provide names for micro-organisms established by international committees; examples are *Escherichia coli* and *Saccharomyces cerevisiae*. The classification attempts to place micro-organisms into structural groups closely related to each other based on similarities in their characteristics (phenotype). The identification of micro-organisms follows the Linnaean system using dichotomous keys or diagnostic keys. The figure shows Whittaker's system of five kingdoms.

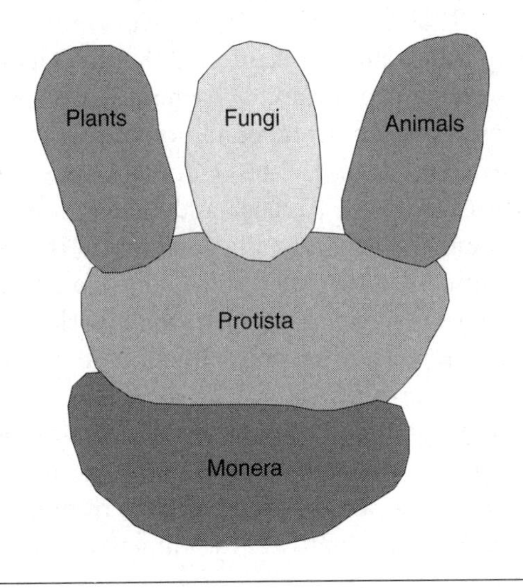

were different from plants and animals and they were therefore placed in a separate kingdom, the Protists. Later the number of kingdoms was increased to five in Whittaker's system with an extra kingdom for the fungi and the Protists divided into Protista and Monera (Box 2.2). The sequence of taxonomic hierarchy is domain, kingdom, division, class, family, genus, and species, based on similarities in characteristics (phenotype).

2.2.1 Ribosomal RNA-based classification and identification

Until recently the only way to describe and identify a micro-organism was to isolate it as a pure culture using techniques like the streak plate. These types of techniques rely on being able to grow the organism. Samples taken from the environment can be placed on a number of media and under a variety of conditions but there is no guarantee than all microbial types present will grow, and in some case a fast-growing strain will mask another. No growth under the conditions used to examine the sample is not necessarily an indication of non-viable cells, as these may just require specific conditions in which to grow. This may be particularly true for samples from extreme conditions.

However, in the last 10 years advances in genetic manipulation have allowed the sequence of small-subunit ribosomal RNA (rRNA) to be determined and used for identification and determination of the relationship between organisms (taxonomy). Woese et al. (1990), by analysing oligonucleotide sequences (8–15 nucleotides) for the 16 S ribosomal subunit, were able to identify common sequences. The 16 S value is a measure in Svedburg (S) units measured as the rate of sedimentation of the ribosome in a centrifuge. As the techniques for the analysis of sequences improved more 16 S sequences were examined and it was shown the 16 S rRNA from methanogens was very different from that of other bacteria. From these data it was inferred that the methanogens were not bacteria and they were named the archaebacteria. Later other archaebacteria were found in extreme environmental conditions and a three-domain classification was proposed (Fig. 2.1). The three-domain system better reflects the evolution of organisms, with the Archaea and Bacteria diverging very soon in evolutionary terms and the origin of life somewhere in the bacterial line of descent. The three-domain system also shows the evolution of higher organisms and shows that the diversity of the microbial world is very much greater than was first thought. It is clear that only a fraction of the microbial organisms have been detected or described.

Even now the biological framework is based on phenotypic data. The phylotype (rRNA) places the organism in an evolutionary framework and the phenotype provides a highly resolved description of the organism. DNA-based taxonomy should provide a framework for the accumulated phenotypic data.

Now, extraction of DNA from the sample and analysis of the 16 S rRNA sequences present can give the number of different organisms present, and their evolutionary relationships or **phylotype**. This method will represent all the organisms present without having to grow them. The sequences of the small-ribosomal-subunit 16 S or 18 S rRNAs have changed very slowly during evolution so that these sequences can be used to classify the organisms and indicate their evolutionary history.

Environmental samples can be analysed for rRNA genes and a survey of the phylotypes made. The simplest way is to extract the DNA from the sample, using the polymerase chain reaction (PCR) to amplify the rRNA using rRNA-specific primers. The amplified DNA can then be analysed by a number of methods.

In order to identify a micro-organism a pure culture is required and the normal method of producing this is the streak plate. However, techniques for the isolation of single microbial cells have been developed. Single cells can be isolated by micromanipulation, which has been improved recently. Micromanipulators have been used since the 1960s but mainly for the larger micro-organisms such as yeasts; bacteria were very difficult to manipulate due to their small size.

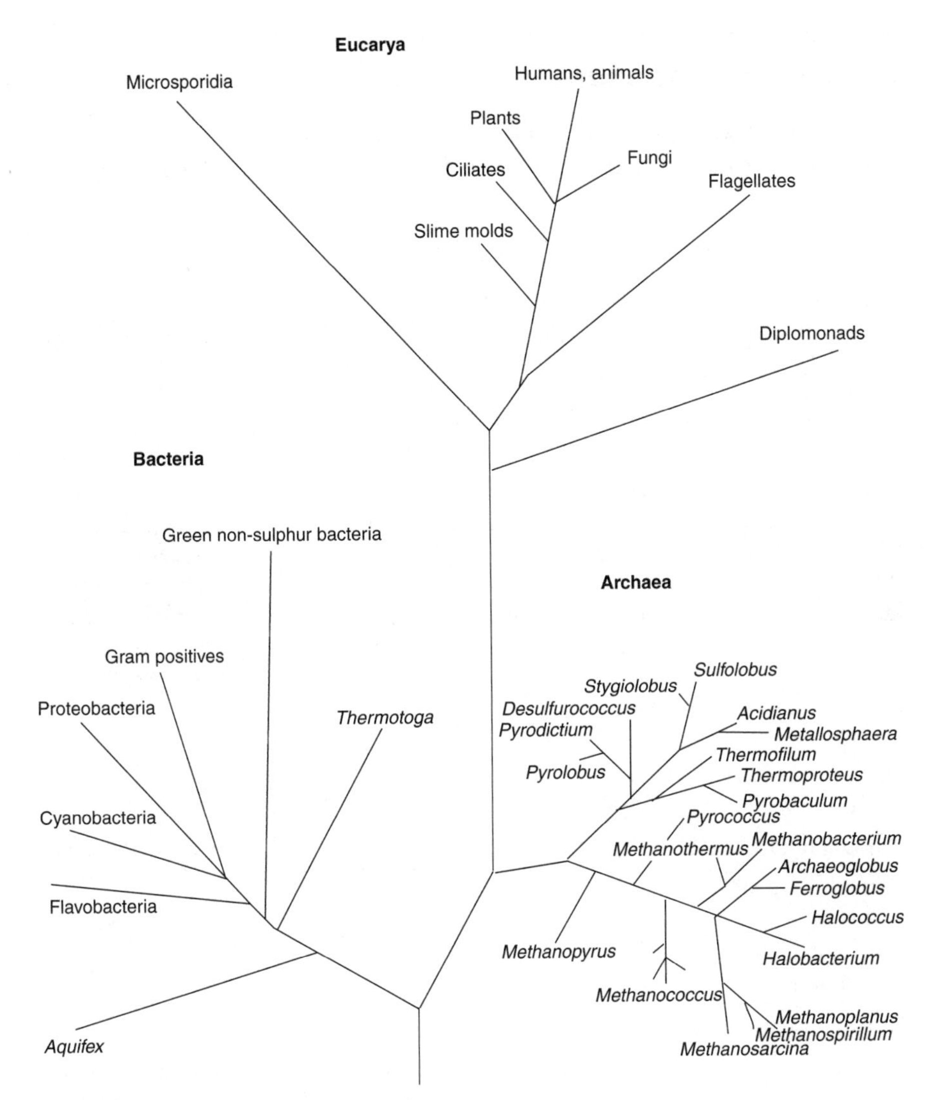

Figure 2.1 The evolutionary relationships between Archaea, Bacteria, and Eucarya (adapted from Hugenholtz and Pace, 1996).

Recently micromanipulators have been fitted with capillary tubes into which single cells can be aspirated (Frohlich and Konig, 2000). Another method is to use optical tweezers where a laser is focused by the microscope objective onto a capillary tube containing medium and cells. A single cell is fixed with the laser beam and separated from the other micro-organisms by moving the tube. The capillary tube is broken at a predetermined point and the single cell transferred to fresh medium. A similar technique has been used to isolate plant cells by tacking selected cells to an adhesive film using a laser beam (Kerk et al., 2003).

2.3 Microbial structure

Most living organisms are complicated and some are highly organized containing numerous structures. However, despite this they all share a number of features:

- they share a common cellular structure,
- all can self-replicate (grow and divide),
- they all perform certain common chemical activities, which are collectively known as metabolism, and
- they all contain DNA, RNA, and proteins.

Table 2.1 Differences between Archaea, Bacteria, and Eukarya

Characteristics	Archaea	Bacteria	Eukarya
Genome	Single, circular	Single, circular	More than one, linear chromosomes
Nuclear membrane (nucleus)	No	No	Yes
Histones bound to DNA	No	No	Yes
Amount of DNA/cell	5×10^{-15} g	5×10^{-15} g	$0.5–3.0 \times 10^{-12}$ g
Extrachromosomal elements	Yes	Yes	Yes
Chloroplasts and mitochondria	No	No	Yes
DNA in organelles	No	No	Yes
Ribosomes	70 S, diphtheria toxin-sensitive	70 S, diphtheria toxin-resistant	80 S, diphtheria toxin-sensitive
Ribosomes in organelles	No	No	70 S
Initiator tRNA*	Methionine	Formylmethionine	Methionine
Flagella (9 + 2 core)	No	No	Yes
Cell membrane	Ester linkage between glycerol and isoprene	Ester linkage between fatty acids and glycerol	Ester linkage between fatty acids and glycerol
Cell wall	Polysaccharide	Peptidoglycan	Polysaccharide
Spore formation	No	In some	No

Note: The ribosome size is given in Svedburg (S) units, which is a measure of how fast ribosomes sediment in a centrifuge.

*tRNA, transfer RNA.

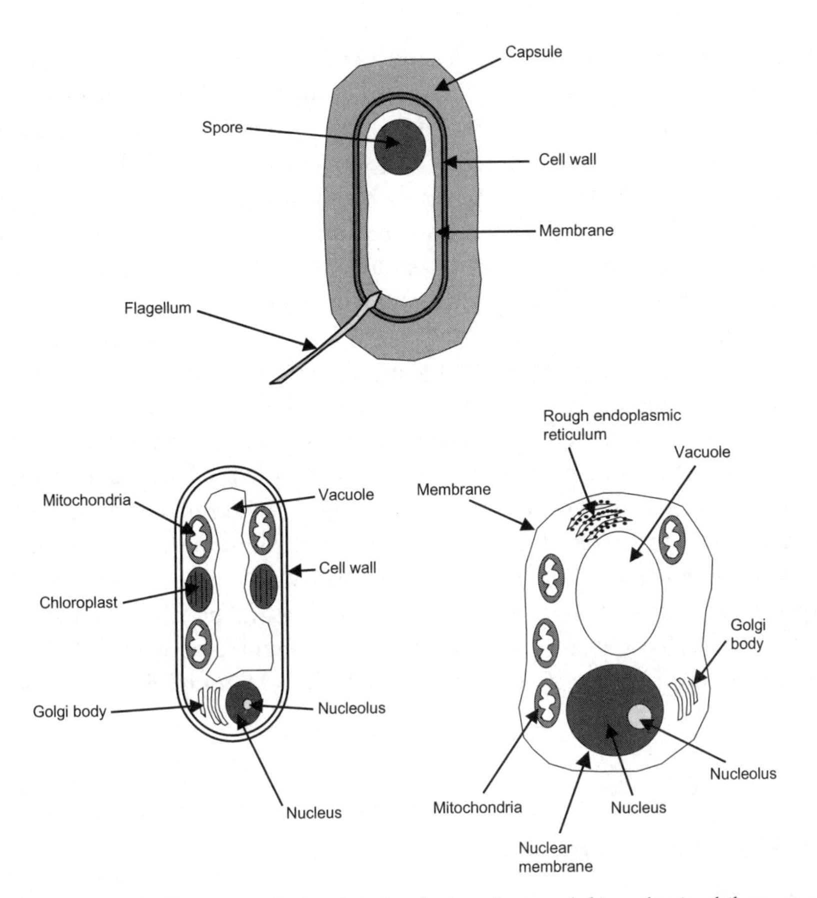

Figure 2.2 Typical cell structure for bacteria (top), plant (bottom left), and animal (bottom right) cells. Few structures can be observed in the bacterial cell except for spores, gas vacuoles, and polyhydroxybutyrate (PHB) granules. In this case a spore is shown. External to the cell wall there can be a capsule or slime layer and with special staining flagella can be observed. Within the rigid cellulose plant cell wall internal structures can be seen which include chloroplasts, mitochondria, the nucleus, and a large vacuole. The same structures apart from the chloroplasts can be seen in the animal cell. Other membranous structures are the rough endoplasmic reticulum and Golgi apparatus.

Although micro-organisms are small, on microscopic examination it was clear that they consist of two distinct cell types. These cell types are the prokaryotes (Archaea and Bacteria) and eukaryotes (Eukarya) and the main differences between these cell types are listed in Table 2.1. The prokaryotes are the smallest and simplest types of organism represented by the Bacteria and Archaea. The prokaryotes generally contain one DNA strand, whereas the eukaryotes have a number of **chromosomes**. Chromosomes are strands of DNA which are attached to proteins (histones). The chromosomes are retained within the nucleus bounded by a nuclear membrane (Fig. 2.2). In addition, the eukaryotes contain other membrane-enclosed organelles including **mitochondria** and **chloroplasts**. These two organelles contain DNA, separate from the nuclear

DNA, which codes for some of the constituents of the organelle. It has been suggested that these organelles originated from bacteria trapped in a larger bacterial cell. The chloroplasts are thought to originate from cyanobacteria and mitochondria from α-protobacteria (Simpson and Stern, 2002). The increase in cell size and complexity is accompanied by an increase in DNA content, and complexity is the main difference between prokaryotes and eukaryotes. Initially the eukaryote/prokaryote division was fundamentally a cytological difference, which is a meaningful cellular organization for eukaryotes, but for the prokaryotes it is really a lack of detail, a negative characteristic. An increased knowledge of the molecular biology of cells made it possible to define prokaryotes on the basis of shared molecular characteristics such as the size of the **ribosomes**; 70 S for prokaryotes and 80 S for eukaryotes. The ribosome is a specialized and complex structure that is responsible for protein synthesis. The bacterial ribosome is about 200 Å in diameter, has a molecular weight of 2.5 million Da and a sedimentation coefficient of 70 S. The 70 S bacterial ribosome can be dissociated into two subunits, 30 S and 50 S. The cytoplasmic ribosomes of eukaryotes are larger at 80 S and have subunits of 60 S and 40 S. The ribosomal subunits contain RNA that is again characteristic of the subunit size, with the 50 S subunit containing 23 S, the 30 S containing 16 S, the 60 S containing 26 S, and the 40 S containing 18 S.

2.4 Shape and size

2.4.1 Prokaryotes: Archaea and Bacteria

Prokaryotes occur in a wide range of shapes, sizes, and arrangements as can be seen in Fig. 2.3. The sizes range from 0.2 to 250 μm (μm means **micron** or micrometer, where 1 μm = 0.001 mm), with a normal range of 1–10 μm. However, an exception is *Epulopiscium fishelsoni*, the largest bacterium found so far at 600 μm in length, which is found in the gut of the surgeon fish in the Red Sea. The simplest shape is the single spherical cell or coccus, which can occur singly, in pairs, and in chains. Not all **cocci** are perfectly spherical and these can vary greatly in shape from spherical to almost square. The next common shape is the rod or **bacillus**, which can also occur singly, in pairs, or in chains. Less common are spiral shapes, the spirillum, and organisms that consist of thin threads. The thin threads are called **hyphae**, and the mass of hyphae is known as a **mycelium**. Typical of this group are the *Streptomyces*, which are found in the soil and often produce antibiotics, such as streptomycin. There are other shapes such as the stalked bacteria like the *Caulobacteria*, but these shapes are less common.

Under the light microscope, which can resolve objects down to about 0.2 μm, there are few structures visible within the bacterial cell. However, depending on the species and growth conditions, storage granules, gas vacuoles,

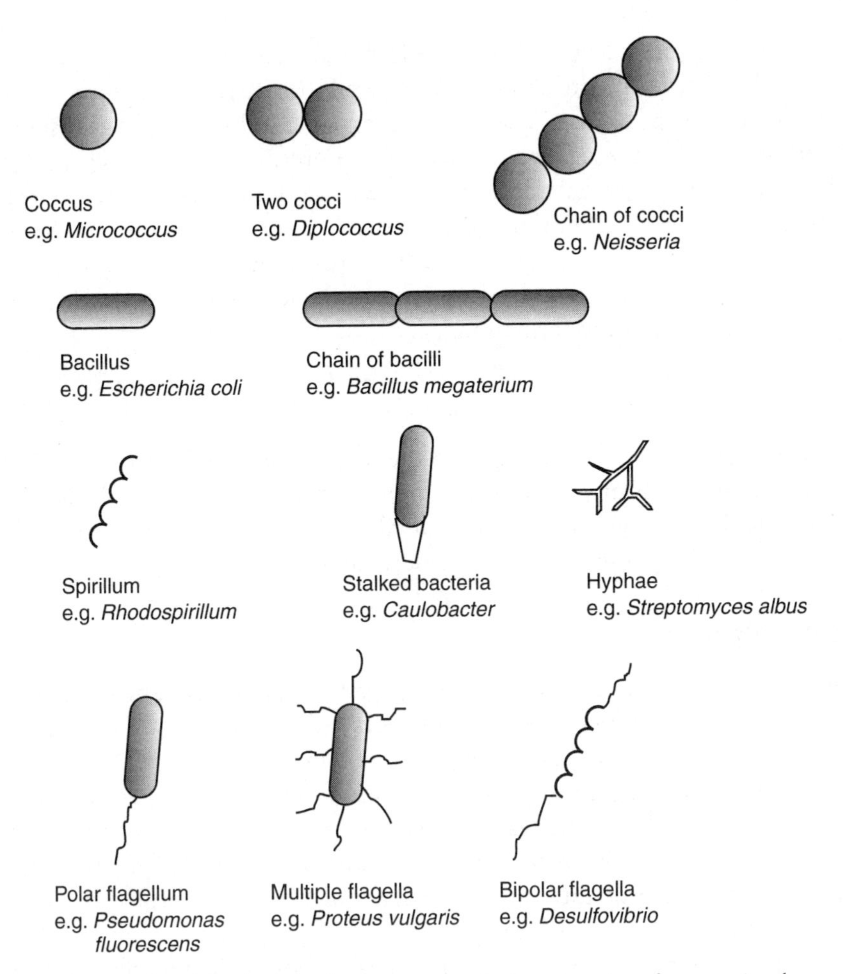

Figure 2.3 Examples of the shapes and arrangements of micro-organisms from cocci and bacillus and the arrangement of the flagella.

and spores may be visible. Outside the cell wall a capsule or slime layer can sometimes be seen (Fig. 2.2). These external structures are often visualized by adding a dye that will show up the capsule as an unstained halo. Capsules and slime layers consist mainly of polysaccharides, some of which have industrial uses. For example, the bacterium *Xanthomonas campestris* secretes a polysaccharide, xanthan, that is used in the food industry as a gelling agent, gel stabilizer, and thickener. Capsules have been assigned a number of functions; the prevention of desiccation, as a permeability barrier, prevention of infection, nutrient reserve, promotion of adhesion to surfaces, and increased pathogenicity. Slime layers are more diffuse than capsules but appear to be important in the adhesion of bacteria to surfaces.

Some bacteria are motile when observed under the microscope or as spreading colonies on solid surfaces. Motility is achieved by using one or more

flagella which are hair-like filaments extending from the cell wall. The flagella can be single, biopolar, or multiple. Flagella (5–20 μm × 0.02 μm) cannot be seen under the light microscope but can be seen by using special staining methods that increase their diameter to the resolution threshold (0.2 μm; Fig. 2.2). Other filaments do occur on bacteria, known as fimbriae and pili, but as these are 2–10 nm in diameter these cannot be seen under the light microscope. Pili are involved with DNA transfer between bacteria and both fimbriae and pili are involved in cell-to-cell adhesion and attachment to surfaces during biofilm formation.

2.4.2 Eukaryotes

The eukaryotic micro-organisms are the fungi, yeasts, and microalgae. Many of these organisms are of major economic importance; yeasts in the food and drink industry, fungi as plant and animal pathogens, and in spoilage, and the protozoa causing diseases such as malaria and sleeping sickness.

2.4.3 Algae

Algae are photosynthetic eukaryotes that can be both micro- and macroscopic and are generally found free-living in fresh and salt water. On land algae are found in the soil, on the surface of plants, and in a symbiotic relationship with fungi to form **lichens**. Most people have encountered algae as the macroscopic seaweeds that inhabit the seashore. However, as some 70% of the Earth is covered with water the microscopic unicellular unattached algae floating free in sea water are probably the most numerous plants and probably fix more carbon dioxide than land plants.

There are seven divisions or phyla of algae, based mainly on the presence of coloured secondary pigments other than the chlorophylls that give the algae their characteristic colour. Storage products and cell-wall structures are also used for identification. Most algae are very versatile in their metabolism. They are photosynthetic but they can utilize a number of organic compounds for growth. Algae generally have a rigid cell wall and show an immense variety in their cell shape from unicellular, to **colonial**, **filamentous**, **parenchymous**, and **coenocytic**. Under the light microscope the various organelles are visible (Fig. 2.2).

2.4.4 Fungi

Fungi can be both micro- and macroscopic, and we are familiar with the macroscopic fungi known as mushrooms. Fungi are most often filamentous organisms consisting of a branching structure (**hyphae**) forming a network known as a **mycelium** (Fig. 2.4). The fungal hyphae are generally branched and may have cross walls separating the hyphae into individual compartments, and growth occurs near the tip of each hypha (Fig. 2.5). Those hyphae without cross walls are known as **coenocytic**. The fungal cell wall contains

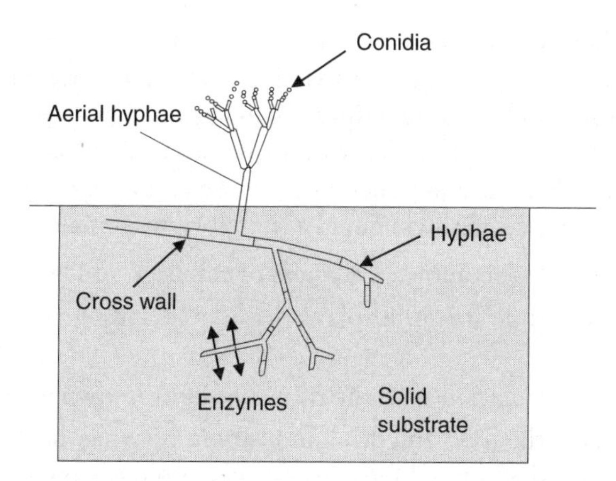

Figure 2.4 The typical structure of a fungus growing on and in a solidified substrate. The aerial production of asexual spores is shown and the colour and form of these is often characteristic of the particular species.

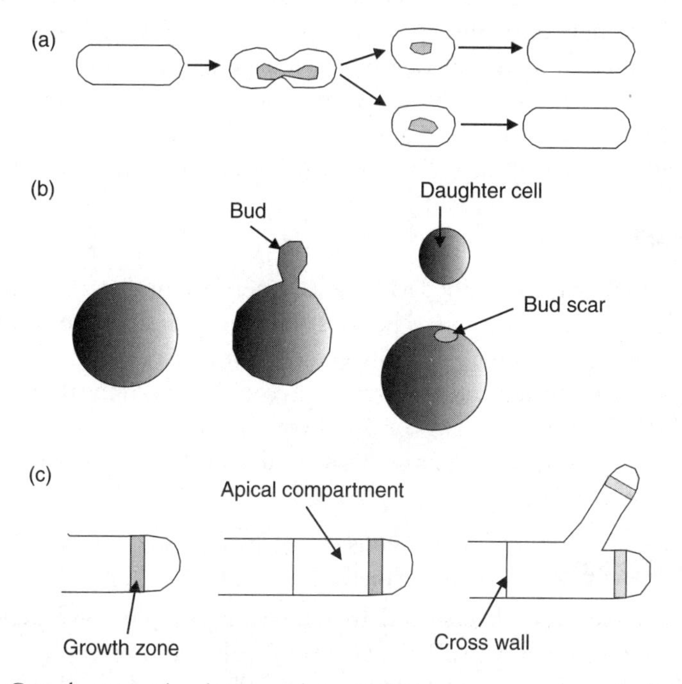

Figure 2.5 Growth patterns in micro-organisms. (a) Division into two identical cells or binary fission, which is the most common method of division. (b) Budding as found commonly in yeast. The bud on separation leaves a bud scar. (c) Fungal growth where there is a peripheral growth zone when growing on solid medium. Extension of the hypha occurs in a zone just behind the hyphal tip.

chitin rather than cellulose as in microalgae. The fungi are not photosynthetic; they derive their energy from either living material (**parasitic**) or dead organic material (**saprophytic**). Thus in nature fungi play an important part in decomposing dead material such as leaves and wood as they have the ability to decompose complex materials such as cellulose and lignin and can carry this out in relatively dry conditions. Fungi are important in many ways.

- They decompose the complex polymers of cellulose and lignin in soils.
- They are responsible for 5000 plant diseases including rusts, blights, and mildews.
- They form an association with the roots of plants known as mycorrhizae.
- They are used in the food and drink industry in brewing, baking, mushroom farming, and production of Quorn (single-cell protein, Marlow Foods).
- They are responsible for the production of many antibiotics such as penicillins and cephalosporins.
- A few can cause human disease.

Fungi are classified by their means of sexual reproduction, the presence of cross wall, chitin in the walls, and non-motile spores. Table 2.2 shows the four true fungal phyla and three phyla that resemble fungi. The fungi grow by elongation of the hyphae just behind the apex of the hyphae and therefore on a solid medium will form a spreading circular colony. In liquid medium the fungus will form a large mycelial mat or pellets, depending on conditions.

2.4.5 Yeast

Yeasts are technically classified as fungi (Deuteromycota) but in practice they are usually treated separately. The most common mode of yeast division is budding where a daughter cell is pinched off from the cell (Fig. 2.5). Buds can form all over the mother cell or in some cases budding is only at the end of the cell. Humans have used yeasts for 3000 years for the production of beer and wine. More recently yeasts have been used widely in genetic engineering, biotechnology, and genetics.

2.4.6 Protozoa

The protozoa are very diverse and are defined as non-photosynthetic, unicellular, and lacking cell walls. Protozoa are divided into four phyla based on their method of locomotion; no locomotion, **cilia, flagella**, and **pseudopodia**. Protozoa have a complex internal structure and many organisms are capable of ingesting particulate food (**phagotrophic**). This ability has allowed the protozoa to develop complex nutritional requirements that are at their greatest in the parasitic protozoa. Most protozoa reproduce asexually, often by simple division (**binary fission**) or more infrequently by multiple fission. Sexual reproduction involves the production of haploid gametes that can combine. The greatest areas of importance for the protozoa are as the cause of waterborne human diseases and as part of aerobic wastewater treatment.

Table 2.2 The four divisions or phyla of true fungi and three resembling fungi

Division or phyla	Characteristics
True fungi	
Zygomycota	Hyphae without cross walls (aseptate); asexual reproduction with non-motile spores in a sporangium; sexual reproduction by fusion of gametangia to form a zygospore, e.g. *Rhizopus* spp.
Ascomycota	Hyphae with cross walls, septa; asexual reproduction by asexual spores or conidia; sexual reproduction by fusion of modified hyphae or male spore with female trichogyne leading to the formation of an ascus containing ascospores, e.g. *Neurospora, Pencillium*.
Deuteromycota	Hyphae with cross walls or yeast forms; asexual spores, conidia formed but no sexual stage found; *Saccharomyces cerevisiae*.
Basidiomycota	Hyphae with cross walls or yeast forms; asexual reproduction rare; sexual reproduction by fusion of hyphae leading to the formation basidiospores in large fruiting body; mushrooms; *Cryptococcus*.
Organisms resembling fungi	
Oomycota	Hyphae without cross walls but contain cellulose rather than chitin; asexual reproduction by motile zoospores; sexual reproduction by fusion of male (antheridium) with female oogonium, producing thick-walled oospores; *Phytophthora infestans* (potato blight).
Chytridiomycota	Unicellular or in chains; forms rhizoids with the substrate; asexual reproduction by motile zoospores from a sporangium; Sexual reproduction by fusion of motile gametes.
Myxomycota	Some fungus-like features but with a cell wall-minus stage; slime moulds, which can form a fungus-like fruiting body; *Dictostelium discoideum*.

2.5 Direct methods for the determination of cell number or mass

The methods used to determine the number or mass of cells present in a sample depends greatly on the nature of the sample and the growth conditions being used. Microbial growth in a simple liquid medium enables a number of direct methods to be used to determine cell numbers or biomass concentrations. Samples from situations where non-microbial solids and cell aggregates are present, such as activated sludge, require more indirect methods to estimate cell numbers or biomass concentrations. A rapid, accurate method of

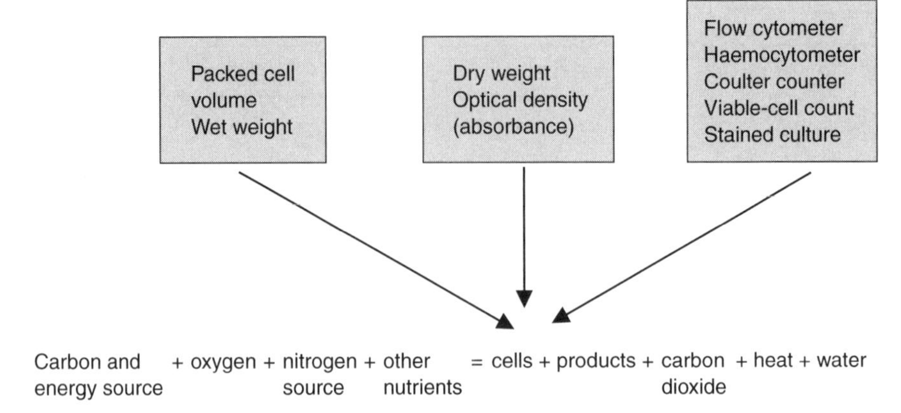

Figure 2.6 Direct methods of determining the mass or number of microbial cells.

estimating cell size and cell number is of great importance in environmental biotechnology. The direct methods are outlined in Fig. 2.6.

2.5.1 Dry weight

A popular direct method for determining the mass of the cells present is to weigh a sample taken from the culture. The cells can be separated from the medium by centrifugation (10 000 **g** × 15 min) using preweighed centrifuge tubes and the weight of biomass determined after drying the pellet in an oven. The disadvantage of this method is that it is slow and requires fairly large samples. An alternative is to filter a known volume of culture through a preweighed filter (0.2 μm pore size) and then dry in an oven or under an infrared lamp. Alternatively the filter can first be weighed wet, and then after drying, thus giving both the wet and dry weights of the biomass. The filter method is faster than centrifugation and can handle smaller samples, although wet-weight determination can be inaccurate.

2.5.2 Cell number

The number of cells in liquid culture can be determined directly by counting the cells. The simplest method is to use a counting chamber (**haemocytometer** or Petroff–Hausser chamber), which is a special microscope slide, etched with a grid (Fig. 2.7). The grids (normally two) are set 0.1 mm below the level of the slide in a trough cut in the slide. The grid is divided into 400 squares, each 1/400 mm^2, and as the depth is 0.1 mm the volume of each square is 0.00025 mm^3 or 2.5×10^{-7} ml. A small drop of the culture is placed in the trough and a glass coverslip placed over the trough. To ensure that the depth is correct the coverslip is pressed down until Newton rings (coloured rings) are observed. After 20 min to allow the cells to settle, the slide is placed under a microscope and the number of organisms per square is counted. To obtain a statistically meaningful result

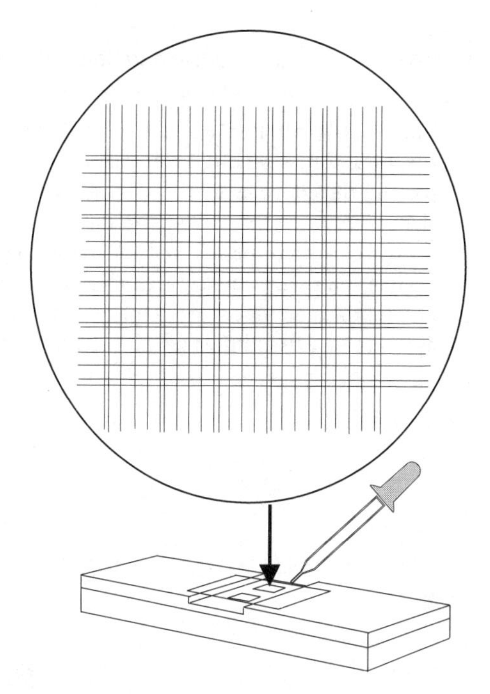

Figure 2.7 The haemocytometer slide or Petroff–Hausser chamber used for the counting of microbial cells. The inset shows the grid system at the base of the trough. Each small square is 1/400 mm^2 and the depth is 0.1 mm, which gives a volume of 2.5×10^{-7} ml. Thus a mean of 20 cells per small square will give a cell density of 8×10^7 cells/ml.

200–500 cells should be counted in total. The mean value per square is calculated and multiplied by 4×10^6 to give the number of cells/ml.

Fluorescent dyes (**fluorochromes**), such as 4′,6-diamidino-2-phenylindole (DAPI) and propidium iodide (DNA dyes) can be used to stain the cells prior to counting. The organisms are counted in an **epifluorescence** microscope where ultraviolet (UV) light is shone on to the specimen from above and any visible fluorescence observed in the microscope.

Another direct counting method is to filter a known volume of the sample through a sterile filter with a pore size of 0.2 μm. The micro-organisms collected on the filter can be either counted directly or stained before counting.

There are two mechanical methods of counting cell numbers, which are the **Coulter counter** and **flow cytometer**. In the electronic particle counter or Coulter counter the micro-organisms are drawn through a small aperture across which an electric field is applied (Box 2.3). A cell passing through the aperture causes a pulse in the field proportional to the cell's size. Thus by passing a fixed volume through the aperture, the number and size distribution of the cells can be determined. The disadvantage of this technique is that the aperture size has to be selected for the cell size expected and the sample has to be diluted sufficiently so that only one cell at a time passes through the aperture. Also, any aggregates in the sample will be counted as one cell or may clog the aperture. In addition, the diluting liquid has to be free of particles as these will give false readings or block the aperture.

The other mechanical method of cell counting is the flow cytometer (cell sorter). In flow cytometry the cell suspension is introduced to a stream of sheath

BOX 2.3 Coulter counter and flow cytometer

The Coulter counter draws a fixed volume of culture through an aperture of a fixed size across which an electric field is applied (see the first figure, below). A cell passing through the aperture causes a pulse in the field proportional to the cell size. Thus the number and size distribution of the cells can be determined. The aperture needs to be selected for the size of cells to be measured and the culture sample diluted in particle-free medium so that only one cell passes through the aperture at one time. Any cell aggregates or solids in the medium will block the aperture so that cultures of this type cannot be used.

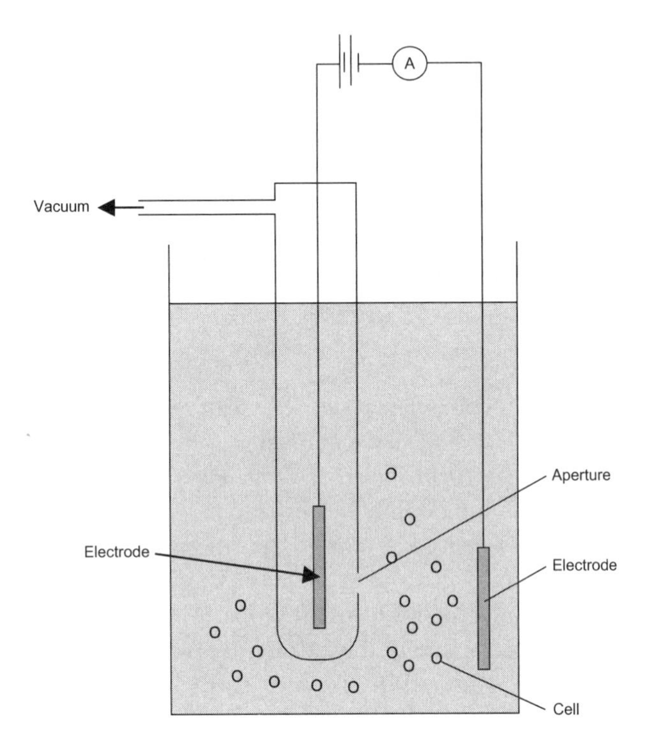

The flow cytometer was developed for animal cells in order to separate and count cells with different DNA contents (haploid, n; diploid, $2n$). A stream of liquid containing the cells is broken up into a series of droplets at a concentration that gives about one cell per droplet. These droplets are passed between a light source and a fluorescence detector (see the second figure, below). If the cells are stained with a flurochrome such as fluorescein or rhodamine, or the DNA with DAPI (4′,6-diamidino-2-phenylindole) they will fluoresce at a specific wavelength which can be detected. The table gives some of the stains available for flow cytometry. If required, the detector will electrostatically deflect the droplet and in this way cells can be sorted as well as counted.

BOX 2.3 ***(Continued)***

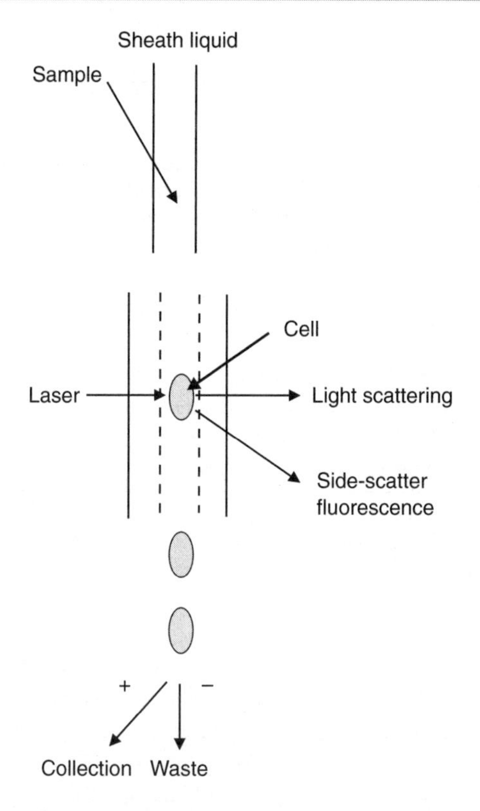

Stains for use with flow cytometry

Component stained	Stain
DNA	Ethidium bromide
	Propidium iodide
	DAPI*
	SYTO®
	Hexidium iodide
	PicoGreen
	YOYO series, based on benzothiazolium-4-pyridinium
	Acridine orange
RNA	Propidium iodide
	Acridine orange
	Pyronine Y
	Thiazole orange
Protein	FITC†
	Rodamine 101 isothiocyanate (Texas Red)

BOX 2.3 ***(Continued)***

(Continued)

Component stained	Stain
Enzyme	Substates linked to coloured and fluorescent compounds; naphthoyl, fluorescein, umbelliferyl, coumaryl, rhodamine
	Galactosidase
	Dehydrogenase, tetrazolium substrate
Antigen	Fluorescently labelled, FITC, TRITC‡
Lipid	Nile red
Ca^{2+}	Aequorin
Membrane	Rhodamine 123
	Oxonol
Carbohydrate polymer	Calcofluor white

*DAPI, 4′,6-diamidino-2-phenylindole.
†FITC, fluorescein isothiocyanate.
‡TRITC, tetramethylrhodamine isothiocyanate.

liquid at a slower flow rate, which ensures that drops are formed that contain only one cell or no cells at all. The drops pass a measuring window where a light or laser is shone into the drops, the light is scattered, and the fluorescence of the sample measured (Box 2.3). These data can be used to estimate the number of cells, measure their DNA content, and sort the cells. The flow cytometer is rapid, sorting 10 000 cells/s at a volume of 10–100 l/min. The incorporation of a range of stains such as propidium iodide, calcofluor white, and fluorescently labelled antibodies can be used to select cells, determine their viability, and identify micro-organisms (Davey and Kell, 1996). Many of the stains, such as propidium iodide, are specific for DNA and RNA (Box 2.3).

Antibodies against specific proteins can be made fluorescent by reacting with fluorescein isothiocyanate (FITC). The flow cytometer was initially developed for animal cells and has been used for plant cell suspensions. Sorting bacteria is more difficult as they are considerably smaller, not always spherical, can contain cell aggregates, and autofluorescence can cause problems with this type of cell counting. Other disadvantages are that flow cytometers are complex and expensive machines. However, flow cytometry has been used to determine the number of cells in aqueous samples (Hoefel et al., 2003) and bacterioplankton in tropical marine samples (Andrade et al., 2003).

2.5.3 **Viable cell count**

One of the main problems with all of the direct counting methods is that they cannot distinguish between viable and non-viable cells. The most commonly used technique for the determination of viable cells is the plate count. Here the sample is serially diluted in a sterile medium (often 0.9 M NaCl) and 0.05–0.2 ml of the diluted suspension spread on to a solid nutrient medium in a Petri dish (Fig. 2.8). The plates are incubated for 24–72 h at the appropriate temperature. If diluted sufficiently each colony that appears on the medium arises from a single viable cell. If the number of colonies formed is counted and the dilution taken into consideration, the number of viable cells can be found. Here again the number of colonies counted has to be sufficient to give a statistically sound result. The percentage viability can be determined by comparing the total cell count with the viable count. The number of cells that grow and form

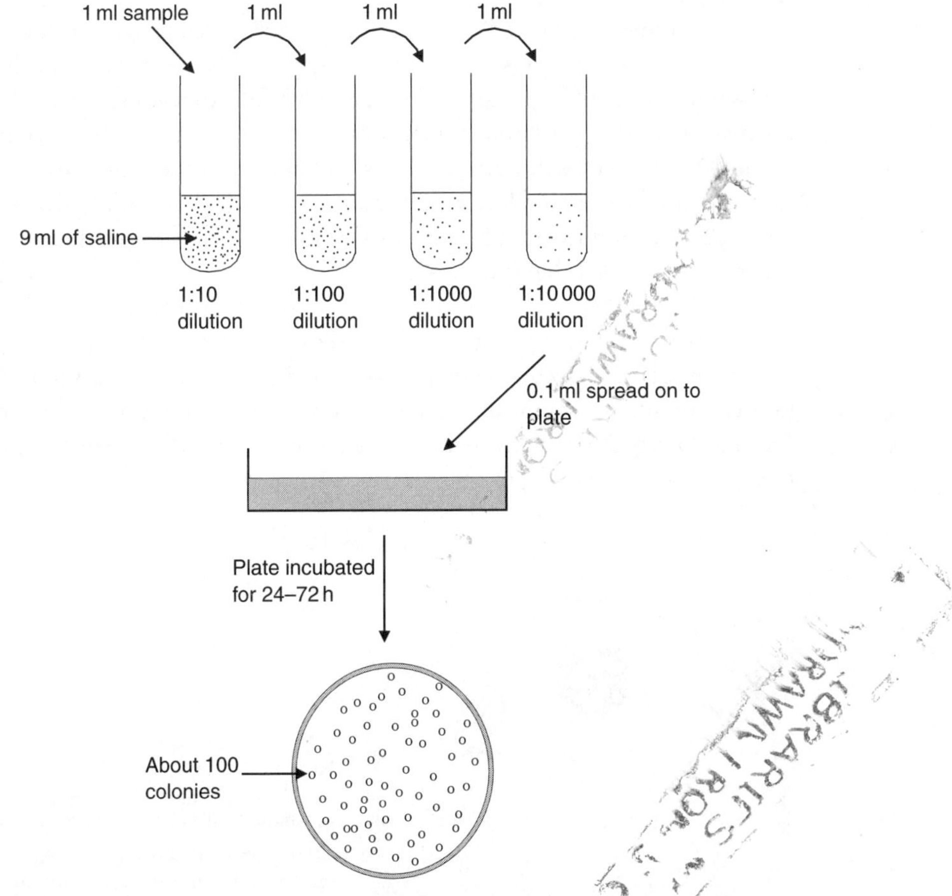

Figure 2.8 The process for the estimation of viable cells in a sample. The sample is serially diluted in order to obtain about 100 cells in 0.1 ml, which can be spread on to the agar plate. The serial dilution is carried out by placing 1 ml into 9 ml of sterile 0.9% saline, mixing well, and passing on 1 ml to the next step. If the initial cell concentration was 1×10^8 cells/ml then five steps will give a concentration of 1000 cells/ml and 100 cells/0.1 ml. Assuming 100% viability spreading 0.1 ml on the plate will give 100 colonies; 0.1 ml of the diluted culture is dropped on to the agar plate and the liquid spread across the plate using a sterile glass spreader. The plates are incubated at the appropriate conditions and the number of colonies counted after incubation.

colonies will depend on using the correct medium and growth conditions, and there may be cells in the sample that are viable but non-culturable.

Variations on the plate technique are the pour-plate method and the Miles and Misra method. In the pour-plate method the liquid sample is mixed with molten agar (at 42–45°C; low enough not to kill the cells) and poured into a Petri dish where colonies develop within and on the plate. The Miles and Misra method (Black, 2002) involves dropping a known volume (generally 0.05 ml) of the serial dilutions of the sample around the circumference of a plate (Fig. 2.9). The plate is not spread but each drop will spread out to some extent on the surface as the medium soaks into the agar, and colonies will develop within this area. In this way the number of plates required is considerably reduced.

If the sample is expected to contain few micro-organisms, too few cells to count accurately, the **most probable number** (MPN) technique can be used. The sample is diluted progressively using 10-fold dilutions. These dilutions are placed in tubes containing medium and a small inverted tube. If viable cells are present growth will occur and gas will collect in the tube. Normally five tubes are used for each dilution. The number of cells in the original sample can be estimated from the number of positive tubes at the various dilutions using the most probable table (Box 2.4).

With dilute samples a large volume of liquid or gas can be filtered through a sterile 0.2 μm filter and the filter placed on the surface of a suitable medium in a Petri dish. Colonies will develop on the surface of the filter and the number of micro-organisms can be determined in relation to the volume filtered. Often these filters will have been marked with a grid to help with the colony count.

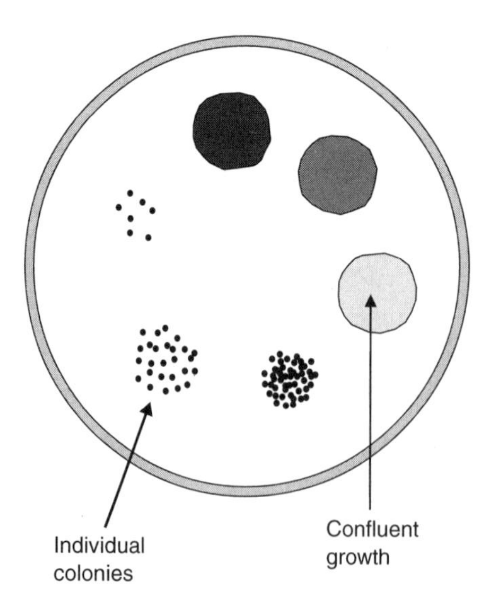

Figure 2.9 The Miles and Misra variation on the viable count. In this case the 0.5 ml of the diluted culture is dropped around the circumference of a plate. The drops are not mechanically spread out but just allowed to spread out themselves and soak into the agar medium. After incubation at the optimum conditions the pattern of growth obtained will be as shown, where six serial dilutions have been added to the plate. In the more concentrated samples the growth is confluent but at the higher dilutions individual colonies can be seen.

BOX 2.4 Most probable number (MPN)

A typical MPN determination will consist of five tubes for each dilution of the culture, usually 10 ml, 1 ml, and 0.1 ml. After incubation those tubes in which there are viable cells will have a cloudy appearance and produce gas that is collected in a small inverted tube. The tubes containing gas is scored as positive. Using the number of positive tubes and the dilution the MPN can be estimated from the table. The table is based on the statistical probability and gives the number of organisms in the original culture.

MPN index, in cells per 100 ml, when five tubes are used with 10 ml, 1 ml, and 0.1 ml dilutions

Dilution				Dilution			
10 ml	1 ml	0.1 ml	MPN index	10 ml	1 ml	0.1 ml	MPN index
0	0	0	<2	4	3	1	33
0	0	1	2	4	4	0	34
0	1	0	2	5	0	0	23
0	2	0	4	5	0	1	30
1	0	0	2	5	0	2	40
1	0	1	4	5	1	0	30
1	1	0	4	5	1	1	50
1	1	1	6	5	1	2	60
1	2	0	6	5	2	0	50
2	0	0	4	5	2	1	70
2	0	1	7	5	2	2	90
2	1	0	7	5	3	0	80
2	1	1	9	5	3	1	110
2	2	0	9	5	3	2	140
2	3	0	12	5	3	3	170
3	0	0	8	5	4	0	130
3	0	1	11	5	4	1	170
3	1	0	11	5	4	2	220
3	1	1	14	5	4	3	280
3	2	0	14	5	4	4	350
3	2	1	17	5	5	0	240
4	0	0	13	5	5	1	300
4	0	1	17	5	5	2	500
4	1	0	17	5	5	3	900
4	1	1	21	5	5	4	1600
4	1	2	26	5	5	5	>1600
4	2	0	22				
4	2	1	26				
4	3	0	27				

There are a number of dyes that can be used to distinguish between dead and living cells both under the microscope and in the flow cytometer (Singleton, 1999; Vives-Rego et al., 2000). The use of these is based on the following parameters.

- Exclusion of dyes by the cell membrane (*Bac*Light™, Evans Blue, Trypan Blue). Live cells have an intact membrane.
- Enzyme activity is characteristic of live cells. Dyes that can cross the cell membrane are converted by internal esterase activity into a compound that is coloured or fluorescent. If the cells are dead the membrane will not be intact so that any coloured product formed will diffuse out of the cells. In some cases extracellular enzyme activity has been used as a marker for viability.
- Addition of antibiotics that will kill or distort any viable cells while leaving dead cells untouched (SimPlate™).

The LIVE *Bac*Light™ stain kit can distinguish between Gram-negative and Gram-positive bacteria using a live culture. Actively growing cells are treated with two dyes, SYTO®, which stains DNA in Gram-negative cells, and hexidium iodide, which stains DNA in Gram-positive cells. Under the epifluorescence microscope hexidium iodide gives an orange colour and SYTO® a green colour. Using a similar technique living and dead cells can be distinguished. In this case two dyes, SYTO® and propidium iodide, are used. The SYTO® dye will stain the DNA in living and dead cells providing they are Gram negative, and propidium iodide will only enter dead cells because of their damaged membranes. The propidium iodide is red under blue-light fluorescence, so the dead cells are red and live cells are green.

Dyes like Evans Blue and Trypan Blue only enter cells with damaged membranes, and have also been used with a number of microbial eukaryotes such as algae and fungi to determine viability.

The dyes based on the internal esterase activity of viable cells cleave the dye into a coloured product that can be seen under the epifluorescence microscope. Examples are fluorescein diacetate (FDA) and carboxyfluorescein acetate (CFA), which are converted to fluorescein by esterase enzymes that will be retained by live cells but will leach out of dead cells. Under an epifluorescence microscope viable cells will appear bright green and dead cells colourless. Other similar stains are ChemChromeB, calcien acetomethyl ester (Calcien-AM), and tetrazolium reduction (α-5-cyano-2,3-ditolyltetrazolium chloride).

Direct plate assays incorporating substrates that can be converted into a coloured product have been developed. The SimPlate™ incorporates a compound that on hydrolysis releases 4-methylumbelliferone, which can be seen under UV light (365 nm).

Another direct counting method uses a cocktail of antibiotics (nalidixic acid, piromidic acid, pipemidic acid, cephalexin, and ciprofloxacin) to kill any viable cells that grow. The difference between the number of normal cells on the antibiotic-containing plates and the control gives the number of viable cells. Soil samples stained with DAPI (a DNA stain) after collection on filters did suffer from interference but this was improved with aluminium oxide filters and another DNA stain SYBR Green 1 (Weinbauer et al., 1998).

The growth techniques used for differentiating between living and dead cells are an important parameter with environmental samples but should be regarded with caution. A cell that does not grow and produce a colony could be regarded as dead, but it may be simply that it has been incubated on the wrong medium or under unsuitable conditions. There are no universal media and conditions suitable for all organisms. This would appear to be particularly true for environmental samples, as they contain such a wide range of micro-organisms. Recent DNA-based analysis has revealed the presence of many more and varied organisms than those detected by growth analysis. It has been estimated that 90–99% of the environmental species were not culturable (Macnaughton et al., 1999). Therefore, the growth characteristics of the organisms of interest need to be known before their numbers can be estimated accurately.

An example of the difficulties of estimating microbial populations which use specific substrates or cultural requirements is the estimation of petroleum-degrading organisms (Mesarch and Nies, 1997). Petroleum-degrading bacteria are abundant in soil and numerous plating techniques have been used for their estimation, including plates containing petroleum, polyaromatic hydrocarbons, and crude oil. However, these have been shown to be difficult to use, as some of the substrates are volatile and the results variable. It has been shown that a medium containing benzoate was better at discriminating petroleum-degrading organisms and could show differences in the microbial populations between clean and petroleum-contaminated soil (Mesarch and Nies, 1997).

2.5.4 Packed cell volume

The **packed cell volume** (PCV) can be determined by pelleting the cells by centrifugation in a calibrated centrifuge tube. The volume of the packed cells can be determined from the calibrated tube, giving a value known as the PCV. PCV is a quick method but the results can be inaccurate.

2.5.5 Spectrophotometry

Another direct method is to measure the absorbance of the culture using a spectrophotometer. Micro-organisms in suspension absorb or scatter light and the amount absorbed is proportional to the cell concentration, within limits, but at high absorbance the relationship ceases to be linear. To be able to determine cell number or mass the individual spectrophotometer has to be calibrated for the organism and medium being used. In many cases the samples will have to be diluted before the absorbance can be read. The spectrophotometric method is much more rapid than filtering or centrifugation but is not suitable for media and samples containing particles, coloured materials, and compounds that adsorb at the wavelengths used.

2.6 Methods for the analysis of the microbial population *in situ*

Modern techniques of analysis have been developed that can estimate the microbial population in the environment *in situ* or after removal.

2.6.1 Confocal scanning laser microscopy (CSLM)

Direct microscopic analysis of the micro-organisms in aggregates and biofilms can only really determine cell morphology. Recently samples have been investigated using techniques such as **scanning** and **transmission electron microscopy**, and **confocal scanning laser microscopy** (CSLM), which can give much higher resolution (Stams and Oude Elferink, 1997). Biofilms appear to be much more complex than was first thought, with the micro-organisms forming separate microcolonies. Thus biofilms vary considerably in terms of depth, density, porosity, and diffusivity which affects the supply of nutrients and removal of wastes. Sampling of aggregates and biofilms cannot be carried out unless some form of disruption can be used. Unless the disruption is complete microbial numbers can be underestimated. In addition, the conditions in an aggregate or biofilm may be difficult to reproduce on a selective medium and although slow-growing species may be the dominant ones in nature they may be underestimated in a plate assay. Normal microbial methods are suitable for suspended microbial cultures consisting of single cells but in most environmental situations the micro-organisms exist as aggregates or as biofilms (Costerton et al., 1995). Anaerobic and aerobic wastewater treatment use mixed microbial populations, most of which occur in aggregates, whereas trickle filters, rotating biological contactors, and fluidized bed reactors (see Chapter 4) use biofilms containing a mixed population.

2.6.2 Enzyme-linked immunosorbant assay (ELISA)

Microbial populations can be detected and specific bacteria localized by the use of antibodies. Bacterial envelopes are strong antigens and both monoclonal and polyclonal antibodies have been prepared against specific bacterial antigens. The antibodies can be linked to fluorescent dyes or enzymes and these labels used to estimate the numbers and positions of specific bacteria. **Enzyme-linked immunosorbent assay** (ELISA) has also been used to study biofilms. These techniques are very specific, inexpensive, and easy to use provided the antibodies are available. One example of the use of immunological techniques has been the detection of methanogens in anaerobic bioreactors. The ELISA technique is a widely used format that can handle a large number of samples. Polyclonal antibodies against eight methanogens (Sorensen and Ahring, 1997) have been used in the ELISA technique to study their levels during anaerobic digestion of cow and pig manure.

2.6.3 Phospholipid fatty acids (PLFAs)

Membrane lipids and associated fatty acids are essential to maintain viability and include **phospholipid ester-linked fatty acids** (PLFAMEs) as well as **phospholipid fatty acids** (PLFAs). Since PLFAs are structurally diverse they can be used to identify groups of micro-organisms and employed as biomarkers. Fig. 2.10 indicates how the lipid could be extracted and the residue used for ribosomal DNA analysis. The profiles of fatty acids can provide a rapid and inexpensive alternative for describing complex microbial populations acting as a fingerprint. On cell death PLFA is rapidly hydrolysed by a phosphodiesterase, forming a diglyceride. Therefore, the ratio of diglyceride to PLFA provides an estimate of viability of the microbial population.

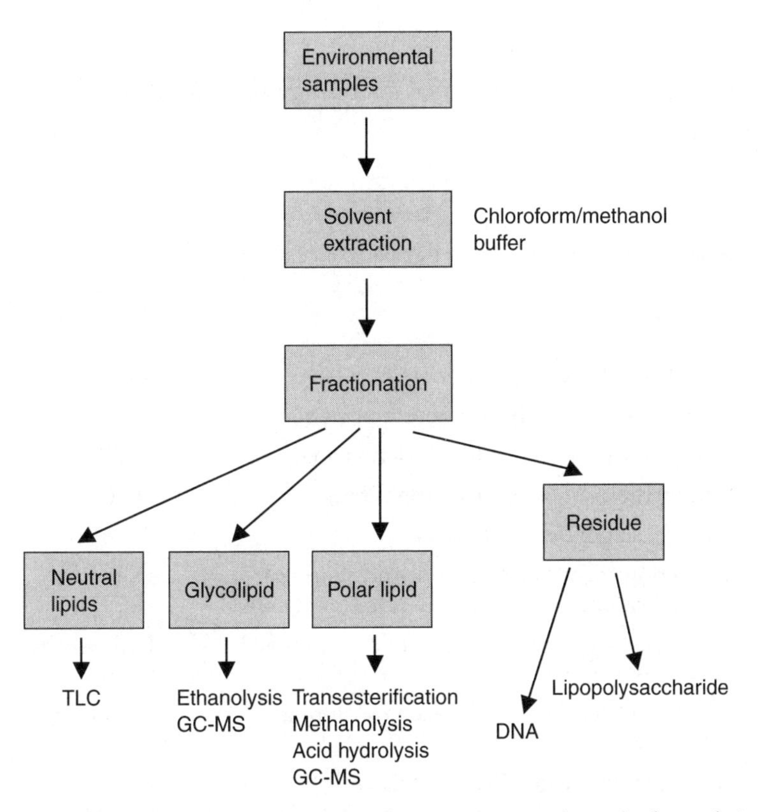

Figure 2.10 The extraction of lipid components from environmental samples for analysis combined with DNA extraction. TLC, thin-layer chromatography; GC-MS, gas chromatography/mass spectrometry.

Bacteria can also be detected and identified by the analysis of their patterns of PLFAMEs or PLFAs. This method has been used to study methanogens and sulphate-reducing bacteria but the technique is not as sensitive as nucleic acid-based techniques and the patterns can be altered by changes in environmental conditions (Bottger, 1996).

2.7 Recombinant DNA technology

Recombinant DNA technology has reached a level where it can now be used to follow genes or specific DNA sequences, and therefore micro-organisms, within the environment. Table 2.3 outlines some of the techniques that can be used to study some of the following environmental systems:

- the microbial ecology of soils and water, in particular extreme habitats such as hydrothermal vents;
- anaerobic and aerobic digesters;
- contaminated ground water;
- bioleaching process;
- evaluation of diversity;
- release of genetically manipulated micro-organisms (GEMs);
- bioremediation processes; and
- drinking-water quality.

Table 2.3 Some of biomolecular techniques used in analysis

Technique	Abbreviation
Amplified (DNA) fragment polymorphism	AFLP
Arbitrary primed polymerase chain reaction	AP-PCR
Amplified ribosomal DNA restriction analysis	ARDRA
Denaturing-gradient gel electrophoresis	DGGE
Pulsed-field gel electrophoresis	PFGE
Restriction fragment length polymorphism of polymerase chain reaction-generated amplicons	PCR-RFLP
Polymerase chain reaction of random-amplified polymorphic DNA	RAPD-PCR
Pulsed-field gel electrophoresis, rare cutting endonuclease	RC-PFGE
Ribosomal intergenic spacer analysis	RISA
Temperature-gradient gel electrophoresis	TGGE

2.7.1 Extraction of RNA and DNA

The first step in the technology is the extraction and isolation of DNA from the samples. A number of methods are available for the extraction of nucleic acids from samples of soil, water, and sediment and these can be divided into two categories (Fig. 2.11). The first methods used were the direct extraction of DNA using chemicals such as alkalis (Ogram et al., 1987), detergents like sodium doedecyl sulphate (SDS), and the enzyme lysozyme to disrupt the bacteria *in situ* (Saano et al., 1995). The DNA can then be extracted with SDS, phenol, or chloroform or a mixture of phenol and chloroform. The second method involves the extraction of the micro-organisms before lysis (Bakken and Lindahl, 1995), and the cells are broken using conventional methods and the DNA extracted (Torsvik et al., 1995). Bead mills (Smalla et al., 1993) and lysozyme combined with freeze–thawing (Tsai and Olson, 1991) have been

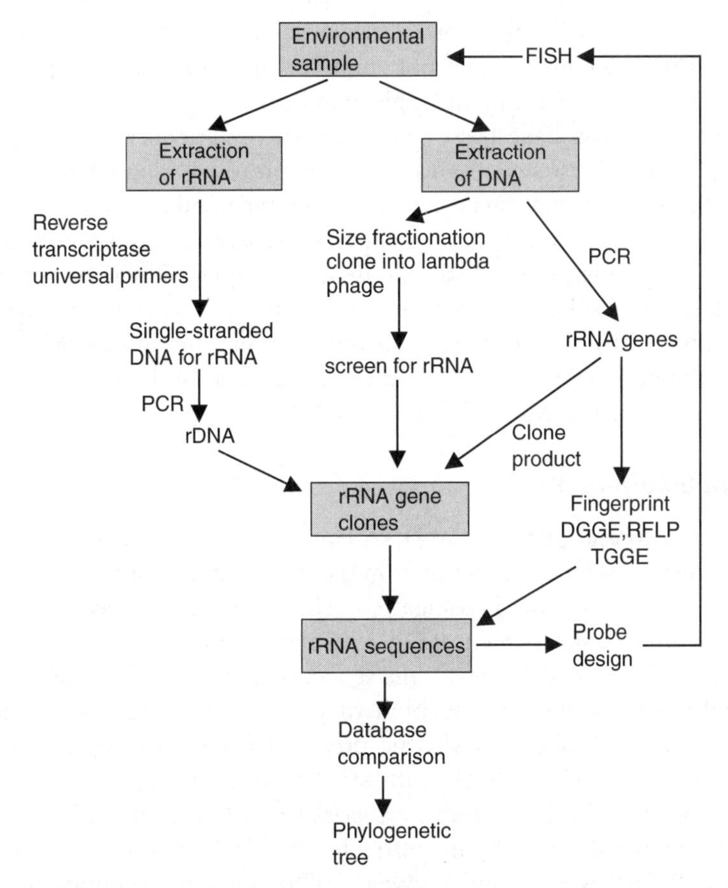

Figure 2.11 Extraction of DNA and rRNA from environmental samples in order to isolate and sequence the rRNA genes present. This will give an estimate of the number of different micro-organisms present and a phylogenic tree can be prepared from databases. The sequences will also enable the production of probes for a specific micro-organism using fluorescence *in situ* hybridization (FISH); DGGE, RFLP, and TGGE are defined in Table 2.3.

used to extract DNA directly with yield of up to 90%. Both extraction techniques are lengthy, require large samples, and can have low yields due to the complex nature of the samples and the need for highly pure DNA. DNA-purification techniques currently used include treatment with cetyltrimethylammonium bromide (CTAB), spin columns, spin cartridges, agarose gels, CsCl precipitation, and ion-exchange chromatography.

2.7.2 Polymerase chain reaction (PCR)

The **polymerase chain reaction** (PCR) can amplify genes or DNA sequences present in low concentrations and small numbers. Thus PCR can improve the detection of a gene manyfold (Graham, 1994). The development of the technique of PCR means that only small amounts of DNA need to be extracted from environmental samples. The PCR technique can be used with very specific primers to detect minor genes or with less-specific primers to detect groups of genes. PCR can be used to detect the presence of specific genes associated with the presence of pollutants or the degradation of pollutants. An outline of the PCR technique is given in Box 2.5.

Samples from soil have been quantified for the presence of the *atz* gene, involved with the first step in atrazine mineralization (Clausen et al., 2002). The copies of the atrazine gene *atz*A exceeded the number of colony-forming units (CFUs), which may be due to non-culturable cells, multiple copies of the gene, or uneven distribution and handling of samples. Others have reported differences between PCR and plate counts of one or two orders of magnitude. PCR has been used to detect cyanobacterial blooms and distinguish between some of the genera present (Baker et al., 2002), and to detect hydrocarbon-degrading organisms (Milcic-Terzic et al., 2001).

2.7.3 16 S-subunit RNA

Once extracted and purified the DNA can be probed for a specific gene or DNA sequence, which can act as a marker for a micro-organism or group of micro-organisms under investigation. The probe can be constructed as an oligonucleotide (a short chain of DNA of about 20 nucleotides, or RNA) or polynucleotide (50 nucleotides) if the sequence of the specific gene is known, and probes can be designed to bind to very specific targets or to a wide range of species. Probes can also be prepared from cloned genes, or from complementary DNA (cDNA) synthesized from extracted messenger RNA (mRNA). One of the most promising series of probes are those based on the gene coding for the rRNA, in particular the 16 S subunit (small-subunit rRNA; ssrRNA) (Amann and Ludwig, 2000). The 16 S subunit rRNA has highly conserved regions and variable regions, and a high copy number (about 1000), so that if the probes are made to the conserved region the probe will detect a group of related organisms whereas if the variable region is used the

BOX 2.5 **Polymerase chain reaction (PCR)**

Polymerase chain reaction is a technique that was first reported in 1985 for the amplification of specific genes. Once pure DNA has been extracted the double-stranded DNA is separated into two stands by heating to 95°C for around 2 min. Short single-stranded oligonucleotides (primers) and *Taq* polymerase (temperature-tolerant DNA polymerase) is added and the mixture cooled to 60°C. This lowering of the temperature allows the primers to anneal to specific sections of the single-stranded DNA complementary to their base sequence. As the mixture contains nucleotide triphosphates the *Taq* now synthesizes new strands of DNA complementary to the DNA template and extending from the primer for up to 10 kb in length (see the figure below). The lower temperature is normally held for about 2 min and then the reaction mixture raised to 95°C. This separates the original and new strands. Another set of binding sites is now available for the primers and when cooled to 60°C after 2 min the *Taq* polymerase starts another round of synthesis. This process is repeated for a number of cycles until the primers or nucleotide run out or the cycling is stopped.

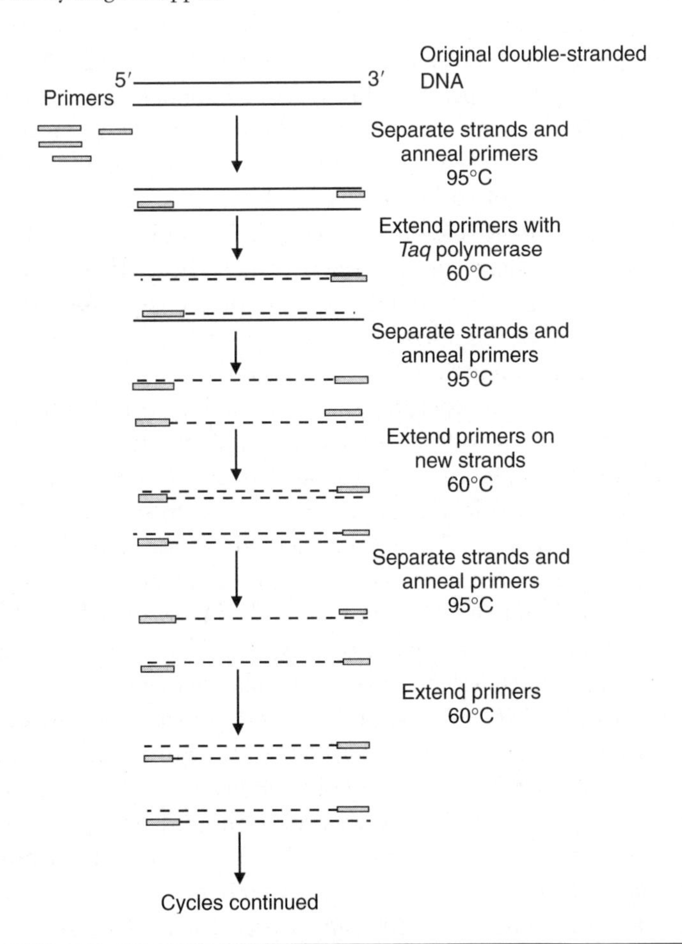

probe will be much more specific. At present there are some 3000–5000 full or partial sequences for the ssrRNA and this is constantly increasing, and these sequences are available in databanks like GenBank and EMBL, so that a considerable number of probes can be designed. The probes can be used with water and soil *in situ*.

For these probes to be able to indicate a gene or sequence they all need some form of label. The incorporation of radioactive labels into the probes can be carried out by nick translation, primer extension, or end labelling. The radioactive labels can be replaced by non-radioactive labels based on biotin, which when incorporated into nucleotides can react with streptavidin which itself can be linked to antibodies or enzymes. The antibody or enzyme reaction can be used to indicate the presence of the probe. The probes once labelled can be used to detect those specific sequences in the extracted DNA or in some case in a bacterial colony. The most commonly used methods of using probes are dot-blot and Southern-blot (Box 2.6) techniques. The use of recombinant DNA techniques for the study of bacterial communities particularly those in environments difficult to reproduce allow the assessment of bacterial diversity without cultivation. An example is the use of whole-cell hybridization with domain- and kingdom-specific fluorescent-oligonucleotide 16 S-derived probes to study the diversity of the thermophilic micro-organisms in deep-sea hydrothermal vents (Harmsen et al., 1997). The variation in microbial populations has been investigated by extracting the nucleic acids, amplifying the 16 S rRNA gene by PCR, cloning the fragments, and comparing the gene sequences (Fig. 2.11). The resulting sequences have shown that the majority of micro-organisms in the explored habitats were previously unknown and that only a small fraction of the microbial diversity has been detected by cultural methods (Polz and Cavanaugh, 1997). This technique has been used to investigate anaerobic fixed-bed reactors, activated sludge, and denitrifying sand beds. Quantitative hybridization can also provide an estimate of the dominance of individual populations. Microbial diversity in the natural environment has also been investigated by using variation in the 16 S RNA and a taxonomy based on this data has been presented (Woese et al., 1990).

DNA technology has been used to follow the bacterial population in activated sludge and the population dynamics in a trickle-bed bioreactor. It has been clear that culture techniques are not capable of providing full details of the population in activated sludge, and 16 S rRNA probes have shown that the dominant organisms were those of the beta subclass of the class Proteobacteria whereas culture methods had shown that the gamma subclass of the Proteobacteria was dominant (Snaidr et al., 1997). The trickle-bed study used fluorescence *in situ* hybridization using 16 and 23 S rRNA probes showed results different from those obtained by culture methods and revealed a very diverse population.

Recombinant technology can be used to estimate the degree of genetic variation in the determination of biodiversity (Karp et al., 1997) and have been

BOX 2.6 Southern blotting

Southern blotting is a technique used to detect specific sequences in DNA fragments. The DNA under investigation is cut into fragments of various sizes by **restriction enzymes**. These enzymes cut the DNA at specific base sequences, which are unique for each enzyme. The fragments of DNA can then be separated on the basis of size by agarose gel electrophoresis. Once separated the DNA fragments can be transferred to nitrocellulose or nylon membranes by overlaying the gel with the membrane for up to 24 h under moderate pressure. This will transfer the DNA from the gel to the membrane, forming a precise replica of the DNA fragments as they have been separated on the gel (see the figure below). Once transferred the DNA fragments can be fixed in place by drying at 80°C for 2 h and the DNA denatured. Denaturation of the DNA allows the probe to anneal with the complementary sequences in the DNA strands, which can then be detected by autoradiography or enzyme activity.

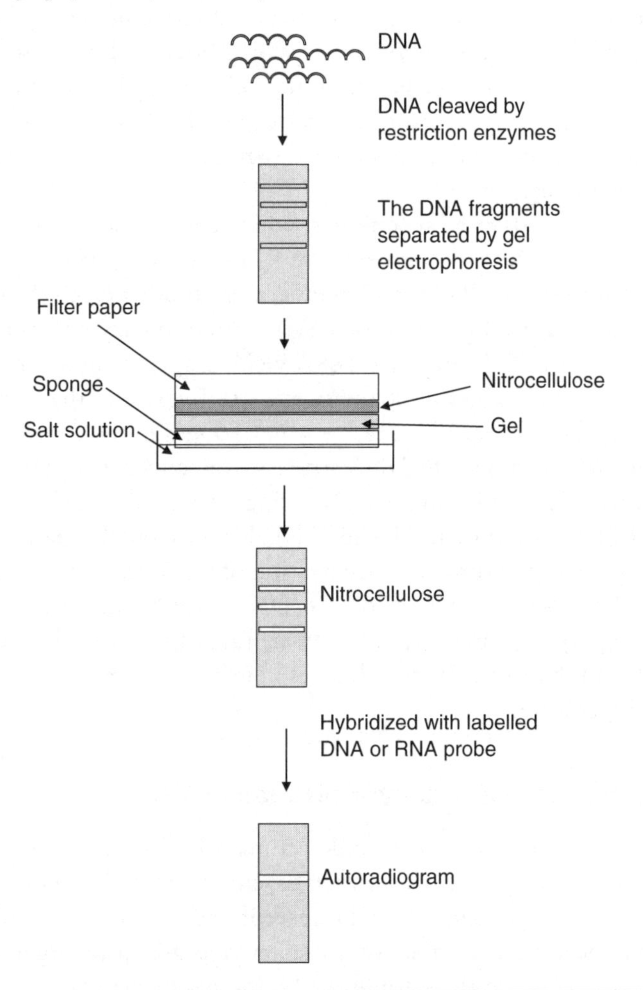

used to detect bioleaching bacteria (De Wulf-Durand et al., 1997). In the case of bioleaching PCR was used to amplify the 16 S rRNA obtained from the bacterial population associated with an acidic mining environment. The monitoring of the release of GEMs is essential if the risk factors associated with this are to be evaluated. GEMs can be followed on the basis of their unique DNA sequences (Jansson, 1995).

Many of the molecular probes are based on the ssrRNA. The rRNA genes can then be detected by fractionating the DNA, cloning fragments into the bacteriophage lambda, and screening the clones for rRNA genes. This method is slow but can be improved by the use of PCR. The rRNA genes can be amplified directly from the isolated DNA by using rRNA primers. The techniques allow the amplification of DNA and thus the identification of rare genes and DNA at very low levels. Clones can then be dot or colony blotted with specific probes, or subjected to **restriction fragment length polymorphism (RFLP)** analysis. Phylotypes can be visualized *in situ* using fluorescently tagged oligonucleotide probes and microscopes fitted with fluorescence detectors. Fluorescence can also be linked to flow cytometry to count cell numbers and CSLM to follow the arrangements of the cells. However, olignucleotide probes require sequence data for their design and at present there are many sequences available.

Other nucleic acid-based techniques, which can be used to follow changes in a population in the environment, are RFLP analysis, **denaturing-gradient gel electrophoresis** (DGGE), and **temperature-gradient gel electrophoresis** (TGGE) (Table 2.3). In RFLP analysis restriction-enzyme degradation of extracted DNA or PCR-amplified DNA yields a specific number of bands of DNA fragments after gel electrophoresis. At this stage the bands can be probed using the Southern-blot technique. Changes in the distribution of the bands indicate changes in the DNA structure or sequence. In DGGE analysis two DNA fragments differing in only a single base substitution, deletion, or insertion can be separated. The method depends on the pattern of DNA melting as the temperature or concentration of denaturant (formamide, urea) increases. The melting of part of the DNA affects the rate of migration through the polyacrylamide gel and the melting of the DNA is affected by its base composition (Box 2.7), and changes in its base composition will alter the melting temperature.

2.7.4 Fluorescence *in situ* hybridization (FISH)

FISH uses specific rDNA probes to detect microbial groups in environmental samples without the need to extract or concentrate the cells. The FISH-labelled cells can be visualized by fluorescent and confocal microscopes. This method has been used to analyse a denitrifying group of organisms where *Azoarcus* and *Thauera* were found to be the major genera. A FISH system with six probes of different colours were used with CLSM to study biofilms

BOX 2.7 Denaturing-gradient gel electrophoresis (DGGE)

If the temperature or the concentration of denaturant (formamide or urea) is increased then blocks of DNA called domains melt at discrete temperatures (T_m). These domains can be from 25 to several hundred base pairs in length and thus DNA fragments of 100–1000 base pairs can have two to five melting domains. The melting temperatures are between 60 and 80°C and the melting temperature can be affected by the base sequence of the DNA region. In practice the DNA is electrophoresed at a high temperature, typically 60°C, in a polyacrylamide gel containing a gradient of denaturant equivalent to a temperature gradient of about 10°C. A DNA fragment which reaches its T_m value will have a domain melt and the DNA mobility will be slowed, and the mobility will be further slowed as other domains melt when their T_m values are reached. Thus DNA fragments with small differences in base composition will have different T_m values and can therefore be separated. The system has been improved by attaching a high-melting-point domain (GC clamp) to the test DNA so that the whole DNA can be melted before the clamp and therefore all the DNA can be analysed. A second improvement has been to use the system to analyse heteroduplexes between control and test DNA samples as any mismatch between the two DNA fragments will cause the DNA to melt at a lower temperature. In addition the control DNA can be radioactively labelled so that the gel can be autoradiographed directly, thus avoiding gel blotting.

(Thurnbeer et al., 2004). FISH can also be used for detection of microalgae, and autotrophic and heterotrophic bacteria. FISH relies on the cells having sufficient rRNA and in some cases the signals were low. Techniques to increase FISH include single probes with multiple fluorochromes and treatment with nalidixic acid to prevent cell division. Standard FISH cannot be used to detect single copies of functional genes at the single-cell level. However, *in situ* PCR has been developed to increase these functional genes and this method has been used to detect the ammonia monooxygenase gene (*amo*A) in a biofilm from a nitrifying reactor that could not be detected normally by FISH.

2.7.5 Proteomics

Proteomics can be defined as the systematic analysis of the protein population in a tissue, cell, or subcellular compartment. The ability to separate and identify proteins from cells allows the study at the protein level of the effects of pollutants and early effects of environmental contamination. The analysis of the protein starts with the one- or two-dimensional separation of the proteins and then the digestion of individual spots or bands. The resulting peptides are extracted and their masses determined by **matrix-assisted laser desorption ionization** (MALDI) and **electrospray ionization** (ESI) **mass spectrometry** (MS). Modern machines give accuracy, mass resolution, and sensitivity which allows for the identification of picomolar–femtomolar amounts of proteins and peptides if matching genomic sequences are available.

2.8 Indirect methods for the determination of microbial numbers and activity

Not all samples taken for microbial identification and cell numbers are suitable for analysis. Cells that form aggregates cannot be plated out successfully for viable counts and the same is true for direct counting using a Coulter counter. There are methods that have been used to break up aggregates, such as sonication, but it is difficult to obtain 100% disaggregation and the sonication process may damage the cells. Aggregates in a sample can also contain more than one type of micro-organism, as with activated sludge. Upon plating such a sample one species may outgrow all the others. Samples that contain solids other than micro-organisms are also difficult to process, as the micro-organisms cannot be distinguished from the inorganic particles by methods like spectrophotometry, and the cells may also be attached to the particles. Therefore, in these cases indirect methods have to be applied which seek to estimate the number of micro-organisms from their metabolic activity by following what they produce or use during growth (Fig. 2.12). These methods are as follows.

- Assay of cellular components—assay of DNA, RNA, protein, and lipids.
- ATP assay—determination of ATP.
- Radiometry—use of radioactive substrates.
- Assay of medium components—sassay of substrates like glucose.
- Viscosity—determination of culture viscosity.
- Product formation—assay of specific products.
- Gas production—determination of the production or use of gases.
- Impedance and conductance—measurement of culture impedance or conductance.
- Heat production—determination of the heat produced by microbial metabolism.

Many of these methods are only suitable for cultures growing in **bioreactors** (Box 2.8). In a bioreactor the growth conditions can be both controlled and monitored, a situation which is not possible with many other methods of cultivation. For example the inlet and exit gas composition, dissolved oxygen, and other parameters can be monitored online in real time in bioreactors.

Cell components such as DNA, RNA, protein, lipids, chitin, ergosterol, and glucosamine can be determined chemically and used as a measure of the mass or number of cells present in a sample. However, this assumes that the proportion

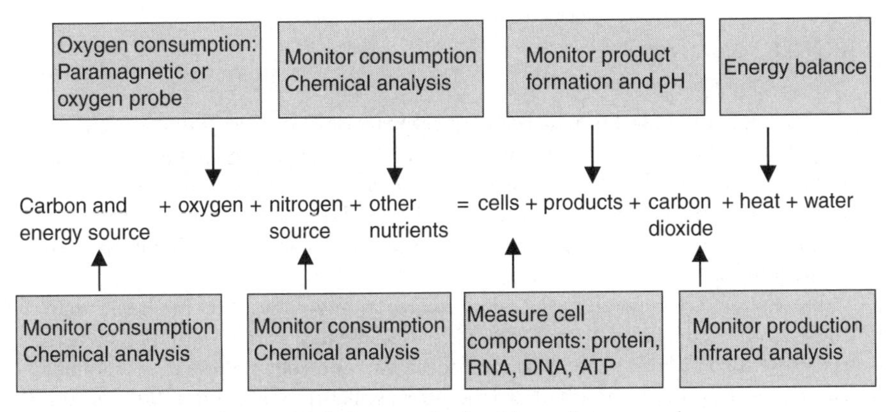

Figure 2.12 Indirect methods of determining growth.

of the cell components remains constant, which is not always true, and that there are no other sources of these components in the sample. Another universal cellular component, adenosine 5-triphosphate (ATP), the main energy currency of the cell, can be measured. Here again it has to be assumed that cells contain a constant level of ATP and all viable cells should contain ATP. Very small concentrations of ATP can be measured using the reaction of ATP with an enzyme called **luciferase**. When combined under the correct conditions ATP causes luciferase to produce light and even very small amounts of light can be measured accurately by a photomultiplier. Incidentally, the enzyme was originally extracted from tails of fireflies (glowworms), where it is responsible for light production. This method of detection and estimating micro-organism numbers has been used to detect microbial contamination in UHT milk.

The nutrient (substrate) level in the culture medium can be determined by enzymatic or physical methods and the reduction in substrate can be used to indicate levels of growth. The utilization of nutrients such as sugar and nitrate are common parameters used and in bioreactors the medium dissolved oxygen level can also be used. In bioreactors dissolved oxygen can be monitored online, as there are accurate, stable, and sterilizable oxygen probes available so that growth can be followed without having to remove a sample. Often carbon dioxide is released during cell growth on sugars and heat generated, and both of these parameters can be followed. Carbon dioxide production in bioreactors can be measured in the exit gas stream using an infrared analyser, and oxygen with a paramagnetic analyser. An example is the determination of fungal growth in a complex medium containing wheat flour using carbon dioxide evolution in a 10-litre bioreactor (Koutinas et al., 2003). The formation of a product by the micro-organism can also be used to indicate growth and if this product is an acid or alkali monitoring the pH is possible. The formation of polysaccharides can be detected by increases in medium viscosity. However, not all products are suitable for following growth, as some are only formed after growth has ceased. The production of heat is an indication of an active

BOX 2.8 Bioreactors

A **bioreactor** is a vessel in which biological reactions are carried out, which can include the cultivation of micro-organisms and reactions involving enzymes and immobilized cells. The term bioreactor has taken over from the term **fermenter,** as it is a more general term and not just a vessel used to cultivate micro-organisms. In the strictest sense a fermenter should only describe a vessel involved with fermentation.

Bioreactors are generally used when the reaction volumes become larger than those easily used in laboratory vessels such as shake flasks and in situations where the growth conditions need to be monitored and controlled. Bioreactor volumes can range from 0.5 l, the smallest laboratory scale, to above 2 000 000 l in the activated-sludge process. Small bioreactors of 0.5–20 l are mainly constructed from glass but above 20 l the construction is stainless steel. There are very many bioreactor designs but the most common design is the stirred-tank design, where the contents are mixed with an impeller. The most common impeller design is the Rushton turbine impeller (see the figure below). The function of the bioreactor design is to provide optimum conditions for the reactions carried out and in the aerobic cultivation of micro-organisms is the provision of a sterile homogeneous culture, oxygen, and maintainance of correct temperature and pH. In addition, a number of sterilizable probes are available for monitoring pH, dissolved oxygen, temperature, and in some cases biomass. The inlet and exit gases can also be monitored for carbon dioxide and oxygen contents. The supply of oxygen is perhaps the most critical feature of bioreactor design, as oxygen is only sparingly soluble in water. To improve oxygen supply the air bubbles are broken up by the impeller and distributed throughout the vessel. The temperature is monitored and controlled by a platinum resistance thermometer using a water jacket.

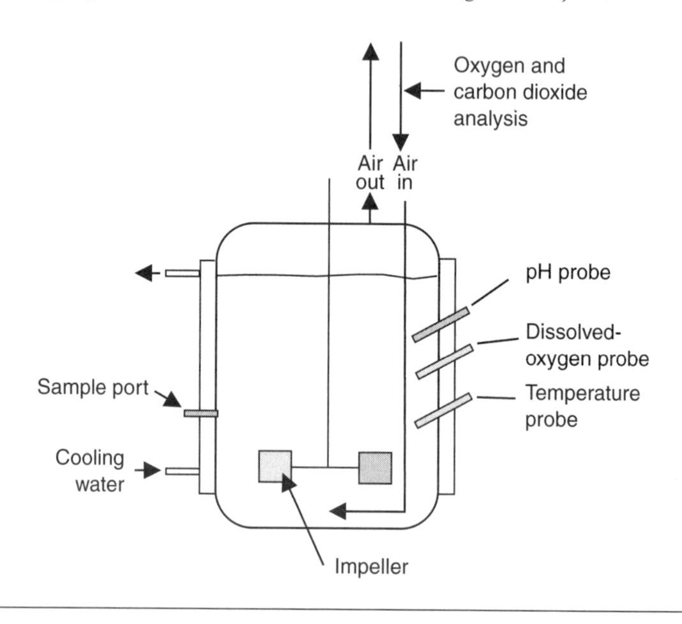

metabolism but is difficult to follow as the culture has to be insulated to avoid incidental heat loss, and cooling is required to maintain an even temperature.

The response of biological material to an applied electrical field is described by its conductivity and permittivity (Markx and Davey, 1999). At low radio frequencies (0.1–10 MHz) the microbial cell membrane behaves as an electrical capacitor. Thus the size of the dielectric permittivity is proportional to the membrane-bound volume of the biomass. Thus by measuring the capacitance at radio frequencies the biomass can be determined and an online steam-sterilizable capacitance probe is now available (Aber Instruments). This allows the online determination of biomass in bioreactors, for example *Streptomyces clavuligerus* (Neves et al., 2000).

2.9 Medium and conditions

In the discussion of viable and non-viable cells it was indicated that the correct medium and conditions were required before any viable cell could grow. In general, the requirements for the growth of micro-organisms are:

- nutrients,
- a source of energy,
- available water,
- the correct temperature,
- the correct pH, and
- appropriate levels or the absence of oxygen.

2.9.1 Nutrients

Elemental analysis of micro-organisms shows that they consist of carbon, hydrogen, oxygen, nitrogen, phosphorus, sulphur, and a variety of trace elements. All these components have to be obtained by the micro-organisms from their environment or medium. The lack of one of the components will restrict the growth of micro-organisms, a condition often found in the environment. Micro-organisms exhibit considerable diversity in the compounds that they use for growth.

2.9.2 Source of energy

The simplest energy source is that used by the photosynthetic organisms, where the power in sunlight is collected. The remaining micro-organisms use a vast range of organic and inorganic compounds to obtain energy for growth. The majority of organisms grown in the laboratory use simple sugars like glucose as their carbon and energy source. In the environment the carbon source can be complex, varied, and limited in its supply.

2.9.3 Available water

Micro-organisms consist of 80–90% water and therefore they will only grow in conditions where there is sufficient free water. The availability of water in solutions is measured as **water activity**, with distilled water having a value of 1.0. In contrast, concentrated sugar solutions have low water activities of 0.6 and below. Humans have used this property of sugar syrups to preserve food for thousands of years. Most bacteria require a water activity of 0.9 to grow but there are some yeasts and fungi that can grow at water activities as low as 0.6. For this reason yeast and fungi are often found growing in low-water conditions such as on bread and on the surface of low-sugar jams. In the atmosphere the amount of water present is expressed as **relative humidity** and at high relative humidities micro-organisms, fungi in particular, can grow on many surfaces. The concentration of salt in a liquid has an important effect on the availability of water. Clearly marine micro-organisms can tolerate moderate levels of salt (3%) but there are organisms that require salt for growth and these are the **halophiles** (Fig. 2.13). Halophiles can grow in salt concentrations above 15% and can be found in salt lakes and pickling brines. Most halophiles belong to the domain Archaea.

2.9.4 Temperature

Temperature influences the rate of growth of micro-organisms and the majority have an optimum at around 30–37°C. There are temperatures above or below this at which microbes cannot grow. For this reason we refrigerate food to slow down or stop bacterial growth and at the other end of the scale a short

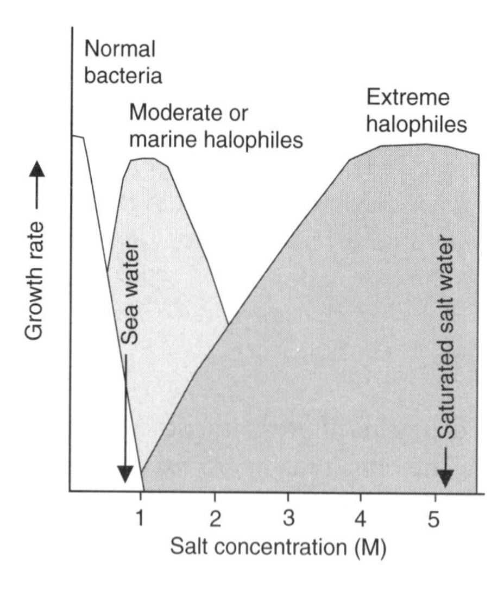

Figure 2.13 The growth conditions for moderate and extreme halophiles.

exposure to 60°C is used to **pasteurize** milk as this kills most enteric bacteria. However, in the environment there are areas where the temperatures are extreme, ranging from cold of the Arctic and Antarctic regions to hot volcanic springs. Initially, it was thought that no life could exist under these conditions, but this is not true as micro-organisms have been found to survive and grow in all these environments. The ranges of temperatures that micro-organisms can grow at have been named and are shown in Fig. 2.14. Those micro-organisms that grow best at or below 15°C are known as **psychrophiles** and have a lower limit of 0°C or below. Some organisms can grow below 0°C if liquid water is available. These bacteria are commonly found in the Arctic and Antarctic, but can also cause food spoilage in fridges and cold stores. The characteristics that allow the cells to grow at these low temperatures are related to changes in the membranes, which contain more unsaturated and short-chain fatty acids, enzyme changes, and ribosomes that remain active at the lower temperatures. Some of these micro-organisms can grow at 0–5°C, have their optimum above 15°C but cannot grow above 20°C. Those micro-organisms that have their optimum temperature between 15 and 45°C are known as **mesophiles** and are the most studied group as they contain both animal and human pathogens. The pathogens mostly have an optimum of 37°C, which is the normal human's body temperature. Yeasts are generally grown at 30°C, although brewing is often carried out at 13–18°C. An interesting case is the organism *Mycobacterium leprae*, which causes leprosy; it cannot grow well at 37°C which is why it only infects the extremities. The only other animal that can be infected by this organism is the armadillo as it has a body temperature lower than 37°C.

Micro-organisms that grow at between 45 and 60°C are known as **thermophiles** and are found at high concentrations in compost heaps, hot springs, and other high-temperature sites. It was thought that these were the most

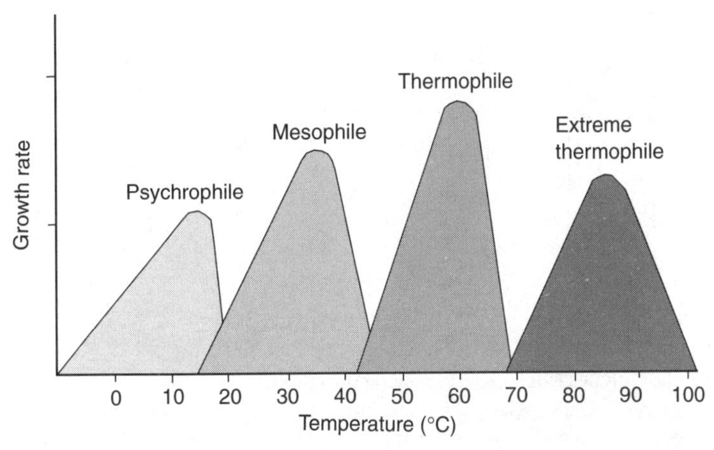

Figure 2.14 The four groups of organisms in relationship to their optimum growth temperature.

extreme micro-organisms but the recent discovery of micro-organisms growing in hydrothermal vents on the ocean floor introduced another group, the **extreme thermophiles** or **hyperthermophiles**. These organisms are capable of growing at 80–110°C. Thermophiles have developed features that allow them to grow at temperatures which would denature the proteins of normal micro-organisms. Changes to proteins, membranes, and DNA polymerase, and DNA that contains more guanine and cytosine have been detected as adaptation to the higher temperatures. Thermophiles are found in areas of volcanic activity and are generally Bacteria and Archaea. The hotter the temperature the higher proportion of Archaea that is found.

2.9.5 pH

Micro-organisms normally grow at around neutrality (pH 7.0) but they can be found growing in both acid and alkaline conditions. The bacteria that can only grow at low pH values of 1–4 are known as **acidophiles** and those that grow above pH 9 are known as **alkaliphiles**. The alkaliphiles can be divided into two groups; the alkaliphiles, including bacteria, yeasts, and fungi, and the haloalkaliphiles where salt is also required for growth (up to 33% salt). The alkaliphiles are found in alkaline lakes and hot springs and the haloalkaliphiles in soda deserts and lakes, where they often have a red pigmentation.

2.9.6 Oxygen

Micro-organisms can be divided into two groups based on their response to oxygen; **aerobes** and **anaerobes**. Aerobes are those organisms that require

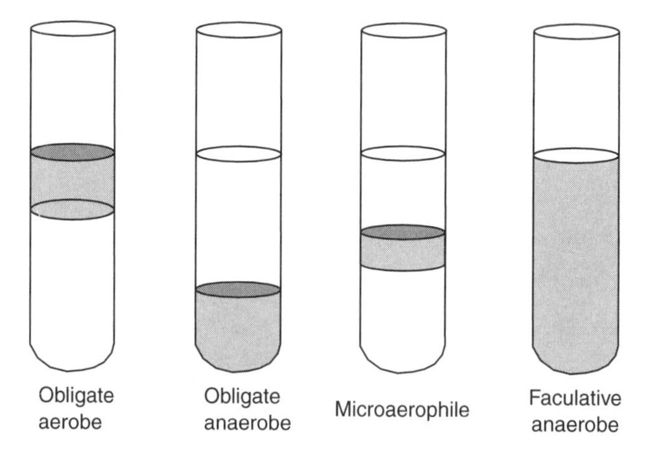

Figure 2.15 The influence of oxygen supply on the growth of micro-organisms. An obligate aerobe will on grow on the surface of the medium in the tube. The obligate anaerobe will only grow at the bottom of the tube where the oxygen levels are low. The microaerophiles will grow somewhere between the first two organisms and the facultative anaerobes can grow throughout the medium.

oxygen for growth in contrast to anaerobes that can only grow in the absence of oxygen (Fig. 2.15). Intermediate between the two are the **microaerophiles,** which will only grow at reduced oxygen concentrations. **Facultative anaerobes** have the ability to grow in the presence or absence of oxygen. Organisms such as *Escherichia coli* and *Saccharomyces cerevisiae* (yeast) can switch between the two conditions and in the absence of oxygen they carry out fermentation. In the case of yeast the main product of fermentation is alcohol, which is the basis of the wine and beer industry. If oxygen were supplied during brewing all that would be produced would be more yeast cells and no alcohol.

2.10 Metabolism

All micro-organisms require energy, carbon, and other molecules to grow and divide. The reactions that are carried out by microbial cells to generate energy and to build up their cellular components are collectively known as **metabolism.** The building of components is known as **anabolism** and the breakdown of materials to produce energy is **catabolism.**

In catabolism large or complex molecules are broken down to simpler products and the energy released collected. In anabolism small molecules are combined to form the cellular components using energy from catabolism. Both processes are carried out in a series of reactions or pathways catalysed by specialized proteins known as enzymes. The pathways of catabolism and anabolism are not the same and in many cases they are not linked in space or time. To function in a coordinated manner catabolism and anabolism are linked by universal high-energy currencies. The main currency is ATP but others include NAD (nicotinamide adenine dinucleotide), PEP (phosphoenolpyruvate) and acetyl-CoA (acetyl coenzyme A) (Fig. 2.16). There are four methods of obtaining energy and the carbon for building the cells and these are (Fig. 2.17):

- **photoautotrophy** (photolithotrophy, photosynthesis),
- **photoheterotrophy,**
- **chemoautotrophy** (chemolithotrophy), and
- **chemoheterotrophy.**

Photoautotrophs obtain their energy from sunlight and use carbon dioxide as the carbon source. The best-known examples are plants and algae but it also includes the less-well-known prokaryotes, the blue-green algae (cyanobacteria) and certain photosynthetic bacteria (purple and green sulphur bacteria). The photoheterotrophs use light as the energy source but build their cell components

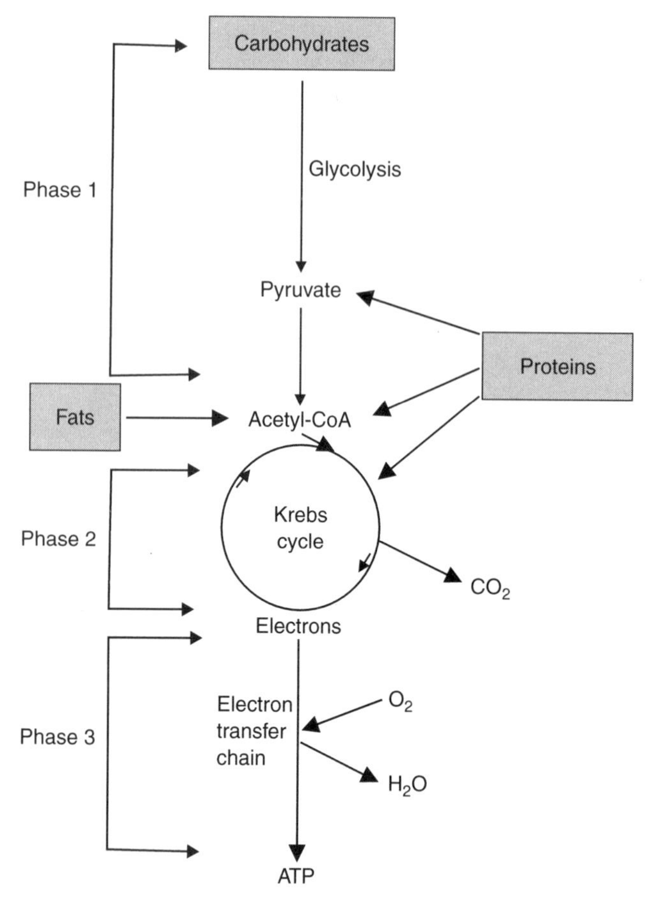

Figure 2.16 The links between catabolism (energy production) and anabolism (construction of the cellular components).

from organic carbon compounds. This is represented by a small group, the non-sulphur green and purple bacteria. Chemoautotrophs use reactions with inorganic chemicals to produce energy and utilize carbon dioxide as their carbon source. The group is represented by the hydrogen-, sulphur-, and iron-reducing bacteria. Finally there are the chemoheterotrophs which use organic compounds as both energy and carbon sources. This is by far the largest group and includes the majority of the bacteria and many eukaryotes.

The catabolic reactions producing energy involve electron transfer where the complex compound is oxidized by the removal of electrons and the electrons are transferred to an electron acceptor that is reduced. During this transfer energy is generated. In photoautotrophs the electron donor is water or inorganic molecules like hydrogen sulphide, and in chemoautotrophs the electron donor is also an inorganic molecule. In photoheterotrophs and chemoheterotrophs the electron donors are organic molecules.

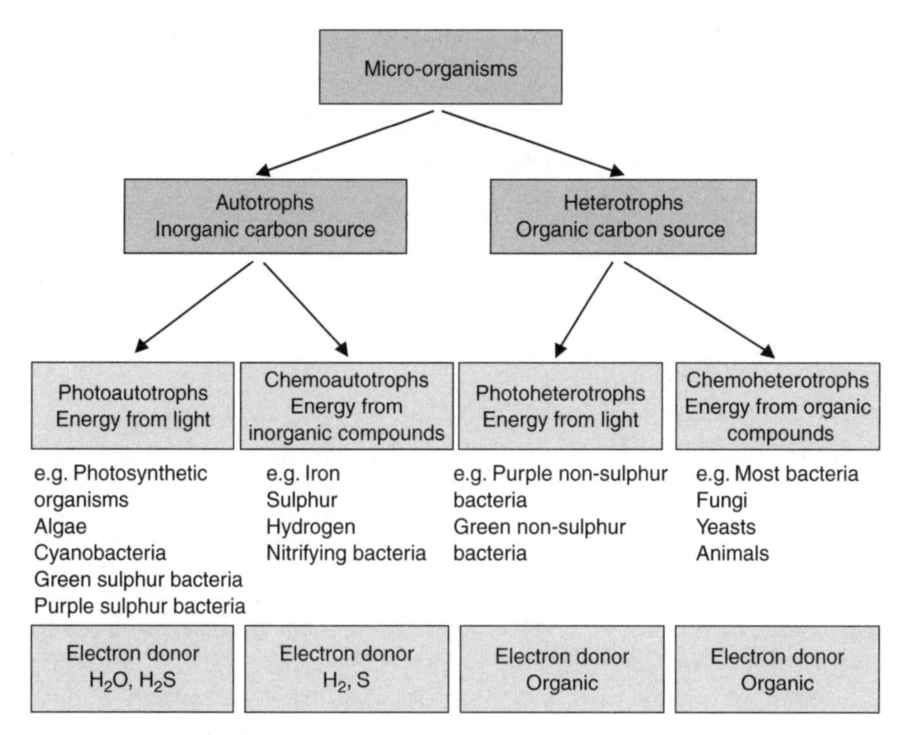

Figure 2.17 The four methods of obtaining energy used by micro-organisms.

2.10.1 Chemoheterotrophs

Catabolism in aerobic **chemoheterotrophs**, which includes most bacteria, fungi, yeast, and animals, can be divided into three phases. If complex carbohydrates are metabolized they are normally broken down to glucose, which is the start point for phase 1 (Fig. 2.16). Phase 1: glucose is broken down to pyruvate via the glycolytic pathway. Phase 2: the pyruvate is converted to acetyl-CoA and passed into the Krebs or tricarboxylic acid (TCA) cycle where it is broken down to carbon dioxide and water. Phase 3: stage 3 is oxidative phosphorylation where the electrons and protons released in glycolysis and the Krebs cycle are used to synthesize ATP. In most cases the final electron acceptor is oxygen, which is reduced to water.

The details of these pathways can be found in the books given in the Further reading list at the end of the chapter. In brief, glycolysis or the Embden–Meyerhof pathway yields a net gain of 2 ATP and 2 NADH molecules per glucose molecule.

$$\text{glucose} + 2\text{ADP} + 2\text{P}_i + 2\text{NAD} = 2\text{ pyruvate} + 2\text{NADH} + 2\text{ATP} \quad (2.1)$$

The ATP in this case is formed by substrate-level phosphorylation that is the simple transfer of a phosphate bond. In the Krebs cycle acetyl-CoA is fed into the cycle and the acetyl-CoA can be derived from the breakdown of fatty acids and amino acids, and glycolysis. The net reaction of the Krebs cycle is as follows.

$$2\ \text{pyruvate} + 2\text{ADP} + 2\text{P}_i + 2\text{FAD} + 8\text{NAD} = 6\text{CO}_2 + 2\text{ATP} + 2\text{FADH}_2 + 8\text{NADH} \quad (2.2)$$

Energy is extracted from NADH by passing electrons from NADH down a series of carrier molecules bound in a membrane, forming the **electron transport chain.**

2.10.2 Fermentation

In aerobic metabolism organic compounds are the source of electrons and oxygen the final electron acceptor. When oxygen levels are low some micro-organisms can switch from oxygen as the electron acceptor to using the products of metabolism as electron acceptors. This process is known as **fermentation.** Fermentation yields less energy than aerobic metabolism, as the compounds cannot be fully oxidized to carbon dioxide and water, as they have to accept electrons. The ATP synthesized in fermentation is produced by substrate-level phosphorylation as found in glycolysis. Thus fermentation begins with an organic substrate, glycolysis and finally the formation of an end product. The type of end product has been used to identify and name various types of fermentation and some of the products are of considerable commercial value. Examples of the fermentation products are given in Table 2.4.

2.10.3 Chemoautotrophs

The **chemoautotrophs** or **chemolithotrophs** are those organisms that obtain their energy by the oxidation of inorganic compounds. These compounds include hydrogen, sulphur, iron sulphide, ammonia, and nitrite ions (Table 2.5). The micro-organisms that are capable of this type of metabolism are a small group of Bacteria and Archaea and each group metabolizes only one substrate. The amount of energy released by this metabolism is small so that the amount of substrate required for growth is large and this makes these organisms important in the environment and in biogeochemical cycling. Some thiobacilli oxidize sulphur to sulphate and sulphuric acid produces acid mine drainage. *Thiobacillus ferrooxidans* oxidizes ferric iron to ferrous iron and sulphur and has also been found in acid mine drainage. The oxidation of ammonia and nitrite are carried out by nitrifying bacteria *Nitrosomonas* and *Nitrobacter* that are very important in soil fertility and activated-sludge systems.

2.10.4 Photoautotrophs

Photoautotrophs are micro-organisms that obtain their energy from sunlight and those that produce oxygen are represented by algae, cyanobacteria

Table 2.4 Types of fermentation in micro-organisms

Pathway	End product	Examples	Process
Homolactic	Lactic acid	*Streptococcus, Lactococcus, Lactobacillus*	Cheese, yogurt
Heterolactic	Lactic acid + ethanol + carbon dioxide	*Leuconostoc* (Embden–Myerhoff pathway not glycolysis)	Sauerkraut
Ethanolic	Ethanol + carbon dioxide	*Saccharomyces cerevisiae* (yeast)	Brewing, wine, fuel
Propionic	Propionic acid + carbon dioxide	*Propiobacterium*	Swiss cheese (with holes)
Mixed acid	Ethanol, acetic, latic, succinic, and formic acids; hydrogen + carbon dioxide	*E. coli, Salmonella, Shigella*	Products used for identification
Butanediol	Butanediol + carbon dioxide	*Enterobacter, Serratia, Erwinia, Klebsiella*	
Butyric acid	Butyric acid, butanol, acetone, carbon dioxide	*Clostridia* spp.	Was used to produce acetone
Amino acid	Acetic acid + ammonia + carbon dioxide	*Streptococcus, Clostridium, Mycoplasma*	
Methanogenesis	Methane + carbon dioxide	Archaea, Bacteria; some are chemoauto-trophs while others are chemo-organotrophs using acetate	Anaerobic digestion, biogas

(blue-green algae), and green and purple sulphur bacteria. These organisms contain a number of pigments including chlorophylls and these trap light energy. Another group of organisms, the purple and green sulphur and non-sulphur bacteria, use light to generate energy but do not generate oxygen. In the sulphur bacteria the inorganic compounds H_2S and H_2 are used as electron donors and in the other group organic compounds such as malate are used as electron donors.

2.10.5 Photoheterotrophs

Photoheterotrophs (photo-organotrophs) are organisms that can generate ATP from light but use organic molecules such as malate as their electron and carbon source. These are represented by the purple and green non-sulphur

Table 2.5 Chemoautotrophic (chemolithotrophic) growth

Electron donor	Electron acceptor	Reaction	Genera
S0	O_2	$S0 + 1.5\ O_2 + H_2O = H_2SO_4$	*Thiobacillus, Sulfolobus, Aquifex, Acidianus*
$S_2O_3^{2-}$	O_2	$S_2O_3^{2-} + 2O_2 + H_2O = 2SO_4^{2-} + 2H^+$	*Sulfolobus, Thiobacillus, Thiomicrospira*
H_2S	O_2	$H_2S + 0.5\ O_2 = S + H_2O$	*Thiobacillus, Sulfolobus, Beggiatoa, Thiothrix*
H_2	O_2	$H_2 + 0.5\ O_2 = H_2O$	*Alcaligenes, Aquifex, Thermocrimis, Acidianus, Hydrogenobacter,*
H_2	NO_3	$4H_2 + HNO_3 = NH_4OH + 2H_2O$	*Pyrolobus*
H_2	NO_3	$H_2 + HNO_3 = HNO_2 + H_2O$	*Aquifex, Pyrobaculum*
H_2	SO_4	$4H_2 + H_2SO_4 = H_2S + 4H_2O$	*Archaeoglobus,*
H_2	CO_2	$4H_2 + CO_2 = CH_4 + 2H_2O$	*Methanobacterium, Methanobrevibacter, Methanococcus, Methanothermus*
Fe^{2+}	O_2	$2Fe^{2+} + 2H^+ + 0.5\ O_2 = 2Fe^{3+} + H_2O$	*Leptospirillum, Siderocapsa, Thiobacillus*
CO	O_2	$CO + O_2 + 2H^+ = CO_2 + H_2O$	*Hydrogenomonas*
NH_4^+	O_2	$NH_4^+ + 1.5\ O_2 = NO_2^- + H_2O + 2H^+$	*Nitrosomonas, Nitrosospira, Nitrococcus, Nitrosolobus*
NO_2^-	O_2	$NO_2^- + 0.5\ O_2 = NO_3^-$	*Nitrobacter, Nitrococcus, Nitrospina, Nitrospira*

bacteria and one example is the extreme halophile *Halobacterium salinarium*, a member of the Archaea.

2.11 Growth and growth kinetics

Most bacterial cells when they reach the correct size divide into two equal cells in a process known as **binary fission**. At the start of division a cross wall forms, the DNA doubles and separates into the new cells, and finally the two separate (Fig. 2.5). In some cases like the chain-forming streptococci the cells fail to separate, forming chains. Not all micro-organisms divide by binary fission, some form buds which pinch off from the mother cell. Rather than dividing fungi elongate their hyphae in an area just behind the hyphal tip. This combined with branching gives fungi a spreading growth on solid medium and the formation of pellets or mats in shaken-liquid cultures. Most eukaryotic micro-organisms, yeast, algae, and protozoa, divide asexually after mitosis.

Micro-organisms, when placed in a suitable medium and under favourable conditions, will increase in size and either divide into two or form a bud. This division will continue indefinitely until some component of the medium becomes exhausted or conditions become unfavourable. Normally growth ceases because of the exhaustion of the carbon source, generally sugar, but changes in conditions such as a rise in temperature can also stop cell division. The growth of a micro-organism in such conditions is known as the **growth cycle** and has a number of recognizable phases (Fig. 2.18).

- The **lag phase** occurs immediately after inoculation of a micro-organism into medium where no growth may occur for a period of time. The length

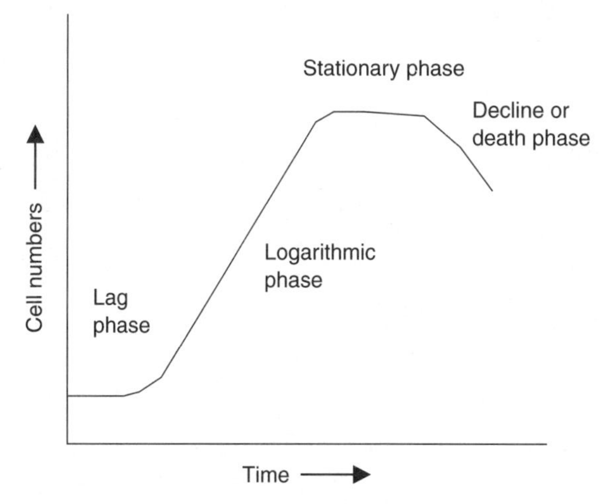

Figure 2.18 The growth phases of micro-organisms.

of this lag phase is variable and dependent on the previous growth history of the inoculum. It represents a period of adaptation to growth in the new environment, and can involve the synthesis of new enzymes.

- The phase following the lag phase is the **exponential** or **logarithmic phase**. In this phase the increase in cell number or cell mass is constant so that the population doubles at regular intervals.
- Once a nutrient becomes limiting, growth slows and ceases. This is the **stationary phase** where there is no growth. In fact in some cases growth may be occurring but any gain is balanced by cell death. In the stationary phase cells may remain viable for long periods.
- The **decline phase** is a condition where the death rate exceeds growth and the cell number or cell mass declines.

2.11.1 Growth kinetics

When microbial cells are in the exponential phase growth is constant and proportional to the initial number of cells. Thus the number or mass of cells will increase exponentially as shown in Fig. 2.18, where X is the initial inoculum. Growth has been followed by measuring the absorbance at 600 nm and can be plotted using both linear and logarithmic scales (see Box 2.9). As can be seen in Box 2.9 the logarithmic scale transforms the growth phase into a straight line, hence the name logarithmic growth. As growth is constant the interval between the doubling of the cells is known as the doubling time (t_d) and the slope of the graph is the growth rate. Thus the amount of cells present after time t (X_t) can be related to the initial cells (X_0).

$$X_t = X_0 2^n \qquad (2.3)$$

where n is the number of doublings after time t. This can be represented by the time t divided by the time it takes to double t_d, the doubling time.

$$N = \frac{t}{t_d} \qquad (2.4)$$

Taking logarithms,

$$\ln\left(\frac{X_t}{X_0}\right) = \ln 2^{\frac{1}{t_d}} \qquad (2.5)$$

$$\frac{\ln X_t - \ln X_0}{t} = \frac{\ln 2}{t_d} \qquad (2.6)$$

$$= \frac{0.693}{t_d} \qquad (2.7)$$

The term

$$\frac{\ln X_t - \ln X_0}{t} \tag{2.8}$$

represents the slope of the lines in panels (b) and (d) of Box 2.9, and is also known as the growth rate μ. Thus

$$\mu = \frac{0.693}{t_d} \tag{2.9}$$

The equation describes growth only in the exponential phase where the growth rate is essentially constant.

BOX 2.9 Example of the growth of micro-organisms

The growth of *S. cerevisae* in medium containing 10 g l^{-1} yeast extract, 20 g l^{-1} glucose, and 20 g l^{-1} peptone at 30°C has been followed. Growth was in a 1-litre shake flask containing 300 ml of medium. Samples were taken and the absorbance at 600 nm, cell dry weight (filter), and total cell number (haemocytometer) determined. The data are given in the table.

Time (h)	Absorbance (600 nm)	Cell dry weight (g l^{-1})	Total cell number (l^{-1})
0	0.08	0.22	7.0×10^9
1.5	0.26	0.32	9.7×10^9
3.5	0.68	0.50	21.5×10^9
6.25	1.84	1.06	6.42×10^{10}
8.2	3.78	2.12	11.3×10^{10}
10.5	7.04	3.65	22.6×10^{10}
12.0	8.6	4.40	31.2×10^{10}
22.0	9.1	4.50	36.0×10^{10}
23.5	9.55	4.63	36.6×10^{10}

From these data the specific growth rate, doubling times, and biomass productivity can be determined. Panel (a) of the figure below shows the absorbance (o) and cell dry weight (•) plotted against time and panel (b) the same data plotted on a log/linear graph. Panel (c) shows the absorbance (o) and cell number (•) plotted against time and panel (d) the same data on a log/linear graph. The data give doubling times (t_d) of 1.5, 1.7, and 2.6 days for absorbance, cell number, and dry weight respectively and this gives growth rates (μ) of 0.46, 0.4, and 0.26 days $^{-1}$.

BOX 2.9 **(*Continued*)**

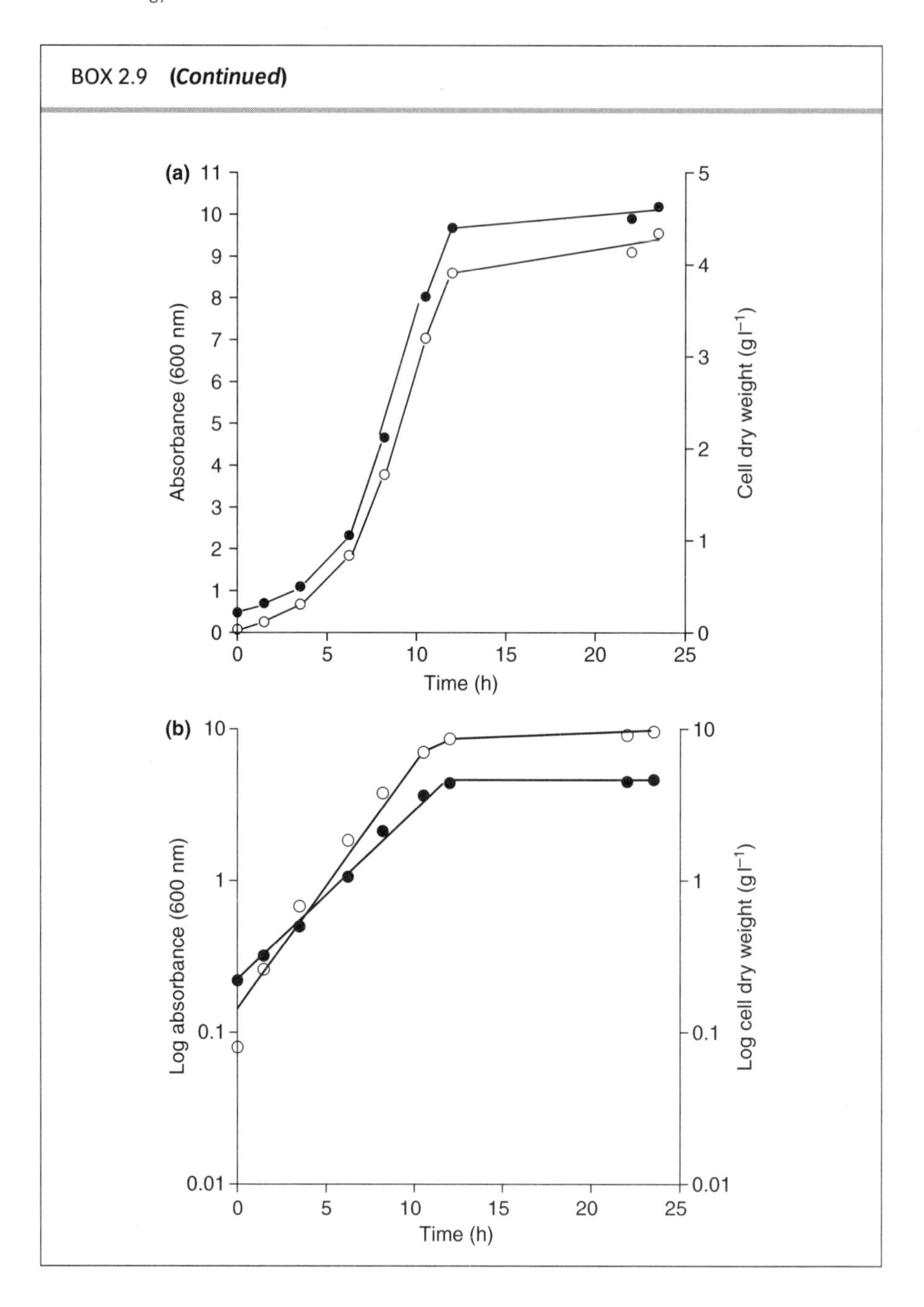

BOX 2.9 ***(Continued)***

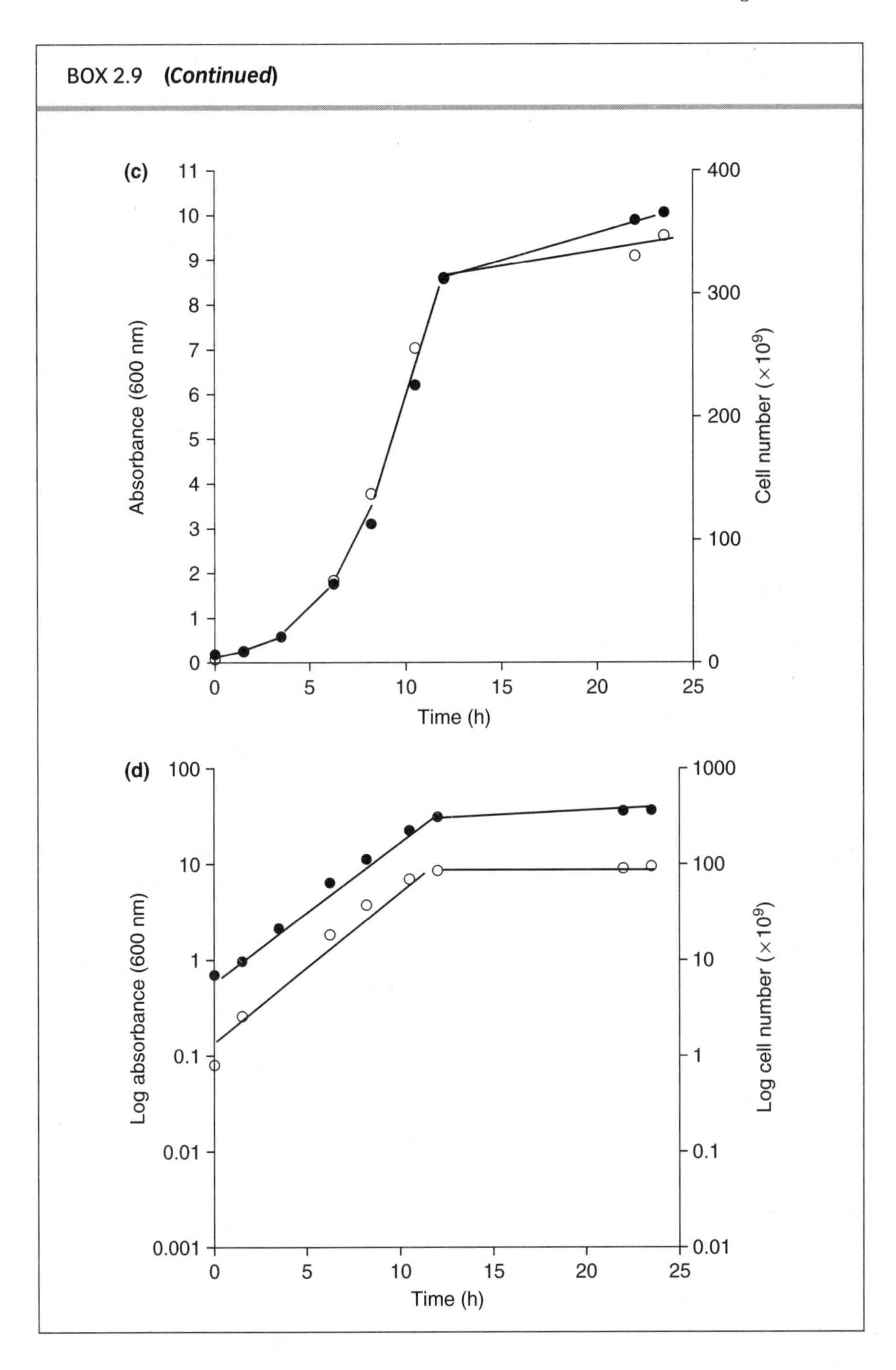

2.11.2 Biofilms

A biofilm is a matrix of microbial cells and extracellular products attached to a biological or non-biological surface. Biofilms consist of single cells, cell aggregates and microcolonies embedded in an exocellular polymer matrix (Winpenny et al., 2000) (Fig. 2.19). Biofilms can be both beneficial, such as in wastewater treatment and activated sludge, and detrimental, such as corrosive biofilms on buildings and dental plaque. Advances in molecular genetics, CSLM, and flow cell technology has revealed the three-dimensional and real-time structure of biofilms. Attachment to the substratum appears to be linked to the possession of pili and flagella. The mature biofilm can be a thick confluent layer or a matrix of microbial colonies. These colonies can take many shapes including towers and mushroom shapes and there are often spaces between the colonies that allow fluid to flow through the matrix.

2.11.3 Eutrophication (blooms)

Eutrophication is the enrichment of natural waters with organic and inorganic material, especially phosphate and nitrogen, which leads to excessive growth of photosynthetic micro-organisms. Eutrophication can occur naturally due to the accumulation of nutrients and organic biomass, droughts, storms, and floods, but the best-known example is that produced by agricultural fertilizer run-off or accidental spills. The nitrates and phosphates in the run-off encourage the prolific growth of photosynthetic organisms, in particular cyanobacteria. The excessive growth known as blooms can produce toxins that kill fish, and in addition on their death the algae sink to the bottom of the lake or river and decompose, which causes a reduction in oxygen content which can also kill fish and other organisms.

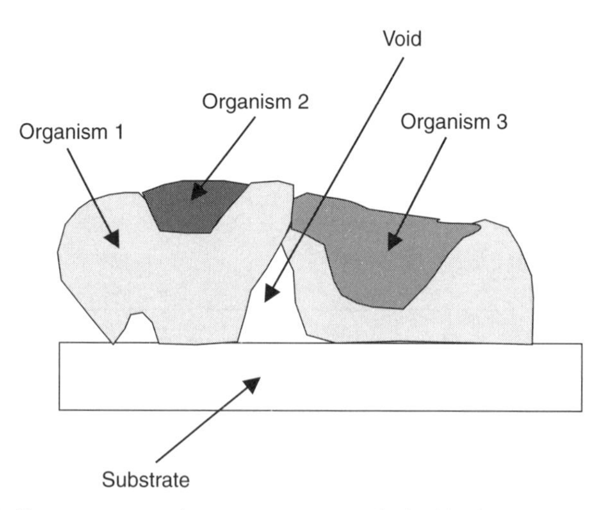

Figure 2.19 Biofilm structure. Each micro-organism in the biofilm forms a colony that can be columnar or mushroom-shaped that leaves spaces or voids that allow the flow of liquids.

2.11.4 Low-nutrient conditions

Most natural ecosystems have a low concentration of nutrients and many micro-organisms have developed strategies to cope with the intermittent nutrient supply. Bacteria have a number of responses to starvation, which include reduction in protein and DNA synthesis at low concentrations of amino acids. Low concentrations of ammonia and phosphate trigger the synthesis of proteins capable of scavenging trace amounts of these nutrients. Low-oxygen conditions can switch facultative anaerobes to fermentation metabolism.

Bacteria that grow specifically at low-nutrient concentrations often have appendages that increase their surface area or form very small cells that have a high surface area/volume ratio. An example of an appendage on a bacterium is the stalk found on *Caulobacter* species (Fig. 2.3).

2.11.5 Spore formation

A few bacterial genera form endospores when nutrients are low or exhausted. The energy for sporulation comes from cellular protein and **polyhydroxybutyrate** (PHB), and takes about 8 h. Once the process of sporulation has started it cannot be stopped (Fig. 2.20). The endospore formed is very resistant

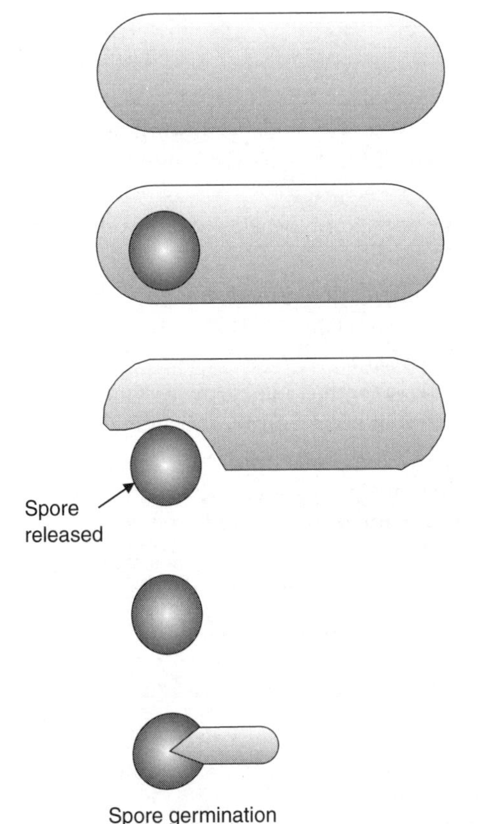

Figure 2.20 The formation of an endospore in a bacillus and subsequent germination.

to desiccation and high temperatures and can remain viable for many years. When conditions become favourable the spore can germinate rapidly and vegetative growth can continue.

Another type of spore can be formed by the myxobacteria. On starvation the cells fuse and form a fruiting body that contains myxospores that can survive for long periods. Eukaryotic micro-organisms frequently form spores as part of their life cycle. Many protozoa form cysts and the slime molds form a fruiting body containing spores. Asexual (conidia) and sexual (conidiospores) spores characterize the fungal formation and many yeasts form ascospores. Algae have both sexual and asexual reproduction.

2.12 Relationship to the environment

Many micro-organisms are free living but in the environment they often form mixed communities in a wide variety of habitats. If these micro-organisms

Table 2.6 Various forms of population interaction

Interaction	Effect
Positive, negative, and neutral	Reduction or stimulation of one population by another. At low population density little effect, i.e. antibiotic production.
Commensalism	One population benefits and the other is unaffected.
Cometabolism	One population growing on one substrate produces as its product a substrate for another population.
Epiphytes	Bacteria can be found growing on the surface of plants, algae, and animals.
Syntrophism	Two populations supply each other's nutritional needs, e.g. archaeal methane production depends on hydrogen transfer.
Rhizosphere	The microbes in soil in contact with plant roots is influenced by root exudates.
Mutualism	A relationship between two organisms that benefits both, i.e. lichens, mycorrhizae, endosymbionts, bioluminescence.
Predation	Many protozoa prey on bacteria.
Parasitism	Some micro-organisms are obligate parasites.

form a stable community they are referred to as **microflora**. The micro-organisms within the microflora may affect each other within this community in a number of ways, as they all have to compete for scarce nutrients. Table 2.6 shows the variation in population interactions that can occur in the environment. In some cases the micro-organisms can actively discourage growth by producing antibiotics, for example *Streptomyces* species. These antibiotics are often produced under low-nutrient conditions. The reverse can also be true, where some organisms can ensure that conditions such as pH are maintained at a level where other organisms can survive. If undisturbed the community will stabilize, a condition in which an alien or introduced organism will find it difficult to establish itself.

Many micro-organisms form associations where each member contributes to the breakdown of a particular substance, known as **cometabolism**. These are known as **consortia** and have been found in populations exposed to pesticides and xenobiotics. Consortia forming on solid surfaces are known as **biofilms** (Fig. 2.19).

2.13 Identification of micro-organisms

Traditionally organisms are classified on their appearance or **phenotype**, which is the observable characteristics of the organism. Traditionally the method of identification of micro-organisms was to isolate a pure culture, which after observation and tests was compared with known organisms using manuals such at Bergey's *Manual of Determinative Bacteriology*. The following characteristics are some of the ones used and are a mixture of morphological and biochemical characteristics. The biochemical characteristics are important as many micro-organisms have few distinguishing morphological features. The preliminary characteristics are

- reaction to Gram stain,
- morphology,
- motility,
- presence of endospores,
- ability to grow under aerobic/anaerobic conditions, and
- the ability to produce the enzyme catalase.

The traditional stain used in microbial identification is the Gram stain, which has been used for over 100 years. Those cells that retain the stain are classified as Gram positive and this characteristic has been linked to their cell-wall structure. Those cells that do not retain the stain are Gram negative and again this is linked to cell-wall structure and antibiotic sensitivity (Box 2.10).

BOX 2.10 The Gram stain

The Gram stain has been used for over 100 years and is generally the first stage in the identification of a bacterium and is probably the most frequently used stain. The Danish scientist Hans Christian Gram discovered the stain that now bears his name in 1884. The process of the Gram stain is given in the figure below. The cells are fixed on the glass slide by heating over a Bunsen flame and the slide immersed in crystal violet solution. This will stain all the cells a purple colour and the dye is fixed by adding a solution of iodine in potassium iodide. This acts as a mordant, fixing the stain within the cells. The slide is then washed with ethanol for 30 s. The ethanol will remove the crystal violet from some cells, which are referred to as Gram negative, while other cells will retain the purple colour and are referred to as Gram positive. However, if the ethanol wash is carried out for too long a time the stain will be lost from all the cells. The Gram-negative cells are shown up by counter staining with another stain, usually safranin, that will stain them reddish pink. In the dichotomous key of bacterial identification one of the first branches is the Gram stain. The Gram stain is also related to the structure of the cell wall of the bacteria. Gram-positive cells have a thick layer of cell-wall material (peptidoglycan) whereas this is thinner in Gram-negative cells.

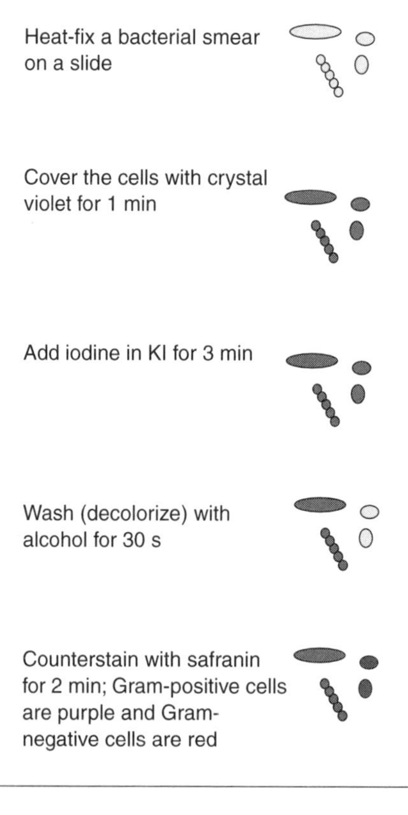

BOX 2.10 ***(Continued)***

More recently, the Gram stain has been used to produce a parameter, the Gram-stain index (GSI),

$$GSI=\left[\frac{B-R}{B+R}\right]$$

Where B is purple/red Gram positive and R is red Gram negative.This index has been used to examine aquatic samples in a study of eutrophication (Saida et al., 2000) using a video camera attached to a microscope and an image analyser.

The colonies formed on solid medium often have a characteristic shape, texture, and colour so this feature could be used as part of the identification of the organism and can be useful if a culture is mixed. The shape of the cells under the microscope can also be used for identification. The main shapes are rods, cocci (circular), and spirals (Fig. 2.3).

The spreading of colonies on agar plates indicates motility due to the presence of one or more flagella. Motility can be seen under the microscope provided that a viscous material is added to the medium to slow the organisms down. With fixed microbial specimens special staining of the flagella with silver will make it possible to see flagella under the light microscope. The presence of an endospore can be seen under the microscope with or in some cases without staining. The ability to grow under aerobic and anaerobic conditions can be determined with stab cultures, pour plates, or covering the culture with paraffin wax. If the cells only grow on the surface they are only aerobic but if growth is throughout the medium then they can grow anaerobically. Anaerobic organisms are notoriously difficult to culture but some indication can be obtained from the stab type of cultures. For those organisms that are very strict anaerobes, incubation in sealed gas jars in which the air has been replaced by nitrogen is required to give good growth.

After the primary tests are carried out a large range of secondary tests can be carried out. Details of these tests can be obtained from most microbiology books (see the Further reading list below). Many diagnostic and medical laboratories use commercial miniaturized multiple test kits to determine the secondary characteristics. Examples of this type of test kit are the Analytical Profile Index (API) and the Enterotube Multitest System.

2.14 Conclusions

This chapter has outlined many of the aspects of micro-organisms that relate to the environment. It should be clear that micro-organisms could and will do almost anything given suitable conditions and do have a great influence on the environment. The discovery of the Archaea and the use molecular biology techniques on environmental samples have shown that the environment contains considerably more organisms than was at first thought and that these can exist in extreme conditions.

2.15 Further reading

Atlas, R.M. (1996) *Principles of Microbiology*. McGraw-Hill, Boston.

Black, J.G. (2002) *Microbiology: Principles and Explorations*. John Wiley, New York.

Dusenberg, D.B. (1996) *Life at the Small Scale*. Scientific American Library, New York.

Heritage, J., Evans, E.G.V., and Killington, R.A. (1996) *Introductory Microbiology*. Cambridge University Press, Cambridge.

Howland, J.L. (2000) *The Surprising Archaea*. Oxford University Press, Oxford.

Hurst, C.J., Crawford, R.L., Knudsen, G.R., McInerney, M.J., and Stetzenbach, L.D. (2002) *Manual of Environmental Microbiology*. ASM Press, Washington DC.

Lester, J.N., and Birkett, J.W. (1999) *Microbiology and Chemistry for Environmental Scientists and Engineers*, 2nd edn. E & RN Spon, London.

Madigan M.T., Martinko M.J., and Parker, J. (1997) *Biology of Microorganisms*, 8th edn. Prentice Hall International, London.

Singleton, P. (1999) *Bacteria in Biology, Biotechnology and Medicine*, 5th edn. John Wiley, New York.

Stanier, R.Y., Ingraham, J.L., Wheelis, M.L., and Painter, P.R. (1998) *General Microbiology*, 5th edn. MacMillan, London.

3 Environmental monitoring

3.1	Introduction	73
3.2	Sampling	83
3.3	Physical analysis	85
3.4	Chemical analysis	86
3.5	Biological analysis	88
3.6	Recombinant DNA technology	88
3.7	Determination of biodegradable organic material	91
3.8	Monitoring pollution	93
3.9	Bioindicators	94
3.10	Biomarkers	96
3.11	Toxicity testing using biological material	100
3.12	Biosensors	102
3.13	Conclusions	110
3.14	Further reading	110

3.1 Introduction

Governments and the public have only over the last few decades become aware of the importance of the environment, and the influence of mankind on the environment. To understand the changes that are occurring in the environment caused by either natural causes or mankind, a large number of environmental parameters have to be taken into consideration. Environmental changes can range from a slow increase in global temperature over years to the rapid accumulation of heavy metals and xenobiotics. Some of the changes are so slow that their determination requires careful monitoring over long time periods. Some of the elements and compounds that accumulate in the environment are present in concentrations so low that they are close to the limits of detection.

Pollution has been defined by Holdgate (1979) as "the introduction into the environment of substances or energy liable to cause hazards to human health, harm to living resources and ecological damage, or interference with legitimate uses of the environment". It is clear that by their very nature human activities produce pollution but the level, and its effect on the balance in the environment, can vary considerably. The first major human influence on the environment was perhaps agriculture, but the industrial revolution, which has left many countries with a legacy of pollution and polluted sites, was also very

Table 3.1 Industrial sites and contaminants

Industry	Sites	Contaminants
Chemical	Acid/alkali works	Acids, alkalis, metals
	Dye works	Solvent, phenols
	Fertilizer and pesticides	Organic compounds
	Pharmaceutical	Organic compounds
	Paint and wood treatment	Chlorophenols
Petrochemical	Oil refineries	Hydrocarbons, phenols, acids, alkali, and asbestos
	Fuel storage	Hydrocarbons
	Tar distilleries	Phenols, acids
Metal	Iron and steel works, foundries, smelters	Metals, especially Fe, Cu, Ni, Cr, Zn, Cd, and Pb
	Electroplating and galvanizing	As above
	Engineering	As above
	Shipbuilding	As above
	Scrap heaps	As above
Energy	Gasworks	Phenols, cyanides, sulphur compounds
	Power stations	Coal and coke dust
Mineral extraction	Mines and spoil heaps	Metals; Cu, Zn, and Pb
	Land restoration	Gas, leachate
	Quarries	Metals
Water supply and sewage	Waterworks	Metals in sludge
	Sewage treatment	Micro-organisms, methane
Miscellaneous	Docks, wharfs	Metals, organic compounds
	Tanneries	Methane, organic compounds
	Rubber works	Toxic, flammable waste
	Military	Explosives, hydrocarbons

Note: Ubiquitous contaminants include hydrocarbons, polychlorinated biphenyls, asbestos, sulphates, and many metals. These may be present on almost any site.

significant. Table 1.1 in Chapter 1 outlines the classes of environmental pollutants that can be found in the environment and Table 3.1 shows some of the industries responsible for the pollutants. The main groups of environmental pollutants which can contaminate land, water, and air are inorganic compounds such as metals and nitrates, organic compounds such as sewage, petrochemicals, and synthetic xenobiotic compounds, micro-organisms including pathogens, gaseous compounds such as volatiles, gases, and particulates. The environmental pollutants can originate from contaminated sites of defunct industries such as gasworks, from existing domestic and industrial sources such as sewage and metals from industries like electroplating, and from accidents and spillages as in the cases of Seveso and the tanker *Amoco Cadiz*.

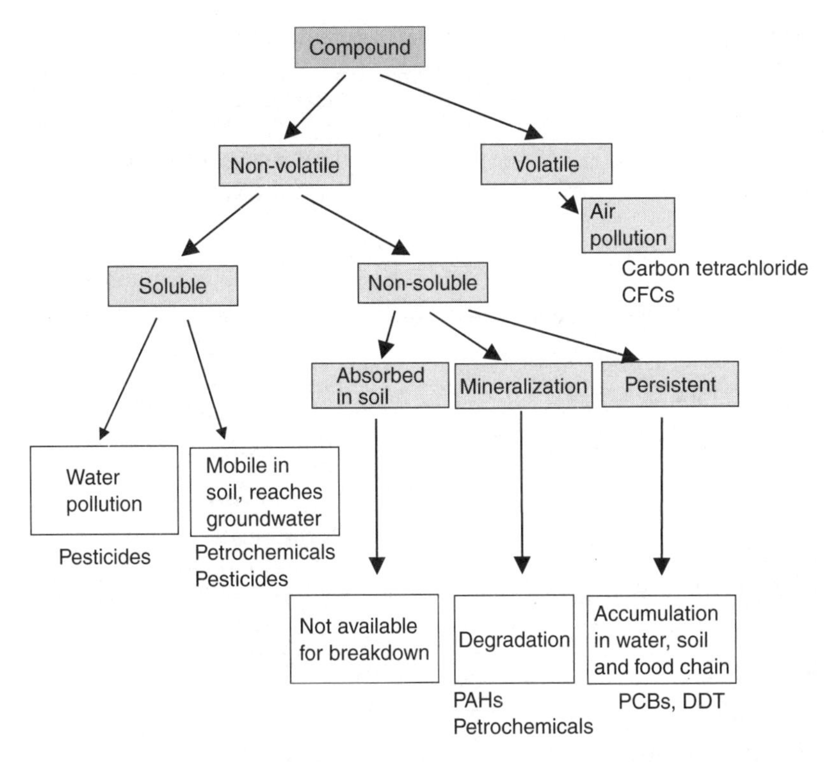

Figure 3.1 An outline of the fate of a compound released into the environment, which depends on its volatility and solubility. CFC, chlorofluorocarbon; PAH, polyaromatic hydrocarbon; PCB, polychlorinated biphenyl; DDT, 1,1,1-trichloro-2,2-*bis*(4-chlorophenyl)ethane.

The fate of pollutants in the environment depends on the properties of the compound and the environmental conditions. The possible fate of compounds released into the environment is given in Fig. 3.1. A full discussion of all the possible routes taken by environmental pollutants is given in Chapter 5. The properties of the compounds that affect their fate are their water solubility, volatility, and reactivity. Often the contaminants are not restricted to their site of introduction, but depending on conditions can migrate and contaminate other parts of the environment such as lakes, rivers, reservoirs, estuaries, and the sea. In addition, the pollutants are not always single compounds but can be complex mixtures containing a number of toxic compounds.

The level and type of environmental contaminant outlined in Table 1.1 (Chapter 1) are regulated by environmental legislation. National laws are normally associated with a number of acts covering many types of pollutants, industries, and conditions (McEldowney and McEldowney, 1996). Of the possible pollutants indicated in Table 1.1 the European Union places the most dangerous compounds on its priority list (Table 3.2). In the USA the Environmental Protection Agency (EPA) has a list of 12 persistent, bioaccumulative, and toxic (PBT) chemicals (Table 3.3) and more extensive lists

Table 3.2 Priority substances under the water framework directive for Europe

Substance	Substance
Alachlor	Lead and its compounds
Anthracene*	Mercury and its compounds
Atrazine*	Naphthalene*
Benzene	Nickel and its compounds
Brominated diphenylethers	Nonylphenols*† (4-(para)-nonylphenol)
Cadmium*	Octylphenols*† (para-t-octylphenol)
C_{10-13}-Chloralkanes*	Pentachlorobenzene*
Chlorfenvinphos	Pentachlorophenol*
Chlorpyrifos*	Polyaromatic hydrocarbons† (benzo(a)pyrene) (benzo(b)fluranthene) (benzo(g,h,l)perylene) (benzo(k)fluoranthene)
1,2-Dichloroethane	Simazine*
Dichloromethane	Tributyltin compounds† (tributyl-cation)
Di(2-ethylhexyl)phthalate*	Trichlorobenzenes† (1,2,4-trichlorobenzene)
Diuron*	Trichloromethane (chloroform)
Endosulphan*	Triflurain
Fluoranthene	
Hexachlorobenzene*	
Hexachlorobutadiene*	
Hexachlorocyclohexane*	
Lindane	
Isoproturon*	

Source: working document ENV/19100/01, European Commission, 2001 (**www.europa.eu.int/comm/environment**).
*Identifies the substance as a priority hazardous substance.
† Where a group of substances has been identified, typical representatives are given underneath in parentheses.

of primary standards for drinking water including micro-organisms and inorganic and organic compounds are given in Tables 3.4 and 3.5. Many of these compounds have strict legal upper limits on land and water. Thus any monitoring must take into consideration the type or types of contaminants present, their availability and the possibility that biomagnification and bioaccumulation may occur. In addition, the environment must be monitored for the presence of plant and animal pathogens. Thus environmental monitoring must be able, in many cases, to detect with accuracy and consistency contaminants present at very low levels. In the determination of the pollutants present, their fate, and their effect on the environment, biotechnology can be of considerable value, especially as molecular biology techniques are increasingly employed.

Table 3.3 US Environmental Protection Agency (EPA) first 12 persistent, bioaccumulative, and toxic (PBT) chemicals

PBT chemicals	PBT chemicals
Aldrin/dieldrin	Mercury and its compounds
Benzo(a)pyrene	Mirex
Chlordane	Octachlorostyrene
DDT*, DDD†, DDE‡	PCBs§
Hexachlorobenzene	Dioxins and furans
Alkyl-lead	Toxaphene

Source: US Environmental Protection Agency, **www.epa.gov**
*DDT, 1,1,1-trichloro-2,2-*bis*(4-chlorophenyl)ethane.
†DDD, 1,1,1-trichloro-2-*o*-chlorophenyl-2-*p*-chlorophenylethane.
‡DDE, 1,1-dichloro-2,2-*bis*(4-chlorophenyl)ethane.
§PCB, polychlorinated biphenyl.

Table 3.4 US national primary standards for drinking water

Contaminant	Maximum contaminant level (MCL; mg/l)	Effects	Sources
Micro-organisms			
Cryptosporidium	99% removal	Gastrointestinal illness	Faecal and human waste
Glardia lamblia	99.9	Gastrointestinal illness	Human and faecal waste
Legionella	NA*	Legionnaires disease	Multiplies in heating systems
Total coliforms	5%	An indicator for the presence of harmful bacteria	Faecal coliform from human and animal waste
Turbidity	5 units	Measures cloudiness, indicating the presence of disease-causing bacteria	Soil run-off
Viruses (enteric)	99.9	Gastrointestinal illness	Human and animal wastes
Disinfection byproducts			
Bromate	0.01	Increased risk of cancer	Drinking-water disinfection
Chlorite	1.0	Anaemia, affects nervous system	Drinking-water disinfection
Haloacetic acids	0.06	Increased risk of cancer	Drinking-water disinfection

Table 3.4 (*Continued*)

Contaminant	Maximum contaminant level (MCL; mg/l)	Effects	Sources
Total trihalomethanes	0.1	Liver, kidney, or central nervous system problems	Drinking-water disinfection
Disinfectants			
Chloramines (as Cl_2)	4.0	Eye/nose irritation, anaemia, stomach discomfort	Water additive to control microbes
Chlorine (as Cl_2)	4.0	Eye/nose irritation, anaemia, stomach discomfort	Water additive to control microbes
Chlorine dioxide (as ClO_2)	0.8	Anaemia, effects on nervous system, infants and young children	Water additive to control microbes
Inorganic chemicals			
Antimony	0.006	Increase in blood cholesterol, decrease in blood sugar	Petroleum refineries, fire retardants, ceramics, electronics, solder
Arsenic	0.01	Skin damage, increased risk of cancer, problems with circulatory system	Natural deposits, run-off from orchards, glass and electronics production
Asbestos	7 MFL†	Increased risk of cancer	Decay of asbestos cement, natural deposits
Barium	0.2	Increase in blood pressure	Discharge of mining wastes, metal refineries, natural deposits
Beryllium	0.004	Intestinal lesions	Metal refineries, coal burning, discharge from electrical, aerospace, and defence industries
Cadmium	0.005	Kidney damage	Corrosion of galvanized pipes, natural deposits,

Table 3.4 (*Continued*)

Contaminant	Maximum contaminant level (MCL; mg/l)	Effects	Sources
			metal refineries, run-off from waste batteries and paints
Chromium	0.1	Allergic dermatitis	Discharge from steel and pulp mills, natural deposits
Copper	1.3	Gastrointestinal illness, long-term liver and kidney damage	Corrosion of plumbing systems, natural deposits
Cyanide	0.2	Nerve damage, thyroid problems	Steel/metal/plastic/fertilizer factories
Fluoride	4.0	Bone disease	Water additive, natural deposits, fertilizer and aluminium factories
Lead	0.015	Affects mental development in children and infants	Corrosion of pipes, natural deposits
Mercury	0.002	Kidney damage	Natural deposits, refineries, run-off from landfills and crops
Nitrate	10	Infants below 6 months old, blue baby syndrome	Fertilizer run-off, leaching from septic tanks, sewage systems, natural deposits
Nitrite	1	Same as nitrate	As above
Selenium	0.05	Hair/fingernail loss, circulatory problems	Petroleum refineries, mines, natural deposits
Thallium	0.002	Hair loss, liver problems	Ore processing, discharge from electronic, glass and drug factories

Source: US Environmental Protection Agency, **www.epa.gov**
*NA, not applicable.
†MFL, million fibres /l.

Table 3.5 List of organic chemicals that can contaminate drinking water

Pollutant	Maximum contaminant level (MCL; mg/l)	Health hazard	Source
Acrylamide	–	Affects nervous system and blood, increased risk of cancer	Added during sewage treatment
Alachlor	0.002	Eye, liver, and kidney problems, anaemia, risk of cancer	Run-off from crops, herbicide
Atrazine	0.003	Cardiovascular problems	Herbicide used on crops
Benzene	0.005	Anaemia, increased risk of cancer	Factories, gas storage, and landfill
Benzo(a)pyrene	0.0002	Reproduction difficulties	Leaching from linings of water tanks, incineration
Carbofuran	0.04	Problems with nervous system	Soil fumigant for rice and alfalfa
Carbon tetrachloride	0.005	Liver problems, increased risk of cancer	Chemical plants
Chlordane	0.002	Liver or nervous problems, increased risk of cancer	Residue from banned termiticide
Chlorobenzene	0.1	Liver/kidney problems	Chemical and agricultural factories
2,4-Dichlorophenoxy acetic acid (2,4-D)	0.07	Kidney, liver, and adrenal problems	Herbicide run-off
Daliapon	0.2	Minor kidney changes	Herbicide run-off
1,2-Dibromo-3-chloropropane	0.0002	Reproductive difficulties	Soil fumigant, run-off
o-Dichlorobenzene	0.6	Liver, kidney, and circulatory problems	Chemical factories
p-Dichlorobenzene	0.075	Anaemia, liver, kidney, and spleen damage	Chemical factories
1,2-Dichloroethane	0.005	Increased risk of cancer	Chemical factories

Table 3.5 (*Continued*)

Pollutant	Maximum contaminant level (MCL; mg/l)	Health hazard	Source
1,1-Dichloroethylene	0.007	Liver problems	Chemical factories
cis-1,2-Dichloroethylene	0.07	Liver problems	Chemical factories
trans-1,2-Dichloroethylene	0.1	Liver problems	Chemical factories
Dichloromethane	0.005	Liver problems, risk of cancer	Drug and chemical factories
1,2-Dichloropropane	0.005	Risk of cancer	Chemical factories
Di(2-ethylhexyl) adipate	0.4	Weight loss, liver problems	Chemical factories
Di(2-ethylhexyl) phthalate	0.006	Reproduction problems, liver and cancer risk	Rubber and chemical factories
Dinoseb	0.007	Reproduction difficulties	Herbicide
Dioxin (2,3,7,8-TCDD)	0.00000003	Risk of cancer, reproduction problems	Emissions from waste incineration and other combustion
Diquat	0.02	Cataracts	Run-off from herbicide use
Endothall	0.1	Stomach problems	Herbicide use
Endrin	0.002	Liver problems	Banned insecticide residues
Epichlorohydrin	0.002	Increased cancer risk	Chemical factories, waste-treatment chemicals
Ethylbenzene	0.7	Liver and kidney problems	Discharge from refineries
Ethylene dibromide	0.00005	Liver and stomach problems	Refineries
Glyphosate	0.7	Kidney problems	Herbicide
Heptachlor	0.0004	Liver damage, risk of cancer	Banned termiticide
Hexachlorobenzene	0.001	Liver and kidney problems	Metal works, agricultural chemical factories

Table 3.5 (*Continued*)

Pollutant	Maximum contaminant level (MCL; mg/l)	Health hazard	Source
Hexachloro-cyclopentadiene	0.001	Kidney and stomach problems	Chemical factories
Lindane	0.0002	As above	Insecticide
Methoxychlor	0.04	Reproductive problems	Insecticide
Oxamyl (Vydate)	0.2	Nervous sytem	Insecticide
Polychlorinated biphenyls (PCBs)	0.0005	Thymus damage, immune problems, cancer risk	Run-off from landfills, waste chemicals
Pentachlorophenol	0.001	Liver and kidney problems	Wood preservation
Picloram	0.5	Liver problems	Herbicide
Simazine	0.004	Blood problems	Herbicide
Styrene	0.1	Liver, kidney, and circulatory problems	Discharge from rubber and plastic factories, landfills
Tetrachloroethylene	0.005	Liver problems, risk of cancer	Factories, dry cleaning
Toluene	1	Nervous system, liver, kidney	Petroleum factories
Toxaphene	0.003	Kidney, liver, and thyroid problems	Insecticide use
2,4,5-TP (Silvex)	0.05	Liver problems	Banned herbicide
1,2,4-Trichlorobenzene	0.07	Changes in adrenal glands	Textile finishing
1,1,1-Trichloroethane	0.2	Liver and nervous problems	Metal-degreasing sites
1,1,2-Trichloroethane	0.005	Liver, kidney, and immune system	Chemical factories
Trichloroethylene	0.005	Risk of cancer, liver	Metal-degreasing sites
Vinyl chloride	0.002	Risk of cancer	Leaching from PVC pipes, plastics factories
Xylenes	10	Damage to nervous system	Petroleum and chemical factories

Source: US Enviromental Protection Agency, **www.epa.gov**

This chapter describes the methods used for determining contaminants present in the environment using microbial assays, biosensors, and genetically manipulated organisms. The effect of the contaminants on the organisms in the environment can be determined using a wide range of biomarkers which are organisms chosen as good indicators of pollution. The population of organisms, their viability, and their nutritional status can now be determined using molecular biology techniques without the restriction of culturing the organisms.

3.2 Sampling

In order to analyse any component of the environment, samples need to be taken and a number of features need to be considered. These features include consideration of what part of the environment needs to be sampled (e.g. water, soil, or air), how many samples need to be taken to cover the area of interest, and the relevant statistical analysis. Also the timing of sampling can introduce variation and seasonal variations need to be taken into consideration. The part of the environment to be sampled will depend on the questions to be answered about the pollutant(s). For example, a chemical spill may contaminate the soil but depending on conditions it may also reach and contaminate the groundwater or adjacent lakes or rivers. In this case a broad spread of sampling will be required to determine whether this has occurred. Samples will have to be taken from the surface and subsurface soil and from any contaminated water. The number of samples is important as results may vary between samples and this variation needs to be estimated. A high degree of variation would indicate that pollution is localized and that further sampling is needed. The collection of a representative sample from a homogeneous source is no problem, but with soil, air, and water samples this is rarely the case. Often soil contamination may be very localized and waste-stream contamination may be intermittent and poorly mixed.

3.2.1 Land (site) sampling

Contaminated sites often require some form of survey, as all the information as to the pollution present cannot be determined from the site history and management. Some directives (British Standards Institute, 1988) suggest that 17 samples should be taken for a 5-hectare site, at three different depths, which should be decreased for sites below 0.5 hectares. However, contamination is frequently patchy so that the number of samples taken will depend on how small an area the contamination covers. Table 3.6 gives the probability of locating contamination using random-sampling grids. These grids can be rectangular, square, diamond-shaped, or herringbone-shaped. The drawback of these types of grid is that they can generate a large number of samples, which

Table 3.6 Probability of locating contamination using random-sampling grids

Contamination as % of total area	Percentage probability (%)		
Number of samples taken ...	10	30	50
1	10	26	39
5	40	79	92
10	65	96	99
25	94	100	100

Source: Adapted from Cairney (1987).

can be expensive to analyse. Therefore, a site history can be of great use in directing the sampling process, and reducing costs.

3.2.2 Water sampling

Sampling of most wastewater and contaminated water is difficult due to their highly variable nature (Keith, 1988). To obtain an accurate assessment, samples will have to be taken over a time period, over different sections of the waterway, and at different depths. There are various automatic methods of taking samples that can be used. Some industrial discharges into waterways are intermittent, which will extend the time that sampling must be carried out. Where to sample in the waterway depends on any in-flow and out-flow of water; stratification and the whole waterway may need to be assessed. If groundwater is to be monitored wells will have to be drilled and the very process of drilling can alter or contaminate samples. Contamination can come from the drilling method, the casing material, and the sample process. These types of consideration have to be evaluated when choosing the sampling method and analysing the results.

When a specific organism is to be surveyed in the environment in order to assess contamination, the samples have to be as representative as possible. One example is the sampling of the edible mussel *Mytilus edulis*, which accumulates metals and can therefore be used as an indicator of metal pollution. The following points had to be taken into consideration when sampling the mussels:

- time of the year; late winter is favoured as metal concentrations are stable at that time;
- size or age; dominant size is taken;
- position on shore; mussels are taken from rocks to avoid contamination by silt; and
- sample size; a minimum of 25 samples.

After sampling the mussels would be washed in fresh water, the soft tissue removed, as this is the site of accumulation, and the tissue extracted and assayed. Alternatively the soft tissue can be stored at –20°C before assay.

3.2.3 Air sampling

Air sampling has much the same problems as water but is also influenced by wind direction and strength (Colls, 1997). The purpose of sampling is to obtain a representative sample and in general there are three sampling systems used for air: pumped systems, pre-concentration, and grab samples. In the pumped system the sample is pumped directly from the air into the analyser. In cases where the pollutant is present at very low concentrations pre-concentration is required before analysis. An example of the type of system is the adsorption of the contaminant on to activated charcoal for removal and analysis at a later date. Grab samples involve the capture of samples of air in bottles, syringes, bellows, and bags for analysis later. Under normal conditions the air is well mixed and samples are normally taken at a height of 2 m, but if boundary layers form then stratification of the air may occur and towers will be required to take samples above 2 m.

If they cannot be analysed immediately samples taken from soil, water, or air can be stored but care is required, as pollutants can be lost or can change during storage. In a number of cases some form of extraction will be needed to prepare and concentrate the sample for analysis and there are a number of liquid/liquid, solid/liquid systems that can be used. Often the extraction methods will also concentrate the sample, helping analysis.

3.2.4 Analysis

Three parameters need to be known before the analysis can be carried out.

1. The limit of detection of the compound, which will give the minimum concentration that can be detected.
2. The variation found with the method of analysis, which will give an estimate of the variation expected between identical samples.
3. How stable is the compound once the sample has been taken?

The number of methods available for the analysis of environmental pollutants and conditions is very great but can be divided into physical, chemical, and biological. Clearly it is in the biological methods that biotechnology will have the greatest influence.

3.3 Physical analysis

Physical methods are often used to determine the conditions that the compounds are exposed to as well as the contaminants, and are in general as follows.

- Gravimetric; used to determine suspended solids (SS), total or volatile solids, and sulphate levels.
- pH; very acid or alkaline conditions will be corrosive and restrict biological activity. Easily measured with a pH electrode.
- Colorimetric; colour and turbidity are important in water quality. These can be determined using comparison tubes, colour discs, colorimeters, and spectrophotometers.
- Dissolved oxygen; this can be measured using an oxygen electrode. Oxygen levels are very important in water quality in order to maintain aerobic biological organisms.
- Ion-specific electrodes; these electrodes can be used to determine the levels of ammonia, nitrate, nitrite, calcium, sodium, and other ions. The determination of nitrate and nitrite is important, as minimum levels have been set for water quality.

The oxygen- and ion-electrode technology allows the possibility of automated analysis, remote sensing, and monitoring. However, many of the newer electrodes suffer from instability and all are prey to fouling and damage.

3.4 Chemical analysis

There are a number of texts (Mason, 1996; Tebbutt, 1998; Gray, 1989; Shaw and Chadwick, 1998) that outline standard chemical methods for the determination of environmental components such as chloride, nitrate, nitrite, and phosphate. An indication of the range of techniques available is given in Table 3.7.

Metals can be determined by chemical methods or by atomic adsorption spectrophotometry. Organic compounds can be assayed by a number of techniques such as high-performance liquid chromatography (HPLC) and gas/liquid chromatography (GLC), which can be linked to mass spectrometers in order to identify the compounds separated (Table 3.7).

The oxygen demand in a waterway can be estimated by measuring the chemical oxygen demand (COD). COD is determined by refluxing the sample with potassium dichromate in concentrated sulphuric acid with silver sulphate as a catalyst (mercuric sulphate is also added to complex any chlorides present). The sample is refluxed for 2 h and the potassium dichromate remaining is determined by titration against ferrous ammonium sulphide. This will give a measure of the oxygen required for the oxidation of the contents in the sample. A different measurement is the total organic carbon (TOC) which measures the organic carbon present and not the oxygen requirement. TOC is measured by electrical combustion of the sample and the

Table 3.7 Methods of chemical analysis of environmental samples

Method	Comments
TLC (thin-layer chromatography)	Replaced the original paper chromatography. Flat plate coated with a variety of media. Compounds move at various rates when placed in a solvent.
GLC (gas/liquid chromatography)	Volatile compound separated on a long column over which a mixture of gases is passed. A number of detection methods can be applied.
HPLC (high-performance liquid chromatography)	The HPLC column consists of fine particles with various groups attached. The liquid phase is passed through the column at high pressure. Again a number of detectors can be used.
MS (mass spectrometry)	The sample is bombarded in a vacuum with electrons or positive ions and the fragments produced are separated by their charge and mass.
NMR (nuclear magnetic resonance)	The resonance frequency of atomic nuclei in a strong magnetic field will vary with their chemical environment.
IR (infrared spectroscopy)	The sample is placed in the beam of infrared radiation. Particular bonds making up groups absorb radiation resulting in a spectrum characteristic for specific groups.
AAS (atomic absorption spectrophotometry)	Used for determining metals. A solution is sprayed into a flame and the emission spectrum measured, which is characteristic for individual metals.
ELISA (enzyme-linked immunosorbant assay)	The very specific reaction between antibody and antigen is used to measure the compound (antigen). Quantification of the binding is estimated by linking the antibody to an enzyme.
RIA (radioimmunoassay)	The same as ELISA except the antibody is radioactive, usually with ^{131}I.
GLC-MS	In this GLC is combined with MS to identify the peaks eluted.
HPLC-MS	In this HPLC is combined with MS to identify the peaks eluted.

carbon dioxide formed determined by infrared analysis. Both measurements are useful in defining the waste in terms of its total carbon content and how much of the carbon could impose a COD.

3.5 Biological analysis

3.5.1 Microbiological determination of cell numbers

Traditional microbiological methods for the detection and estimation of microbial numbers involve a number of techniques both direct and indirect. The details of the methods used for direct estimation of micro-organisms are given in section 2.5 (Chapter 2) and are outlined below.

- Dry weights produced by centrifugation or filtering.
- Cell numbers; haemocytometer, Coulter counter, flow cytometer.
- Viable cell counts, plate counts, most probable number (MPN), Miles and Misra, filter and culture, specific fluorescent dyes (BacLight™, FDA or fluorescein diacetate, SimPlate™).
- Packed cell volume.
- Spectrophotometry; optical density (absorbance) at various wavelengths.

These microbiological methods have limitations with environmental samples. Organisms attached to sediments and particles are difficult to quantify and the results probably grossly underestimate the cell numbers. The viable count methods relies on the premise that each colony formed has been derived from a single cell and by using selective media the number of specific bacteria such as *Salmonella* and coliforms can be estimated. These selective techniques tend to favour the faster-growing species and some of the important slower-growing species can be missed. Pure culture isolation and estimation by viable counts or MPN cannot reproduce conditions found in the environment and the interactions between the micro-organisms, and therefore many cells will not grow. It has been estimated that 90–99% of the environmental species are not culturable (Macnaughton et al., 1999). Therefore, the growth characteristics of the organisms of interest need to be known before their numbers can be estimated accurately.

3.6 Recombinant DNA technology

Some of the techniques used in molecular biology and recombinant DNA technology are described in Chapter 2. These developments have enabled the

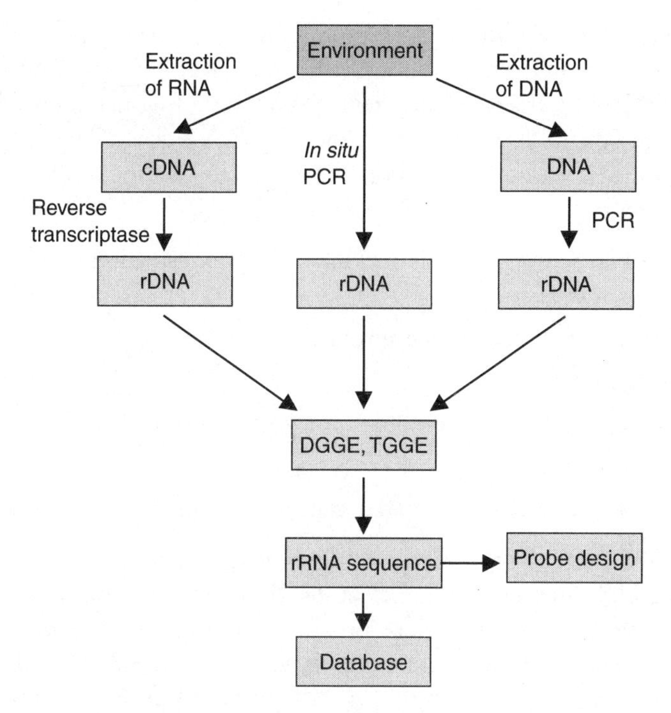

Figure 3.2 Some of the molecular biology techniques that can be used to determine the microbial population in either environmental samples or *in situ*. Both DNA and RNA can be extracted and the DNA produced amplified by PCR. The PCR products can be analysed by gel fractionation using DGGE or TGGE (denaturing-gradient gel electrophoresis, temperature-gradient gel electrophoresis) giving a number of bands that can be extracted. The extracted DNA can be further amplified by PCR and sequenced. The sequence can be used to identify the micro-organism and to design probes.

study of the environmental microbial population both in the laboratory and *in situ* in considerable detail which was not possible with conventional methods based on the culture of micro-organisms.

The possible methods and the sequence of their use are given in Fig. 3.2, where DNA can be amplified by the polymerase chain reaction (PCR) from extracted DNA or RNA treated with reverse transcriptase. The most frequent primers used in the PCR technique are those that select for the 16 S ribosomal RNA (rRNA) sequences. Denaturing-gradient gel electrophoresis (DGGE) or temperature-gradient gel electrophoresis (TGGE) can separate the product of PCR and the bands obtained can be amplified by PCR and sequenced. Once a sequence has been obtained it can be compared with the database of rDNA sequences to give an identification. The sequence will also allow the design of probes, which can be used in fluorescence *in situ* hybridization (FISH).

If specific primers are chosen these will amplify low-copy or single genes present in the sample, which would not have been detected otherwise.

The use of recombinant DNA techniques for the study of bacterial communities, particularly those in environments difficult to reproduce, allows the assessment of bacterial diversity without cultivation. Many of the molecular probes are based on the small-subunit rRNA (ssrRNA). The rRNA genes can then be detected by fractionating the DNA, cloning fragments into the bacteriophage lambda, and screening the clones for rRNA genes. The following are examples of the use of recombinant DNA technology in the study of microbial communities.

- The microbial ecology of soils and those from extreme habitats such hydrothermal vents.
- Aerobic and anaerobic sewage systems.
- Biodiversity.
- Microbial populations in contaminated groundwater.
- Bioremediation.
- Release of genetically manipulated micro-organisms.

An example is the use of whole-cell hybridization with domain- and kingdom-specific fluorescent-oligonucleotide 16 S-derived probes to study the diversity of the thermophilic micro-organisms in deep-sea hydrothermal vents (Harmsen et al., 1997). Microbial populations in the deep sub-seafloor as far down as 800 m have been shown to be low in culturability although 16 S rRNA techniques have shown a significant biomass (Webster et al., 2003). The variation in microbial populations has been investigated by extracting the nucleic acids, amplifying the 16 S rRNA gene by PCR, cloning the fragments, and comparing the gene sequences. The resulting sequences have shown that the majority of micro-organisms in the extreme habitats are unknown and only a small fraction of the microbial diversity had been detected by cultural methods (Polz and Cavanaugh, 1997). Quantitative hybridization can also provide an estimate of the dominance of individuals in populations. Microbial diversity in the natural environment has also been investigated by using variation in the 16 S RNA and a taxonomy based on these data has been adopted (Woese et al., 1990).

DNA technology has been used to follow the bacterial population in activated sludge and the population dynamics in a trickle-bed bioreactor. It has been clear that culture techniques are not capable of providing full details of the population in activated sludge and 16 S rRNA probes have shown that the dominant organisms were those of the beta subclass of Proteobacteria, whereas culture methods had shown that the gamma subclass of Proteobacteria was dominant (Snaidr et al., 1997). The trickle-bed study used FISH with 16 and 23 S rRNA probes and gave results different from those obtained by culture methods. The FISH technique showed that there was a very diverse population in the trickle-bed filter.

Recombinant technology can be used to estimate the degree of genetic variation in environmental samples (Karp et al., 1997) and has been used to detect bioleaching bacteria (De Wulf-Durand et al., 1997). In the case of bioleaching

PCR was used to amplify the 16 S rRNA obtained from the bacterial population associated with an acidic mining environment.

Gene expression can be measured at the protein level (proteomics) or at the messenger RNA (mRNA) level. In the latter case the molecular biology technology of DNA microarrays is being applied to environmental samples. DNA microarrays can be used to examine expression of thousands of genes simultaneously. DNA assays have been used to identify genes that are affected by xenobiotics and metals.

3.6.1 Proteomics

Proteomics is "the systematic analysis of the protein population in a tissue, cell or subcellular compartment". The ability to separate and identify proteins from cells allows the study at the protein level of the effects of pollutants and early effects of environmental contamination (Liebier, 2002; Steen and Pandley, 2002). Analysis of the protein starts with the one- or two-dimensional separation of the proteins and then the digestion of individual spots or bands (Fig. 3.3). The resulting peptides are extracted and their masses determined by matrix-assisted laser desorption ionization (MALDI) and electrospray ionization (ESI) mass spectrometry (MS). Modern machines give accuracy, mass resolution, and sensitivity which allows for the identification of picomolar–femtomolar amounts of proteins and peptides if matching genomic sequences are available. The integration of these new mass-spectrometry methods, data-analysis algorithms, and information databases of protein and gene sequences has enabled the characterization of the protein profile of an organism.

3.7 Determination of biodegradable organic material

The Royal Commission on sewage disposal established biological oxygen demand (BOD) in 1912 as an important parameter of water quality. Since then it has been a key parameter in the monitoring of water quality and treatment. The BOD is a measure of the oxygen demand in a sample as a result of its metabolizable organic content.

The oxygen demand of wastewaters has been determined traditionally by incubating a sample in a sealed bottle in the dark at 20°C. After 5 days of incubation the oxygen consumed is measured and this is the BOD_5 value. Sometimes the sample is diluted, if necessary, with a volume of activated sludge. Activated sludge is a mixture of aerobic organisms produced by the treatment of waste. The precise test procedure is given in a number of handbooks on water analysis. The test does give reasonable results with normal sewage but the test is biological and will be affected by the time of incubation, temperature, and micro-organisms present. For typical waste this is not a

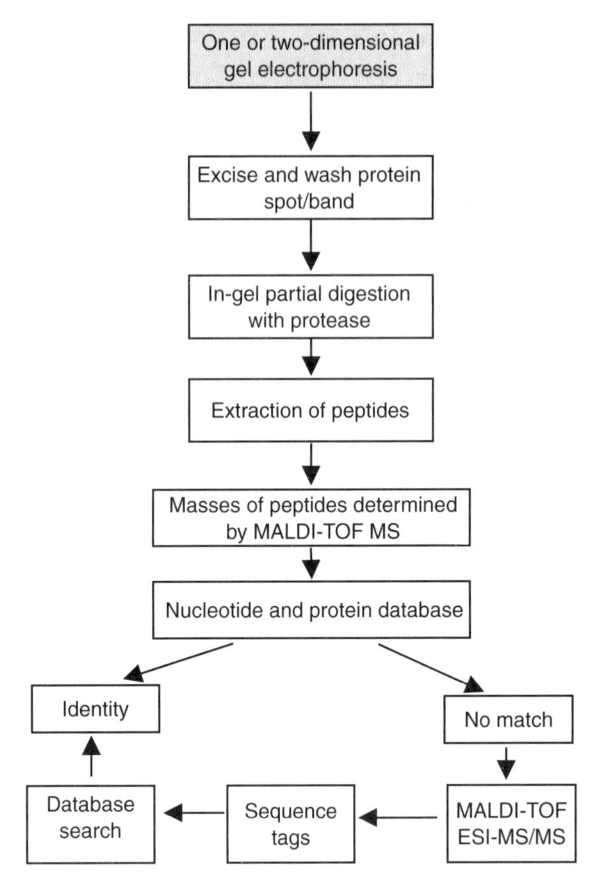

Figure 3.3 Proteomics, some of the methods used when studying protein expression in micro-organisms. The proteins are separated by one- or two-dimensional electrophoresis and the spots or band are extracted. The extracted proteins are partially digested and the masses of the peptides determined by matrix-assisted laser desorption ionization-time of flight (MALDI-TOF) mass spectrometry (MS). This may give sufficient information to identify the protein from the databases. If no match is found further analysis using other techniques such as electrospray ionization (ESI) tandem MS (MS/MS).

problem but for waste which contains high levels of nitrogenous material longer than 5 days will be required for complete oxidation. If the waste is too strong the oxygen will be depleted before oxidation is complete, and certain compounds require the presence of certain micro-organisms for oxidation and are in fact toxic to many micro-organisms. Variations in the BOD assay are to extend the incubation to 7 days, to oxidize ammonia by addition of allythiourea, and dilution of the sample. Determining BOD is a standard method that requires only simple equipment but although used widely the BOD assay has a number of deficiencies:

- slow, taking 5 days before results are obtained;
- time-consuming and expensive for a large number of samples;
- it does not fully represent the natural conditions;

Table 3.8 Typical BOD_5 and COD values (mg/l) for waste streams

Waste	BOD_5	COD
Sewage	200–400	400–600
Cattle waste	16 000	150 000
Pig waste	30 000	70 000
Poultry waste	2 400	170 000
Whey	45 000	65 000
Brewing	2 000	17 000
Sugar beet	2 000	3 000
Potato processing	2 000	3 500
Distillery	7 000	10 000

- oxidation may not be complete in 5 days;
- difficult to interpret with wastes containing high levels of nondegradable organic material; and
- imprecise, particularly with low levels of organic material.

Due to these concerns alternative methods have been investigated, including absorbance and fluorescence (Comber et al., 1996). It was shown that there is a correlation between BOD and absorbance but this is not true for fluorescence. As yet BOD still remains the main indicator of the level of organic material in wastes. BOD is clearly related to the COD; where COD measures all the oxidizable material in the waste, and BOD measures only the biodegradable organic material. Thus the two values will not be the same for a given sample as some wastes contain organic materials such as lignins which are degraded so slowly as to be regarded as non-biodegradable. The ratio of COD to BOD provides a useful guide to the types of organic material in the waste, but because of the rapidity of obtaining the results TOC is often used to define highly contaminated wastes.

In Table 3.8 it can be seen that cattle waste has a COD value considerably higher than the BOD value due to the high levels of fibre, cellulose, and lignin in the waste that degrade only very slowly. This is a feature of ruminant waste whereas with wastes from sugar beet and potato processing the BOD and COD values are much closer.

3.8 Monitoring pollution

The most frequent approach used to determine the effect of pollution on the environment is to determine the concentrations of the pollutants in any environmental situation. To do this there are a wide range of chemical methods and analyses available. However, this does not determine the real effect of the pollution on the organisms that make up the environment as the effects can be modified by availability, degradation, and transport of the pollutants. Alternatives to the chemical methods are to use a biological system to measure the pollutant or effects *in situ* and include bioindicators, biomarkers, and specific test organisms.

- The effects of pollutants on whole organisms representative of the environment, known as **bioindicators.**
- The effects of pollutants on physiological, biochemical, and molecular characteristics of organisms in the environment, known as **biomarkers.**
- The effect of the pollutant on test organisms in the laboratory.

Bioindicators and biomarkers have the advantage that they measure the action of the pollutants in the real and complex environment where there may be many and complex interactions at sublethal levels.

3.9 Bioindicators

Bioindicator organisms are those that can be used to identify and quantify the effects of pollutants on the environment and ideally

- are distributed widely,
- are abundant and not mobile,
- are available all year,
- are easy to collect,
- have a high concentration factor for the specific pollutant, and
- are sensitive to the pollutant.

Examples of the types of changes that can be observed are ecological, behavioural, and physiological. Ecological factors involve changes in population density, key species, and species diversity. Behavioural changes can be feeding activities, bacterial mobility, and web spinning. Physiological changes can be the accumulation of heavy metals, carbon dioxide production, BOD, and microbial activity. Some examples of bioindicators are shown in Table 3.9.

Lichens are a symbiotic relationship between an alga and fungus and have many of the properties required of a bioindicator. Both lichens and mosses are abundant, not mobile, easy to collect, and lichens are particularly sensitive to air pollutants and have been used widely used as bioindicators of air quality (Conti and Cecchetti, 2001; Onianwa, 2001).

Table 3.9 Some examples of bioindicators

Organism	Pollutant	Character
Earthworms (*Eisenia fetida*)	2,4,6-Trinitrotoluene (TNT)	Toxicity and avoidance
Honey bee	Metals	Extracted from bees and honey
Lichens	Air pollution	Physiological parameters
Periphyton*	Acid mine drainage	Chlorophyll content
Mosses	Air pollution	Metal content
Dab (*Limanda limanda*)	DDT	Malformations
Semaphore crab (*Heloecius cordiformis*)	Heavy metals	Extracted from organs
Mussels	Metals	Metal content of soft tissues

*Periphyton is the biomass on surfaces in streams.

Honey bees have been used as a bioindicator for the contamination of the atmosphere by heavy metals (cadmium, chromium, and lead). Honey bees, honey, pollen, and wax were assayed for the presence of the heavy metals and differences were found between reference sites and those in cities (Conti and Botre, 2001).

Another group of immobile organisms, which are suitable for use as bioindicators, are the bivalve molluscs. The filter-feeding mussels are known to accumulate heavy metals, making them ideal for the monitoring of coastal waters. The Mediterranean mussel *Mytilus galloprovincialis* was used successfully to monitor heavy metal contamination using both metal accumulation and the response of certain enzymes to pollution (Regioli and Principato, 1995). Measurements of periphyton chlorophyll content, biomass, and ash-free day weight are commonly used as indicators of pollution. This type of system has been used to assess the impact of acid mine drainage on streams (Vinyard, 1996). Marine pollution has been monitored by following increases in malformations in embryos of dab (*Limanda limanda*) in the North Sea, which have been linked to an unusually high input of DDT and polychlorinated biphenyls (PCBs) from the River Elbe (Von Westernhagen et al., 2001).

Behavioural, sublethal, and lethal tests have been applied to concentrations of 2,4,6-trinitrotoluene (TNT) in contaminated soils using earthworms (Schaefer, 2004). The semaphore crab (*Heloecius cordiformis*) has been shown to be a good indicator of lead in Australian estuaries (MacFarlane et al., 2000).

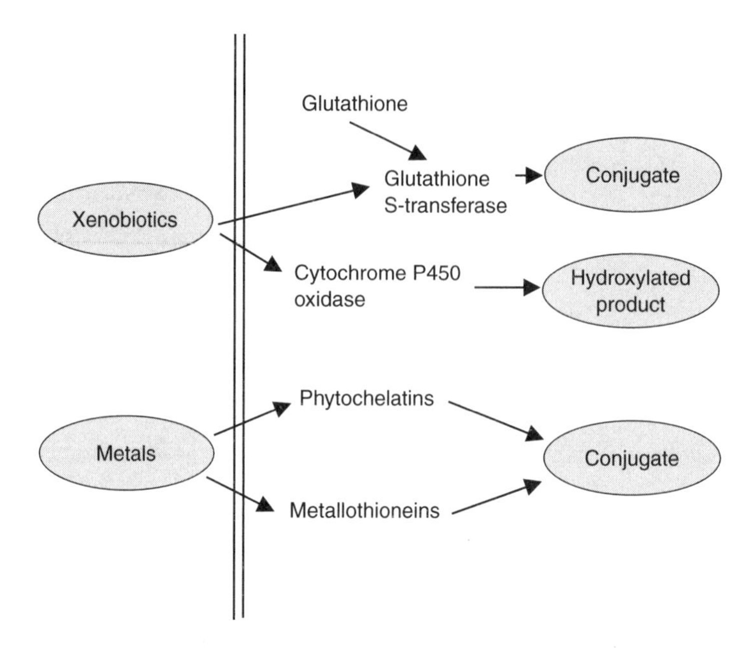

Figure 3.4 The initial stages of some of the processes for the detoxification of xenobiotics and sequestration of heavy metals. Xenobiotics can be oxidized by the cytochrome P450 complex or conjugated with glutathione. Metals are conjugated with either phytochelatins or metallothioneins.

3.10 Biomarkers

Biomarkers can be defined as quantitative measures of changes in the biological system that respond to exposure to metals and/or xenobiotic substances that lead to biological effects. The advantages of biomarkers are as follows.

- The determination can be made over a period of time and not the single sample often used in chemical analysis.
- They can indicate the risks of exposure to a particular chemical.
- By determining the effects in various habitats the different routes to exposure can be established.
- They can provide information on the toxicity of single compounds or mixtures in the real and complex environment at sublethal levels.

Biomarkers can be placed into three groups; biochemical, immunochemical, and genetic indicators.

3.10.1 Biochemical indicators

Biochemical indicators are based on the ability of the pollutant to generate a response at the gene level, inducing or increasing specific enzymes involved

Table 3.10 Biochemical biomarkers

Organism	Assay	Pollutant
Oyster (*Crassostrea gigas*)	Metallothioneins	Metals
Mussel (*Mytilus edulis*)	Glutathione	
Plants	EROD*, AHH†	Dioxins
Freshwater mussel (*Unio tumidus*)	GSH‡ enzymes	Copper, thiram
Sea star (*Asterias rubens*)	Cytochrome P450, benzopyrene hydroxylase	PAHs§

*EROD, ethoxyresorufin-*o*-dethylase.
†AHH, aryl hydrocarbon hydroxylase.
‡GSH, glutathione.
§PAH, polyaromatic hydrocarbon.

with detoxification of contaminants (Fig. 3.4). The extent of this expression serves as a measure of the available concentration of the pollutant. Table 3.10 gives some examples of biochemical indicators used to follow the effect of a number of pollutants. The detoxification of xenobiotics is carried out by the hydrolytic activities of esterase and amidase enzymes but the major reactions are oxidations catalysed by cytochrome P450, which occurs in multiple forms. Exposure to xenobiotics will induce an increase in the enzyme activity. Another process of detoxification found in both mammals and plants is the conjugation with glutathione, catalysed by the enzyme glutathione S-transferase. One of the numerous mechanisms for the detoxification of heavy metals is sequestration by metallothioneins in bacteria and phytochelatins in plants. These are short cysteine-rich peptides with the general structure (γ-glutamic acid-cysteine)$_n$-glycine, where $n = 2$–7. These peptides are synthesized from glutathione by the enzyme glutathione synthase. Other enzymes used to measure the effects of pollutants are ethoxyresorufin-*o*-dethylase (EROD) and aryl hydrocarbon hydroxylase (AHH), which are found to be induced by exposure to hydrocarbons and polyaromatic hydrocarbons (PAHs).

3.10.2 Immunochemistry

The specific reaction between antigens and antibodies can be used to determine the presence of xenobiotics in environmental samples. Antibodies against PCBs, PCDDs (polychlorinated dibenzo-*p*-dioxins), and PCDFs (polychlorinated dibenzofurans) have been developed and used in an ELISA system to determine PCBs in samples (Hahn, 2002). ELISA has also been used to determine methanogens in samples from an anaerobic digester.

Table 3.11 Molecular biology biomarkers

Selection	Assay
Antibiotic-resistance gene, e.g. *npt*II	Selective plates, resistance to kanamycin
Heavy-metal-resistance gene, e.g. *mer*	Selective plates, e.g. resistance to mercury
Chromogenic marker gene, e.g. *lac*ZY	Plate assay with X-gal converted to a blue-coloured product
Bacterial luciferase gene, e.g. *lux*AB	Bioluminescence with or without luciferin, plate count or luminometry
Eukaryotic luciferase gene, e.g. *luc*	Bioluminescence with plate or luminometry
Green fluorescent gene, e.g. *gfp*	Green fluorescence, flow cytometry, microscopy, plates

3.10.3 **Genetic indicators**

There are a number of molecular techniques that can be used to follow the effect of pollutants in the environment and these include the introduction of genetically engineered indicator organisms, and antibiotic- and heavy-metal-resistance genes.

One of the most rapid and sensitive methods of screening for gene expression is to fuse the relevant promoter sequences to those for a product that is easily detected. The detection can be by antibiotic and heavy metal resistance but the best is one where light is produced. A number of examples of these are given in Table 3.11.

The *gus*A gene product will cleave the colourless substrate X-gal (5-bromo-4-chloro-3-indolyl β-D-galactoside) to give a blue colour and this gene can be fused to a promoter that responds to the pollutant. Another example is the insertion into an *E. coli* strain of a plasmid containing the *lac*Z gene, which codes for the enzyme β-galactosidase, which has been linked to the *ars*R gene that codes for the regulatory protein of the *ars* operon (Scott et al., 1997). The *ars* operon is involved in the removal of antimony and arsenic from the cell. Thus when the cells containing the plasmid are exposed to antimony or arsenate the enzyme β-galactosidase is induced and this can be assayed by the addition of *p*-aminophenyl β-D-galactopyranoside. The product of the reaction, *p*-aminophenol, can be determined electrochemically. Thus a bacterial sensing system has been developed that responds to antimony and arsenate.

There is another group of genes where the product does not require substrate addition, as these genes code for light-producing proteins; the luciferase enzyme *lux*AB from prokaryotes and the *luc* gene from eukaryotes,

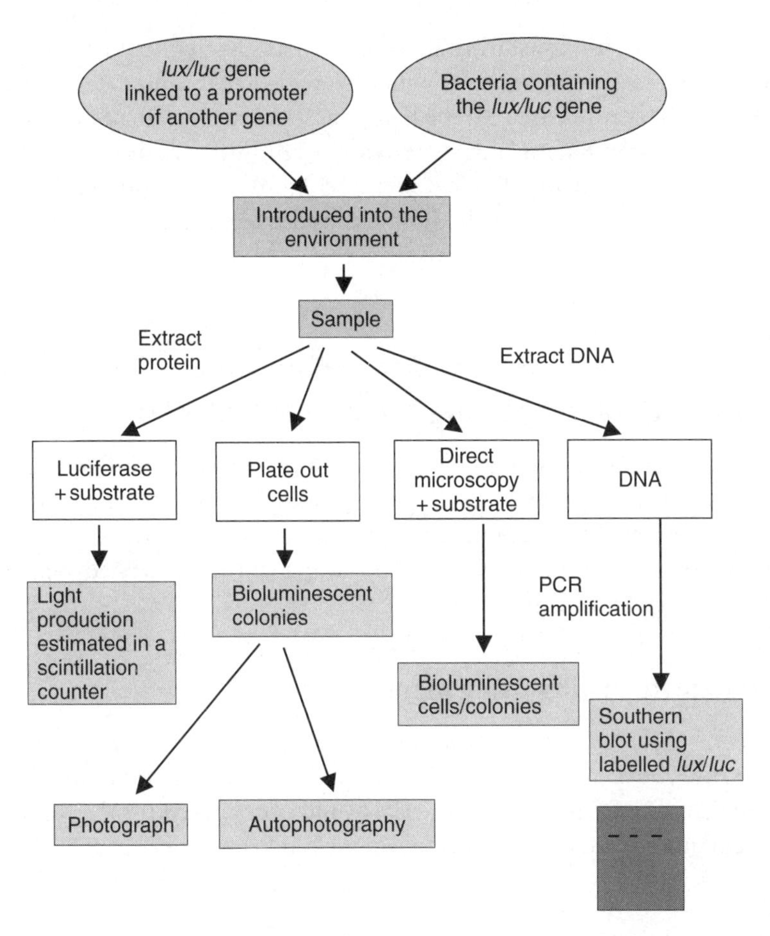

Figure 3.5 The use of genes coding for protein that can emit light in the detection of effects of pollution in the environment.

and more recently a green fluorescent protein coded by the *gfp* gene from a jellyfish, *Aequorea victoria* (Misteli and Spector, 1997; Kendall and Badminton, 1998). These types of marker gene have been placed under the control of the promoters of genes associated with a response to toxic compounds. Thus if the engineered bacteria are exposed to a toxic compound it will trigger light production, which can be detected easily (Fig. 3.5). An example is the construction of an *E. coli* strain which contained a promoter (*alk*B) from *Pseudomonas oleovorans* and the *lux*AB genes of *Vibrio harveyi* (Sticher et al., 1997). The cells responded to alkanes by the production of light and were used to detect soil contamination by heating oil. Another *E. coli* construct was used to determine the levels of metals and *Vibrio fischeri* was used to screen the effect of TNT (Frische, 2002; Ren and Frymier, 2003). Bioluminescence was used to track genetically engineered *Xanthomonas campestris* after its release into the environment (Shaw et al., 1992). *X. campestris* was engineered to contain the *lux* gene, which under the conditions used was constantly bioluminescent. The emission of light was detected using a special camera, which provided real-time measurement of the organisms after release into the field. Bacterial strains are not the only cell lines used; a recombinant mouse hepatoma cell line containing a luciferase gene under the control of a cytochrome P450 1A1 promoter has been constructed. The cytochrome P450 is induced when the cells are exposed to dioxins, PAHs, and petrochemicals.

3.11 Toxicity testing using biological material

Toxicity bioassays often rely on measurement of the lethal effects of a compound on a specific biological system and this is usually carried out with a single pollutant. Some examples of the use of biological material to detect the effect of a compound on a specific organism are outlined below and examples are given in Table 3.11.

3.11.1 Toxicity testing using plants and algae

The use of algae for toxicity testing was introduced in 1964 when the alga *Selenastrum capricornutum* (now known as *Pseudokirchneriella subcapitata*) was applied to pollution (Skulberg, 1964) and is now used for the assessment and evaluation of water quality (Environmental Protection Agency, 1971).

Table 3.12 Trophic levels representing toxicity testing

Trophic level	Type	Example
1	Primary producers	*Chlorella vulgaris*, *Selenastrum capricornutum*, *Senendesus quadricauda*, *Lemna* spp.
2	Primary consumers	Daphnia water fleas, brine shrimp (*Artemia salina*)
3	Secondary consumers, Carnivores	None
4	Tertiary consumers	Fathead minnow, zebra fish, guppy, rainbow trout, blue-gilled sunfish
5	Quaternary consumers, Birds of prey	Pigeon, quail, pheasant, kestrel, buzzard

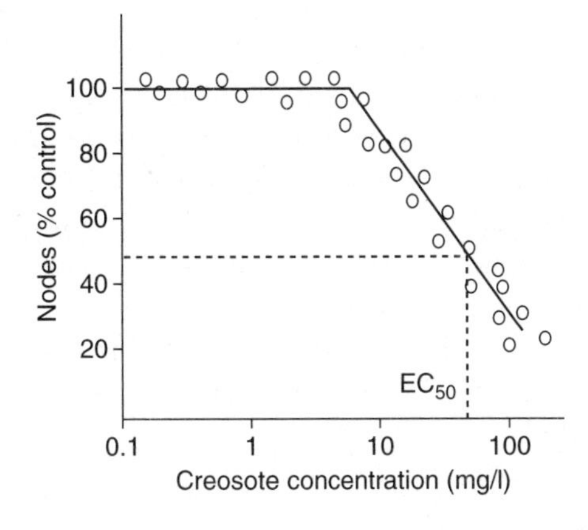

Figure 3.6 The effect of creosote on the growth of *Myriophyllum spicatum* L (aquatic macrophyte) and the determination of the EC_{50} value. Data from McCann et al. (2000).

Toxicity testing using plants and animals can be represented by a number of trophic levels (Table 3.12). The common freshwater algae *Chlorella vulgaris*, *S capricornatum*, and *Senendesmus subspicatus* are often used to represent trophic level 1. Toxicity testing with algae has been developed into standard tests with a number of agencies. Most of the tests are the exposure of the algal culture to the pollutant for 3–4 days and the reduction in growth rate used to calculate the EC_{50} value. The EC_{50} value is the concentration of pollutant that reduces growth by 50% (Fig. 3.6). The EC_{50} value gives the level of toxicant that will affect the environment. The result of the toxicity tests can be affected by the culture conditions such as test species, culture volume, and light intensity.

Other aquatic plants have been proposed as alternatives to microalgae, including duckweed *(Lemna* spp.) *Eloidea canadensis*, periphyton, and phytoplankton (Lewis, 1995; Cleuvers and Ratte, 2002). Marine algae have also been proposed, including filamentous algae *Ceramium strictum* and *Ceramium tenuicorne* (Bruno and Eklund, 2003), *Phaedactylum tricornutum*, and *Champia parvula* (Eklund and Kautsky, 2003).

In trophic level 2 the main organism used is the water flea (*Daphnia* spp.), although the brine shrimp (*Artemia salina*) has also been used. The effect of the pollutant is estimated by the number of organisms immobilized after 24 h of exposure. In trophic level 3 no test organisms exist, possibly due to their similarity in response to level 4 organisms. The tertiary organisms are carnivores such as fish like trout, blue-gilled sunfish, zebra fish, fathead minnow, and guppy. The final group, level 5, represents the top of the food chain and includes a number of bird species.

3.11.2 Luminescent organisms

MICROTOX™, BIOTOX™, ToxAlert®, and LUMISTox® are systems that use the reduction in light emission by luminous bacteria such as *Photobacterium fisceri* and *Photobacterium phosphoreum* as a measurement of the toxicity of a compound over a 15-min exposure period.

3.11.3 Ames test

The Ames test was developed to test substances for their ability to produce mutations in bacteria. The test consists of the treatment of a *Salmonella typhimurium* mutant, which requires the amino acid histidine for growth (*His*$^-$), with the test compound along with an extract of rat liver. The rat liver is added because many non-toxic compounds can be converted to mutagens by the enzymes present in the mammalian liver, which are not found in bacteria. If the compound is a mutagen then reverse mutations will occur and colonies will form on a medium lacking histidine. The number of colonies will give a measure of the mutagenic potential.

3.11.4 Molecular biology biomarkers

Another development using genetic manipulation of biological material for the estimation of toxicity has been the generation of a transgenic strain of the nematode *Caenorhabditis elegans* (Candido and Jones, 1996). The strain contains the *lac*Z gene from *E. coli* fused to the *hsp*16 gene. The *lac*Z gene codes for the enzyme β-galactosidase and the *hsp*16 is an inducible gene that is induced when the nematode undergoes a heat-shock response. Thus when the nematode is stressed the enzyme is induced in the worm. The enzyme can be detected by the addition of a substrate such as *o*-nitrophenyl-β-galactopyranoside (ONPG), which will produce a blue colour when cleaved by the enzyme.

Often in the testing of pollutants a combination of many of the tests mentioned are used to determine the effect of the pollutant on the environment. Wild-type and recombinant cell lines have been used to assess aryl hydrocarbon and oestrogen receptor activity of dioxin-like chemicals and the oestrogen receptor activity of a wide range of chemicals (Giesy et al., 2002). Dioxin-like chemicals are of importance as they cause a range of toxicities and are carcinogenic. The actions of these chemicals are mediated through the aryl receptor (AhR). Compounds that mimic oestrogen activity disrupt normal reproduction and developmental processes, which leads to cancers and reproductive problems. Thus it is important to monitor these chemicals.

Another example is the range of biomarkers and bioindicators used to study the effect of pentachlorophenol (Table 3.13). Pentachlorophenol has been widely used as a wood preservative and because of its highly chlorinated nature is resistant to microbial attack and is toxic, and therefore requires close monitoring.

3.12 Biosensors

A biosensor can be defined as a device that incorporates a biological sensing element in close proximity or integrated with a signal transducer in order to

Table 3.13 The effect of pentachlorophenol on different biomarkers and bioindicators

System	Type	Indicator	Exposure period (h)	EC_{50} (μM)
Chlorella vulgaris	Microalga	Inhibition of growth	24	59
Daphnia magna	Cladoceran	Immobilization	24	2.1
Allium cepa	Plant	Mitotic index	24	10
Blue-gilled sunfish	Fish	Mortality	96	0.2
Fathead minnow	Fish	Mortality	96	1.76
Rainbow trout	Fish	Mortality	48	0.35
*Vibrio fischeri**	Bacterium	Luminescence	1	1.9
RTG-2 cell line	Rainbow trout	Neutral red uptake	24	90
Vero cell line	Monkey	Cell growth	24	34
Vero cell line	Monkey	MTT† reduction	24	37.6

Source: Data derived from Repetto et al. (2001).
* *V. fischeri* from a MICROTOX ™ test kit.
†MTT, 3-(4,5-dimethylthiazol-2-yl)-2,5-diphenyl-2*H*-tetrazolium bromide.

quantify a compound or conditions. The basic components of a biosensor are shown in Fig. 3.7. A biosensor essentially uses the specific reactions of biological materials such as enzymes and antibodies with molecules in order to detect and quantify these molecules, often in complex mixtures. The differences between a biosensor and a simple enzyme assay are that the reaction is linked to a transducer that produces a signal proportional to the numbers of molecules present. Thus the biosensor can give an online continuous determination of the compound present. The biological material carrying out the reaction can be enzymes, antibodies, hormone receptors, proteins, organelles, lectins, DNA, and whole cells. The biological reaction can be converted into a signal by a very extensive range of transducers. To function correctly the biological material has to be kept in close contact with the transducer and this is often achieved by immobilizing the biological material. There are a large number of methods for the immobilization of biological material and the method used depends on the properties of the material to be immobilized. The techniques of immobilization are outlined in Box 3.1.

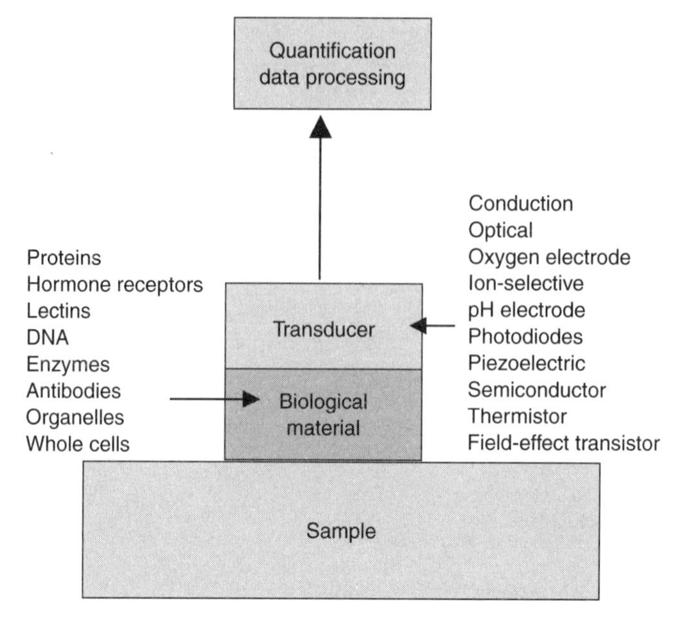

Figure 3.7 The basic components of a biosensor.

BOX 3.1 **Immobilization of enzymes**

The immobilization of enzymes, the attachment of the enzyme to a surface, or trapping of the enzyme in a matrix or behind a membrane confers some advantages. In the first case the enzyme is not free in the medium and therefore once the substrate is converted to the product the enzyme is not lost and can be reused. This also allows a continuous process to be used. The extended use of the enzyme and continuous process means that the process is less costly and that more-expensive enzymes can be used. Immobilization also increases the stability of the enzyme. Enzymes lose their activity over time and in some cases very rapidly (in a matter of hours) due to thermal degradation and protein breakdown. The rate of breakdown is often measured as the half-life, which is the time that an enzyme takes to lose half its original activity (see the figure below).

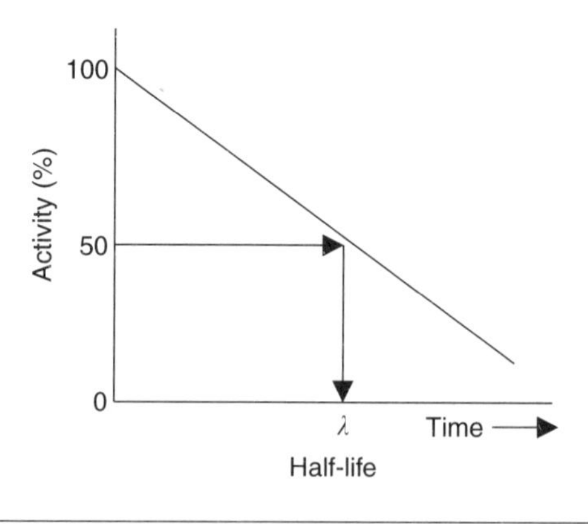

BOX 3.1 ***(Continued)***

Immobilization increases the half-life of enzymes considerably with values of up to 40 days. This means that sensitive and expensive enzymes can be used in continuous processes.

The methods used for the immobilizing both enzymes and microorganisms is shown in the figure below. Entrapment can be in either preformed or *in situ*-formed polymers such as alginate, behind semipermeable membranes, and in small semipermeable capsules. Polymers often have large pore sizes and are therefore more suitable for the immobilization of cells rather than enzymes. Enzymes can be bound to a range of surfaces such as charcoal, sand, and glass using a number of techniques. Some substrates such as charcoal will adsorb enzymes to their surface, others will bind using metals as intermediates, and ionic binding involves attachment through the amino or carboxyl groups on the enzyme. Covalent binding uses the amino and carboxyl groups on the enzyme to form a chemical link to a surface.

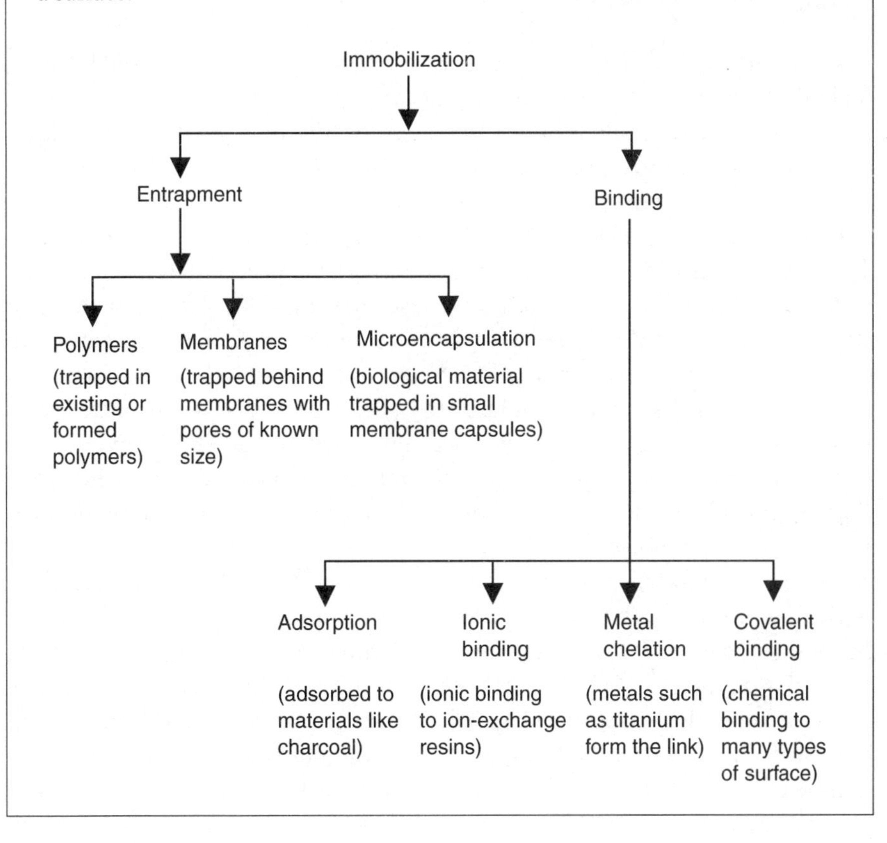

Biosensors were first developed for the lucrative medical industry and one of the most successful was the glucose biosensor based on the enzyme glucose oxidase. The advantages of biosensors for environmental analysis and monitoring include

- online sensing, real-time monitoring,
- fast response,
- operation with complex mixtures,
- very specific response,
- response to a class of compounds, allowing the biosensor to monitor toxicity, mutagenicity, carcinogenicity, and genotoxicity,
- high sensitivity in some cases,
- lower cost and quicker operation than conventional assays, and
- portability in some cases.

The disadvantages of using biosensors are those limitations imposed by using biological material. The biosensor cannot be steam-sterilized and therefore cannot be used in sterile bioreactor operation, the biological material has a finite life, and it can be poisoned by some compounds and metals.

However, biosensors are beginning to be used for environmental monitoring (Denizen and Turner, 1995) and some examples are given in Table 3.14. The oxygen demand of wastewaters and effluents has been determined traditionally by measuring the BOD_5 value (see section 3.7). Although this is a standard method and requires only simple equipment, 5 days are required to obtain a result. However, with a BOD biosensor a fast, online determination is possible. BOD biosensors can be either of the biofilm or respirometer type. In a biofilm BOD biosensor a microbial film is sandwiched between a porous membrane and the oxygen-permeable membrane of a Clark oxygen electrode (Fig. 3.8). A change in oxygen levels in the microbial culture will be proportional to the metabolizable content of the material to be measured. In some cases the electrical signal from the oxygen electrode has been replaced by optical signals using the luminous bacterium *P. phosphoreum*. The signal can be monitored as a steady-state value or an initial rate of change, which gives the BODst value. The BODst is not identical to BOD_5 but there is good correlation between the two values (Fig. 3.9). The respirometer method uses a small bioreactor fitted with an oxygen electrode containing micro-organisms. The sample can be added and the response followed by the electrode or added continuously. Another BOD sensor design is shown in Fig. 3.10, which is based on a microbial fuel cell (Chang et al., 2004). In this system the anode and cathode are separated by a cation-exchange membrane and the electrodes are graphite felt. The chambers have a volume of 20 ml and are seeded with activated sludge where a stable current was generated after 4 weeks.

Table 3.14 Examples of biosensors for use in environmental monitoring

Measurement	Biological component	Transducer
BOD_5	Single and multiple bacterial species, adsorbed or entrapped	Oxygen electrode, electric
BOD_5	Mediator-less microbial fuel cell	Amperometric
BOD_5	*Photobacterium phosphoreum* luminescence	Light intensity
Atrazine, estrone, and isoproturon	Antibodies	Solid-phase fluorescence
E. coli contamination in water	Membrane-based DNA/RNA hybridization with liposome amplification	Reflectometer
Phenols	Horseradish peroxidase enzyme absorbed on to graphite electrode	Amperiometric
Nitrate	Immobilized whole cells of *Agrobacterium radiobacter*	Electrochemistry
Nitrate	Nitrate reductase from *Alicaligenes faecalis*	Amperiometric
Nitrate	Enzyme from denitrifying *Thiopaera pantotropha*	Optical
Phosphate	Immobilized pyruvate oxidase (POD)	Amperiometric
Heavy metals	Immobilized phytochelatin	Capacitance
Herbicides	Immobilized whole cells of *Chlorella vulgaris*	Chlorophyll fluorescence,optical
Two pesticides	Immobilized acetylcholinesterase and choline oxidase	Oxygen electrode
Pesticides	Immobilized acetylcholinesterase	Electrochemistry

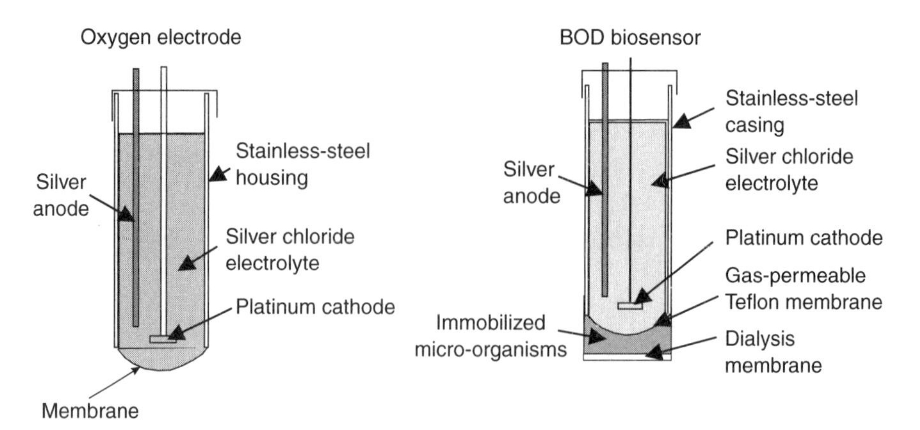

Figure 3.8 The construction of a BOD biosensor based on an oxygen electrode. Micro-organisms are immobilized between the gas-permeable electrode membrane and a medium-permeable dialysis membrane. The more metabolizable material in the medium the higher the metabolic activity of the cell and therefore the more rapid the reduction in oxygen.

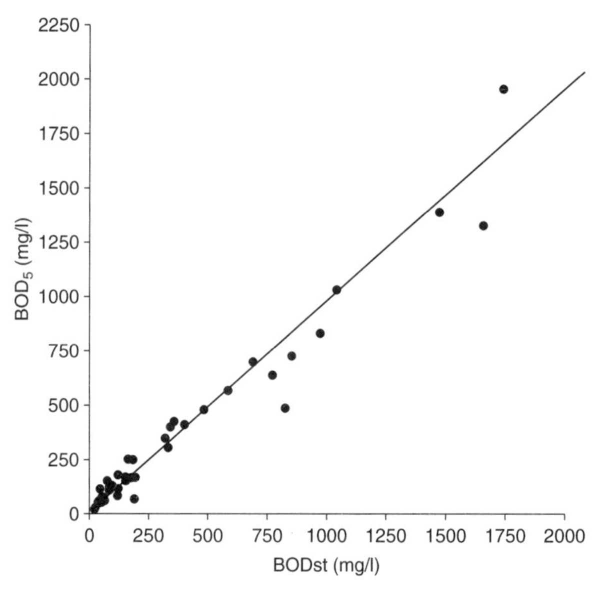

Figure 3.9 The correlation between the BOD_5 and BODst values derived from a number of wastewaters and micro-organisms (derived from Liu and Mattiasson, 2002).

The wastewater was run through the anode and the cathode has air-saturated water, and a steady state was reached after 60 min which was proportional to the wastewater's strength.

Biosensors that can measure pesticide concentrations have been developed using the enzyme acetylcholinesterase and also acetylcholinesterase combined with choline oxidase. Acetylcholinesterase converts acetylcholine to choline

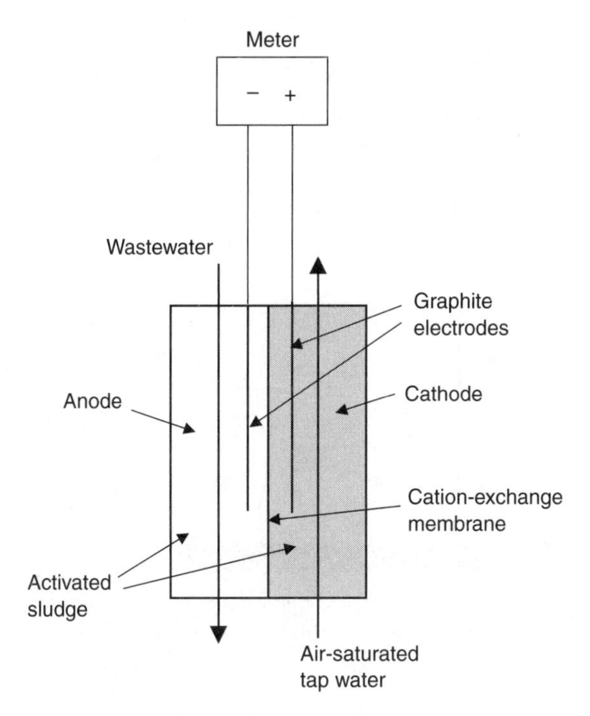

Figure 3.10 A BOD sensor based on a mediator-free fuel cell. The anode and cathode are separated by a cation-exchange membrane and the compartment stabilized with activated sludge. After 6 weeks the current stabilized and wastewater was run into the anode and air-saturated water into the cathode. The current generated was proportional to the BOD value of the wastewater. Source: Chang et al. (2004).

and choline oxidase converts the choline to betain and hydrogen peroxide. Hydrogen peroxide formation can be monitored amperometrically or by an oxygen electrode. Pesticides inhibit the action of acetylcholinesterase and therefore produce a reduction in peroxide formation, which can be used to estimate toxicity.

There have been a number of reports of biosensors for phenols, which were based on the oxidation of phenols to catechols and quinones by the enzyme tyrosinase where the reaction requires oxygen. If the enzyme tryosinase is linked to an oxygen electrode levels down to about 50 parts per billion (ppb) could be detected. Metals such as lead and cadmium can be detected by their inactivation of oxidases and dehydrogenases linked again to an oxygen electrode. Combining bacterial cultures with disposable print-screen electrodes (Fig. 3.11) has produced biosensors for metals and hydrocarbons. The detection of gases has been an area of considerable interest but to date the only gas to be detected with a biosensor is carbon dioxide, using bacteria linked to an oxygen electrode. Thus biosensors can provide cheap, reliable, and accurate monitoring of the environment which will also give real-time analysis.

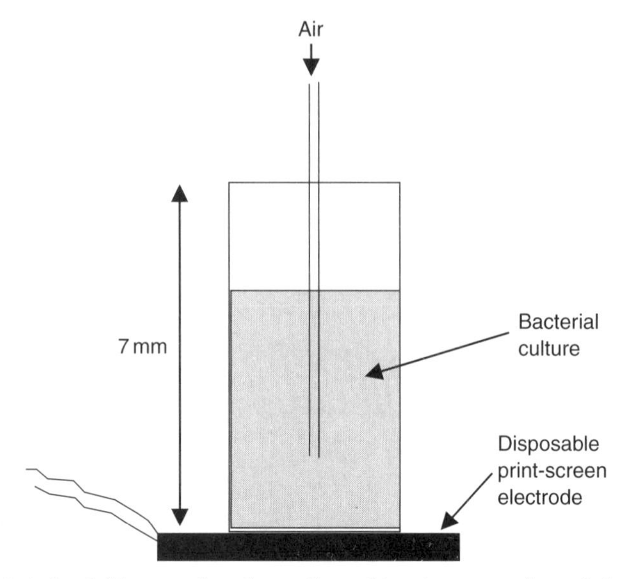

Figure 3.11 A simple biosensor based on a disposable print-screen electrode in contact with a microbial culture, which can be used to monitor heavy metals and hydrocarbons.
Source: Paitan et al. (2003).

3.13 Conclusions

Biotechnology in the form of recombinant DNA technology will influence the monitoring of the environment in the following areas.

- The monitoring of the environment will see the continued expansion of the uses of molecular biology and recombinant technology. The technology has allowed the study of the microbial population in environmental samples and *in situ* without the need to culture the micro-organisms. This has revealed a hitherto unknown population of Bacteria and Archaea living in some of the most hostile and unusual environments. Recombinant technology will also influence the application of bioindicators and biomarkers.
- Biomarkers using micro-organisms with the green fluorescent protein (*gfp*) gene inserted will be used to monitor pollution *in situ*.
- Recombinant DNA technology will be used to follow the introduction of genetically modified micro-organisms into the environment.
- Biosensors will be able to determine the levels of contaminants in the environment and perhaps provide real-time online monitoring.

3.14 Further reading

Atlas, R.M. and Bartha, R. (1993) *Microbial Ecology*. Addison Wesley Longmann, Harlow.

Gray, N.F. (1989) *Biology of Waste Water Treatment*. Oxford University Press, Oxford.

Hall, E.A.H. (1990) *Biosensors*. Open University Press, Milton Keynes.

Huggett, R.J., Kimerle, R.A., Mehrle, P.M., and Bergman, H.J. (1992) *Biomarkers*. SETAC Special Publications Series, Lewis Publishers, Ann Arbor.

Mason, C.F. (1996) *Biology of Freshwater Pollution*, 3rd edn. Longman Scientific and Technical, Harlow.

Reeve, R.N. (1994) *Environmental Analysis*. John Wiley, Chichester.

Shaw, I.C. and Chadwick, J. (1998) *Principles of Environmental Toxicology*. Taylor and Francis, London.

Tebbutt, T.H.Y. (1998) *Principles of Water Quality Control*, 5th edn. Butterworth-Heineman, Oxford.

Trevors, J.T. and Van Elasa, J.D. (1995) *Nucleic Acids in the Environment: Methods and Applications*. Springer Verlag, Heidlberg.

4 Sewage treatment

4.1	Introduction	112
4.2	Function of the waste-treatment system	116
4.3	Sewage-treatment methods	117
4.4	Modifications to existing processes	140
4.5	Removal of nitrogen and phosphorus	147
4.6	Sludge treatment and disposal	154
4.7	Anaerobic digestion	165
4.8	Agricultural waste	170
4.9	Industrial waste	170
4.10	Conclusions	171
4.11	Further reading	172

4.1 Introduction

Domestic and institutional liquid waste, and industrial and agricultural wastes make up the bulk of the biodegradable pollutants that can be released into the environment (Table 1.1). The use of micro-organisms for the disposal of sewage was developed around 1910–14 in Manchester and a number of European cities. In earlier days, sewage was buried or run into rivers and waterways. In the 1800s the population of most cities expanded greatly due to industrialization, which meant that the volume of sewage produced was too much for this type of disposal and rivers and canals became very polluted. In the UK rivers like the Thames became anaerobic, devoid of aquatic life, producing unpleasant smells and were responsible for the spread of diseases like typhoid and cholera. Sewage disposal was so acute in London that it was the practice for those who could afford it to move to the country in the summer. The treatment of sewage is linked to the provision of clean water, as polluted water is the cause of many diseases. The elimination of or reduction in water-borne diseases by the introduction of sewers and sewage disposal probably did more for the health of the population than the introduction of antibiotics later in the century.

This chapter covers the biological methods of treating sewage, the processes involved, and the modifications made to the basic processes. The disposal of waste sludge is discussed along with the treatment of industrial and agricultural wastes.

4.1.1 **Pollution caused by biodegradable material**

Before the industrial revolution the sewage disposed of in canals, rivers, and at sea was sufficiently diluted so that the organic content of the waterway did not reach too high a value. Natural waterways contain indigenous populations of micro-organisms that can use the dissolved organic compounds, and are in turn part of the food chain for protozoa, insects, worms, and fish. Under normal conditions in a waterway, the population of micro-organisms is part of a balanced ecosystem and small quantities of organic compounds will not disturb this balance. However, this balanced system can be destabilized if excess metabolizable organic materials, such as high levels of sewage, are released into the waterway. The addition of metabolizable organic compounds will cause a considerable increase in the growth and metabolism of the aerobic microbial population of the waterway. This excess growth will use up all the available oxygen dissolved in the water, causing the waterway to become anaerobic. Figure 4.1 illustrates the effect of the addition of a biodegradable organic material to a river. The dissolved oxygen level drops rapidly at the point of addition (often known as the biological oxygen demand (BOD) sag), but provided the amount of organic material added is not too high the oxygen level will rise slowly as the material is mixed and moves downstream from the site of addition. The subsequent rise in dissolved oxygen is due in part to mixing in the river, dilution with unaffected water in the waterway, and metabolism of the organic material. Stagnant waters such as lakes and ponds become anaerobic more rapidly on the addition of excess organic material as they have no flow to mix the system and to add clean water (Table 4.1). However, if the

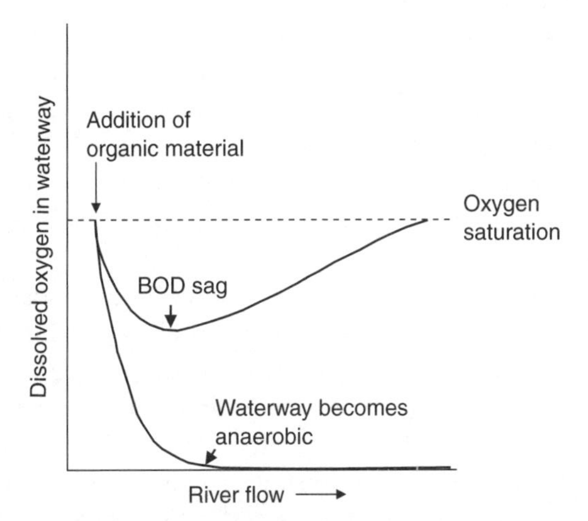

Figure 4.1 The consequences of adding organic material to a river or stream in terms of the dissolved oxygen. The dissolved oxygen will dip close to the site of addition but with mixing the dissolved oxygen will rise again to the saturation value. This feature is known as the 'BOD sag'. However, if the organic addition is large the dissolved oxygen may not recover and the waterway will become anaerobic.

Table 4.1 Typical values for the oxygen-exchange coefficient, *f*

Type of water	*f* (cm/h)
Stagnant water	0.4–0.6
Water flowing at 0.6 m/min	1
Sluggish polluted river	2
Sluggish clean water 51 mm deep	4
Water flowing at 10 m/min	7.5
Open sea	13
Water flowing at 15 m/min	30
Turbulent stream	30–200
Water flowing down a 30° slope	70–300

Source: Data from Gray (1989) and Tebbutt (1998).

organic addition is too great the increased microbial metabolism will keep the conditions in the waterway anaerobic. Under anaerobic conditions the aerobic micro-organisms will decline or die and the anaerobes will increase. Anaerobic metabolism is slower than aerobic metabolism so that the rate of degradation of the organic material decreases, which can lead to build up of excess organic material, or sediment, in the water. The continued lack of oxygen will lead to the death of other aquatic organisms such as fish and crustacea. Anaerobic metabolism in the sediment will produce gases such as hydrogen sulphide and methane so that the production of gas bubbles can be an indication of anaerobic conditions. In addition to sewage, industrial wastewater from processes such as brewing, baking, paper mills, metal processing, and agriculture can run into waters; these wastes are in general much stronger than sewage. Sewage-treatment plants normally only accept domestic wastewater and industrial plants have to treat their own waste prior to returning it to the waterways or to the domestic wastewater system. If an industrial waste is discharged into the sewers, the local water company normally makes a charge for its treatment.

Waterways are not only affected by the addition of biodegradable wastes (nutritional pollution), but can be affected by chemical and physical pollution. Chemical pollution is mainly industrial, with the release of acids, alkalis, and toxic compounds, which can poison the living organisms in the waterway. Physical pollution is the release into the waterway of materials that can change the water's physical conditions, and consists mainly of the release of large quantities of warm water from power stations, which use the water for cooling. The change in temperature of the water can encourage the excess growth of native organisms or the growth of a new organism, thus changing the balance of the normal population.

4.1.2 Sewage

Domestic waste comes in two forms; liquid or solid. This chapter concentrates on the treatment of the liquid domestic waste, sewage. Domestic wastewater is derived from private houses, commercial buildings, and institutions such as schools and hospitals.

Sewage is some 99.9% water by weight, containing dissolved organic material, suspended solids, micro-organisms (pathogens), and a number of other components. The suspended solids can range from >100 μm to colloidal in nature. The composition, concentration, and condition of the sewage may differ widely depending on the origin, time, or weather conditions. The strength can vary daily and by season and can be diluted by rainwater. An example of the composition of domestic sewage is given in Table 4.2.

In typical sewage, 75% of the suspended solids and 40% of the dissolved material are organic. The dissolved organic materials are a mixture of proteins (65%), carbohydrates (25%), fats (9%), and detergents (1%). There are substantial inorganic components in sewage, including sodium, calcium, magnesium, chlorine, sulphates, phosphates, bicarbonates, nitrates, and ammonia with traces of heavy metals. The oxygen uptake by the micro-organisms when they metabolize the organic material in sewage is known as the oxygen demand and this is usually expressed as BOD_5, in g/m^3 or mg/l. The BOD_5 value is calculated by seeding a sample of the waste with a mixture of micro-organisms and incubating at 25°C for 5 days. The oxygen concentrations at the beginning and end of this assay are measured and the difference is the BOD_5 value. The oxygen demand of sewage can also be estimated with a chemical technique using an oxidizing agent and this is known as the chemical oxygen demand (COD).

Table 4.2 Typical composition of domestic sewage

Component	Concentration (mg/l)
Total solids	300–1 200
Suspended solids (SS)	100–350
Total organic carbon (TOC)	80–290
Biological oxygen demand (BOD_5)*	110–400
Chemical oxygen demand (COD)*	250–1 000
Total nitrogen	20–85
Ammonia	12–50
Nitrites	0
Nitrates	0
Total phosphorus	4–15

*See Chapter 3.

Sewage normally has a BOD_5 value of between 110 and 400 mg/l (Table 4.2), whereas wastes from some industrial and agricultural processes such as tannery waste and animal slurry can have much higher BOD_5 values of up to 50 000 mg/l.

4.2 Function of the waste-treatment system

The main function of domestic waste-treatment systems is to reduce the organic content, pathogenic micro-organisms, and suspended solids as far as possible in order to be able to return the water to rivers and coastal waters without causing pollution, especially those rivers used as a source of drinking water (Fig. 4.2).

The quality of the treated waste released from the treatment system depends on the volume and condition of the receiving water and its ability to dilute the waste and whether there is to be water abstraction further downstream. In general, a 20:30 standard is adopted, which means an effluent level of 20 mg/l BOD_5 and 30 mg/l suspended solids (SS) when the effluent is to be diluted 8:1. The secondary drinking-water standards for water supplies in the UK and the USA are given in Table 4.3. The primary drinking-water standards are given in Chapter 3 (Table 3.4).

The volume of domestic wastewater from toilets, baths, washing machines, etc. in the UK is about 9 million m^3/day, industry uses about 7 million m^3/day, and with run-off due to rainfall this gives a total of about 18 million m^3/day,

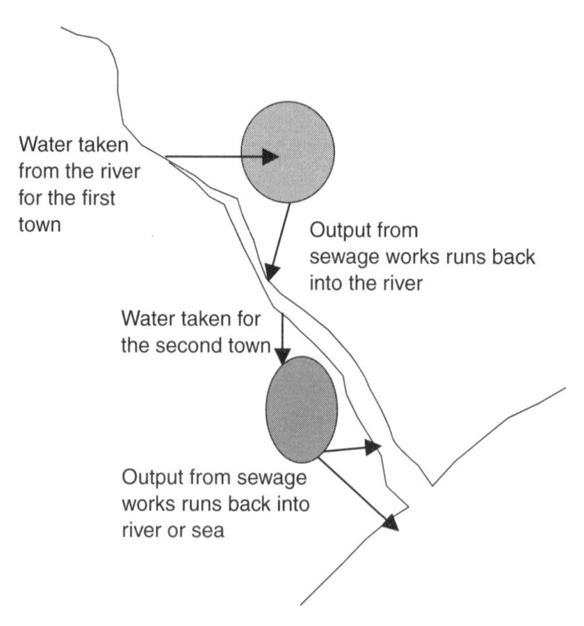

Figure 4.2 An illustration of the need to treat wastewater when it is returned to the river as it is extracted further down the river for drinking water.

Table 4.3 National secondary drinking-water regulations for the USA and UK

Contaminant	USA (mg/l)	UK (mg/l)
Aluminum	0.05–0.2	NA*
Chloride	250	250
Colour	15 colour units	15 color units
Copper	1.0	1.0
Corrosivity	Noncorrosive	NA
Fluoride	2.0	NA
Foaming agents	0.5	NA
Iron	0.3	0.3
Manganese	0.05	0.1
Odour	3 threshold number	Nil
pH	6.5–8.5	7.5 ± 1
Silver	0.1	NA
Sulphate	250	400
Total dissolved solids	500	1 000
Zinc	5.0	5.0
BOD_5	NA	20 (30 mean/month)

Source: USA data from the US Environmental Protection Agency, **www.epa.gov**

*NA, not available.

all of which needs treating before it can be returned to the rivers or canals. It has been calculated that each person in the UK contributes on average some 230 l of sewage/day. Thus although the scale of sewage treatment is enormous it is a very dilute growth medium and its value per tonne is very low compared with other biotechnological processes.

4.3 Sewage-treatment methods

The first attempts at using micro-organisms to treat sewage used single cultures but as sewage is a complex mixture processes developed which used naturally occurring mixed microbial populations. In the treatment process potentially polluting materials are brought into contact with the microbial population under conditions where they are broken down and metabolized. As the nature and composition of wastewater can vary so much there is no universal waste-treatment process but four stages of treatment are generally incorporated; the preliminary, primary, secondary, and tertiary stages (Fig. 4.3).

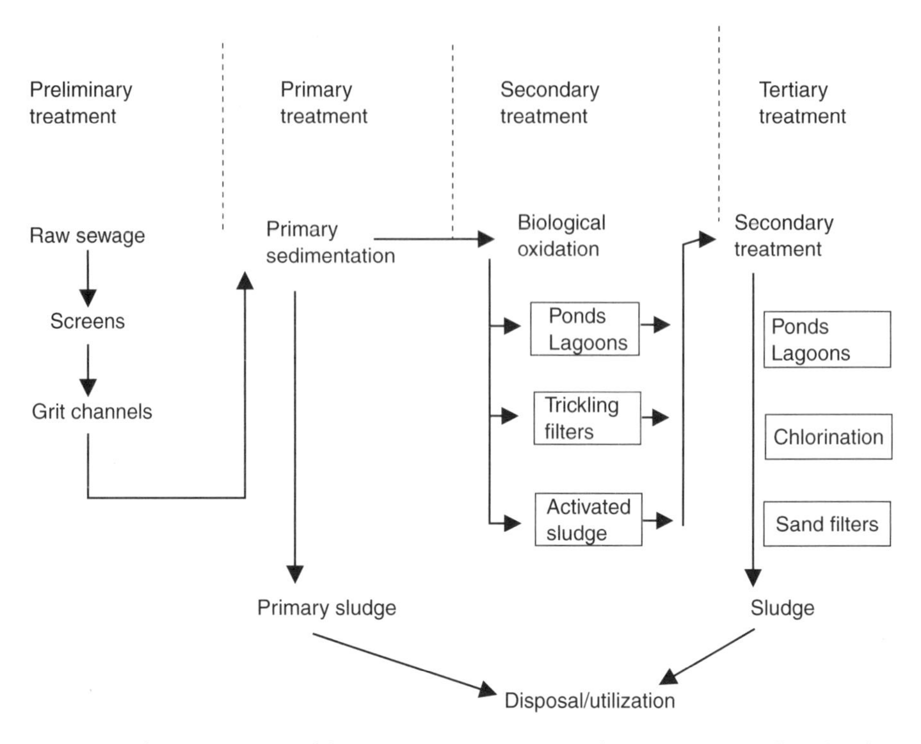

Figure 4.3 The various stages of the sewage-treatment system. The raw sewage is collected and large debris and grit removed by screens and grit channels. Large suspended solids are settled out in the primary sedimentation tank forming the primary sludge. It is the effluent from the first sedimentation tank that is treated by a number of biological processes including ponds, trickling filters, and activated sludge. After treatment in these systems the suspended solids which are mainly microbial flocs are removed by a secondary sedimentation step. The liquid from this secondary sedimentation can be run into a waterway or treated by some form of tertiary treatment such as maturation ponds, sand filters, and chlorination. The sediment (sludge) is either recycled back to the activated-sludge system or combined with the primary sludge for disposal.

- Preliminary treatment: this is the removal of large debris and grit. The large debris is normally collected on screens and often macerated or broken up with various grinders and added back to the system. Other solids such as grit, derived from the run-off from roads, are removed in grit channels and can be recycled after washing.
- Primary treatment: the sewage is allowed to settle for 1.5–2.5 h, removing the suspended solids which flocculate easily, and thus reducing the BOD_5 load by about 40–60%.
- Secondary treatment: the effluent from the primary treatment still contains the dissolved organic materials and 40–50% of the suspended solids. It is at this stage that biological action is used to remove the organic material. The biological reactions can be either aerobic or anaerobic, although the aerobic treatment is the most widely used as it is more rapid. A number of processes can be used in the aerobic or anaerobic treatments, including lagoons or ponds, trickling filters, activated sludge, rotating biological contactors, and anaerobic digesters.

- Tertiary treatment: this treatment may be required to remove phosphate, nitrate, and pathogenic micro-organisms to produce potable water and to prevent eutrophication. The processes can involve chemical precipitation, disinfection with chlorine, filtration through sand filters, and the use of maturation ponds.

4.3.1 Lagoons or ponds

Ponds or lagoons were used for sewage treatment long before the controlled use of micro-organisms was developed at the beginning of the last century, and ponds are popular alternatives to aerobic sewage treatments in countries where land is freely available and sunshine plentiful. Ponds or lagoons can be operated aerobically, anaerobically, or by a mixture of the two and Fig. 4.4 shows the various options available.

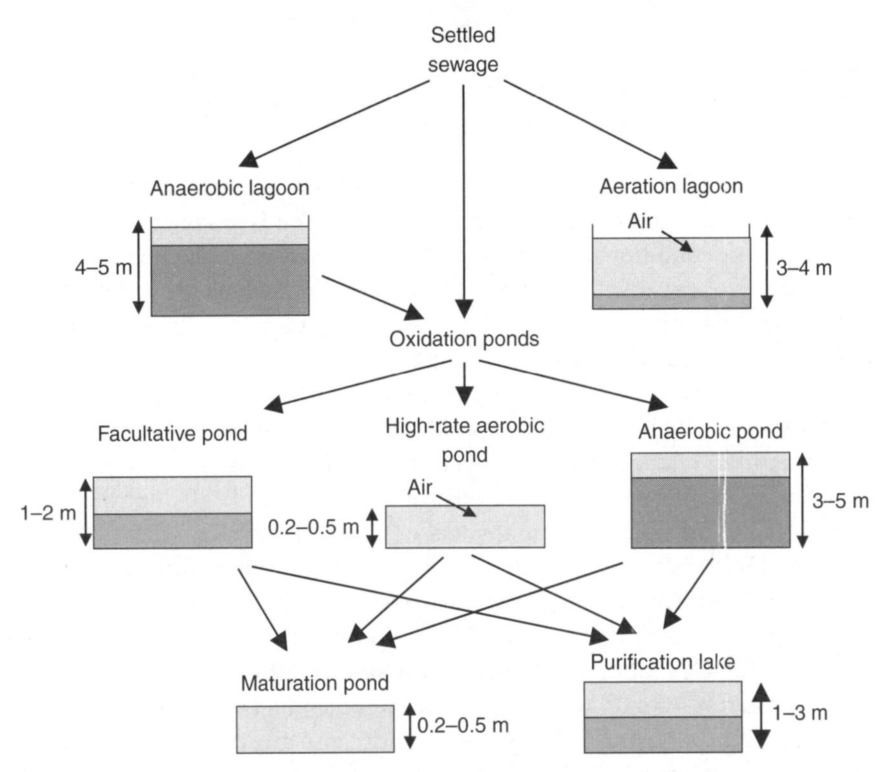

Figure 4.4 The types of ponds and lagoons which can be used for waste treatment. Settled sewage or other wastes can be treated in anaerobic lagoons with a depth of 4–5 m, aeration lagoons, which are much shallower and can be aerated on the surface, and oxidation ponds. The oxidation ponds can be of three types depending on their depth and therefore aeration. High-rate ponds are shallow (0.2–0.5 m) and can be stirred, facultative ponds (1–2 m) have both anaerobic and aerobic conditions, and anaerobic ponds (3–5 m) are mainly anaerobic. The effluent from these types of pond can be further purified in purification lakes and maturation ponds where the particulates can settle out and the organic material fully removed.

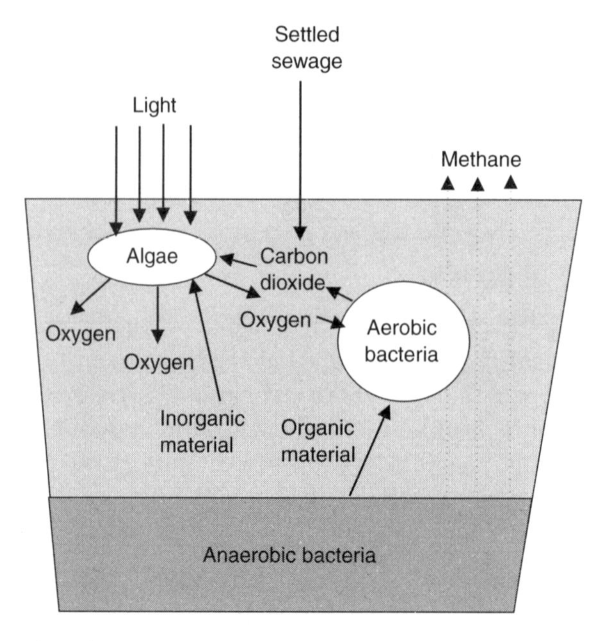

Figure 4.5 An outline of the population that can occur in a facultative pond. Anaerobic organisms are found in the lower regions and aerobic organisms in the surface regions, and additional oxygen is supplied by the algal population at or near the surface.

Facultative ponds are generally shallow (1–2.5 m) and the biological processes which occur are represented in Fig. 4.5. The pond combines both aerobic and anaerobic conditions. The organic matter in solution or the fine suspended material is broken down by aerobic micro-organisms, which are supplied with oxygen by diffusion and mixing from the surface, and by the algae growing on or near the surface. The algae use sunlight, carbon dioxide, and the inorganic compounds present in the sewage for their growth. Much of the suspended solid settles to the bottom of the pond where the conditions are anaerobic. The anaerobic micro-organisms break down these solids to yield methane, nitrogen, and carbon dioxide and this can represent some 30% of the BOD_5 load. Some of the operating parameters of facultative ponds, where 70–95% of the BOD_5 load can be removed in 7–50 days, are given in Table 4.4.

Aerobic ponds are much shallower than facultative ponds, up to 1 m in depth, so that light can reach the bottom. The shallow pond means that more oxygen is supplied by algal photosynthesis, in addition to diffusion from the surface. High-rate aerobic ponds are similar to the aerobic ponds, except that they are operated to ensure maximum algal metabolism and growth. Light is generally the limiting nutrient for algal growth in ponds so that the ponds are much shallower, 0.2–0.5 m deep to avoid shading, and oxygen production by the algae is often enhanced by some form of mixing. The algae produced in the high-rate ponds are often harvested and used for both animal and fish food.

Maturation ponds are generally of similar construction to the facultative ponds but are used as a tertiary treatment with longer retention times of

Table 4.4 Some of the parameters for facultative ponds

Parameter	Units
Depth (m)	1–3
Retention time (days)	7–50
BOD_5 loading (kg/acre/day)	9–22
BOD_5 removal (%)	70–95
Microalgal concentration (mg/l)	10–100
Effluent suspended-solids concentration (mg/l)	100–350

7–15 days, which allows suspended solids to settle before the water is discharged into a waterway. Anaerobic lagoons or ponds are mainly used for the pretreatment of wastes before they are passed on to facultative ponds. The lagoons and ponds are suitable for high-strength wastes with BOD_5 values much above 300 mg/l. Anaerobic conditions are maintained by increasing the depth of the pond to 1–7 m and increasing the BOD_5 load. The retention times can be between 2–160 days with the removal of 70–80% of the BOD_5 added. Purification lakes are large lakes used to remove the pollution from a river and as such function as large maturation ponds. Aeration lagoons, unlike the other ponds, are used in the primary treatment of sewage or industrial waste. The lagoon is generally deeper than high-rate ponds, at 3–4 m, and oxygen supply is provided by mechanical means by diffusion aeration units or surface aerators that also mix the waste. Simplicity, low construction costs, and cheap operation are some of the advantages of ponds and lagoons.

All of these various ponds and lagoons can be used in isolation but frequently they are combined or mixed with other systems (Chapter 5). High-BOD_5 wastes over 300 mg/l are normally treated first in an anaerobic lagoon or pond for 1–5 days that will remove 50–70% of the BOD_5 load. The effluent is then passed to the second stage, a facultative pond with a retention time of 20–40 days (Fig. 4.6). The facultative pond can take low-strength wastes directly without passing through the anaerobic pond. The final stage is the maturation pond, where the suspended solids will settle out over a period of up to 7 days. This type of sequence can give clean final effluents of <25 mg/l BOD_5 and <75 mg/l COD. Some lagoons have been linked to marsh plant ponds, sand and gravel filters and in one case these have been combined into one system.

4.3.2 **Trickling filters**

Most micro-organisms in nature are associated with or attached to solid surfaces and these are known as biofilms. The trickling filter is based on a randomly packed solid medium, which will act as a surface on which a mixed culture of micro-organisms will attach and grow. If sewage is run through

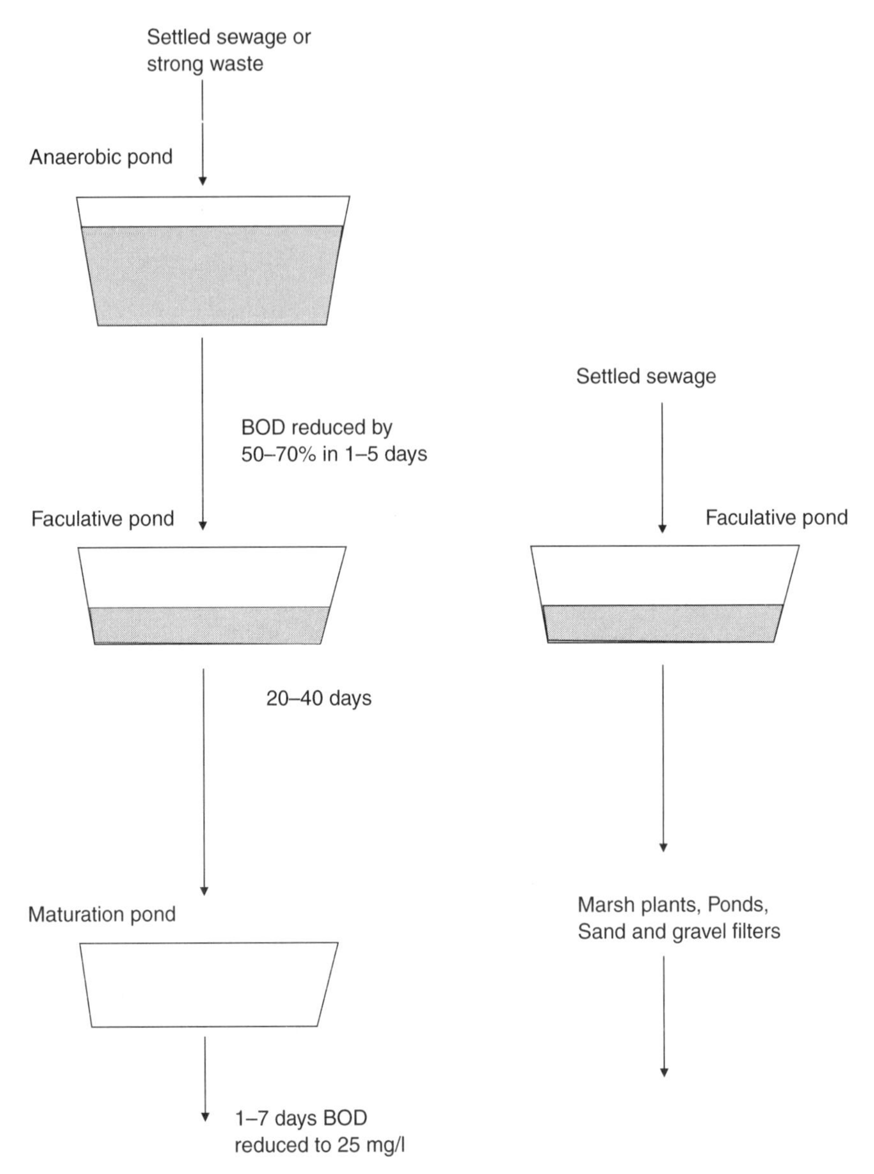

Figure 4.6 A possible sequence of ponds for the treatment of waste. Strong wastes with a BOD_5 over 300 mg/l are usually treated in an anaerobic pond followed by treatment in a facultative pond with final polishing in a maturation pond. Settled sewage can be run directly into the secondary stage, the facultative pond.

such a filter the attached biomass will metabolize the organic content. The biofilm, which develops on the filter packing, will give the filter a high biomass content as the solid packing has a high surface area, and so the high biomass content should be able to metabolize the sewage rapidly.

In its early form the trickling filter consisted of a bed 1–3 m deep packed with stones, clinker, or slag of between 40 and 60 mm in diameter (Fig. 4.7).

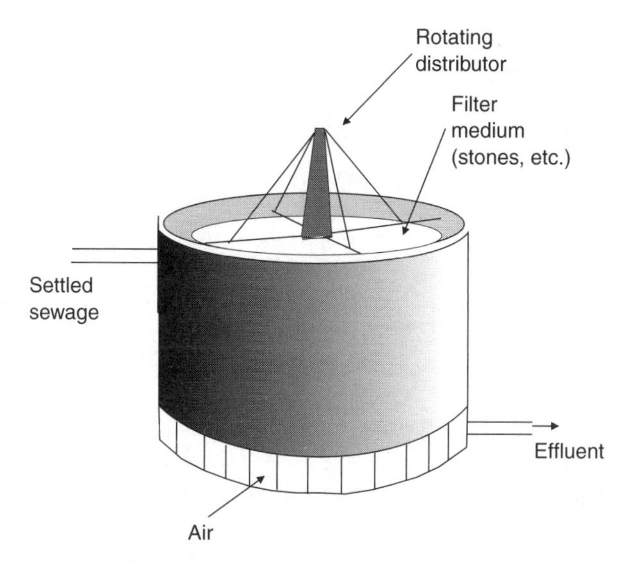

Figure 4.7 The design of a trickling filter. The circular vessel 1–3 m in depth is filled with stones, clinker, or slag and settled sewage is run from the rotating arms on to the bed. The organic material in the sewage is metabolized by the microbial biofilm, which develops on the stones. The clean effluent is collected at the bottom of the filter. Circulation of air through the filter bed is encouraged by air vents at the base of the vessel.

The gaps between the stones allow air to penetrate the bed and this is improved by having ventilation at the base. The wastewater is applied by a rotating distribution system, and the treated water is collected by a drain at the base. Such a filter does not require inoculation but develops its own population, which is a complex mixture of bacteria, fungi, protozoa, algae, and larger organisms like insects and worms. A biofilm develops on the packing and the biofilm is made up principally of bacteria and fungi. The biofilm is in a dynamic equilibrium, which can alter as the content of waste stream changes, and with seasonal changes. Once the biofilm reaches a certain thickness, layers will be detached and these aggregates are collected in a settling tank.

Filter systems are widely used for secondary treatment as they have the advantage that once constructed they require little or no maintenance; they are also economical and tolerant to changes in wastewater composition. As the efficiency of the filter depends upon the biomass contained in the biofilm any increase in the solid-phase area will increase the load potential of the filter. Filter design has been improved by the incorporation of a plastic solid phase, either random or modular in design, which has increased the biofilm area significantly, and thus allows the filter to handle higher BOD_5 loads (Table 4.5). Trickling filters have been used in a single-pass process producing an effluent of a high standard. However, if the load on these filters increases over time, due to an increase in population or the addition of stronger industrial effluents, the system will need more than one filter unit. The most common modification to deal with these types of change is to have a degree of recirculation

Table 4.5 Some of the properties of materials used in filter beds

Material	Size (mm)	Specific surface area (m^2/m^3)
Slag	50	125
Clinker	62	120
Rounded gravel	25	150
	62	65
PVC shapes	–	85–220
Polypropylene shapes	–	120–190

Source: Derived from Gray (1989).

which will even out fluctuations using a two-stage system, where the first filter is used at a high rate and the partially treated waste passed on to the second filter for final treatment. In some cases the two filters can be alternated to allow the primary one to recover from the effects of a high loading. In this way filter units can treat more waste without the need to construct new filter units.

4.3.3 Activated-sludge process

The activated-sludge process is basically a very large tank (bioreactor) containing a mixed population of micro-organisms that are living on the organic material in the sewage. The sewage flows from one end of the bioreactor to the other in sufficient time for all of the organic material to be removed. The basic process is given below.

sewage + micro-organisms + oxygen = growth of micro-organisms (biomass) + carbon dioxide

The microbial population is allowed to develop naturally so that it contains micro-organisms capable of metabolizing all the components of the sewage. In order to get a high rate of organic-material removal the microbial biomass should be as high as possible and for this some 20% of the final microbial biomass is returned to the start of the bioreactor. In this process the waste is brought into contact with a high concentration of micro-organisms present as aggregates under aerobic conditions (Fig. 4.8). The effluent from this continuous stream is run into a settling tank where the biomass and nondegraded solids settle out. The clarified effluent is virtually free of solids and, apart from a final polishing or tertiary treatment, can be run into a river. The normal operation is a plug flow system in a rectangular tank, typically 6–10 m wide, 30–100 m long, and 4–5 m deep (volume 5000 m^3). In a plug flow system there is a continuous flow of sewage and culture through the rectangular bioreactor without mixing along the length. There is however mixing from top to bottom, which aids aeration.

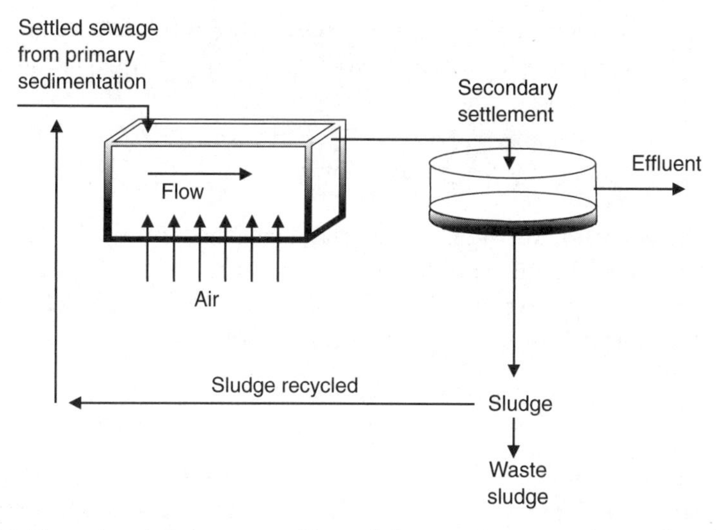

Figure 4.8 The activated-sludge process. The settled sewage is run into a long tank at the same time as recycled sludge (20%) is added. The mixture is passed down the tank where it is aerated. At the end of the tank the mixture is passed into a settlement tank where the microbial aggregates and any remaining solids settle out. Some of this sludge is recycled and the remainder collected for disposal.

4.3.4 Operating parameters

The growth phases of a microbial culture have been described in Chapter 2 where the lag, logarithmic, and stationary phases have been identified. Activated sludge contains a very mixed population but many of the characteristics of single microbial cultures can be applied to activated sludge. The microbial metabolism of simple carbohydrates such as glucose follows the reaction

$$C_6H_{12}O_6 = 6CO_2 + 6H_2O \tag{4.1}$$

The elemental composition of typical domestic sewage has been estimated and the products of its oxidation are

$$C_{10}H_{19}O_3N + 12.5O_2 + H^+ = 10CO_2 + 8H_2O + NH_4^+ \tag{4.2}$$

The energy obtained by the mixed population from the oxidation of the sewage is used for growth and cell maintenance. Briefly, the growth of micro-organisms is autocatalytic and in the exponential phase can be described by the following equation (Chapter 2).

$$\frac{dX}{dt} = \mu X \tag{4.3}$$

Where X is the biomass concentration (mg/l); t is the time in days and μ is the specific growth rate (days^{-1}). The relationship between the specific growth rate and the substrate concentration is expressed by the Monod equation.

$$\mu = \mu_m \frac{S}{K_s + S} \tag{4.4}$$

where μ_m is the maximum growth rate (days^{-1}), S the substrate concentration (mg/l), and K_s the saturation constant (mg/l). The two equations can be combined.

$$\frac{dX}{dt} = \mu_m \frac{SX}{K_s + S} \tag{4.5}$$

The concentration of organic material that can be used for growth in sewage treatment is in general the BOD_5 value. The yield of cells or biomass from any substrate is given by the yield value $Y_{X/S}$. The yield is defined as the mass of cells formed per unit mass of substrate consumed.

$$\frac{dX}{dt} = -Y_{X/S} \frac{dS}{dt} \tag{4.6}$$

The normal method of measuring yield is to measure the accumulation of biomass and substrate consumed over a period of time. These values represent a global yield over the time period measured and yield values can change depending on the state of the culture. The energy required for cell maintenance can alter, which will affect the yield value. Thus the substrate will be used for both growth and maintenance.

$$\text{substrate} = \text{cell growth} + \text{maintenance}$$

In conventional sewage treatment systems the continuous flow of sewage through the bioreactor (plug flow) is similar to a continuous culture system. In a continuous culture the microbial population is maintained in a continuous state of balanced growth, by continuously removing some of the culture and replacing it with fresh medium at the same rate (Box 4.1). In the activated-sludge system a high concentration of cells is maintained by recycling some of the cells which flow from the bioreactor.

However, sewage is a weak substrate at 110–400 mg/l compared with the usual concentration of sugars used to cultivate micro-organisms at 20 g/l. Sewage can therefore only support a low-density culture but the waste-treatment system needs to treat very large volumes of this weak substrate. To treat this rapidly needs high cell numbers, and in order to get this some of the micro-organisms (sludge) are recycled. The recycling of sludge gives a high biomass and allows a steady-state biomass level while having a flow rate about the same as the maximum growth rate. If the biomass is retained in the bioreactor for an increased time compared with the conventional system the substrate concentration will be very low. As a consequence the cells will grow more slowly and many cells may not grow at all. Under these conditions the maintenance energy increases and some cells will die and lyse so that the biomass levels will drop (Fig. 4.9). In contrast, if the flow rate is increased to a high level there will be insufficient time for all the substrate to be removed and there will be a drop in the biomass level as the growth rate cannot keep up with the increased flow.

BOX 4.1 Continuous culture

Continuous culture is a system for growing cells in an open bioreactor, where the microbial population is maintained in a continuous state of balanced growth, by continuously removing some of the culture and replacing it with fresh medium. There are two types of continuous culture, the chemostat and the turbidostat. The chemostat is the most commonly used system and is operated by supplying an essential growth-limiting nutrient at a constant rate with the result that culture density and growth rate adjust themselves to the supply. The chemostat is shown schematically below.

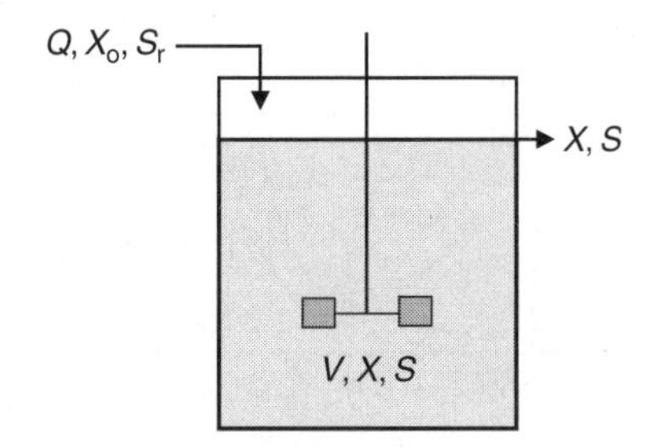

The self-regulatory state in the chemostat and the steady state can be expressed by a series of equations. These equations relate cell density and limiting substrate concentrations by developing material balances.

Thus for cells

Cell accumulation = cells in − cells out + growth − death

$$\frac{dX}{dt} = \frac{QX_o}{V} - \frac{QX}{V} + \mu X - \alpha X$$

where Q = flow rate of medium supply (l/h), V = constant working volume (l), X_o = cell concentration in medium supply (g/l), X = cell concentration in bioreactor (g/l), μ = specific growth rate (h^{-1}), and α = specific death rate (h^{-1}).

With a single chemostat the medium supply is sterile (no cells) and the death rate is negligible ($\mu >> \alpha$). Consequently the equation can be simplified to,

$$\frac{dX}{dt} = \frac{-QX}{V} + \mu X$$

The term Q/V is termed the dilution rate (D), the number of culture volumes passing through the bioreactor per hour, so that the equation can be rewritten as

$$\frac{dX}{dt} = -DX + \mu X$$

$$\frac{dX}{dt} = X(\mu - D)$$

At a steady state growth is zero, therefore $dX/dt = 0$, and the equation simplifies to

$$\mu = D$$

BOX 4.1 **(Continued)**

This holds until the dilution rate exceeds the growth rate, and the culture reaches its maximum growth rate for the conditions. If the dilution rate exceed the growth rate cells extracted from the bioreactor cannot be replaced by growth so that the culture concentration will be reduced in a condition known as washout, as shown in the graph below.

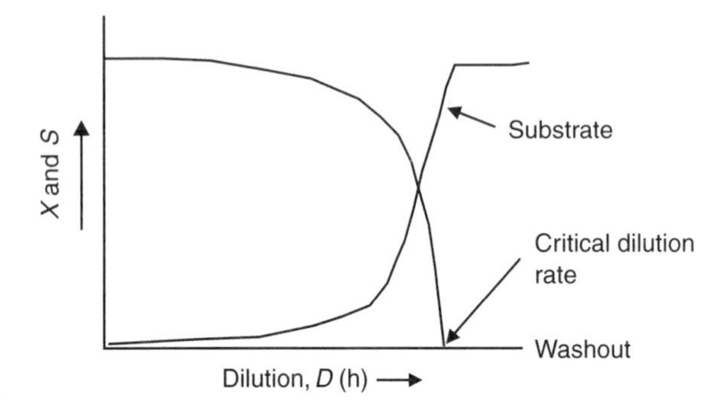

Dilution rate and substrate concentration

Again the relationship can be examined by making a material balance.

Nutrient accumulation = nutrient in − nutrient out − consumption − maintenance − product formation

$$\frac{dS}{dt} = \frac{QS_r}{V} - \frac{QS}{V} - \frac{\mu X}{Y_{X/S}} - mX - \frac{qpX}{Y_{P/S}}$$

where S_r = limiting substrate in medium flowing into bioreactor (g/l), S = substrate concentration in bioreactor (g/l), $Y_{X/S}$ = biomass yield coefficient (g of cell/g of substrate consumed), $Y_{P/S}$ = product yield coefficient (g of product/g substrate consumed), m = maintenance requirement (g of substrate consumed/g of cells/h), μ = specific growth rate (h^{-1}), and qp = specific rate of product formation (g of product formed/h).

In general mX is very small compared with that used for growth ($mX \ll \mu X/Y_{X/S}$) and can be ignored, and product formation can also be discounted. Therefore

$$\frac{dS}{dt} = \frac{QS_r}{V} - \frac{QS}{V} - \frac{\mu X}{Y_{X/S}}$$

At a steady state Q/V is the dilution rate D, and $dS/dt = 0$.

$$\frac{\mu X}{Y_{X/S}} = D(S_r - S)$$

At a steady state $\mu = D$: therefore

$$X = Y_{X/S}(S_r - S)$$

BOX 4.1 **(Continued)**

The relationship between X, S, and the dilution rate D can be carried out using the Monod equation:

$$\mu = \mu m \frac{S}{K_S + S}$$

At a steady state $dX/dt = 0$ and $dS/dt = 0$.

$$D = \mu m \frac{S}{K_S + S}$$

Solving for S,

$$S = \frac{K_S D}{\mu m - D}$$

$$X = Y_{X/S}(S_r - S)$$

Substituting for S in these equations,

$$X = Y_{X/S}\left(S_r - \frac{K_S D}{\mu m - D}\right)$$

Chemostat with cell recycling

The recycling of cells generated from continuous culture will enable the system to be operated at dilution rates higher than the maximum growth rate.

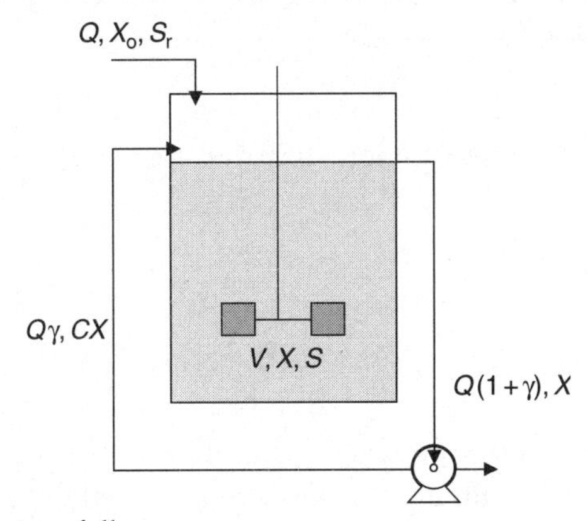

The cell balance is as follows:

cell accumulation = cells in + recycled cells + cell growth − cells out − cell death

$$\frac{dX}{dt} = \frac{QX_o}{V} + \frac{FCX}{V} + \mu X - \frac{(1 + \gamma)FX}{V} - \alpha X$$

BOX 4.1 (*Continued*)

Where C = recycled biomass concentration factor and γ = recycle ratio. Assuming cell maintenance and cell death are negligible, $dX/dt = 0$ and $D = Q/V$.

$$\mu = D(1 + \gamma - \gamma C)$$

For the substrate

Substrate accumulation = medium substrate + substrate in recycle − consumption − substrate out

$$\frac{dS}{dt} = \frac{QS_r}{V} + \frac{(1+\gamma)Q_r}{V} - \frac{\mu X}{Y_{X/S}} - \frac{QS}{V}$$

Under steady-state conditions $dS/dt = 0$ and $Q/V = D$, and the equation can be simplified.

$$X = \frac{D}{\mu} Y_{X/S}(S_r - S) \quad (4.3)$$

Substituting for μ:

$$X = \frac{Y_{X/S}(S_r - S)}{(1 + \gamma - \gamma C)}$$

This equation shows that the biomass X will always be higher with recycling as the lower expression will always be less than 1 as C will be greater than 1.

The yield coefficient, when combined with the growth equation, gives

$$\mu = -Y\frac{dS}{dt}X \quad (4.7)$$

$$\frac{dS}{dt} = -\frac{\mu}{Y}X \quad (4.8)$$

However, under conditions where a significant amount of substrate is used to maintain the cells, maintenance needs to be added to the equation

$$\frac{dS}{dt} = -\frac{\mu}{Y}X - K_m X \quad (4.9)$$

where K_m is the maintenance coefficient.

4.3.4.1 *Operating factors*

The population of the activated sludge is much the same as that which develops on the trickling filter, except that it is a little less heterogeneous and is therefore less adaptable to changes in waste composition. Most of the biomass

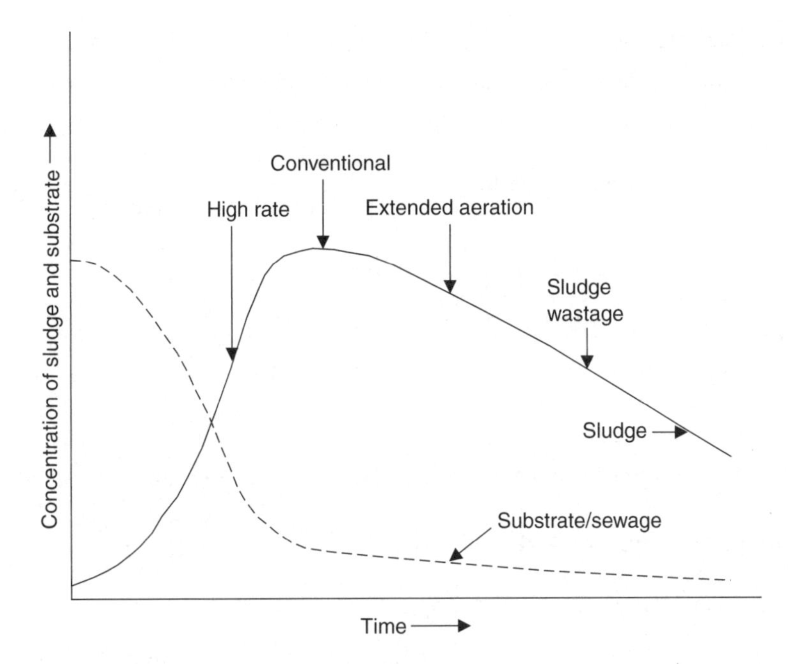

Figure 4.9 Changes in the concentration of biomass (sludge) and substrate (BOD_5) with time as the mixed culture growing. The conventional activated-sludge system maintains the biomass at the highest level and the substrate low by recycling some of the sludge to the beginning of the system. A high-rate system will still contain rapidly growing micro-organisms but some substrate will remain. Under these conditions there will be a rapid throughput but incomplete BOD_5 removal. Extended aeration is the reverse, where the micro-organisms are retained in the system for an extended period with no or low substrate. As a consequence some of the cells die and are lost and others stop growing, and the biomass levels drop as these are washed out of the system.

occurs as flocs or aggregates and the concentration of biomass in the tank is referred to as the mixed-liquor suspended solids (MLSS). Normal MLSS levels range from 200–300 to 800 mg/l in high-rate systems, and this level can be controlled by the amount of activated sludge recycled. The MLSS is determined by drying samples after filtering at 105 °C.

4.3.4.2 *Sludge-residence time (SRT) or sludge age*

SRT is the mean time that the sludge spends in the bioreactor. This is determined by the flow rate into and out of the reactor and the biomass concentration.

$$\text{SRT} = \frac{VX}{(Q_w X_w) + (Q_c X_c)} \tag{4.10}$$

where V = the reactor volume (m^3), X the MLSS (mg/l), Q_w the sludge-wastage rate, death rate (m^3/day), X_c the MLSS exit concentration (mg/l), Q_c the effluent-discharge rate, and X_w is the concentration of waste sludge (mg/l). Under normal conditions the MLSS remains constant; then the SRT is the mean cell residence time.

4.3.4.3 *Plant loading*

The volumetric plant loading is the flow of wastewater in relation to reactor volume.

$$\text{Volumetric loading (h)} = \frac{\text{Volume (m}^3) \times 24 \text{ h}}{\text{Input flow rate (m}^3\text{/day)}} \tag{4.11}$$

$$\text{Hydraulic retention time (HRT)} = \frac{VX \times 24}{Q} \tag{4.12}$$

where Q is the sewage flow rate (m^3/day). The hydraulic retention time (HRT) does not take into account the recycling of activated sludge.

4.3.4.4 *Organic loading*

This is the loading of the reactor with BOD as kg of BOD/day.

$$\text{Organic loading (kg BOD}_5\text{/m}^3\text{/day)} = \frac{Q(\text{m}^3\text{/day}) \times \text{BOD}_5(\text{mg/l})}{V\,(\text{m}^3) \times 1000} \tag{4.13}$$

The rate of BOD_5 removal (g of BOD_5/day):

$$\text{Removal rate (g/day)} = \frac{\text{Input BOD} - \text{output BOD}}{\text{HRT}} \tag{4.14}$$

However, there is sludge recycling.

$$\text{BOD concentration inlet (mg/l)} = \frac{(Q \times \text{input BOD}) + (Q_r \times \text{output BOD})}{Q + Q_r} \tag{4.15}$$

where Q_r is the sludge recycling rate (m^3/day).

4.3.4.5 *Sludge loading*

This is the relationship between BOD and activated sludge in the reactor. This is often referred to as the food to biomass ratio (f/m). The effect of increasing f/m on BOD removal is shown in Fig. 4.10.

$$\frac{f}{m} = \frac{\text{BOD loading rate}}{\text{Volume of sludge}} \tag{4.16}$$

$$\frac{f}{m} = \frac{Q \times \text{BOD}}{VX} \tag{4.17}$$

As

$$\text{HRT} = \frac{VX \times 24}{Q} \tag{4.18}$$

$$\frac{f}{m} = \frac{\text{BOD}}{X} \cdot \frac{24}{\text{HRT}} \tag{4.19}$$

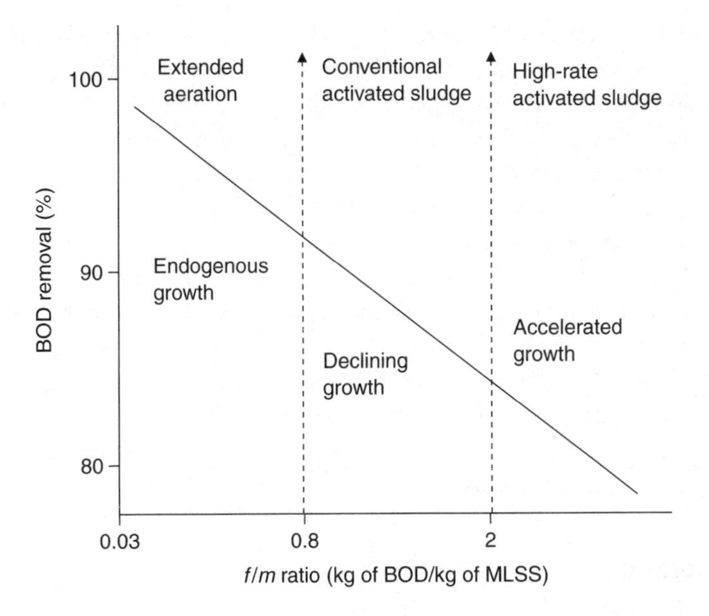

Figure 4.10 The effect of the *f/m* ratio (food/biomass or sludge loading) on the BOD_5 removal. In the high-rate system the ratio of BOD_5 to micro-organisms (MLSS, mixed-liquor suspended solids) is high but the removal low. The reverse is true for the extended-aeration system.

4.3.4.6 *Sludge settling and sludge volume index*

Sludge density index (SDI) and sludge volume index (SVI). Both are calculated by allowing the sludge to settle for 0.5–1 h in a measuring cylinder. Well-settled sludge has an SVI value of 100, but there are settling problems with a sludge that has an SDI value of 150 and above.

4.3.4.7 *Recirculation of sludge*

Sludge is returned to the reactor and, assuming no wastage:

$$\frac{Q_r}{Q + Q_r} = \frac{V}{1000} \tag{4.20}$$

Q_r is the sludge recycling rate (m^3/day), Q is the mean inflow rate (m^3/day), and V is the volume of settled solids after 30 min in a 1000 l cylinder.

$$Q_r = \frac{V \times Q}{1000 - V} \tag{4.21}$$

The solids concentration of recirculated sludge is

$$X_r = \frac{1\,000\,000}{\text{SVI}} \tag{4.22}$$

X_r is the concentration of returned sludge (mg/l). The ratio of returned sludge to the reactor is expressed as a percentage of influent wastewater:

$$\frac{Q_r}{Q} = \frac{V}{1000 - V} \times 100 \tag{4.23}$$

Different systems used for waste treatment are best compared by their ability to treat waste per unit volume. For trickling filters the waste treated per unit volume can be expressed as hydraulic and organic loads. The hydraulic load is expressed as cubic metres of wastewater treated per cubic metre of filter. In general a high load for a filter will be above 3 m^3/m^3 and the upper limit is set by the filter becoming flooded with organic material, the spaces between the filling becoming clogged and the whole system becoming anaerobic. The organic load is expressed as kg of BOD_5 removed/m^3/day. A BOD_5 load of below 0.6 kg/m^3/day is regarded as a low rate. To achieve an exit BOD_5 value of 20 mg/l organic loads of up to 1.0 kg/m^3/day are normally used in a single pass. However, if nitrification, the removal of ammonia, is required the flow and organic load have to be reduced as the organisms responsible for nitrification only grow slowly. A worked example of a conventional activated-sludge system is given in Box 4.2.

4.3.5 Aeration

As activated sludge is an aerobic process considerable effort is made to maximize the supply of oxygen and avoid oxygen limitation of growth. One of the most common methods of providing oxygen is to introduce air at the base of the tank through spargers or porous 'diffusers'. The porous diffusers provide a stream of small bubbles and the smaller the bubbles the better the aeration due to the increase in the bubble surface-area-to-volume ratio. Bubbling with air is generally sufficient to provide all the oxygen needed as sewage is a fairly dilute substrate (200–250 g/m^3) and therefore the oxygen demand is fortunately low. Air flow rates of 7–10 m^3/m^3 of sewage are normally used. If increased aeration is required then mechanical surface aeration can also be used and there are a number of designs, which include a partly submerged paddle (Kessener brush) and turbines rotating at the surface (Simcar) (Fig. 4.11). Air can also be entrained in the activated sludge by withdrawing some of the sludge and pumping it back into the tank using a venturi or a high-speed jet. The activated-sludge tank is operated as a plug flow reactor and as such the oxygen requirement will drop as the BOD is reduced as the material passes down the reactor. In order to even out the varying oxygen requirements a number of operational modes have been developed.

Where constant aeration is provided along the tank this will not match oxygen requirements due to the reduction in substrate along the tank. To avoid this tapered aeration has been developed where a diminishing supply of air is introduced along the tank (Fig. 4.12). This can be achieved in a series of tanks as well as in a long plug flow tank.

In the conventional activated-sludge process the biomass is essentially in the form of flocs due in part to the production of extracellular bacterial polymers. The exact nature of this material is not known but it is responsible for forming aggregates of between 20 and 1000 μm in diameter. There is rapid attachment of suspended and colloidal material to these flocs as soon as the

BOX 4.2 Conventional activated-sludge treatment

In a conventional activated-sludge treatment system settled sludge is run continuously into the bioreactor (see figure). In order to maintain a high biomass in the tank 20% of the activated sludge is recycled. Some of the parameters of a conventional wastewater system are given in the table below.

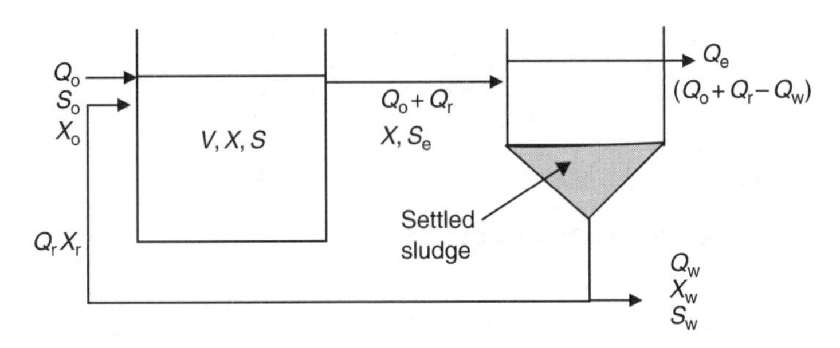

Parameter	Parameter value
Volume of bioreactor (V)	2 000 m^3
Substrate (BOD) (S)	250 mg/l
Exit substrate (BOD) (S_e)	10 mg/l
Steady-state biomass (X)	350 mg/l
Concentration of biomass recycled (X_r)	1 500 mg/l
Yield of cells on substrate (Y)	0.55 g of cells/g of substrate
Recycle ratio (γ)	0.2
Recycled biomass concentration factor (C)	4.28
Rate of sludge wastage (Q_w)	443 m^3/day
Rate of effluent production (Q_e)	1034 m^3/day

In the conventional activated-sludge process run at a steady state the biomass X is given by the following equation.

$$X = Y_{X/S}(S - S_e)$$
$$X = 0.55\ (250 - 10)$$
$$X = 132 \text{ mg/l}$$

However, in the activated-sludge process a proportion of the settled sludge is recycled. Thus

Cells in the bioreactor = cells in + recycle stream + growth − cells out.

If cell maintenance and death are discounted the equation can be simplified to

$$\mu = D\ (1 + \gamma + \gamma C)$$

BOX 4.2 (*Continued*)

Where D the dilution rate is Q/V. This can be converted to the following, where S_0 is the starting sewage concentration (mg/l):

$$X = \frac{Y_{X/S}(S_o - S)}{(1 + \gamma + \gamma C)}$$

$$X = 0.55 \frac{(250 - 10)}{(1 + 0.2 + 0.2 \times 4.28)}$$

$$X = 395 \text{ mg/l.}$$

This is a considerably higher biomass value than without recycling.

Sludge-residence time (SRT) or sludge age

$$\text{SRT} = \frac{VX}{(Q_w X_w) + (Q_e X_e)}$$

$$\text{SRT} = \frac{2000 \times 350}{(433 \times 395) + (1034 \times 10)}$$

$$\text{SRT} = 3.77 \text{ days.}$$

Bioreactor loading (hydraulic retention time, HRT)

$$\text{HRT} = \frac{V \times 24}{Q}$$

$$\text{HRT} = \frac{2000 \times 24}{960}$$

$$\text{HRT} = 50\,\text{h (2.08 days).}$$

Organic loading

$$\text{Loading} = \frac{Q \times \text{BOD}}{V \times 1000}$$

$$\text{Loading} = \frac{960 \times 250}{2000 \times 1000}$$

$$\text{Loading} = 0.12 \text{ kg of BOD/kg/day.}$$

Sludge loading

$$\frac{f}{m} = \frac{Q \times \text{BOD}}{V \times X}$$

$$\frac{f}{m} = \frac{960 \times 250}{2000 \times 350}$$

$$= 0.34 \text{ g/g/day}$$

If the flow rate is increased from 960 m^3/day to 20 000 m^3/day,

$$\text{HRT} = \frac{2000 \times 24}{20000}.$$

$$\text{HRT} = 2.4 \text{ h (0.1 days)}$$

BOX 4.2 **(Continued)**

Organic loading

$$\text{Load} = \frac{2000 \times 250}{2000 \times 1000}$$

Sludge loading

$$\frac{f}{m} = \frac{20000 \times 250}{2000 \times 350}$$

$$\text{Loading} = 7.14 \text{ g/g/day.}$$

Sludge residence time (SRT)

$$\text{SRT} = \frac{2000 \times 350}{2000 \times 1000}$$

$$\text{SRT} = 0.35 \text{ h.}$$

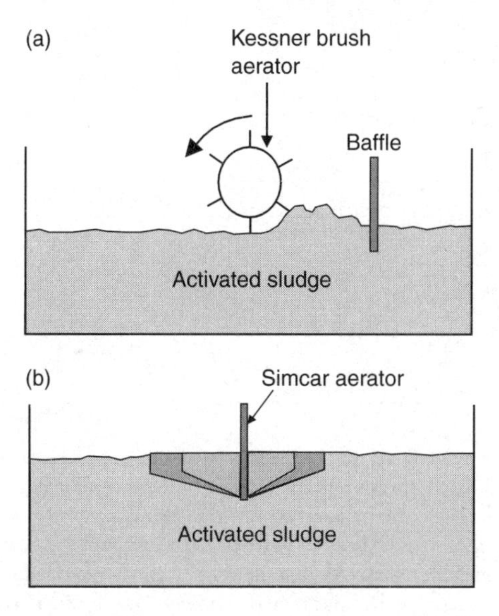

Figure 4.11 The two forms of surface aerator used in the activated-sludge process; (a) the Kessner brush and (b) the Simcar aerator. Both aerators agitate the surface, increasing diffusion of oxygen into the liquid.

activated sludge is mixed with the sewage. Once attached the breakdown of the material takes much longer in a process known as stabilization. In the contact-stabilization process the adsorption and stabilization take place in separate tanks (Fig. 4.12). The contact period is usually between 0.5 and 1 h; the sludge is settled and the returned sludge aerated for up to 5 h to complete oxidation (stabilization). The process can be achieved in a single tank if the waste enters the tank towards the end and is then recycled after a short contact time.

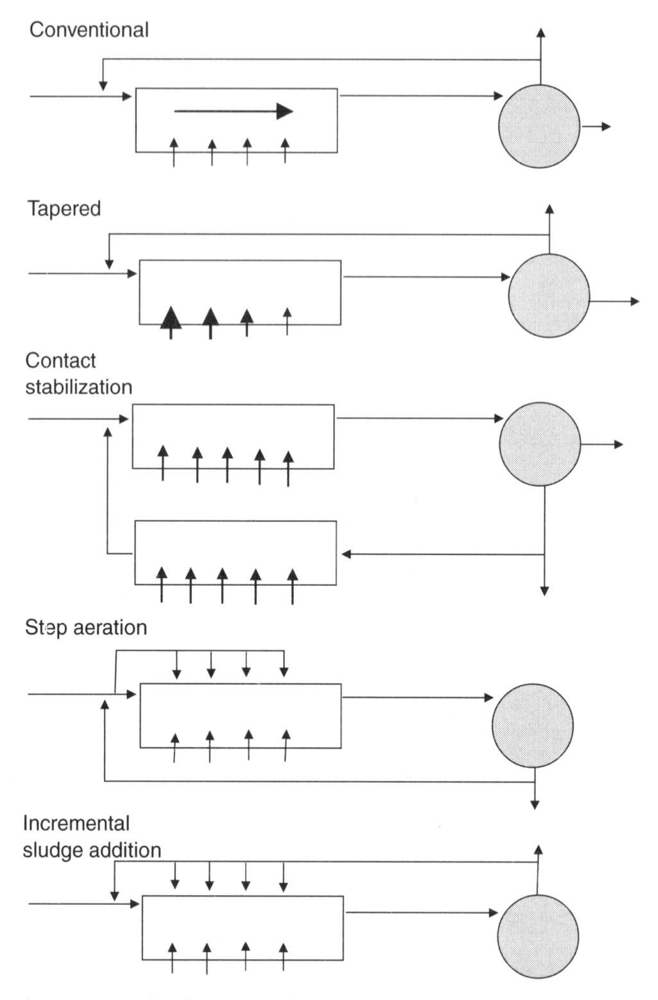

Figure 4.12 Alternative mode of operating the activated-sludge process in order to have better contact between the micro-organisms and the organic material, and improved aeration. In the conventional system the tank is aerated at the same rate along the length of the tank. The tapered aeration mode supplies more air at the start where the organic load is greatest. In contact stabilization the recycled sludge is mixed with the sewage and aerated before returning to the process. In this way the micro-organisms are in contact with organic material for longer. In step aeration sludge and sewage are mixed before they are added along the length of the tank. The incremental-sludge-addition mode adds recycled sludge to the tank along its length.

With step aeration there is an incremental waste and sludge feed along the length of the tank which should go some way to balancing the oxygen supply (Fig. 4.12). Step aeration can be carried out in a series of tanks as well as in a single tank. Incremental sludge feeding has a similar effect to step aeration except that the recycled sludge is fed into the tank along its length, and the waste runs in at a single point.

All the previous systems used a continuous plug flow system but completely mixed systems have been developed similar to those found in fermentation bioreactors where the object is to produce a homogeneous suspension within the reactor. This requires both aeration and mixing to be provided. A settling area can separate the sludge before the effluent is released. The advantages of such a system are increased aeration, which allows increased BOD_5 loading and the ability to withstand BOD_5 shocks. The disadvantages are that the sludge produced is more difficult to settle out due to its smaller floc size.

4.3.6 Mode of operation

Irrespective of the type of system used the rate of supply of sewage has to be regulated in order to achieve an effluent level 20 mg/l BOD_5. However, by altering the operating conditions different loading and conversion rates are possible. The conventional activated-sludge system tries to keep the biomass as high as possible by sludge recycling (Table 4.6). In the extended-aeration variation the biomass and substrate remain in contact for much longer and under these conditions growth is very much substrate-limited, which means that less aeration is required and sludge production is reduced as the cellular metabolism is directed towards energy formation rather than cell growth. The best example of such a process is the Pasverr or oxidation ditch (Fig. 4.13). This is an oval ditch, 1–3.5 m in depth, fitted with gates or paddles that aerate and give it a directional flow. The waste is run into the ditch just before the aerator and travels around the whole ditch at 0.3–0.6 m/s before it is run off into a settling tank. Some of the settled sludge is added back to the system just after the aerator.

High-rate operations involve only the partial treatment of the waste and are often used as the first part of the treatment of industrial waste and sometimes domestic waste.

Table 4.6 Operating parameters for different activated-sludge rates

Rate	Retention time (h)	BOD loading (kg of BOD/ m^3/day)	Sludge loading (*f/m*; kg of BOD/kg/day)	Sludge age (days)	Sludge production (kg/kg of BOD removed)
Low, extended aeration	24–72	<0.3	<0.1	>5–6	0.4
Medium, conventional treatment	5–14	0.4–1.2	0.2–0.5	3–4	0.5–0.8
High, high-rate	1–2	>2.5	>1	>5–6	0.8–1

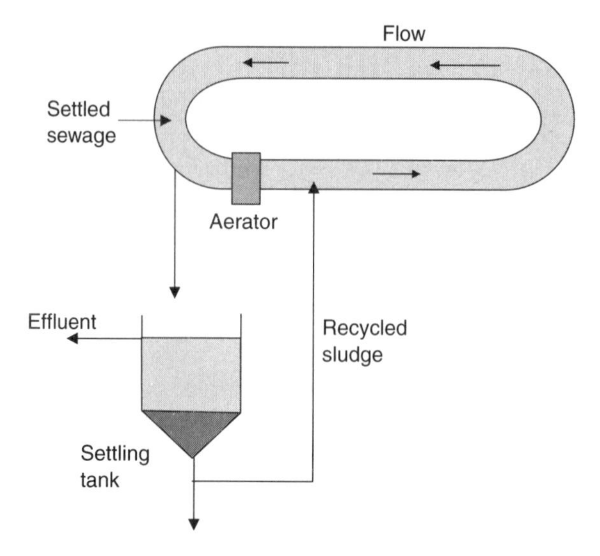

Figure 4.13 The Pasverr oxidation ditch. Settled sewage and recycled sludge are mixed and passed around an oval ditch by an aerator which both mixes and aerates, creating a highly aerated system. After one cycle the sludge is separated from the liquid and some sludge returned to the system.

4.4 Modifications to existing processes

4.4.1 The biotower

The development of plastic media as the packing for trickling filters has been followed by the development of tall tower filters without the need for support. These tower bioreactors, because of their high surface area, can handle a high BOD loading and in some cases forced air is supplied to the base of the tower to increase the rate of degradation (Fig. 4.14). The towers are mainly used to reduce BOD in high-strength wastes and are often added to a conventional system ahead of the aeration tanks to smooth out variations in BOD and to increase plant capacity.

4.4.2 The rotating biological contactor

The availability of plastic media has also seen the another development; the rotating biological contactor. Here a drum of honeycomb plastic or closely spaced discs are slowly rotated (1–2 rev./min) with the base (40%) of the drum in the settled sewage or wastewater (Fig. 4.15). A typical unit may be 5–8 m in length and 2–3 m in diameter and separated into a series of chambers. The chambers help to maintain a form of plug flow. Aeration occurs as the drum rotates free of the liquid. The large area and good aeration means that the rotating contactor can handle a wide range of flows, needs only short contact times, has 4–5 times the capacity of a conventional filter, and no recycling is

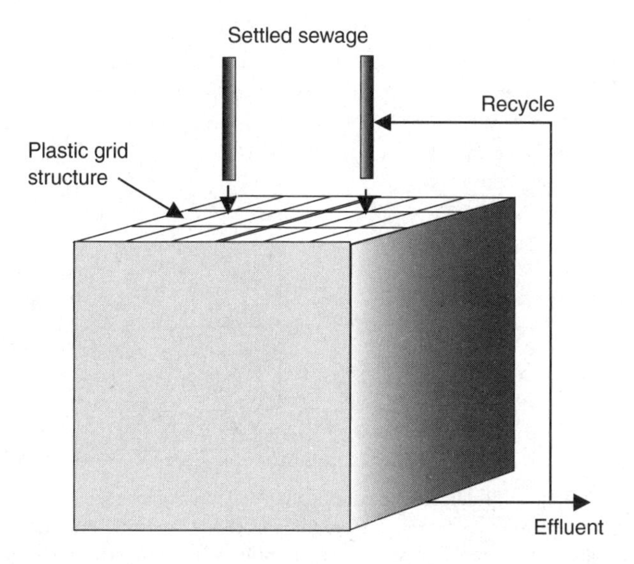

Figure 4.14 A schematic diagram of the construction of a biotower. Modular plastic grids are combined to form a tower structure. The plastic grids are constructed to give a high surface area while maintaining good airflow. A microbial population forms a layer (biofilm) on the plastic surface and this degrades the organic material. The wastewater is distributed across the top of the tower and the effluent collected at the bottom. Depending on the rate of flow and the strength of the wastewater the effluent can be recycled, run into a waterway, or passed on to another treatment system.

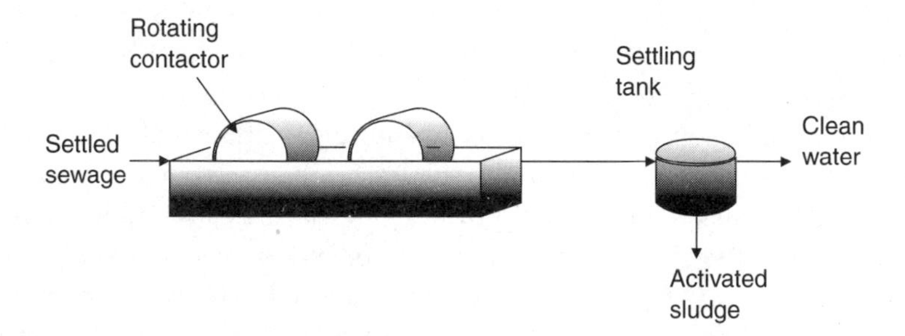

Figure 4.15 The rotating biological contactor consists of one or more rotating drums of porous material which supports a biofilm, a thin layer of micro-organisms. About one-third of the drum is immersed in the sewage so that, as the drum rotates, aeration is provided when the cells are out of the liquid. The rate of rotation is 1–2 rev./min.

required. The disadvantages of the rotating biological contactor are that in cold climates the system needs covering, the cost of running the motor, and maintenance.

4.4.3 The fluidized bed

One method of increasing the area of the support in a biofilm reactor is to use smaller and more robust biofilm supports such as sand. However, sand coated

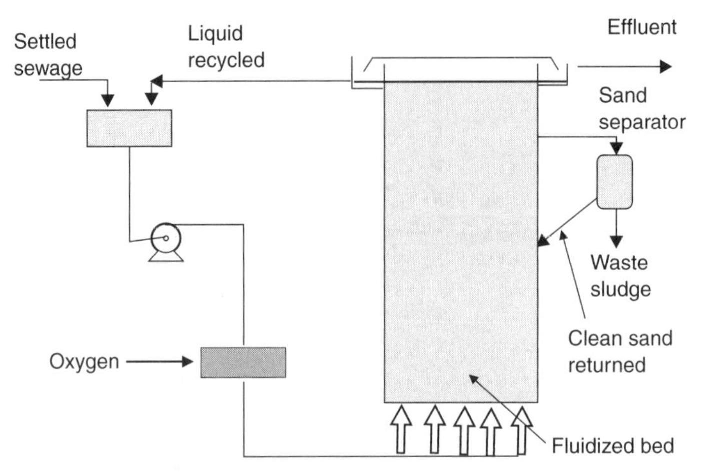

Figure 4.16 The biological fluidized-bed reactor uses small particles such as sand as these have a very large surface area on which the micro-organisms can grow (biofilm). The settled sewage or wastewater is aerated by bubbling air or oxygen, prior to pumping the liquid, into the base of the vessel. The flow of liquid into the base will fluidize the sand giving good distribution of the waste. Excess biomass can be removed by extracting some of the sand and sonication to remove the micro-organisms, after which the clean sand is returned to the vessel.

with a biofilm in a reactor would soon clog and trap particles and become anaerobic. To overcome this problem the bed of sand can be fluidized by the upward flow of liquid. The area for biomass support is 3300 m^2/m^3 compared with 150 m^2/m^3 for rounded gravel which supports a MLSS of 40 000 mg/l compared with 150–350 mg/l for activated sludge. The high biomass clearly has a considerable oxygen demand such that pure oxygen is supplied by injection into the settled sewage prior to entering the fluidized bed (Fig. 4.16). The biomass build up on the sand particles can be controlled as coated sand can be extracted, the biomass removed from the sand in a cyclone, and the clean sand returned to the vessel. This type of system has proved particularly useful for treating high-strength wastes, especially those from industrial sources. The fluidized beds operate with short retention times of around 20 min but the system is expensive to operate due to the use of oxygen and the costs of pumping.

An alternative to the use of fluidized beds is the airlift bioreactor where the biomass forms as a biofilm on small particles of about 0.3 mm in diameter. The bed is fluidized by the introduction of air at the base of the vessel, which also supplies oxygen. The advantage of such a system is the higher biomass coating the particles due to the better oxygen supply, which means that with the same concentration of organic waste there is a lower growth rate and therefore less sludge is generated. Another similar system, known as the moving-bed biofilm, uses polyethylene carriers to retain the biomass (Fig. 4.17) to treat thermomechanical pulping whitewater under thermophilic conditions (55 °C). The reactor is fluidized by a supply of air at the base of the vessel (Jahren et al., 2002) and has been used for other wastes.

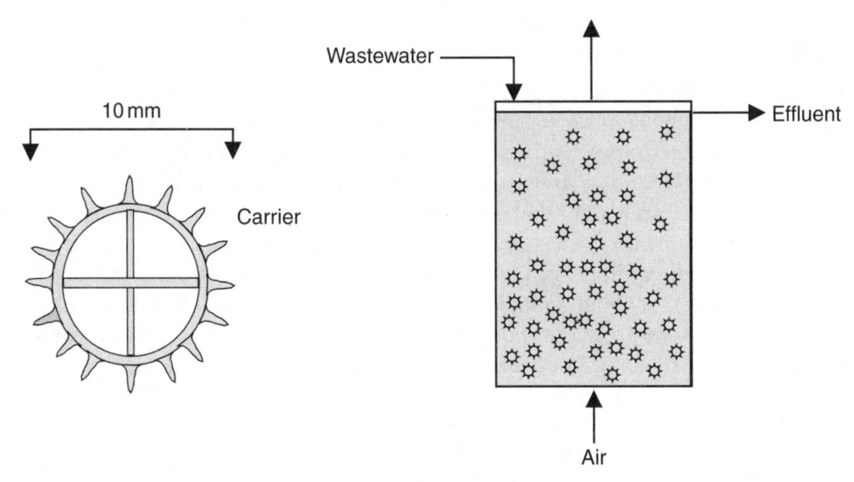

Figure 4.17 A Kaldness moving-bed biofilm bioreactor. Here the biomass is grown on plastic carriers (100 mm diameter) which are mixed by an air supply at the bottom of the vessel. The vessel is supplied continuously with waste without sludge recycling. The system can be used anaerobically if mechanical stirring is provided and such a system has been used in full-scale treatment of wastewaters. Modified from Jahren et al. (2002).

4.4.4 The deep-shaft process

The deep-shaft process was a development from ICI's single-cell protein work and was based on the airlift design of bioreactor. The airlift design of bioreactor operates by introducing air at the bottom of the vessel. The introduced air reduces the overall density of the liquid and the air bubbles rise; these factors combined cause a flow of water upwards. If this upward flow is separated from the rest of the vessel by a partition (draft tube), a circulating flow can be generated so that both mixing and aeration can be achieved by sparging air (Box 4.3). The airlift bioreactor is normally a tall narrow vessel so that at a height of 100 m or more a pressure of about 10 atm will be found at the base. The high pressure will force more oxygen into solution, improving aeration considerably. In practice the deep-shaft bioreactor is sunk into the ground, either as concentric pipes or divided vertically (Fig. 4.18), and because of its increased aeration has often been installed to treat high-BOD_5 industrial wastes. Once a flow has been started air can also be injected into the downcomer to be carried downwards to the base of the vessel. A deep-shaft process has been installed at Marlow Foods, single-cell protein (Quorn) plant to treat the waste from the cultivation process, and others have been installed worldwide. The system has the advantage that it requires only a small space compared with conventional systems and due to the high aeration rate will deal with high-BOD_5 wastes, 3–6 kg of BOD_5/m^3/day with a 90% removal. This treatment rate is intermediate between high-rate and conventional sewage treatment, but the process produces less sludge. Figure. 4.18 shows the details of a full-scale installation for the treatment of sewage that is capable of treating 30 000 m^3 of waste/day at 600 mg/l BOD_5 where the sewage is retained in the vessel for 1 h.

BOX 4.3 Airlift bioreactor

The figure below shows a schematic diagram of an airlift bioreactor, which is the principal alternative to the stirred-tank bioreactor. In the airlift the stirrer (impeller) is replaced by a stream of air to mix the culture. Air is injected (sparged) at the base of a tube (draft tube) held in the centre of the bioreactor. The addition of air lowers the density of the liquid in the draft tube and rising bubbles cause the liquid to flow upwards. As a consequence liquid is drawn in at the draft-tube base and a circulation of liquid around the bioreactor is established. This form of mixing has the advantage of not requiring a stirrer system; it has no moving parts, simple construction, and good mixing from top to bottom of the bioreactor.

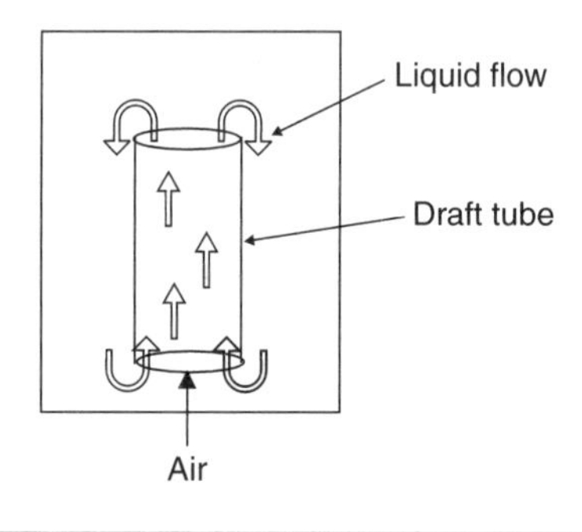

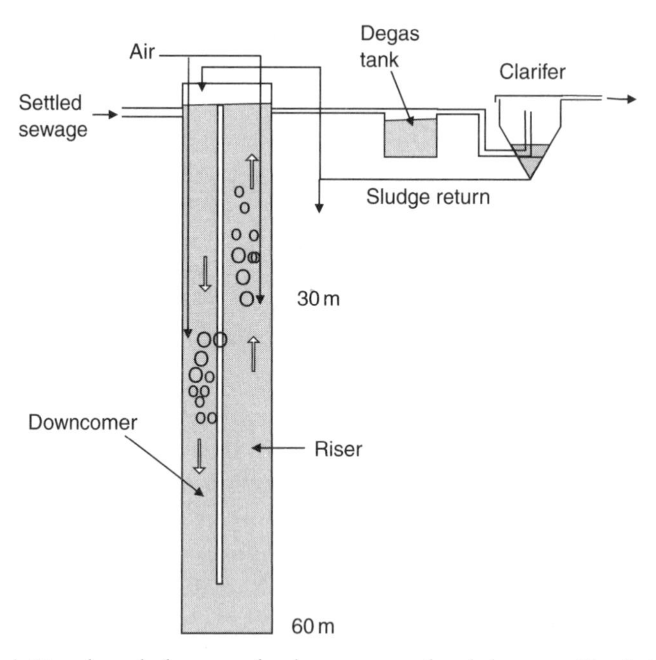

Figure 4.18 A 60-m deep-shaft process for the treatment of settled sewage. The design is based on an airlift bioreactor (Box 4.3). The depth of the shaft (60 m) ensures that oxygen is forced into solution. Once a liquid flow has been established, air is added to the downcomer. This provides more air to the system and produces a highly aerobic process.

4.4.5. Addition of pure oxygen

The aeration of the activated sludge can be improved by the addition of pure oxygen to a closed system or to open tanks. The closed system has the advantage that the oxygen is not lost to the atmosphere but the presence of high oxygen concentrations does constitute an explosive hazard requiring strict safety precautions. The high aeration also causes an accumulation of carbon dioxide that can reduce the pH and thus reduce nitrification (Fig. 4.19). A system like this was marketed in the UK as the Unox system. The tanks are divided into a series of compartments, each of which is mixed by a surface aerator. This type of system can be used to sustain a higher biomass (increased MLSS), with lower sludge production and double the loading rate.

4.4.6 The captor process

In order to maintain a high biomass at the start of the plug flow process of waste treatment by activated sludge a modification has been used. Here the activated-sludge biomass is immobilized in reticulated plastic pads of 25 mm × 25 mm × 12 mm in dimension, similar in nature to washing-up pads. The activated-sludge micro-organisms from aggregates readily colonize these pads. The pads are retained in the early part of the aeration tank by screens and have been shown to give higher biomass levels of 6–8 g/l (Fig. 4.20). To maintain an active biomass some of the pads are stripped of

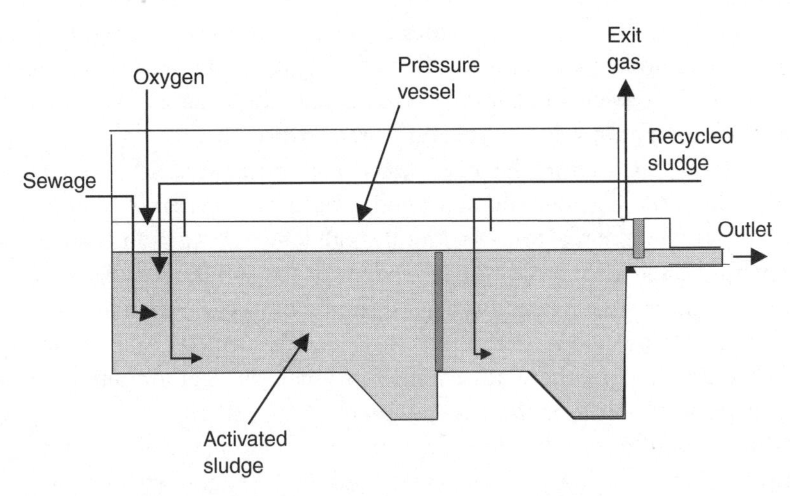

Figure 4.19 An activated-sludge process using pure oxygen to improve waste removal. Oxygen has a low solubility in water (7 mg/l) so that the application of pure oxygen will improve oxygen transfer in this system. An example is the Unox system, which is sealed to avoid the loss of expensive oxygen and to reduce the fire hazard of pure oxygen. Oxygen applied under pressure is forced in the activated sludge in a number of chambers arranged in a cascade. The carbon dioxide and residual oxygen is collected and treated at 1000 °C. Adapted from Crueger and Crueger (1990).

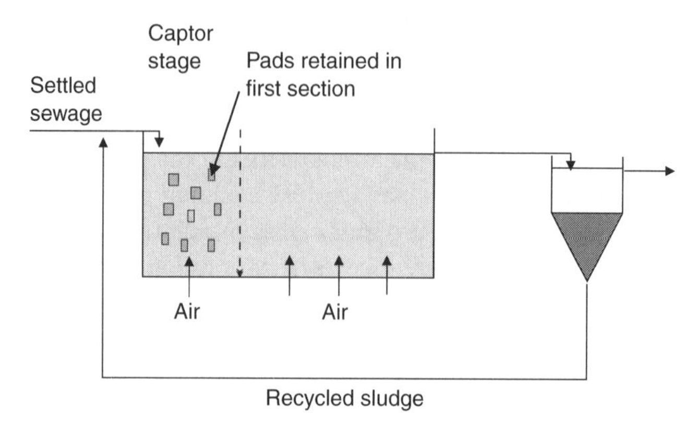

Figure 4.20 The Captor process. A high biomass is retained at the start of the activated-sludge process where the organic material is highest by immobilizing the biomass in plastic pads. The pads are small (10 cm) plastic meshes that form a large area on which micro-organisms can grow. When the pads become full the biomass can be squeezed out and the pads returned to the system.

excess biomass by a system that removes the pads, squeezes out the sludge, and returns the empty pads to the tank.

4.4.7 Membrane bioreactors

The development of ultrafiltration and microfiltration membranes for biological separations has allowed this technology to be applied to wastewater treatment (Fig. 4.21). The membrane allows the passage of small molecules while retaining the micro-organisms that make up the activated sludge (Brindle and Stephenson, 1996). A membrane bioreactor was first used to treat landfill leachates but since this time three types of membrane systems have been developed; solid/liquid separation, gas permeable system, and extractive process.

Membrane bioreactors have been used in both aerobic and anaerobic modes where the high biomass retained gives a rapid breakdown of organic compounds, and the reactors can handle high loads. As the solids are retained the hydraulic retention time is independent of the solids. Because of the high biomass in the system it requires a good supply of oxygen but a low substrate-to-biomass ratio reduces the amounts of sludge formed. Apart from many experimental systems there are over 20 full-scale membrane units that have been installed in the Netherlands and Germany.

As the membrane will allow gases to pass through while retaining the biomass reactors of this sort can provide bubble-free aeration with a high surface area for oxygen transfer. The membrane also provides a support for biofilm formation. In the extractive process the membrane allows the chemical pollutant to pass through the membrane into the biomass where they can be degraded. The membrane system can also separate the biomass from a biologically hostile wastestream and this type of system has been tested with

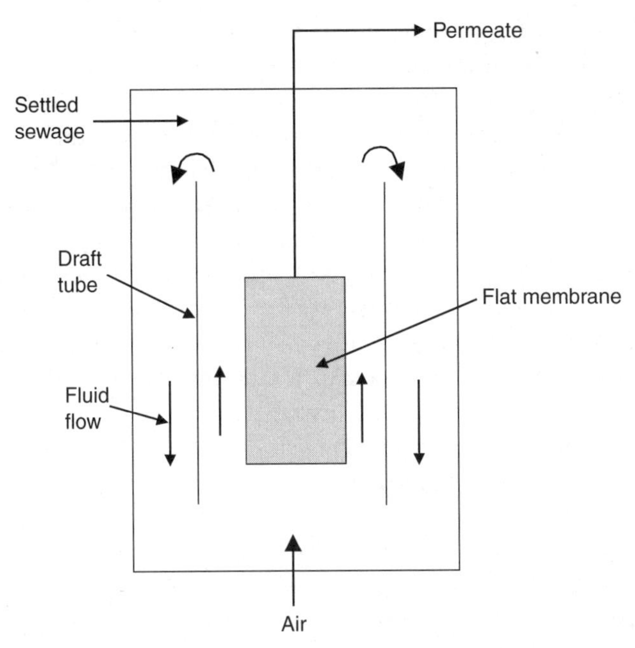

Figure 4.21 A system for the use of flat membranes to treat wastewaters. The advantages of a membrane bioreactor to treat waste are that the biomass concentration can be very high as there is complete separation of biomass and effluent. The main problem is clogging of the membrane. This can be reduced by creating a rapid flow across the membrane. In this system an airlift configuration directs air over the membranes which are situated in the draft tube. The clean permeate is removed from the membrane by a pump. Adapted from Shim et al. (2002).

pollutants such as nitrobenzene, benzene, and dichloroanaline with over 99% removal. Membrane bioreactors are more expensive than conventional activated-sludge and trickling-filter processes but have the advantage that less sludge is generated, a higher COD removal, good oxygen supply, and a suitability for small plants and where high-quality effluent is required.

4.5 Removal of nitrogen and phosphorus

Waste streams not only contain biologically metabolizable organic materials but also nitrogen- and phosphorus-containing compounds. The nitrogen-containing compounds in sewage are ammonia, proteins, and amino acids (Table 4.2). Increasingly nitrates are being found in wastewater as a result of run-off from agricultural land. Ammonia is formed from urea, a major constituent of urine, and can also be formed during the natural breakdown of proteins and is found in some industrial wastes. Ammonia has an offensive smell, is poisonous to aquatic life at concentrations as low as 0.5 mg/l, and increases the chlorine dosage needed for the treatment of drinking water.

The European Inland Fisheries Advisory Commission (EIFAC) has recommended that the maximum un-ionized ammonia concentration should be 0.025 kg/m^3 (20 g/m^3 total ammonia).

Nitrogen-containing compounds are required for the synthesis of proteins by all types of micro-organism during their growth. The micro-organisms normally utilize ammonia first but nitrates and urea can also be used. In waste-treatment systems a third of the total nitrogen content is removed by assimilation during growth, and the rate of removal depends on the biomass level and rate of growth in the waste-treatment system.

Nitrogen compounds can be removed by physical, chemical and biological treatment: ion exchange, reverse osmosis, ammonia stripping, chemical denitrification, chemical precipitation, assimilation, and biological treatment. Despite the number of chemical and physical methods, increasingly biological methods are used to remove nitrogen.

The potential for nitrification is limited at lower temperatures (<15 °C) due to the slowing of growth but this can be improved by bioaugmentation with nitrifying bacteria. The nitrifying bacteria are supplied by a bioreactor in the sludge return line, fed with a nitrogen-rich flow from sludge treatment (Salem et al., 2003).

4.5.1 Nitrification

The levels of ammonia in domestic wastewater are about 25 mg/l (25 g/m^3) but industrial wastes can contain up to 5 g/l ammonia with the recommended level at 25mg/l (25 g/m^3). Ammonia is oxidized rapidly to nitrate in the environment and in wastewater-treatment systems in a process known as nitrification. The conversion is carried out by two groups of chemoautotrophic bacteria, which use the oxidation of ammonia as a source of energy. The first stage of ammonia oxidation is carried out mainly by the genera *Nitrosomonas, Nitrosococcus, Nitrosospira, Nitrocystis,* and *Nitrosogloea.* The reaction is as follows, although the oxidation of ammonia is more complex than given in the equation.

$$2NH_4^+ + 3O_2 \rightarrow 2NO_2^- + 4H^+ + 2H_2O + (\text{energy } 480\text{–}700\,\text{kJ}) \quad (4.24)$$

The energy released is used by the organisms to synthesize cell components from inorganic sources. The release of hydrogen ions can cause a drop in pH and it is clear that a good supply of oxygen is required. The growth of nitrifying bacteria is very slow (μ_{max} 0.1–1 day^{-1}) compared with that of heterotrophic bacteria (μ_{max} 0.46–2.2 day^{-1}).

The nitrite formed is converted to nitrate by the genera *Nitrobacter, Nitrocystis, Nitrosococcus,* and *Nitrosocystis,* but *Nitrobacter* has been the most studied. The reactions is as follows:

$$2NO_2^- + O_2 \rightarrow 2NO_3^- + (\text{energy } 130\text{–}180\,\text{kJ}) \quad (4.25)$$

As the oxidation of nitrite to nitrate yields less energy than the oxidation of ammonia the cell yield of *Nitrobacter* is less than *Nitrosomonas* and the growth rates are also slow with a μ_{max} of 0.28–1.44 day^{-1}. The characteristics of the organisms involved in nitrification affect the wastewater treatment as follows:

- The growth rate is slower than for heterotrophic organisms so that the organic load has to be balanced to their slower growth rate, otherwise the organisms will be washed out.
- There is a low cell yield per unit of ammonium oxidized.
- The organisms require significant amounts of oxygen, 4.2 g/g of NH_4^+ converted.
- The system may need some form of buffering due to the acid conditions produced by the hydrogen ions.

If nitrification is not required in the sewage process then a higher rate of flow can be used. Recently bioaugmentation with nitrifying bacteria has been shown to be effective in maintaining nitrification in stress situations such as low temperatures and high SRT values.

4.5.2 Denitrification

Nitrification in treatment plants and soils combined with the nitrates from agricultural run-off can give rise to high nitrate levels (above 50 mg/l) in waterways, which are used to supply drinking water. High nitrate levels are associated with one disease, methaemoglobinaemia, which affects children below the age of 6 months. The children have an incomplete digestive system and the intake of nitrate leads to the accumulation of nitrite ions, which enter the blood system and block oxygen transport by haemoglobin. Thus the EU have set an absolute limit of 50 mg/l for nitrate, and a recommended limit of 25 mg/l, although a number of UK water-treatment systems are working at 80 mg/l. Nitrate can be converted to nitrite in the human stomach and nitrite has been shown to be converted to carcinogenic nitrosamines and leads to concern over the development of stomach cancer on consumption of high nitrate water.

Nitrate removal, ion exchange, or biological processes can carry out denitrification. The ion-exchange process depends on the resin's affinity, which on a conventional anion-exchange resin is

$$SO_4^{2-} >> NO_3^- > Cl^- \geq HCO_3$$

Any sulphate in the waste will bind in preference to nitrate, but once this has occurred nitrate will exchange with chloride. Once the resin is exhausted it will require regeneration with excess sodium chloride which yields a solution containing high concentrations of sodium sulphate, sodium nitrate, and sodium chloride, which will need disposal. Adding these high-salt solutions to waterways is unacceptable and in practice it is passed on to sewage works for treatment.

The biological conversion of nitrate to nitrite and eventually nitrogen occurs under conditions where oxygen is very low or absent. The process of oxidation involves the loss of electrons and in normal conditions oxygen acts as an electron acceptor but when oxygen levels are low inorganic ions such as nitrate, phosphate, and sulphate can act as electron acceptors. In wastewater where nitrification has occurred, combined with nitrate from agricultural run-off the concentration of nitrate will be higher than sulphate or phosphate. A number of facultative heterotrophic micro-organisms occur in sewage-treatment systems which are capable of converting nitrate to nitrogen provided an electron donor is present.

$$3NO_3^- + 6H^+ \rightarrow 3NO_2^- + 3H_2O \quad (4.26)$$

$$2NO_2^- + 8H^+ + 6e^- \rightarrow N_2 + 4H_2O \quad (4.27)$$

The electron donor is usually an organic compound and in some cases methanol as been used to supplement the normal organic source. The reactions with methanol are as follows:

$$3NO_3^- + CH_3OH \rightarrow 3NO_2^- + CO_2 + 2H_2O \quad (4.28)$$

$$2NO_2^- + CH_3OH \rightarrow N_2 + CO_2 + H_2O + 2OH^- \quad (4.29)$$

The process of denitrification requires low oxygen levels (anaerobic), an organic carbon energy source, a level of nitrate of 2 mg/l or above, and a pH of 6.5–7.5.

4.5.3 Nitrification and denitrification processes

In order to meet increasing European standards, the removal of ammonia and other nitrogenous compounds from wastewater has been developed. Within the sewage system the processes of biological nitrification and denitrification can be organized in a number of ways. In general, the first step of the removal of ammonia by nitrification can be carried out in parallel with the removal of organic material, provided that the hydraulic retention time is not too short (Table 4.7). Denitrification, in contrast, requires a change in growth conditions from aerobic to anaerobic and an organic carbon source. Both nitrification and denitrification can be achieved by partitioning the sewage-treatment system (Fig. 4.22) (Lee et al., 1997) or providing separate reactors which can be used with both suspended and fixed-film cultures (Fig. 4.23) (Tchobanoglous and Burton, 1991). In the single vessel the anoxic zone is situated at the start of the aeration tank where the carbon levels are high, and the anoxic conditions are achieved by stopping aeration. The process using separate vessels is much easier to use and control.

Another system for combined nitrification and denitrification is the sequencing batch reactor (SBR) where a single vessel is used but a programmed

Table 4.7 Parameters for biological nitrification and denitrification processes

Process	Hydraulic retention time (HRT; days)	Retention time (h)	MLSS (mg/l)	pH
Carbon removal	2–5	1–3	1–2	6.5–8
Nitrification	10–20	0.5–3	1–2	7.4–8.6
Denitrification	1–5	0.2–2	1–2	6.5–7

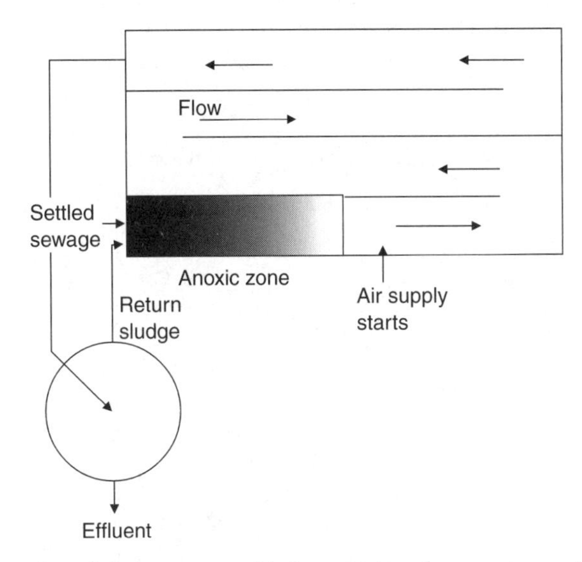

Figure 4.22 An activated-sludge process with the provision of an anoxic zone (anaerobic) at the start in order to achieve denitrification. The anoxic zone is formed by stopping aeration at the first stage where the organic material is highest.

sequence of operations is applied, which can be feeding, anaerobic conditions, aerobic conditions, sludge settling, and effluent removal. This type of operation has been used for the treatment of a number of wastes such as agricultural run-off and landfill leachates but it has the potential for the combining nitrification and denitrification. An example of nitrate, nitrite, and ammonia levels in a sequencing batch reactor are given in Fig. 4.24. The ammonia levels drop during the initial anaerobic phase and the subsequent aerobic phase. Nitrate concentration in contrast is low at the start but rises due to nitrification in the aerobic phase. Both nitrate and nitrite are denitrified during the anoxic phase. The sequencing mode can also be applied to anaerobic reactors (Ndon and Dague, 1997) and in two-stage processes (Ra et al., 1998) where the second

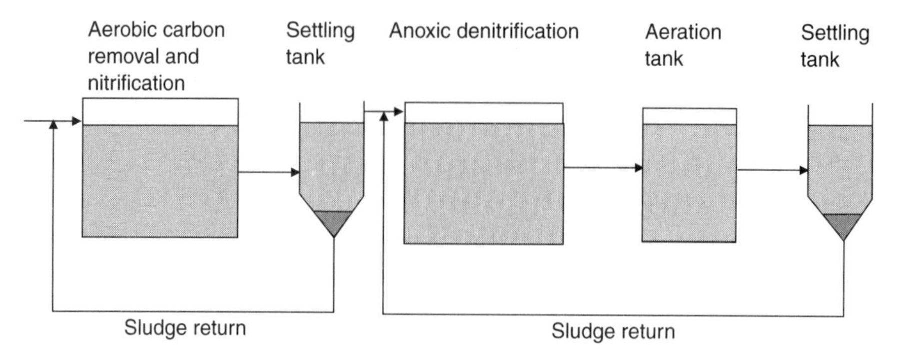

Figure 4.23 An outline of a two-stage or sequencing process for both nitrification and denitrification. In the first stage the normal activated-sludge process occurs with the removal of organic materials and nitrification. In the second stage anoxic conditions cause denitrification which is followed by an aeration tank to strip out the nitrogen formed to ensure precipitation in the settling tank.

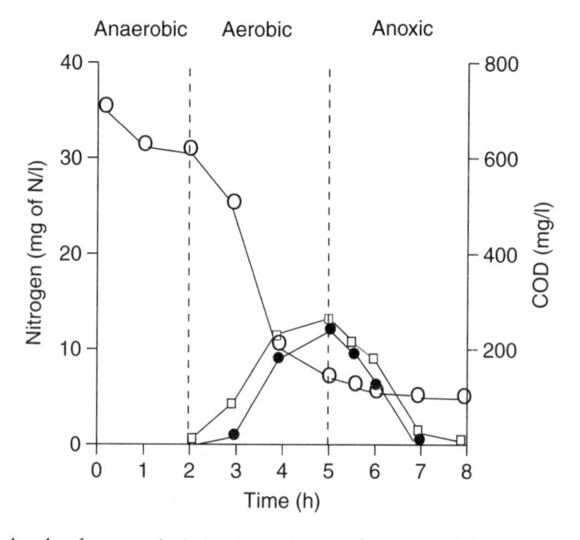

Figure 4.24 The levels of ammonia (o), nitrate (□), and nitrite (●) in a sequencing batch reactor. The three phases, anaerobic, aerobic, and anoxic, are followed by a 1.5-h settlement phase and 0.5-h decantation phase. From Rhee et al. (1997).

stage is the anoxic phase. Some new systems have been developed based on partial nitrification of ammonia to nitrite in anaerobic ammonium oxidation. The systems are based on the following reaction carried out by planctomyces-like bacteria.

$$NH_4^+ + NO_2^- \rightarrow N_2 + 2H_2O \qquad (4.30)$$

Table 4.8 gives three examples of this type of system which have the advantages that they require no carbon addition, produce little sludge, and use less

Table 4.8 Comparison of different ammonia-removal technologies

System	SHARON	Anammox	CANON	Conventional
Bioreactor number	1	1	1	2
Feed	Wastewater	Ammonia/nitrite mixture	Wastewater	Wastewater
Product	Ammonia and nitrite	Nitrate and nitrogen	Nitrate and nitrogen	Nitrate, nitrogen, and dinitrogen oxide
Conditions	Oxic	Anoxic	Oxygen-limited	Oxic, anoxic
Oxygen requirement	Low	None	Low	High
pH control	None	None	None	Yes
Biomass retention	None	Yes	Yes	None
COD requirement	None	None	None	Yes
Sludge production	Low	Low	Low	High
Bioreactor capacity (kg of N/m^3/day)	1	6–12	1–3	0.05–4
Bacteria	Aerobic ammonia oxidizers	*Planctomycetes*	Aerobic ammonia oxidizers and planctomycetes	Nitrifiers and heterotrophs

Source: Jetten et al. (2002).

Note: SHARON, Andmmox, and CANON systems are explained in the text.

energy and oxygen (Jetten et al., 2002). The use of such a system would reduce costs and the prime targets would be wastewater that contain high amounts of ammonia and low organic compounds such as landfill leachates and sludge liquor. The CANON (completely autotrophic N removal over nitrite) process combines aerobic and anaerobic ammonium-oxidizing bacteria. The SHARON (single-reactor system for high-ammonia removal over nitrite) process converts ammonia to nitrite using *Nitrosomonas eutropha* but could be linked to the Anammox process, which converts mixtures of ammonia and nitrite to nitrogen with ammonia as the electron donor. This system is carried out by *Brocadia anammoxidans* and *Kuenenia stuttgartiensis*, which are planctomyces-like bacteria.

Excess phosphorus is one of the causes of eutrophication, the excess growth of algae, which is a problem in water bodies. Evidence suggests that one of the main sources of phosphorus is wastewater-treatment plants. Conventional treatment plants are designed to remove organic carbon compounds and latterly nitrogen compounds. The removal of phosphorus is achieved by encouraging the microbial population to accumulate phosphorus as polyphosphate granules. The enhanced biological phosphorus removal (EBPR) system started with the observation that an anaerobic reactor gave phosphate removal provided that the nitrate concentration was limited (Seviour et al., 2003).

4.6 Sludge treatment and disposal

Conventional activated-sludge and filter processes produce large volumes of primary sludge in addition to the excess settled secondary sludge (activated-sludge). Almost 1 million tonnes (dry weight) are produced each year in the UK. In the activated-sludge process this secondary sludge is mainly the microbial biomass produced by the metabolism of the organic material. The microbial yield on settled sewage is about 50%. A proportion of this biomass is recycled (20%) and the remainder is combined with the primary sludge for disposal. In the trickling filters the same problem applies except with a lower loading the sludge produced is less but there is no recycling. Therefore, large volumes of sludge are formed which have a solids content of about 1–4% and represent one of the main problems of disposal in wastewater treatment.

4.6.1 Reduction in sludge production

One solution to the large amount of sludge produced is to reduce the levels of sludge produced. This can be achieved by the following (Liu and Tay, 2001):

- oxic, anaerobic digester,
- high-dissolved oxygen,
- uncouplers, and
- ozonation.

4.6.2 Disposal of excess sludge

The waste sludge is a mixture of organic material and microbial cells, which can be degraded by other micro-organisms. There are a number of methods employed to dispose of excess sludge. In the UK until recently 67% of the sludge was disposed of on land, 29% dumped at sea, and 4% removed by incineration. It was agreed at the 1990 North Sea Conference to phase out disposal of sewage sludge at sea by 1998 to comply with EC Urban Wastewater

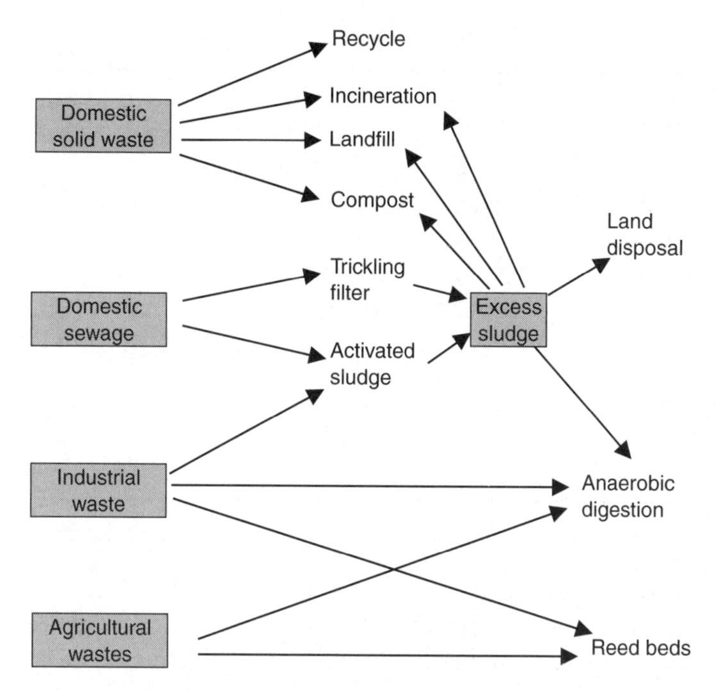

Figure 4.25 Activated-sludge disposal in relationship to the treatment of domestic solid, industrial, and agricultural wastes. Disposal at sea has now been discontinued.

Treatment Directive (91/271/EEC). The banning of sludge dumping at sea was a measure to improve coastal water quality, although there was no substantial evidence of environmental damage other than chromium accumulation. Thus there will be a need to find a replacement for dumping at sea. The possible fates of the excess sludge are given in Fig. 4.25 and are as follows:

- landfill,
- incineration,
- spray irrigation (agricultural disposal),
- drying,
- composting, and
- anaerobic digestion.

4.6.3 Landfill

Domestic solid waste is the solid refuse produced by households and institutions consisting of glass, metal, paper, and organic material (Table 4.9). The trends in the treatment of domestic solid waste have changed over the years. In the USA in 1985 83% was dumped in landfill sites, 5% combusted, and 12% recycled, and by 1993 landfill disposal had been reduced to 62%, 16% was combusted,

Table 4.9 Composition of solid domestic waste (%)

Composition	UK waste	US waste
Paper	35–60	32.0
Garden waste	2–35	14.8
Food waste	2–8	8.5
Metals	6–9	6.3
Glass	5–13	6.4
Plastics	1–2	11.8
Textiles/wood	1–3	19.3 (Textiles/wood, Rubber/leather and Miscellaneous combined)
Rubber/leather	1–3	
Miscellaneous	2	

4% composted, and 18% recycled. Although there has been an increase in recycling and incineration the bulk of the domestic waste is still disposed of in landfill sites in the USA and UK. The elimination of domestic waste is difficult as packaging produces much of the paper. Domestic waste could be reduced by recycling of glass, metal, and paper, and a reduction in packaging.

Once wastes have been produced there are a number of methods that can be used to treat or dispose of wastes that include domestic refuse, excess sewage sludge, and agricultural and industrial wastes.

Perhaps the oldest method of disposing of waste material is to bury it and this method has been used for both industrial and domestic waste for some considerable time. Initially most landfill sites were unsealed so that any leachate derived from the degradation of the contents could be dispersed in the surrounding soil but in practice the leachate often contaminated the groundwater. The main problem of landfill disposal is finding suitable sites in terms of geology: a non-porous substratum is needed to prevent contamination of the groundwater, siting must be away from habitation to avoid odours, and there are now stricter regulations as to what can be disposed of in a landfill site. The most suitable site for a landfill is an abandoned quarry or open-cast site although a site can be constructed above ground by building retaining walls. The site will need to be lined with an impermeable barrier to avoid contamination of the groundwater. Materials such as clay, plastics such as polyethylene and polyvinylchloride, and rubber have been used based on compacted soil. The site, once constructed, can be filled over a wide area or the fill can be confined to cells or terraces (Fig. 4.26). In all cases the waste is compacted by purpose-built vehicles and each day's waste must be covered with soil or ash. Compaction controls the air and water transfer, reduces the volume and reduces the possibility of spontaneous combustion by reduction in oxygen. The overall dimensions of the refuse cell will vary with the

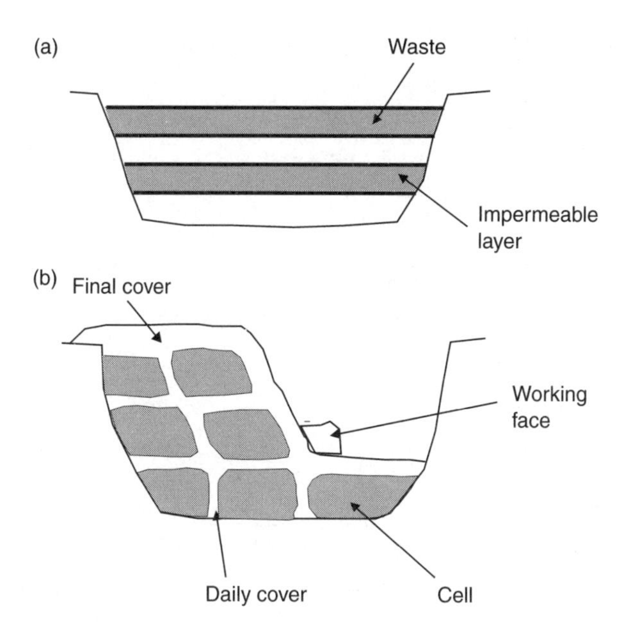

Figure 4.26 Methods of constructing landfill sites. (a) The use of thin layers separated by impermeable layers will not allow the easy collection of gases or the penetration of water. (b) In the second method the waste is laid down in compacted cells which are covered daily by soil or ash.

characteristics of the waste, the cover soil, the availability of land, and the topography of the site. During construction permeable horizontal trenches or perforated pipes may be installed in order to extract gas and to collect leachate (Fig. 4.27). Once the landfill has reached its working level the top is sealed with a layer of clay, an impervious lining topped by a drainage layer and soil. The capping prevents rainwater entering the site, reducing the leachate formed. Once capped the organic content of the landfill will be degraded anaerobically in a similar manner to the anaerobic digester (Chapter 4) and will produce methane, and a leachate. The rate of degradation is dependent on moisture content and as the site has been capped water may have to be added. The leachate will have a high BOD_5 level due to the limited degradation of the organic content and may also contain metals and toxic chemicals particularly when industrial waste is dumped at the same site (co-disposal), and will require treatment. Examples of landfill leachates are given in Table 4.10. The capping of the site will restrict gas extraction so that pipes or gravel vents are inserted during construction or inserted after capping (Fig. 4.28). Landfill gas is of variable composition but contains methane, carbon dioxide, and hydrogen and has a calorific value of about 50% that of natural gas. The wells are normally inserted after 1–2 years of capping the site and give at best yields of ~100 m^3 of gas/tonne of refuse (Fig. 4.28). The yield is only 25% of the possible yield but the rate of gas formation and migration within the site makes complete extraction difficult.

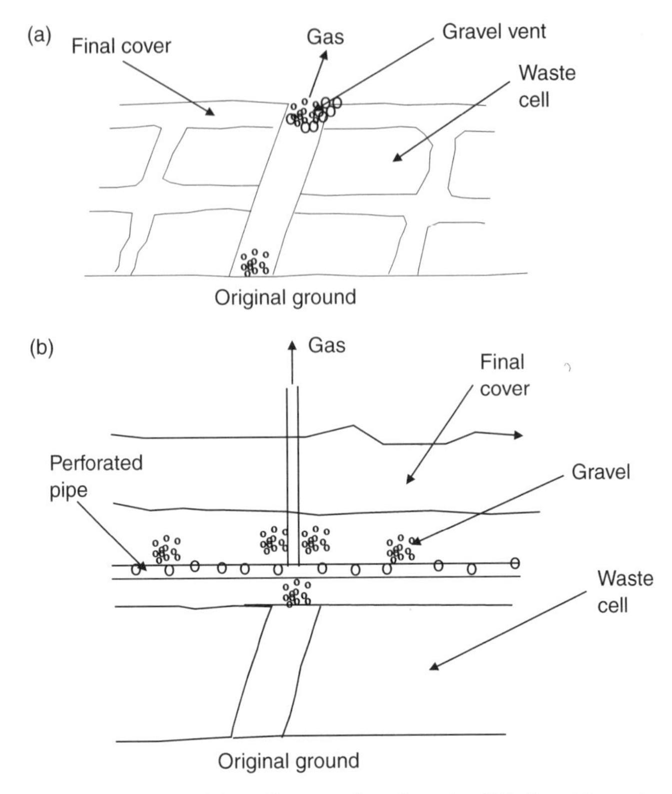

Figure 4.27 The management of the collection of gas from landfill sites either using (a) gravel vents or (b) perforated pipes, which are laid when the landfill is constructed.

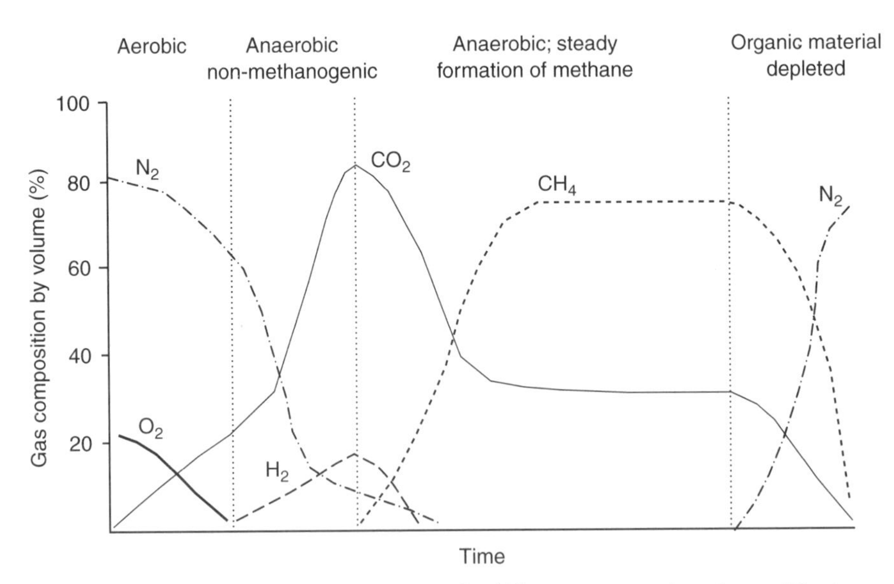

Figure 4.28 The changes in gases produced in a landfill site over a number of years. The time taken to produce a steady supply of methane is typically 2 years.

Table 4.10 Composition of the leachate from recent and old landfill sites

Parameter	Recent leachate (mg/l)	Old leachate (mg/l)
COD	23 800	1 160
BOD	11 900	260
TOC	8 000	465
Fatty acids	5 688	5
Ammoniacal N	790	570
Oxidized N	3	1
Chloride	1 315	2 080
Sodium	960	1 300
Magnesium	252	185
Potassium	780	590
Calcium	1 820	250
Manganese	27	2
Iron	540	23
Nickel	0.6	0.1
Copper	0.12	0.3
Zinc	21.5	0.4
Lead	8.4	0.14

Source: Attewell (1993).

Despite the prevention and minimization of waste, landfill will continue to be one of the main methods of solid-waste disposal. Of the 67% sludge disposed of on land in the UK some 16% goes into landfill sites either solely or co-disposed with domestic wastes. A much higher proportion is landfilled in the UK than in European countries where incineration and composting are used more widely. In the USA in 1993 62% of domestic waste was landfilled, 16% combusted, and 4% composted.

Landfill sites under construction will be active for up to 100 years before the sites are suitable for alternative use and are issued with a Certificate of Completion. To obtain this certificate the biodegradable content of the landfill must have reached a steady state (stabilized) and metals and toxic contaminants flushed out. Thus the management of landfill sites needs modification and landfills should be regarded as anaerobic reactors rather that just sealed disposal systems. The problem with landfill is that although the biodegradable material is broken down the process is very slow, probably due to the lack of water and nutrients in the system. However, if the leachate, which is rich in organic materials, is recycled the degradation is faster, more gas will be produced, and the whole system will be stabilized faster. In the UK a landfill tax has just been introduced which will probably reduce landfill use. Alternatives to landfill are being investigated in Europe and the USA. If the leachate is not recycled it will require treatment before it can be released into waterways and there are a number of technologies available. One treatment is aerobic tanks or lagoons. This contrasts with the anaerobic digestion in the landfill and it has been successful in reducing leachate BOD_5 and COD levels, especially from the early leachates. Often sequencing batch reactors are used for leachate treatment where the sequence steps are feed, aerobic reaction, settle, and decant, and these steps can run over a 24-h period. Other methods of leachate treatment are anaerobic treatment, spray irrigation, and reed beds.

4.6.4 Incineration

Incineration can be used for industrial wastes, domestic wastes, and toxic wastes, and in some cases is preferable to land disposal. The critical factors for effective combustion are the temperature, length of time at high temperature, and the effective mixing of the waste with air. There are a number a combustion systems which will operate at 1500–3000°C, the temperature required to break down organic wastes, which include rotary kilns, liquid injection, fluidized beds, and multiple-hearth designs (Fig. 4.29). The advantages of incineration are the reduction in volume, the ability to treat toxic materials, and the construction of incinerators in areas where landfill disposal is not available.

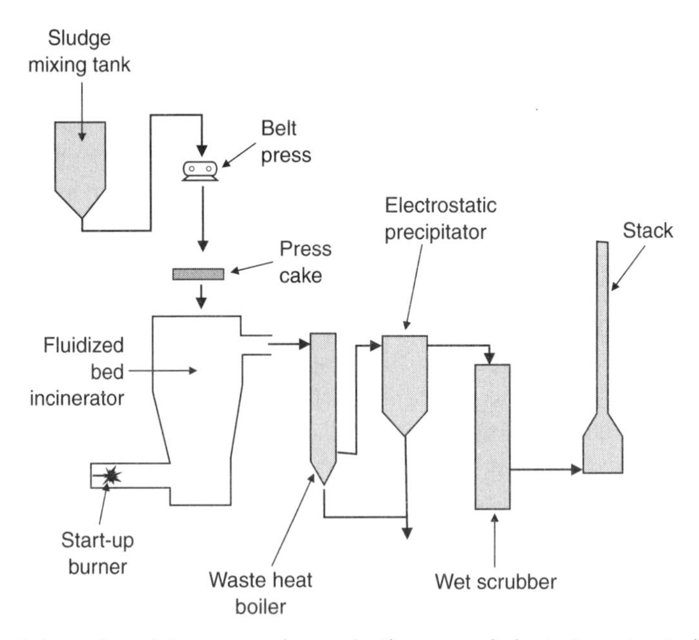

Figure 4.29 An outline of the process of autocalorific sewage sludge incineration, including an electrostatic precipitator and wet scrubber to clean the flue gases.

The disadvantages of incineration are as follows: the process can be expensive to operate and to construct, it is not a complete disposal as the resulting ash will need to be disposed of in a landfill site and this ash may contain high levels of metals, and incineration also produces particulates and flue gases. The flue gases from incinerators can contain hydrogen chloride, sulphur oxides, and nitrogen oxides, which can be removed by wet scrubbing, carbon monoxide, which can be formed by incomplete combustion, and carbon dioxide, which when formed increases the level of greenhouse gases. The incineration of chlorinated compounds at low temperatures, which can happen with excess air, will allow the production of dioxins and furans, which are toxic and carcinogenic. The formation of particulates is also to be avoided and electrostatic precipitation will need to be incorporated in an incinerator system.

The large-scale incineration of sewage sludge is expensive, with high capital costs, and is only a partial disposal option, as the ash formed needs disposal. However, the development of autothermic incineration has made incineration more attractive. In the process primary and secondary sludges are mixed together and water removed by pressing so that a cake of 30% solids is produced which can support autothermic combustion. Advanced combustion systems like fluidized beds working at temperatures of 750–850 °C create more heat than is required to heat the inlet air and remove the water from the sludge. This means that once the process has started no fuel needs to be added as the sludge itself generates sufficient heat (Fig. 4.29). The ash formed can be removed by an electrostatic precipitator and a wet scrubber will remove sulphur dioxide, hydrogen fluoride, and hydrogen chloride. The ash contains the heavy metals that are present in the sludge, represents 30% of the original dry mass and 1–2% of the volume, and is normally disposed of in landfill sites.

4.6.5 Composting

One method of disposal and reuse of organic waste is composting, which can be used for the organic components of domestic waste, excess activated sludge, excess straw and chaff, and some agricultural wastes.

Composting can be defined as "the biological stabilization of wastes of biological origin under controlled aerobic conditions" and is a process that most gardeners are familiar with. In composting the organic component of municipal waste is decomposed by a mixture of micro-organisms under warm, moist, and aerobic conditions (Anderson and Smith, 1987). This is in contrast to the anaerobic landfill conditions where decomposition is slow and methane is formed. In composting the matrix or conditions should be open to allow an adequate supply of oxygen. In the presence of oxygen the rapid metabolic processes produce heat. Not all the heat can be dissipated and the compost will rise in temperature to 50 °C and above. The high temperatures are responsible for the inactivation of pathogens, producing a very acceptable final product.

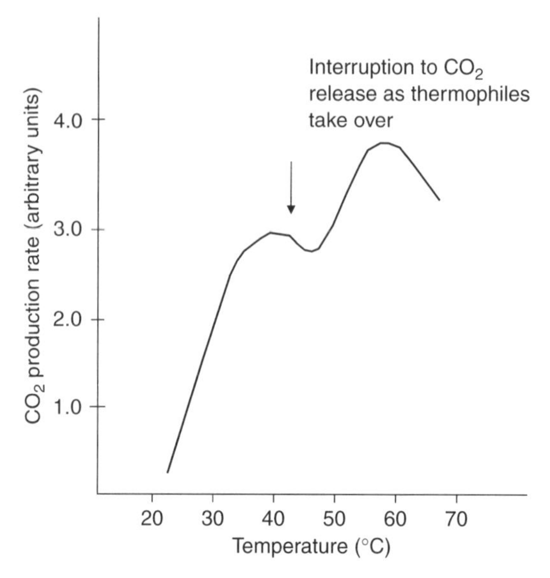

Figure 4.30 The metabolism of the micro-organisms in a compost system where the release of carbon dioxide is used as a measure of microbial activity. The dip in carbon dioxide released at 40 °C indicates the change from mesophilic organisms to thermophiles, as the temperature of the compost heap rises. Adapted from Plat et al. (1984).

Once the compost system has been assembled aerobic mesophyllic micro-organisms start the degradation but as the temperature rises in the compost thermophilic organisms take over. The pattern of growth can be seen in Fig. 4.30, where growth has been followed by the formation of carbon dioxide. Three days at 55°C or above is sufficient to inactivate most pathogens and viruses. The system with sufficient oxygen will be rapid (6–11 weeks) with low energy consumption giving a hygienic standard end product. There are a number of composting systems or processes, which can be divided into two groups, open and closed systems (Table 4.11).

The simplest open system is the Windrow, where the waste to be composted is piled in long heaps, often covered with straw to conserve heat, and aeration is achieved by periodic turning of the heaps. In other static systems aeration is by either blowing air in the base of the heap (Rutgers) or by suction from the base (Beltsville). A cross-section of a typical heap is shown in Fig. 4.31 and the typical temperature cross-section for such systems is shown in Fig. 4.32.

The closed systems are normally situated in buildings and up to seven different systems have been described in which some form of mechanical means gives aeration and mixing in either continuous or discontinuous modes (Anderson and Smith, 1987). An example is the Dano biostabilizer, which is a sloping plug flow bioreactor where mixing and aeration are carried out by rotating the cylinder. This system combines the removal and recycling of

Table 4.11 Types of composting system

Type	Operation
Windrow	Waste piled in long rows outdoors; aerated by turning regularly
Static pile (Beltsville)	A Windrow pile aerated by suction system; outdoors
Static pile (Rutgers)	A Windrow pile aerated by blowing air through the pile; outdoors
Dano stabilizer	Plug flow system mixed by rotating cylinder
Tunnel	Plug flow with waste moved by hydraulic ram through a rectangular reactor
Fairfield–Hardy digester	Solid-bed system mixed using augers
Tower or silo	A vertical plug flow reactor
Continuous composting	Waste composted in vertical plug flow reactor where fresh material is added at the top and compost withdrawn from the bottom
Discontinuous composting	Waste composted on a series of floors on to which the waste is sequentially transferred

ferrous metals with composting. The residence time in the cylinder is about 1–5 days, which is not enough to give complete composting so that the partially composted material is passed on to a Windrow system for 1–6 weeks depending on conditions and starting material. The silo system consists of a vertical plug-flow vessel where air is introduced at the base and the compostable material pumped in at the top, which gives both mixing and aeration. There is a facility for recycling of the compost and the residence time in the first vessel is about 14 days. After this time the compost is moved to the curing vessel where the high temperature developed in the first vessel inactivates any pathogens present. About 25% of the curing vessel is removed after 14–20 days incubation to be replace by compost from the first vessel. The compost formed is used for horticulture. Composting has a lot to commend it although it has not been popular in the UK for the treatment of domestic waste; it may be used more widely in the future as landfill sites become more difficult to find and operate.

Some agricultural products such as chaff and straw are disposed of by composting. Straw used to be burned but as it contributes to atmospheric particlates and CO_2 emissions this practice has been stopped.

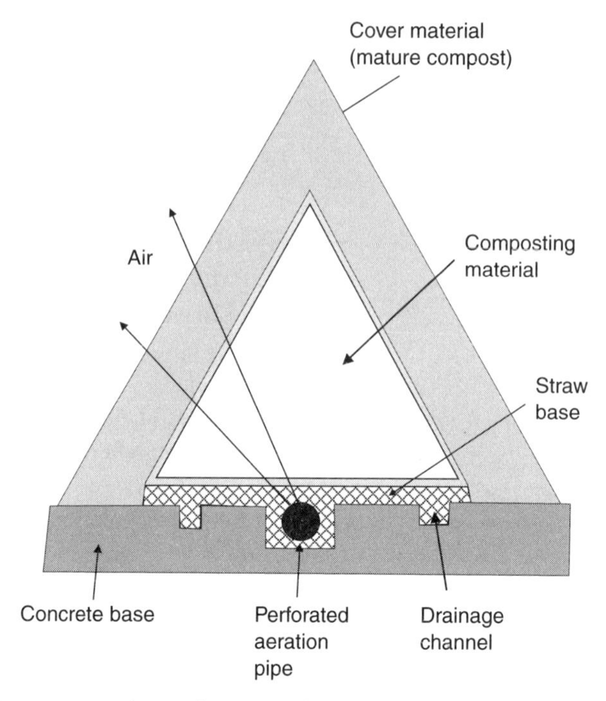

Figure 4.31 Cross-section of a Windrow type of compost heap with aeration. Air is pumped through the perforated pipe and out through the composting material to supply oxygen to ensure aerobic metabolism. Adapted from Anderson and Smith (1987).

4.6.6 Agricultural disposal

Some sludge is disposed of by some form of agricultural use. Any sludge that is applied to agricultural land is required to have some form of biological or chemical treatment to reduce the levels of pathogens unless it is injected below the surface. These treatments are

- pasteurization, heating at 70 °C for 30 min or more,
- anaerobic digestion at 35 °C for 12 days,
- thermophilic aerobic digestion for 7 days at 55 °C,
- composting (see above)
- alkali stabilization, at pH > 12 for 2 h,
- liquid storage for 3 months, and
- drying and storage for 3 months.

All these methods reduce the pathogen content of the sludge before it is applied to the land but there are restrictions on what crops are grown and how soon the land can be used for certain crops. Treated sludge can be used for growing turf but must not be applied to growing fruit or vegetable crops and treated land cannot be used for this purpose for 10 months following

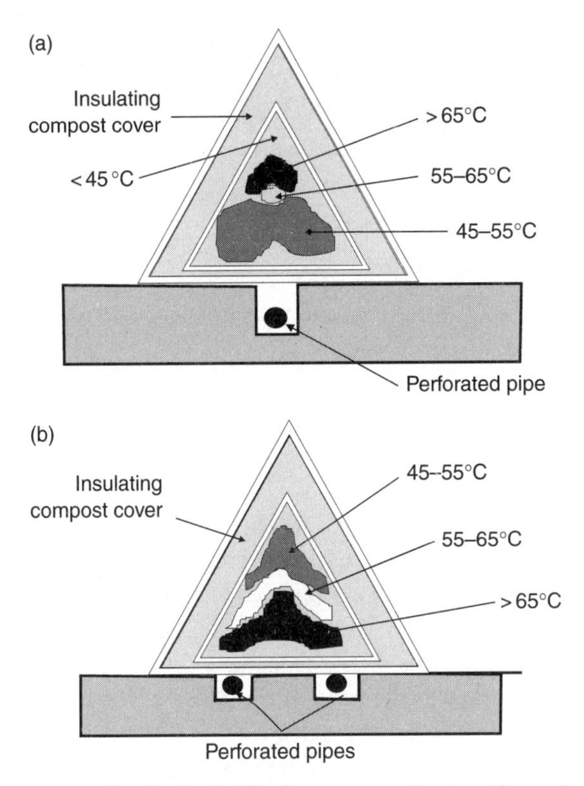

Figure 4.32 Temperature distribution in Windrow compost heaps with aeration by (a) blowing (Rutgers) or (b) sucking (Beltsville). The temperatures produced are due to the aerobic metabolism of the micro-organisms growing on the composting material producing heat which cannot be lost as the heap is insulated.

treatment. Prior to drying the sludge is normally conditioned so that its ability to settle before dewatering is improved. Often polyvalent ions such as Fe^{3+} or Al^{3+}, polyelectrolytes, or soil are added to improve precipitation. Dewatering is often carried out in drying beds, but filtration and centrifugation have also been used to produce a compact cake.

Almost all sludge contains heavy metals as micro-organisms have the ability to sequester metals, so that the application of sludge to soils carries with it the risk of producing high levels of heavy metals in the soil. There are limits for metals in sewage sludge (Table 4.12) and in soil.

4.7 Anaerobic digestion

Liquid wastes had been treated anaerobically in simple anaerobic ponds for thousands of years before the development of anaerobic digestion in closed vessels. One of the best methods for disposing of excess sludge is to subject it

Table 4.12 The limits of metals in activated sludge and soil treated with sludge

Metal	Concentration in sludge (mg/kg)	Concentration in soil (mg/kg)
Cadmium	20–40	1–3
Copper	1 000–1 750	50–140
Nickel	300–400	30–75
Lead	750–1 200	50–300
Zinc	2 500–4 000	150–300
Mercury	16–25	1–1.5

Source: Attewell (1993).

to anaerobic digestion; a process that has been used in the large sewage works for some time. However, more recently anaerobic processes have been used to treat industrial wastes with high contents of insoluble or organic compounds.

The advantages of anaerobic digestion are that the process produces much less biomass or sludge, forms methane (biogas), no aeration is required, and the associated smell is less as the process is enclosed. The methane can be used as a fuel to run boilers or to generate electricity with about 1.16×10^7 kJ produced per 1000 tonnes of COD removed (Speece, 1983). The disadvantages of anaerobic digestion are that the process requires good mixing, a temperature in the region of 37 °C, and a substrate with a high BOD_5 (1.2–2 g/l), and that the process has long retention times, of 30–60 days. The requirement for a high-BOD_5 waste means that anaerobic digestion is suitable for some agricultural and industrial wastes but is not suitable for normal sewage.

Anaerobic digestion is a complex process involving a considerable number of reactions, three main groups of organisms, and can be divided into four main stages. The first stage is the hydrolysis of the fats, proteins, and carbohydrates that make up the main components of sewage (Fig. 4.33) to form fatty acids, alcohols, and ketones, by hydrolytic micro-organisms like *Clostridium, Eubacterium,* and *Bacteroids.* In the second stage, the acidifying phase, fatty acids are then converted to acetate, carbon dioxide, and hydrogen. Other amino acids and sugars are converted to acetate, hydrogen, and carbon dioxide. The reactions are carried out by bacteria like *Peptococcus* and *Propionibacterium.* Acetate and hydrogen can also be formed directly from the primary components of the sludge. In the third phase, the acetogenic phase, organic acids are converted to acetate and carbon dioxide by bacteria like *Syntrophobacter, Desulfovibrio,* and *Syntrophomonas.* The fourth phase is the formation of methane by methanogenic bacteria such as *Methanobacterium, Methanobacillus, Methanococcus,* and *Methanosarcina,* which are some of the most oxygen-sensitive bacteria known. There are about 20 species of methanogenic bacteria known and these are found in intimate contact with the acetogenic

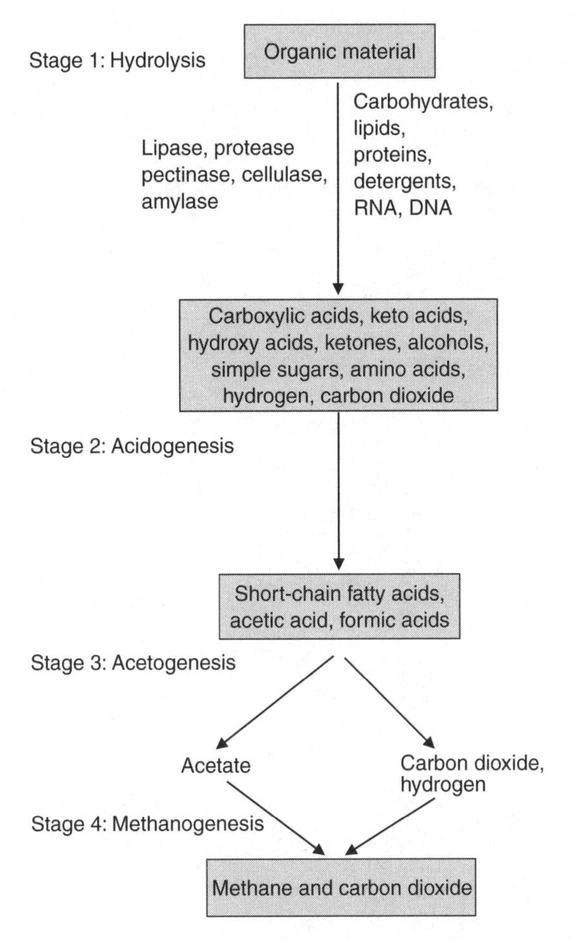

Figure 4.33 The stages of the anaerobic breakdown of sewage to biomass and biogas.

bacteria. The methanogenic bacteria convert hydrogen, carbon dioxide, and acetate to methane as follows (Fig. 4.34):

$$4H_2 + HCO_3^- + H^+ \rightarrow CH_4 + 3H_2O \quad (4.31)$$

$$CH_3COO^- + H_2O \rightarrow CH_4 + HCO_3^- \quad (4.32)$$

4.7.1 The anaerobic process

The first anaerobic digesters were simple sealed vessels which were not mixed or heated but since then a number of different designs have been developed, including upflow sludge blanket, fixed film, fluidized beds, and two-stage processes. The simplest design is the contact process where the waste is mixed with recycled anaerobic sludge and retained in a sealed vessel at 30–37 °C. After 30–60 days digestion is complete, the mixture is separated, the liquid discharged for further treatment, and much of the biomass returned to the vessel (Fig. 4.35). The problem with this type of system is that it subjects the

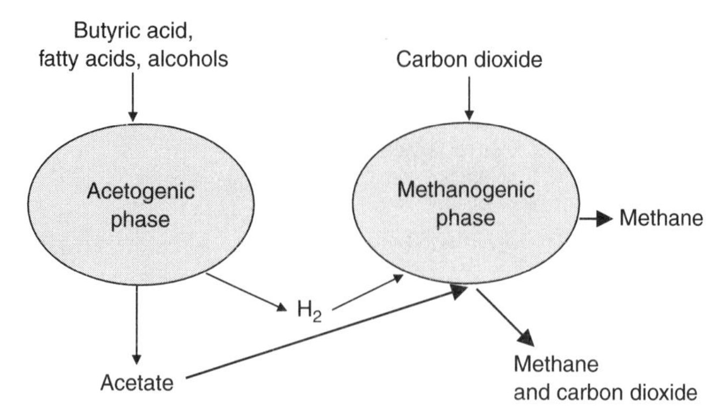

Figure 4.34 The intimate relationship between the acetogenic and methanogenic bacteria.

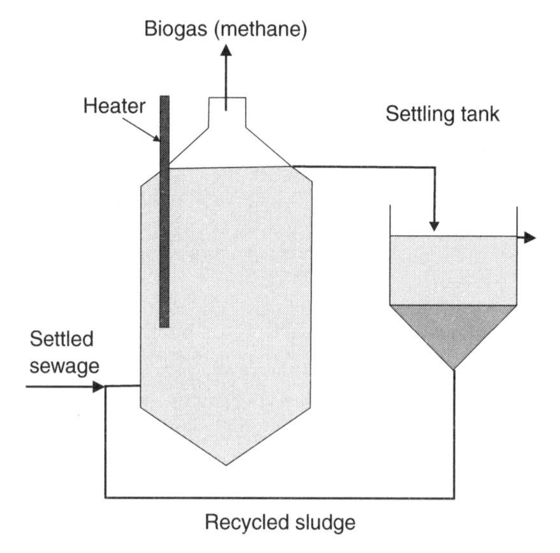

Figure 4.35 Conventional anaerobic-sludge process. The vessel is heated and sealed so that the biogas can be collected.

methanogens, which are strict anaerobes, to an aerobic environment when the vessel is discharged and even if this exposure is only for a short time it will affect their activity. To avoid this disturbance other designs have been developed which are based on continuous-flow systems.

In the upflow sludge blanket the waste is introduced at the bottom of the vessel into the area where the sludge settles (Fig. 4.36). Therefore the waste immediately contacts the sludge, ensuring a more rapid rate of reaction. Mixing is achieved by the incoming flow of waste and by the gas generated and at the top of the vessel the biomass, liquid, and gas are separated in a settling section. This type of vessel has been applied successfully to a number of wastes at high loading rates. In order to maintain a high biomass level and good contact with waste fixed-film or immobilization biomass have been used. Porous material

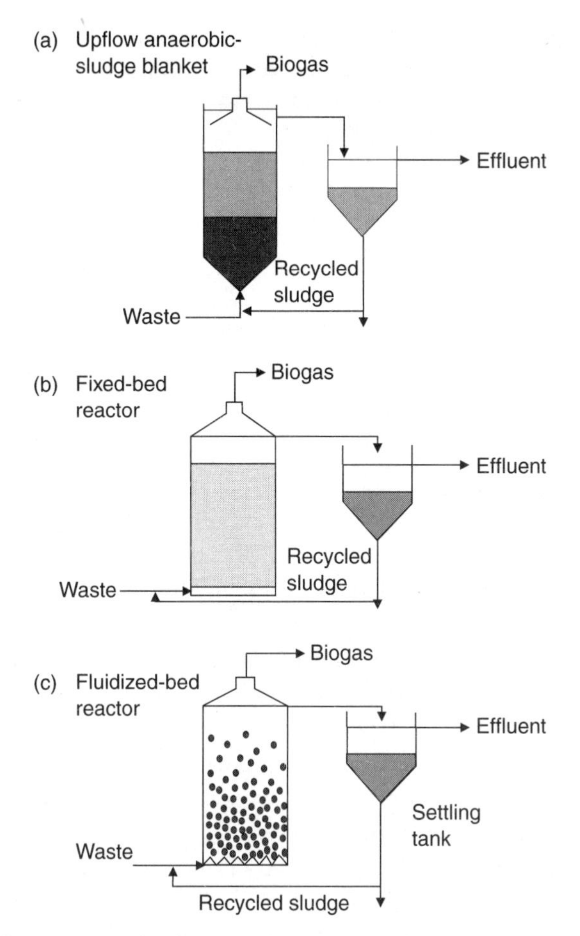

Figure 4.36 Different systems for the anaerobic treatment of wastewaters. (a) In the upflow anaerobic-sludge blanket the wastewater passes through a layer of sludge into an upper layer where the cell aggregates settle out. (b) In the fixed-bed reactor the micro-organisms are immobilized on a matrix and the wastewater passed though this matrix. (c) In the fluidized-bed reactor the micro-organisms are immobilized on and in carriers which are fluidized by pumping in the wastewater.

such as plastics, and solids such as gravel, sand, and glass granules have been used to immobilize the anaerobic micro-organisms (Fig. 4.36). This type of system can be operated in a number of modes and in either an upflow or down-flow mode. In the fixed-bed reactor the waste passes upward through the particles coated with biomass. This system does suffer from channelling and clogging. An alternative is the fluidized-bed reactor where the cells are immobilized on sand particles, which are fluidized by the upflow of waste (Fig. 4.36). A comparison of the function of these various types is shown in Table 4.13 where it is clear that the upflow sludge blanket has the highest loading rate and COD removal. Attempts have been made to separate anaerobic digestion into a two-stage process by separating the acidifying stage from methanogenesis by using two separate vessels run under different condition and flow rates.

Table 4.13 A comparison of the loading and removal rates of some anaerobic digestors

Reactor type	COD loading rate (kg/m^3/day)	COD removal (%)
Contact	1–6	80–95
Upflow filter	1–10	80–95
Fluidized bed	1–20	80–87
Downflow filter	5–15	75–88
Sludge bed	5–30	85–95

Source: Speece (1983).

4.8 Agricultural waste

Agricultural waste can be divided into solid and liquid waste. Some 5–10% of the chaff and straw waste is used to produce compost for mushroom cultivation. Traditionally wheat straw and horse and chicken manure are wetted, mixed, and placed in a Windrow system. Aeration is normally by turning every 2–3 days. After 7–14 days outdoors temperatures of 80 °C have been reached inside the compost. The second stage is carried out indoors where the compost is incubated at 50–60 °C with aeration after which the compost is packed into trays ready for mushroom seeding.

Liquid agricultural waste is mainly cattle, pig, or chicken slurry, a mixture of dung and urine. The quantities are large when associated with large herds, factory farming, and silage run-off. In contrast to domestic sewage these wastes have very high BOD_5 values of 10 000–25 000 mg/l (Table 4.14). Other liquid agricultural wastes are the run-off of nitrates and phosphates from agricultural land and herbicides and pesticides from treatment and spraying. Farmyard slurries have traditionally been disposed of by spraying or spreading on to the land, but in the case of intensive farming there is far too much slurry produced to be disposed in this way. As a consequence large farms have installed anaerobic digesters or facultative pond systems for the disposal of the slurries. Constructed wetlands with subsurface flow have been used to treat low-concentration dairy waste.

4.9 Industrial waste

The organic wastes from various industries can be both liquid and solid and generally have higher BOD_5 values than domestic wastes. The degradable organic wastes are in general the products of food, drink, meat, and vegetable

Table 4.14 Examples of industrial and agricultural wastes

Waste	BOD_5 (mg/l)	COD (mg/l)
Animal wastes		
Cow	16 000	150 000
Pig	30 000	70 000
Poultry	24 000	170 000
Whey	45 000	65 000
Processing wastes		
Abattoir	100–3 000	
Brewing	500–2 000	17 000
Sugar beet	450–2 000	600–3 000
Potato processing	2 000	3 500
Distillery	7 000	10 000
Poultry	500–800	600–1 000
Wool industry	300–600	200–8 000

Source: Data from Gray (1989) and Glazer and Nikado (1994).

processes along with waste from the production of yeast, citric acid, and antibiotics. All these processes and industries produce strong wastes in terms of BOD_5 value (Table 4.14). If the industry is willing to pay the local water company many of these wastes can be released into the sewage system and be treated by the normal methods. In order to avoid these charges a number of biological methods of waste removal have been tested including methods for the upgrading of the waste to an animal food.

One solution for strong waste is to install an anaerobic digestion system to fully or partially treat the wastes, as if the wastes are high in BOD_5 then anaerobic digestion is very suitable. An example of this type of system was the installation of an anaerobic digester at a cheese-production plant (Kemp and Quickenden, 1989). The cheese whey was digested anaerobically, producing methane, which was used to run the boilers and generate electricity (Fig. 4.37). The BOD reduction was about 95% and after denitrification the effluent was of high quality and could be released directly into the river.

4.10 Conclusions

- The aerobic treatment of sewage will continue with any new systems based on improvements in the engineering to provide better mixing and aeration.

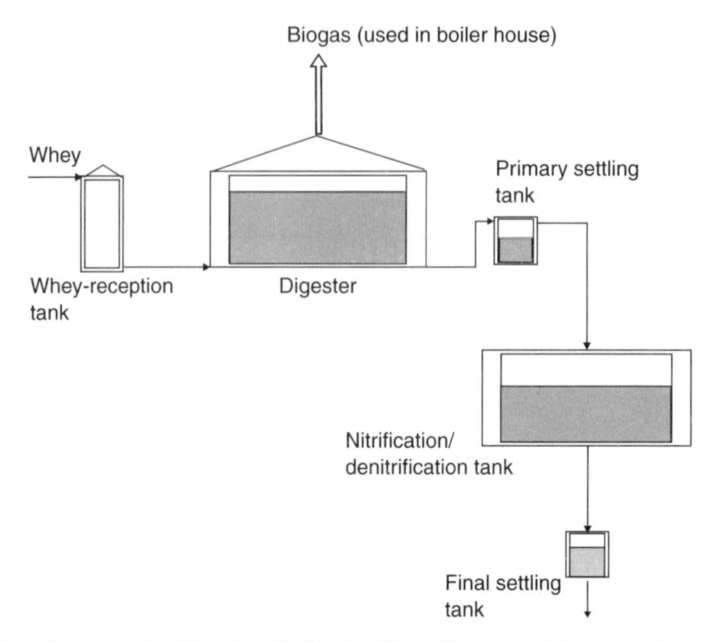

Figure 4.37 The anaerobic digestion of whey in a South Caernavon Creamery. The anaerobic digester produces mainly methane, which is used to fuel the heaters and pumps in the plant. The effluent is first settled and the liquid from the settling tank treated in an anaerobic/aerobic sequencing tank to reduce the level of ammonia and nitrate respectively. Nitrogen-containing compounds will be high in whey, as there is a high protein content compared with many other wastes. Adapted from Stafford (1989).

- Recombinant DNA technology will be increasingly be used to follow the population dynamics in both aerobic and anaerobic digestion systems.
- Better management of landfill for gas production and leachate recycling should increase the rate of degradation and allow the reuse of landfill sites.

4.11 Further reading

Attewell, P. (1993) *Ground Pollution*. E & FN Spon, London.

Clark, R.B. (1992) *Marine Polution*. Clarendon Press, Oxford.

Gerardi, M.H. (2002) *Nitrification and Denitrification in the Acivated Sludge Process*. Wiley-Interscience, New York.

Gray, N.F. (1989) *Biology of Waste Water Treatment*. Oxford University Press, Oxford.

Tebbutt, T.H.Y. (1992) *Principles of Water Quality Control*. Pergamon Press, Oxford.

5 Bioremediation

5.1	Introduction	173
5.2	Synthetic compounds	176
5.3	Petrochemical compounds	179
5.4	Inorganic wastes	183
5.5	Bioremediation strategies	187
5.6	Bioremediation techniques *in situ*	195
5.7	Bioremediation techniques *ex situ*	198
5.8	Phytoremediation	204
5.9	Metals bioremediation	216
5.10	Gaseous bioremediation	220
5.11	Biochemical pathways of biodegradation	224
5.12	Conclusions	228
5.13	Further reading	229

5.1 Introduction

The contamination of soil and water with organic and inorganic pollutants is of increasing concern and the subject of legislation. These pollutants include complex organic compounds, heavy metals, and natural products such as oils and are derived from industrial processing, deliberate release, and accidental release.

The classes of pollutant released into the environment are outlined in Table 1.1, and Chapter 4 deals with the treatment of the biodegradable contaminants such as sewage, industrial waste, and agricultural waste. However, these are not the only pollutants released into the environment; a wide range of inorganic and organic materials are released (Table 5.1). Some of the sources of these contaminants are given in Table 3.1. The contaminants can come from industrial effluents, deposition from flue and exhaust gases, old industrial sites and disused mines, run-off from waste tips and landfill, excess application of herbicides and pesticides, and accidental spills. The source and type of contaminant will influence the nature of the pollution. An accidental spillage may be restricted to a small area on land but at sea this may contaminate a wide area. Deposition from the air may contaminate very wide areas but the concentration may be very low. Air deposition does however contribute a high proportion of the heavy metals that are accumulated in soil in the UK.

Table 5.1 Contaminants released into the environment

Contaminant	Example
Solvents	Chloroform, carbon tetrachloride
Pesticides	DDT*, lindane
Herbicides	Arochlor, atrazine, 2,4-D†
Fungicides	Pentachlorophenol
Insecticides	Organophosphates
Petrochemicals	Benzene, toluene, PAHs, xylene
Explosives	Trinitrotoluene
Polychlorobiphenyl	PCBs
Phenols/chlorophenols	Nitrophenol, chlorophenol
Heavy metals	Cadmium, lead, mercury

*DDT, 1,1,1-trichloro-2,2-bis(4-chlorophenyl)ethane.
†2,4-D, 2,4-dichlorophenoxyacetic acid.

The structure, volatility, and chemical activity of the compound will also greatly influence their fate as shown in Fig. 5.1. If volatile, the compound will evaporate, adding to air pollution, although if reactive it may react with the soil or soil micro-organisms before it evaporates. If the compound is water soluble it may dissolve in rivers, lakes, or groundwater. This will mean that it will be mobile in the soil and therefore may contaminate the water table. However, the microbial population may also degrade the pollutant, as water-soluble compounds are more accessible. If the compound is insoluble (hydrophobic) it will be more difficult to metabolize by micro-organisms but some compounds will dissolve in the cells lipid membranes. If the compound is inert it will accumulate in the organism and may become part of a food chain. Alternatively, it may bind to the soil components and remained fixed within the soil. Metals are a particular case as they cannot be biodegraded, but can be absorbed by micro-organisms that should allow their concentration and disposal. Many of the organic compounds released into the environment are not normally found in the environment and are known as **xenobiotic**. This would apply to the insecticide DDT and the herbicide lindane but not to petrochemicals as these are the products of living material laid down millions of years ago.

A low level of a compound in the environment may not appear at first to be a problem, but some organisms in the environment may concentrate the compound to levels considerably higher than those in the surrounding environment. In this way a toxic level of the compound may be reached and the accumulation will be compounded if the organism is part of a food chain. The increase in successive organism in the food chain in a process known as **biomagnificaton**

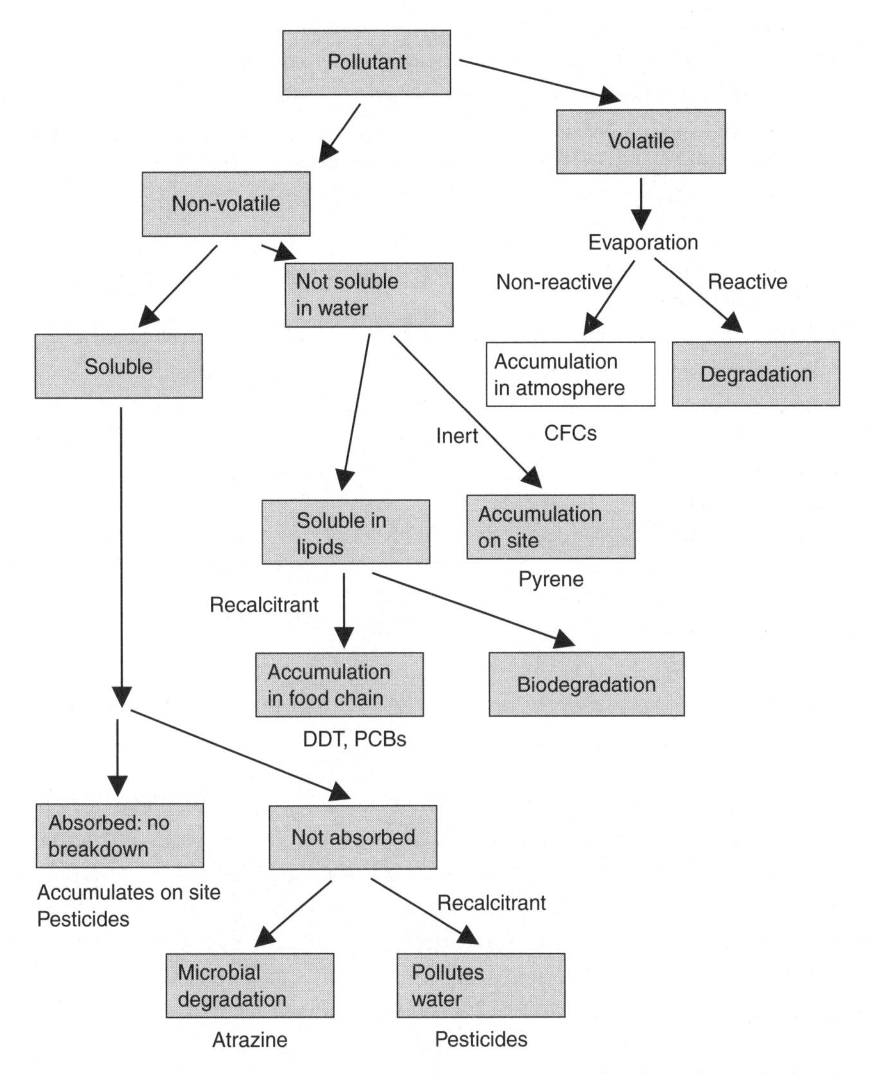

Figure 5.1 The fate of a pollutant released into the environment that depends on whether the pollutant is water soluble and volatile.

(see Box 5.1). An example of this is the accumulation of DDT in grebes that were at the top of the food chain on a lake treated with 0.01–0.02 ppm of DDT to control gnats. The levels found in the grebes were some 100 000 times the DDT level applied. Observations like this lead to the eventual banning of the use of DDT.

This chapter describes the types of contaminant that can be released into the environment, their structure, fate, and the consequences of the pollution. Strategies for the biotreatment of synthetic compounds, petrochemicals, metals, and gaseous wastes are discussed and the bioremediation techniques used for these various wastes outlined.

BOX 5.1 Terms used to describe environmental pollutants

The term biodegradable implies that the compound can be degraded or transformed by a biological process. The term does not however give any indication of the extent of degradation.

- **Mineralization** is complete degradation to carbon dioxide, water, and other inorganic compounds, and partial degradation refers to breakdown to an intermediate stage.
- **Persistent** organic compounds undergo degradation under certain circumstances.
- **Recalcitrant** compounds are not degraded under any conditions.
- **Bioaccumulation** is the increase in a compound in an organism compared with the level found in the environment.
- **Biomagnification** is the increase in a pollutant in tissues of successive organisms of a food chain.
- **Bioconcentration factor** refers to the concentration of a pollutant from the environment, and the factor is the concentration in an organism compared with that in the environment.

5.2 Synthetic compounds

Many thousands of synthetic organic (**xenobiotic**) compounds have been produced and many of these have found their way into the environment. Some of the most commonly found are the pesticides (biocides), herbicides, and preservatives and some of their structures are given in Fig. 5.2. Most biocides and herbicides are released into the environment by direct use although some may be released during manufacture, and spills do occur. The scale of herbicide production can be seen in Table 5.2, where the herbicide atrazine is produced at a rate of 39 000 tonnes per year. Other synthetic compounds such as polychlorobiphenyls (PCBs) are used in hydraulic fluids, plasticizers, adhesives, and lubricants; flame retardants and dielectric fluids in transformers are released during production, from spillage and disposal. Another group of contaminants found frequently in the environment are chlorinated compounds such as trichloroethene, carbon tetrachloride, and pentachlorophenol, which are used as solvents and for wood treatment. Other contaminants such as dioxins and dibenzofurans can be formed during the combustion of polyaromatic hydrocarbons (PAHs) and when heating plant oils. So, depending on the properties, such as solubility and the nature of their release, these contaminants can be localized or widespread.

Many of the xenobiotic compounds released into the environment accumulate because they are only degraded very slowly and in some cases so slowly as to render them effectively permanent. The half-lives (the time required to remove half of the compound present) of some of the halogenated pesticides

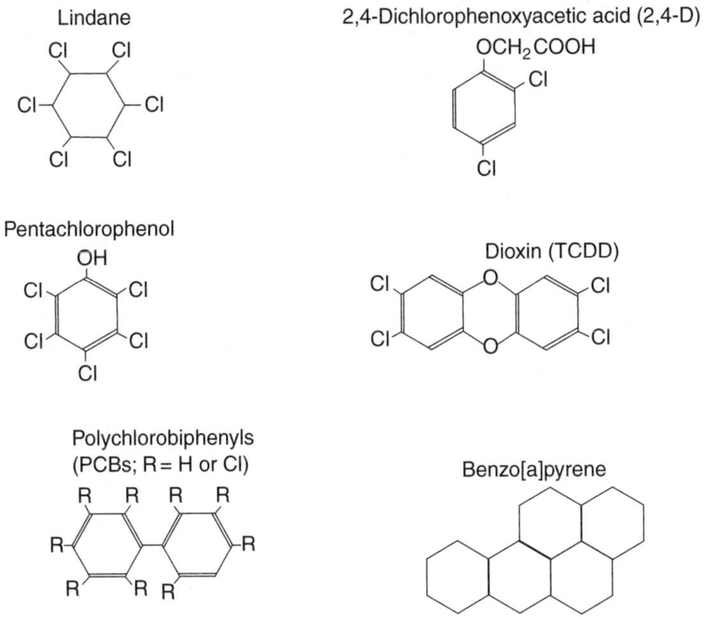
Lindane
Cl Cl
Cl Cl
Cl Cl
2,4-Dichlorophenoxyacetic acid (2,4-D)
OCH2COOH
Cl
Cl
Pentachlorophenol
OH
Cl Cl
Cl Cl
Cl
Dioxin (TCDD)
Cl O Cl
Cl O Cl
Polychlorobiphenyls
(PCBs; R = H or Cl)
R R R R
R R
R R R
Benzo[a]pyrene

Figure 5.2 The structures of some herbicides, pesticides, and some environmental contaminants.

Table 5.2 Annual production of herbicides

Product	Production (tonnes)
Atrazine	39 000
Cyanazine	13 000
Diuron	3 000
2,4-D	30 000
Arochlor	50 000
Glyphosate	12 000

are given in Table 5.3, which can be measured in years. For example, DDT requires from 3 to 10 years for half of the compound to be broken down.

The persistence of xenobiotics in the environment is influenced by the structure and properties of the molecules, their toxicity to micro-organisms, and the environmental conditions. Many natural complex polymeric molecules, such as lignin, will be naturally slow to degrade. Many xenobiotic compounds are also complex in structure and are therefore slow to degrade. For many of the xenobiotics their activity and persistence is also linked to the presence of halogens in the compound (Fig. 5.3). The toxicity and persistence of an organohalogen is influenced by the number of halogen molecules, their

Table 5.3 Persistence of some halogenated biocides

Biocide	Approximate half-life (years)
DDT	3–10
Dieldrin	1–7
Heptachlor	7–12
Atrazine	0.47
Toxaphene	10

DDT

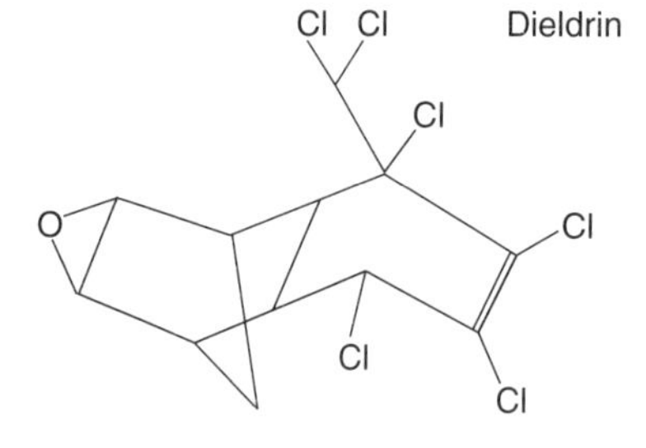

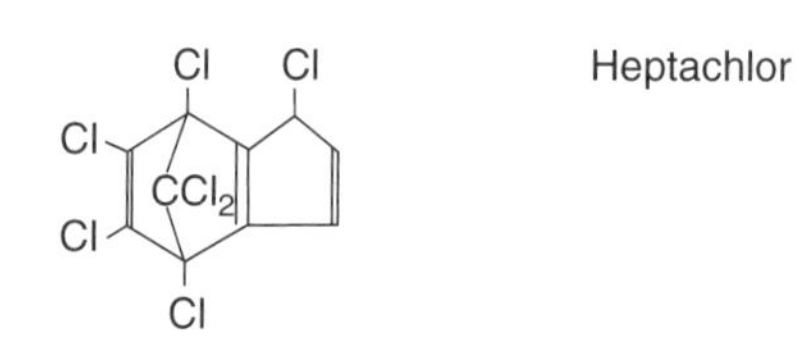

Figure 5.3 The structures of some halogenated xenobiotics.

position, and the type of halogen. Figure 5.4 shows the effect of the addition of another chlorine to 2,4-dichlorophenoxyacetic acid (2,4-D), a biodegradable compound, forming 2,4,5-trichlorophenoxyacetic acid (2,4,5-T), which is recalcitrant. A mixture of 2,4-D and 2,4,5-T is known as Agent Orange, and its persistence was of great concern when it was used as a defoliant in the Vietnam War. The first commercial production of organochlorines was in 1907 with carbon tetrachloride, and later in 1920 trichloroethane was produced. It was found that adding three or more chlorine molecules to phenolic herbicides increased their efficiency. Organochlorines are effective biocides due to their hydrophobic properties, allowing them to pass through or into membranes, disrupting cellular activity and inhibiting oxidative phosphorylation. The solubility of organochlorines decreases as the chlorine content increases, which often increases their toxicity. For these organochlorines to be

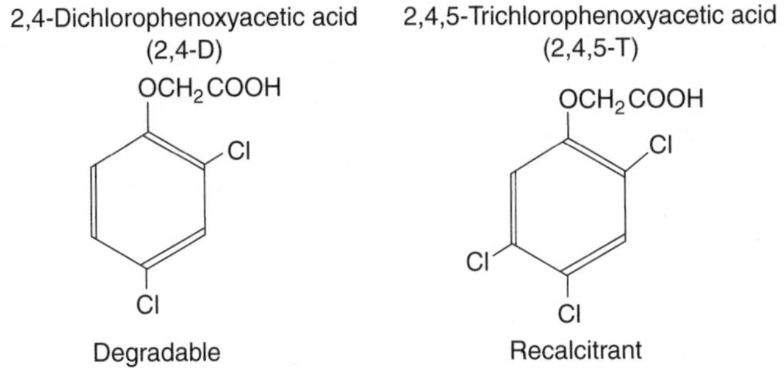

2,4-Dichlorophenoxyacetic acid (2,4-D)
OCH_2COOH
Cl
Cl
Degradable

2,4,5-Trichlorophenoxyacetic acid (2,4,5-T)
OCH_2COOH
Cl
Cl
Cl
Recalcitrant

Figure 5.4 The influence of the addition of chlorine to a molecule that makes it difficult to degrade (recalcitrant).

Table 5.4 The characteristics of some PAHs

Compound	Number of rings	Boiling point (°C)	Molecular weight (Da)	Solubility (μg/mol)
Naphthalene	2	218	128	31.7
Phenanthrene	3	340	178	1.29
Anthracene	3	342	178	0.07
Pyrene	4	404	202	0.135
Benzo[a]pyrene	5	495	252	0.0038
Benzo[g,h,i]perlene	6	500	276	0.00026

degraded by micro-organisms their increased toxicity and low solubility in the aqueous phase have to be overcome. The problem may also be increased as some organochlorine compounds are a mixture of isomers, as found with PCBs, which decreases their rate of degradation. An example of the decrease in solubility and increase in toxicity as the molecules become more complex are the PAHs (Table 5.4 and Fig. 5.5). Another problem with these persistent compounds is that in many cases there is little information on their toxicity and long-term effects.

5.3 Petrochemical compounds

Crude oil is an extremely complex and variable mixture of organic compounds. Crude oil has accumulated underground as a result of the anaerobic degradation of organisms over a very long time. Under conditions of high temperature and pressure the organic material has been converted to natural gas, liquid crude oil, shale oil, and tars. At the underground temperatures shale oils and tars do not flow, but the crude oil is liquid, and unless contained will

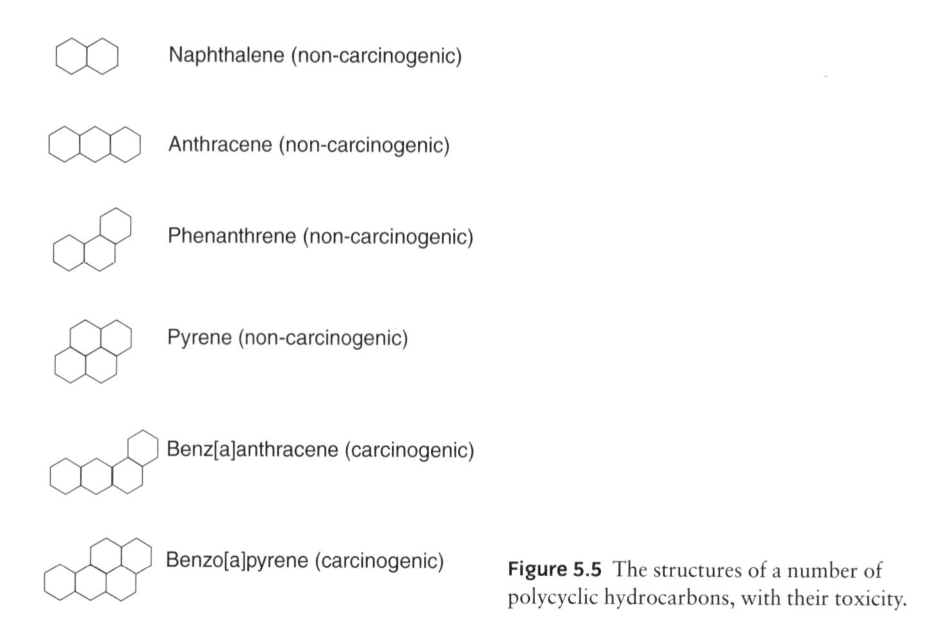

Figure 5.5 The structures of a number of polycyclic hydrocarbons, with their toxicity.

escape to the surface, where the volatiles evaporate, forming a tar bed. The majority of the compounds in crude oil are hydrocarbons, which can range in molecular weight from the gas methane to the high-molecular-weight tars and bitumen. These include the following:

- methane,
- aliphatic *n*-alkanes, pentanes, hexanes, heptanes, and cycloparaffins,
- monocyclic aromatics; benzene, toluene, ethylbenzene, and xylene and ('BTEX'),
- polycyclic hydrocarbons (5–35%), naphthalene, phenanthrene, and anthracene (PAHs),
- heterocyclic compounds (0.05–3.0%) containing sulphur, nitrogen, and heavy metals, and
- tars and bitumen.

These hydrocarbons can also come in a wide range of molecular structures; straight and branched chains, single or condensed rings, and aromatic rings (Box 5.2). The two major groups of aromatic hydrocarbons are the monocyclic hydrocarbons such as benzene, toluene, ethylbenzene, and xylene ('BTEX'), and the PAHs such as naphthalene, anthracene, and phenanthrene. The proportion of each individual compound can vary greatly between crude oil sources and this variation in composition affects the properties of the oil. Oils with a high proportion of low-molecular-weight material are known as 'light oil' and flow easily, while 'heavy oils' are the reverse. In addition to the hydrocarbons, crude oil contains between 0.05 and 3.0% heterocyclic

BOX 5.2 Low-molecular-weight compounds found in crude oil

Some of the structures of low-molecular-weight compounds found in crude oil are shown below: the gases methane, ethane, and propane; the aliphatic alkane pentane; the cycloparaffin cyclohexane; the monocyclic aromatics benzene and toluene; and the PAH naphthalene.

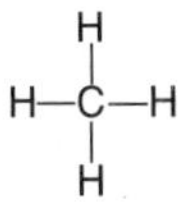

Methane 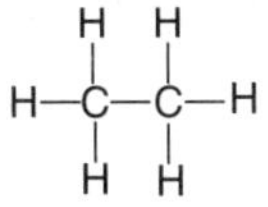 Ethane 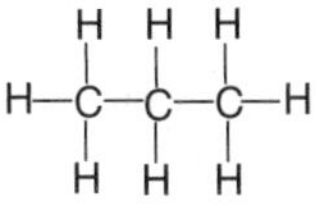 Propane

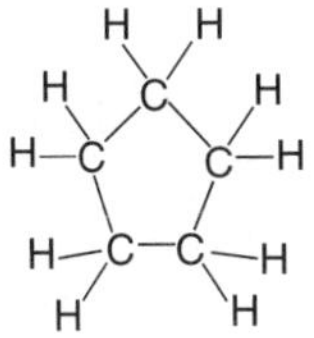

Cyclopentane 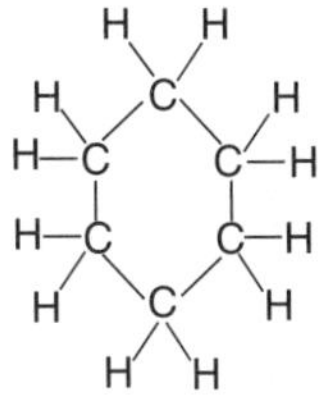 Cyclohexane

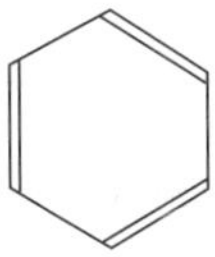

CH_3

Benzene 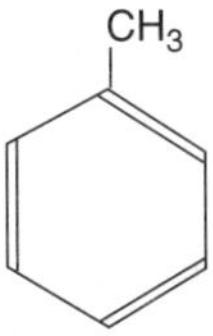 Toluene 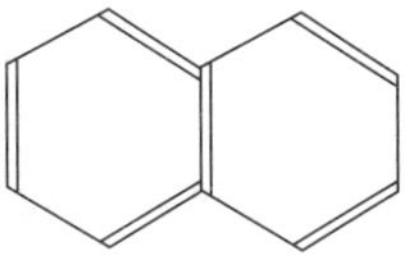 Naphthalene

compounds, containing sulphur, nitrogen, and oxygen, and some heavy metals. When crude oil is refined in a number of processes most of the PAHs are converted into monocyclic aromatic compounds. Typically naphthalenes can constitute 5–35% of the crude oil, and are reduced to 1–7% after refining. The refined oil can be split by distillation into petroleum, diesel, heating oil, and many other products (Box 5.3).

5.3.1 Crude oil and product release on to the land

Crude oil and its refined products can be released into the environment from a number of sources, which can pollute both land and water. Crude oil, which has accumulated underground, can reach the surface if not contained by impermeable rock. Leaks on land have long since been exploited but leaks of crude oil at sea can still be observed.

BOX 5.3 The refined crude oil products separated by distillation after the crude oil has been refined

The main gas methane is removed prior to refining and the refining process breaks down the polycyclic hydrocarbons to monocyclic hydrocarbons such as benzene, toluene, ethylbenzene, and xylene (BTEX). The table below gives the boiling points and range of carbon numbers in the various fractions. The higher the number of carbons the higher the boiling point; thus diesel requires a higher temperature to burn than petroleum.

Crude oil products

Product	Carbon number	Boiling point (°C)	Examples
Methane	1	−160	Methane
Petroleum ether	4–6	20–60	Pentanes, hexanes
Light naphtha	6–10	60–100	Cyclohexanes, cycloheptane
Petroleum (gasoline)	3–8	40–205	BTEX
Paraffin (kerosene)	10–14	165–200	Polycyclic hydrocarbons
Diesel	15–20	175–365	Polycyclic hydrocarbons
Fuel oil	20+	350+	Polycyclic hydrocarbons

Apart from this release of crude oil, the main source of crude oil and oil products, such as petrol and diesel, released on land comes from the disposal of waste motor oil, the leaking of storage tanks, and other spillages and accidents during its transport. Estimates suggest that there are between 100 000 and 300 000 leaking petroleum and petroleum-based product tanks in the USA (Lee and Gongaware, 1997; Mesarch and Nies, 1997). It has been estimated that there are 350 000 contaminated sites in Europe, of which the largest proportion are contaminated by petroleum products (Troquet et al., 2003). The petroleum leaks are of particular interest as petroleum can contain up to 20% BTEX. The BTEX compounds, although not miscible with water, are mobile and can contaminate the groundwater (Bossert and Compeau, 1995). Figure 5.6 shows the possible fate of the components of a petrochemical leak from a storage tank. The volatile components will be lost to the atmosphere if the leak is on or very near the surface, but if the leak is below soil level the mobile components can migrate down through the soil to the water table. Any compounds that are water soluble can also migrate down through the soil and into the groundwater. The higher-molecular-weight components are mostly immiscible with water and may either move slowly through the soil or remain on or near the surface, depending on the soil structure. Insoluble compounds may also be absorbed very tightly on to the soil particles. If the water-immiscible components do

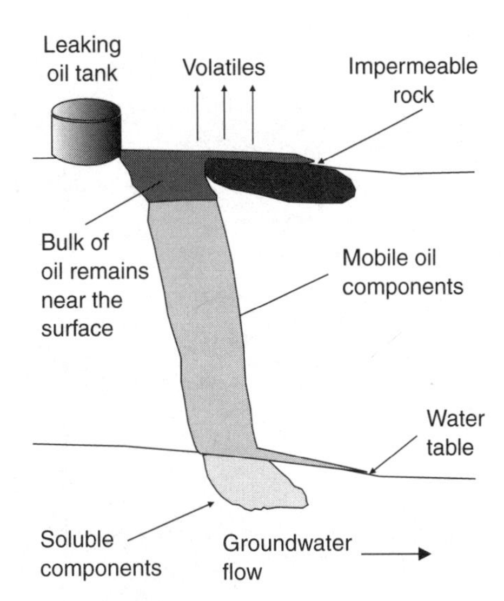

Figure 5.6 The distribution of hydrocarbons in soil from an oil spill on or just below the surface. Adapted from Bossert and Compeau (1995).

migrate through the soil and reach the water table they will form a layer on the surface of the water and spread out in this manner.

5.3.2 Crude oil and product release into the sea

The most spectacular releases of oil are the oil tanker wrecks where very large quantities are released, as in the case of the *Exxon Valdez* and *Sea Empress* oil tankers. Although these spills gain considerable public attention, taken over a period of time the amount spilled is about the same as that released by natural seepage (Prince, 1997). The levels of oil released into the oceans are given in Fig. 5.7 and it has been estimated to be $(1.7–8.8) \times 10^6$ tonnes. It can be seen that the accidental spillage from tankers is far less than that derived from motor-oil replacement and other sources on the land which run into the oceans via rivers and streams. A considerable amount of oil is also released in the process of cleaning the oil tanks of oil tankers. This type of pollution should be decreasing as there is an International Convention for the Prevention of Pollution from Ships (MARPOL) that was signed in 1973 and later amended by protocol in 1978. The cleaning waste should be retained and pumped out when the ship is next in port. However, not all tankers conform to the agreement. The bioremediation of oil spills at sea is described in Chapter 10.

5.4 Inorganic wastes

Metals and other inorganic compounds are discharged into the environment from a number of activities including mining, smelting, electroplating, and

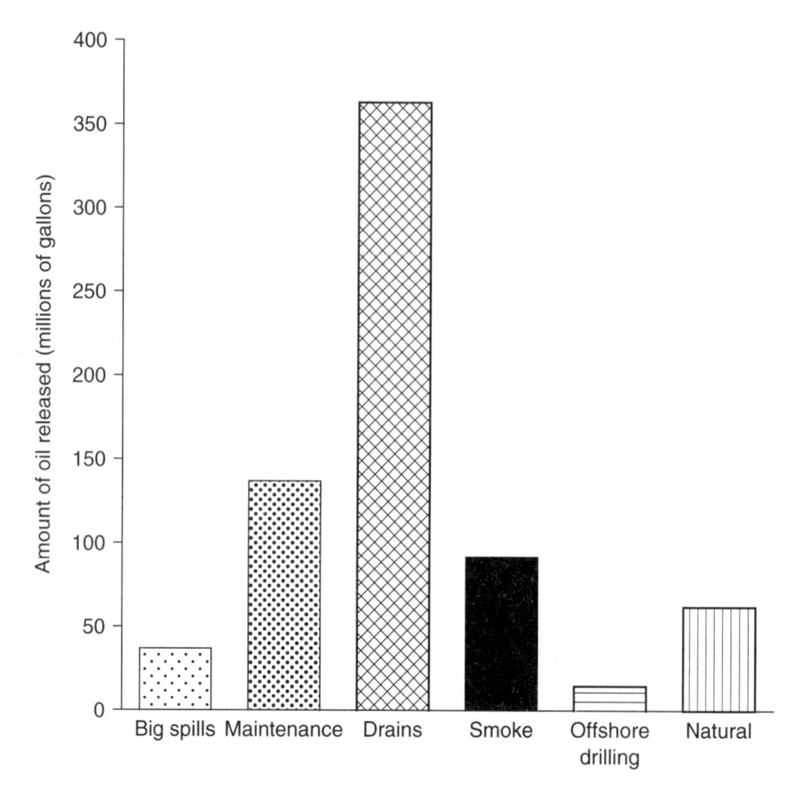

Figure 5.7 The sources of oil released into the oceans.
Source: www.seawifs.gsfc.nasa.gov

farming. Many metals are required by living organisms for their normal function but at high concentrations these can become toxic. Table 5.5 lists some of the sources and effects of inorganic wastes. One example is copper, which is an essential micronutrient (trace element) for plants but an excess of which can cause inhibition of photosynthesis and pigment synthesis, and damage to plasma membranes (Marschner, 1995). The damage caused by copper is due to generation of superoxide anions, hydrogen peroxide, and hydroxyl radicals due to incomplete oxidation of oxygen during respiration. The free radicals and hydrogen peroxide will attack lipids, proteins, and DNA causing mutation and death (Halliwell and Aruoma, 1993). An example of the toxicity of metals to humans was the mercury poisoning at Minamata in Japan that killed 43 people and disabled many others. The mercury originated from a factory producing vinyl chloride and acetaldehyde, which released low levels of methyl mercury into the sea. Marine organisms adsorbed and concentrated the mercury and consumption of the fish passed on the mercury contamination to the local population at very high concentrations in a process of biomagnification.

Table 5.5 The source and effects of inorganic contaminants

Contaminant	Maximum contaminant level (MCL; mg/l)	Effects	Sources
Antimony	0.006	Increase in blood cholesterol, decrease in blood sugar	Petroleum refineries, fire retardants, ceramics, electronics, solder
Arsenic	0.01	Skin damage, increased risk of cancer, problems with circulatory system	Natural deposits, run-off from orchards, glass and electronics production
Asbestos	7 MFL*	Increased risk of cancer	Decay of asbestos cement, natural deposits
Barium	0.2	Increase in blood pressure	Discharge of mining wastes, metal refineries, natural deposits
Beryllium	0.004	Intestinal lesions	Metal refineries, coal burning, discharge from electrical, aerospace, and defence industries
Cadmium	0.005	Kidney damage	Corrosion of galvanized pipes, natural deposits, metal refineries, run-off from waste batteries and paints
Chromium	0.1	Allergic dermatitis	Discharge from steel and pulp mills, natural deposits
Copper	1.3	Gastrointestinal illness, long-term liver and kidney damage	Corrosion of plumbing systems, natural deposits
Cyanide	0.2	Nerve damage, thyroid problems	Steel/metal/plastic/fertilizer factories
Fluoride	4.0	Bone disease	Water additive, natural deposits, fertilizer and aluminium factories

Table 5.5 (*Continued*)

Contaminant	Maximum contaminant level (MCL; mg/l)	Effects	Sources
Lead	0.015	Affects mental development in children and infants	Corrosion of pipes, natural deposits
Mercury	0.002	Kidney damage	Natural deposits, refineries, run-off from landfills and crops
Nitrate	10	Infants below 6 months old, blue baby syndrome	Fertilizer run-off, leaching from septic tanks, sewage systems, natural deposits
Nitrite	1	Same as nitrate	As above
Selenium	0.05	Hair/fingernail loss, circulatory problems	Petroleum refineries, mines, natural deposits
Thallium	0.002	Hair loss, liver problems	Ore processing, discharge from electronic, glass and drug factories

Source: US Environmental Protection Agency, **www.epa.gov** *MFL, million fibres/l.

Because of the toxicity of some metals national and international standards have been set for metal levels in drinking water (Chapter 3). Metals released from some industrial processes often far exceed the levels set for drinking water values, but these wastes are usually greatly diluted when added to waterways or sewage systems.

If metals are released into the environment they cannot be biodegraded and depending on the conditions will follow a number of pathways. Metals can adsorb on to the soil particles, run off into rivers or lakes, and leach into the groundwater. Metals are released into the environment in industrial effluents, landfill leachates, mining, household waste, and spills at hazardous waste sites. Metals can occur in different forms depending on conditions, which is known as speciation. Sequential extraction can be used to determine speciation in an environmental sample. In the case of cadmium the divalent form is soluble and therefore mobile, but it can complex with oxides and organic compounds and is not soluble above pH 7.5.

5.5 Bioremediation strategies

In many cases the clean-up of contaminated sites has been carried out using physical and chemical methods such as immobilization, removal (dig and dump), thermal, and solvent treatments. However, advances in biotechnology have seen the development of biological methods of contaminant degradation and removal, a process known as bioremediation. Potentially bioremediation is cheaper than the chemical and physical options, and can deal with lower concentrations of contaminants more effectively, although the process may take longer.

The strategies for bioremediation in both soil and water can be as follows.

- Use the indigenous microbial population.
- Encourage the indigenous population.
- Bioaugmentation; the addition of adapted or designed inoculants.
- Addition of genetically modified micro-organisms.
- Phytoremediation.

5.5.1 Indigenous micro-organisms

Soils contain a very large number of micro-organisms, which can include a number of hydrocarbon-utilizing bacteria and fungi (Namkoong et al., 2002), representing 1% of the total population of some 10^4–10^6 cells/g of soil. In addition, cyanobacteria and algae have also been found in the soil to degrade hydrocarbons. Hydrocarbon-contaminated soils have been found to contain more micro-organisms than uncontaminated soils (Table 5.6), but the diversity of the micro-organisms was reduced (Bossert and Compeau, 1995; Milcic-Terzic et al., 2001). The indigenous microbial population in the marine environment also contains micro-organisms that can degrade hydrocarbons (Head and Swannell, 1999).

The fate of organic compounds in the environment is affected by a number of factors that can be grouped into two areas; firstly those factors affecting the growth and metabolism of the micro-organisms and secondly those imposed by the compound itself. The factors imposed by the compound are as follows: the chemical structure of the organic compound; its availability and/or solubility; and the effects of photochemistry.

The rate of degradation of an organic compound will be dependent on the structure of the compound. The simpler aliphatic and monocyclic aromatics are readily degradable, but the more complex structures, like PAHs, are not easily degraded and may persist for some time (Table 5.3 and Fig. 5.4). The persistence will be increased if the compound or its breakdown products are

Table 5.6 The number of degrading bacteria detected in diesel-contaminated soil

	Bacterial count (CFU/g)	
	Diesel contaminated soil	Control soil
Heterotrophic	$(2.2 \pm 0.4) \times 10^{7}$	$(3.6 \pm 0.7) \times 10^{10}$
Toluene-degrading	$(7.3 \pm 0.7) \times 10^{4}$	ND*
Naphthalene-degrading	$(5.4 \pm 1.0) \times 10^{4}$	ND*
Diesel-degrading	$(8.3 \pm 1.7) \times 10^{5}$	ND*
Total cells (acridine orange direct counts)	$(6.1 \pm 1.9) \times 10^{12}$	$(3.4 \pm 1.2) \times 10^{12}$

Notes: colony forming units (CFU) were determined by spread plates. Values are the means from three replicates.

Source: Milcic-Terzic et al. (2001).

*ND, not determined.

BOX 5.4 Terms for the degradation of organic compounds

The degradation of organic compounds in the environment can follow a number of routes from mineralization to recalcitrance.

- **Detoxification** is where the compound is converted into non-toxic metabolites without being mineralized.
- **Activation** is the reverse, where non-toxic compounds are converted into toxic compounds.
- **Co-metabolism** is the breakdown of a compound which is not used as a carbon or energy source.
- **Gratuitous metabolism** is the degradation of a compound by an unrelated enzyme. This may be due to the enzyme being not very specific with its substrates.

toxic (Box 5.4). The persistence of organohalogen compounds is influenced by the number and type of halogen, with increasing resistance with more halogen atoms (Figs 5.3 and 5.4). Another crucial factor is the availability of the compound for degradation within the environment. On land the soil structure, its porosity and composition, and the solubility of the compound itself, will affect availability. At sea miscibility and solubility are important in degradation. Some compounds can be adsorbed to clays and are thus rendered invulnerable to degradation. To overcome this problem surfactants have been added to contaminated soils to improve the availability of hydrocarbons (Mihelcic et al., 1993). The addition of surfactant has been shown to increase the degradation of the biocide Aroclor 1242 (PCB) by increasing its availability (Ferrer et al., 2003).

5.5.2 Stimulation of indigenous microbial growth

The biodegradation of organic compounds including hydrocarbons is associated with microbial growth and metabolism and therefore any of the factors affecting microbial growth will influence degradation. If the micro-organisms cannot use the pollutants as their sole source of energy and carbon skeletons, some other growth substrate will be needed. In some cases if another substrate is present the micro-organisms may use this in preference to the hydrocarbons. The micro-organisms may also require supplementation with nitrogen and phosphorus as demonstrated in marine situations.

The following factors can affect the growth of micro-organisms:

- presence of other biodegradable organic material,
- presence of nitrogen- and phosphorus-containing inorganic compounds,
- oxygen levels,
- temperature,
- pH,
- presence of water; soil moisture,
- number and type of micro-organisms present, and
- presence of heavy metals or salt.

The aerobic degradation of hydrocarbons is considerably faster than the anaerobic process (Holliger and Zehnder, 1996), so that a supply of oxygen will be needed to maintain aerobic conditions if rapid degradation is required. A soil with an open structure will encourage oxygen transfer and a waterlogged soil will have the reverse effect. The temperature affects microbial growth, so that at low temperatures the rate of degradation will be slow. Nutrient addition to soils at temperatures of 4–10°C has been shown to have little effect as the low temperature reduces growth to such a low level. However, in other conditions the addition of nitrogen- and phosphate-containing fertilizer greatly increases the degradation of contaminants and hydrocarbons (Fig. 5.8). The pH of the soil will affect both the growth and the solubility of the compound to be degraded. The presence of large numbers of hydrocarbon-degrading micro-organisms in the soil will clearly be an advantage at the start, but as most soils contain these types of organism growth will soon increase the numbers, so that seeding with specific hydrocarbon-degrading organisms will probably not be needed. In some cases hydrocarbon contamination may also be associated with high levels of heavy metals, which may inhibit microbial growth depending on the concentration and type of metals.

The rate of degradation of xenobiotics in soil and water is also dependent on the presence of micro-organisms with the enzymatic capability to degrade the polluting molecules. As xenobiotics are not normally found in nature, the level of the degrading micro-organism may be very low, and in many cases a period of adaptation is required before degradation occurs.

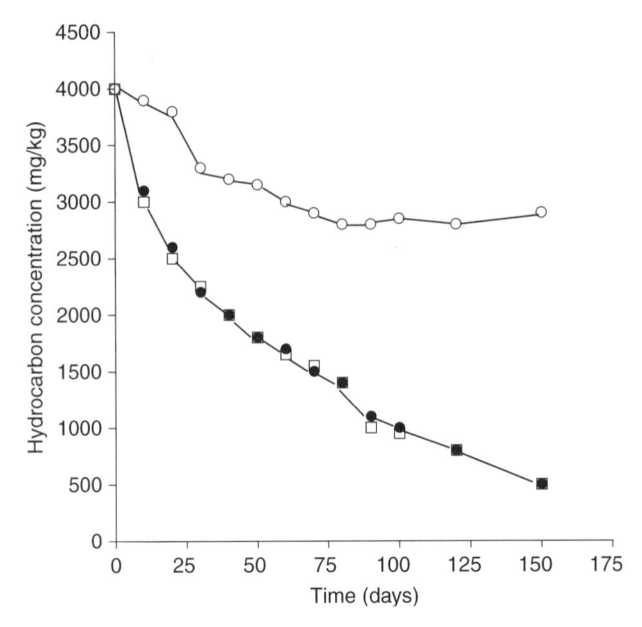

Figure 5.8 The removal of diesel from alpine soils. ○, Control with no additions; •, addition of fertilizer; □, addition of fertilizer and micro-organisms. Adapted from Margesin and Schinner, 1997.

Over the years *in situ* aerobic **co-metabolism** has been demonstrated (Semprini, 1997). This is a situation where the addition of a second carbon source stimulates the degradation of the pollutant. In the case of trichloroethylene removal oxygen and methane, the co-metabolic substrates, were added to the subsoil. The major problem with the addition of a second microbial substrate is to ensure that the substrates added are distributed through the soil and not just close to the injection bore. To reduce the uneven distribution recirculation systems have been investigated.

Not all soil pollutants are easy to remove from the soil or are readily available for degradation. In some cases the pollutant or its breakdown products can be either reversibly or irreversibly bound to the soil particles and contents. An example is trinitrotoluene (TNT), which is degraded in the soil by sequential oxidation to aminodinitrotoluene (ADNA), diaminonitrotoluene (DNAT), and triaminotoluene (TAT) in the presence of oxygen (Fig. 5.9). DNAT and TAT react irreversibly with the clay and humic components in the soil. Therefore, a sequence of anaerobic/aerobic processes is needed to remove TNT from soils.

Many of the soil contaminants are insoluble in the aqueous phase so that if they are to be degraded or removed they have to be made available to the micro-organisms. One solution has been to apply biosurfactants, which are surface-active molecules, which act as detergents and solubilize, emulsify, and disperse the pollutants. Non-ionic surfactants such as Triton X-100 and Triton

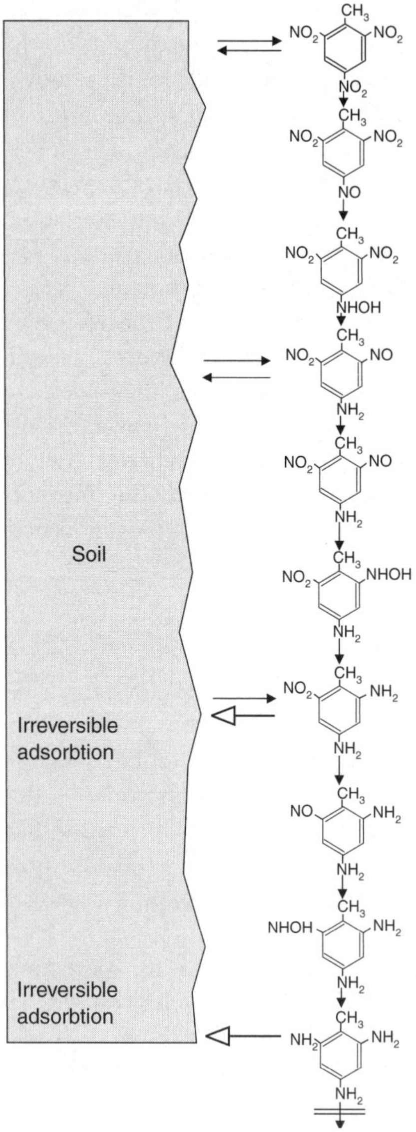

Figure 5.9 The aerobic degradation of trinitrotoluene (TNT) where the degradation products irreversibly bind to clay and humic components in the soil.

X-45 have been shown to increase the breakdown of phenanthrene (Churchill et al., 1995) and pentachlorophenol (Cort et al., 2002). Biologically produced surfactants (biosurfactants) can be produced *in situ* by adding specific micro-organisms, added as an extracted biochemical, or by stimulation of the indigenous population. Biosurfactants have been used to solubilize oils and xenobiotics such as PCBs and organophosphates. *Acinetobacter radioresistens* produces a protein bioemulsifer, alasan, which was shown to solubilize PAHs. The gene for this protein has been cloned and transferred to *Escherichia coli* and expressed. This may make the protein more available for bioremediation. Another method of extraction has been the use of liquid carbon dioxide for the removal of compounds. The method has been used successfully for the removal of diesel from soil (Lee and Gongaware, 1997). Soil washing has been use to remove sparingly soluble pollutants such as pentachlorophenol. The soil is generally removed and washed by a series of scrubbing and physical separation techniques. In this method the contaminants are partitioned into the liquid phase and the fine particles are collected for biotreatment or disposal. The water containing the pollutant can be treated in a number of systems.

Recent studies have shown that the mobility of crude oil can be increased by the addition of biodiesel to the contaminated soil, which dissolves the oil. Biodiesel, which is made of methyl esters from plant oils, has a low toxicity to plants and is readily biodegradable. The addition of biodiesel and nutrients (nitrate and phosphate) to coal tar PAH-contaminated soil has been shown to stimulate the breakdown of PAH which was not found with the nutrients alone (Taylor and Jones, 2001). Some 85% of the PAH was removed in field experiments after 55 days of treatment.

5.5.3 Bioaugmentation

Bioaugmentation is the addition of selected organisms to contaminated sites in order to supplement the indigenous microbial population and speed up degradation. In general, introduced bacteria decline rapidly after introduction and their growth is poor. This decline in introduced bacteria is probably due to their failure to compete with the indigenous population. However, this may be more effective where ecological selectivity is carried out based on supplying a specific substrate that can only be used by the augmenting organism (Atagana, 2003).

Bioaugmentation has been considered for crude oil removal for some time, but the results have been mixed (Venosa and Zhu, 2003). Twelve commercial augmentation cultures were tested on Alaskan crude oil in the laboratory and after 28 days four cultures showed enhanced oil degradation. Bioaugmentation of weathered diesel fuel in Arctic soil had no effect on the rates of removal. Thus, bioaugmentation can be effective in the laboratory but in the field this may not be true. Two field experiments with crude oil and marsh sediments concluded that oil biodegradation was not limited by the population of hydrocarbon degraders and that seeded micro-organisms could not compete with indigenous populations. The fate of introduced organisms has been followed in the environment using molecular techniques such as DGGE and TGGE analysis of 16S rDNA fragments. Other possible causes for bioaugmentation failing are

- rarely, a limiting population of micro-organisms,
- the concentration of the contaminant not sufficient to support growth,

- the environment may contain substances that inhibit growth,
- predation by protozoa,
- the added micro-organisms may use some other substrate in the environment, and
- the introduced micro-organisms may not be able to penetrate the soil to get to the contaminant.

More recently bioaugmentation has had more success using activated soil rather than pure cultures. The activated soils are those containing indigenous microbial populations recently exposed to the contaminant. The technique has the advantages that it introduces naturally developed populations not cultured outside the soil, and that it is a mixed population (Gentry et al., 2004). This technique has been shown to enhance the degradation of pentachlorophenol, atrazine, and chlorobenzoate.

Bacteria are not the only micro-organisms used for bioaugmentation; fungi have also been used as they can grow under low-water conditions, are present in both soil and water, and the hyphae can penetrate the soil, making the pollutant available. A number of fungal inocula have been used to bioaugment soils contaminated with PCP and this removed 80–90% within 4 weeks (Lestan and Lamar, 1996). A selected strain of *Methylosinus trichosporium* was used in a field study where the organism was selected for its high trichloroethylene (TCE) transformation rate under low-copper conditions. Once injected 50% of the cells attached to the soil, forming a biofilter that was efficient for the transformation of TCE (Erb et al., 1997). A fungus, *Absidia cylindrospora*, has been used to degrade fluorene (Garon et al., 2004) and *Cladophialophora* sp. strain T1 degraded BTEX (Prenafeta-Boldu et al., 2004).

The genetic information for a number of degradative pathways is found on plasmids or other mobile elements (transposons). Table 5.7 shows a number of different mobile elements in relation to xenobiotic degradation. The use of these elements to produce new pathways of degradation and addition to polluted water and soil is a possible bioaugmentation strategy (Top et al., 2002). Examples are the addition of *Enterobacter agglomerans* DK3 containing the plasmid RP4 Tn4371, which encodes for biphenyl degradation. The donor strain when added to soil disappears rapidly but the plasmid was transferred to other micro-organisms in the soil which express the gene (de Rore et al., 1994). In a similar experiment 2,4-D-degradative plasmids were introduced to soil in an *E. coli* donor and in the presence of 2,4-D this had a positive effect on the amount of transfer. Some of the novel biodegradation pathways appear to have been formed in a random assembly of horizontally transferred genes and it would appear that these processes occur naturally in polluted soil. The application of plasmid-assisted bioaugmentation may be feasible where the limiting factor is the absence of the suitable catabolic capacity (Top et al., 2002).

Table 5.7 Examples of mobile genetic elements for the degradation of xenobiotics

Mobile element	Bacterial strain	Substrate
Plasmids		
pSS60	*Achromobacter* sp. LBS1C1	4-Chlorobenzoate
pBRC60	*Alcaligenes* sp. BR60	3-Chlorobenzene
pENH91	*Ralstonia eutropha* NH9	3-Chlorobenzoate
pJP4	*Ralstonia eutropha* JMP 134	2,4-D
pTSA	*Comamonas testosteroni* T-2	*p*-Toluenesulphonic acid
pCS1	*Pseudomonas diminuta*	Parathion
pC1-3	*Delftia acidovorans* CA28	3-Chloroanaline
pPS12-1	*Burkholderia* sp. PS12	1,2,4,5-Tetrachlorobenzene
Class 1 transposons		
Tn5280	*Pseudomonas* sp. p51	Chlorobenzene
Tn5271	*Alcaligenes* sp. BR60	Chlorobenzoate
DEH	*Pseudomonas putida* PP3	Chlorinated aliphatic acids
Class 2 transposons		
Tn4651	*Pseudomonas putida* mt-2	Toluene, xylene
Tn4655	*Pseudomonas putida* G7	Naphthalene

Sources: Top et al. (2002) and Top and Springael (2003).

5.5.4 Genetically manipulated organisms

The creation of a superbug by genetic manipulation has been considered for some time. A multiplasmid-containing *Pseudomonas* strain has been produced which is capable of oxidizing aliphatic, aromatic, terpenic, and polyaromatic hydrocarbons. Another multiplasmid organism, *Pseudomonas putida*, has been produced which can degrade both lighter alkanes and aromatics (Venosa and Zhu, 2003). In the 1970s Chackrobarty (Black, 2002) was first to produce a superbug for the degradation of hydrocarbons (Fig. 5.10). Different plasmids were combined to produce a bacterial strain, which could degrade camphor, octane, xylene, and naphthalene. This strain was the first to be granted a US patent for its construction and use. However, the survival of such stains in the environment must be questioned and the issues of release into the environment of genetically manipulated organisms have to be addressed in terms of safety, containment, and public perception.

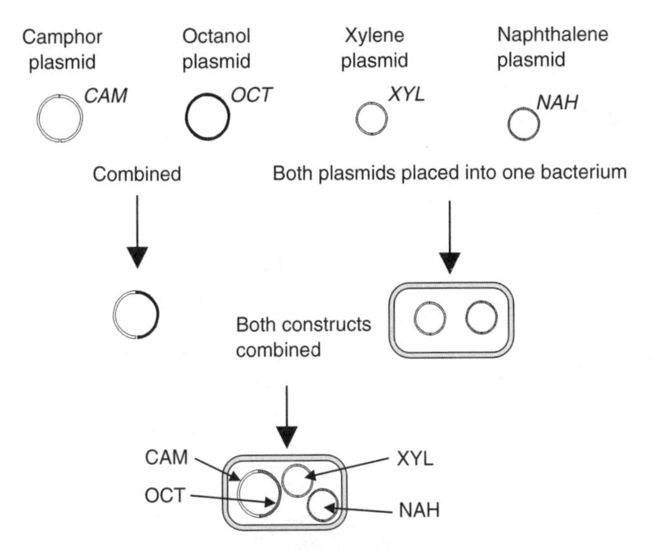

Figure 5.10 The construction of a hydrocarbon-degrading bacterium by the combining of four bacterial plasmids containing the genes for camphor (CAM), octanol (OCT), xylene (XYL), and naphthalene (NAH) degradation. In one case the camphor and octanol genes were combined into one plasmid whereas the xylene and naphthalene genes were added as separate plasmids in a multiplasmid-containing bacterium.

5.6 Bioremediation techniques *in situ*

5.6.1 Bioremediation on land

The bioremediation methods employed will depend on the area contaminated, the properties of the compound(s) involved, the concentration of the contaminants, and the time required to complete the bioremediation. There are a number of options for bioremediation; the contamination can be treated on site or the contaminated material excavated and treated on or off site. The trend in the USA is for removal as this avoids litigation over any contamination not removed by an *in situ* treatment. If the contaminant is water soluble a pump-and-treat technique can be used (Fig. 5.11). In this method water is introduced into the contaminated areas and removed at another site to be treated on or off site. This has been used frequently to treat metal-contaminated soil where the metal is not tightly bound to the soil. The *in situ* and *ex situ* processes are outlined in Fig. 5.12 and include bioventing, biosparging, stimulation, and phytoremediation. The following methods can be used to treat a contaminated site without excavating the contaminated soil.

5.6.2 Land farming

The simplest of the on-site treatments is land farming, which involves mixing of the soil by ploughing or some form of mechanical tilling. Ploughing

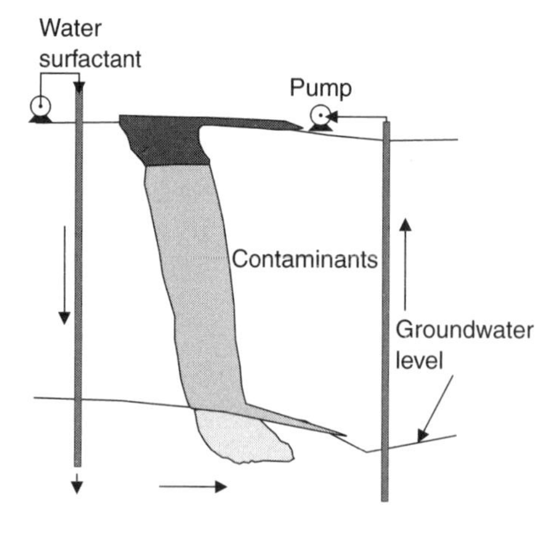

Figure 5.11 The pump-and-treat method for the treatment of contamination of the water table. The contaminated water is pumped out and treated on or off site. The water pumped out is replaced by pumping water and surfactant into an adjacent well(s).

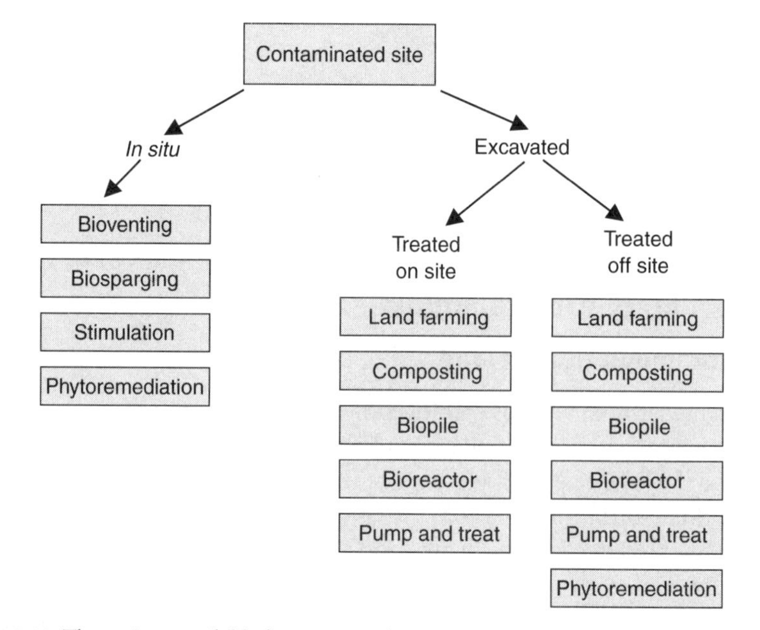

Figure 5.12 The options available for *in situ* and excavated treatments of contaminated soil. The excavated soil can be treated on and off site.

increases the oxygen levels in the soil and distributes the contaminants more evenly, which increases the rate of degradation. In some cases nutrients are added to increase biodegradation. Typically 4–6 months are required to remove contaminants such as PAHs. This method is best suited for shallow contamination of the soil surface. Often the treatment area is lined and dammed to retain any contaminated leachate. The rate of degradation depends on the microbial population, the type and level of contamination, and the soil type. The average half-life for degradation of diesel fuel and heavy oils is in the order of 54 days with this type of system (Bossert and Compeau, 1995).

5.6.3 Bioventing

Bioventing is an *in situ* process, which combines an increased oxygen supply with vapour extraction. A vacuum is applied at some depth in the contaminated soil; this draws air down into the soil from holes drilled around the site and sweeps out any volatile organic compounds (Fig. 5.13). Nutrient supplementation can be provided by running nutrients into trenches dug across the site. The increased supply of air will increase the rate of natural degradation by the aerobic micro-organisms. Clearly this is only effective for reasonably volatile compounds, and where the soil is permeable. However, the vapour extracted may need some form of treatment and one biological solution is the use of biofilters (section 5.10).

5.6.4 Biosparging

Biosparging is a process to increase the biological activity of soil by increasing the supply of oxygen by sparging air or oxygen into the soil. Air injection was tried at first but was replaced by pure oxygen in order to increase degradation rates. The expense of this type of treatment has limited its application to highly contaminated sites but on-site generation of oxygen has reduced costs.

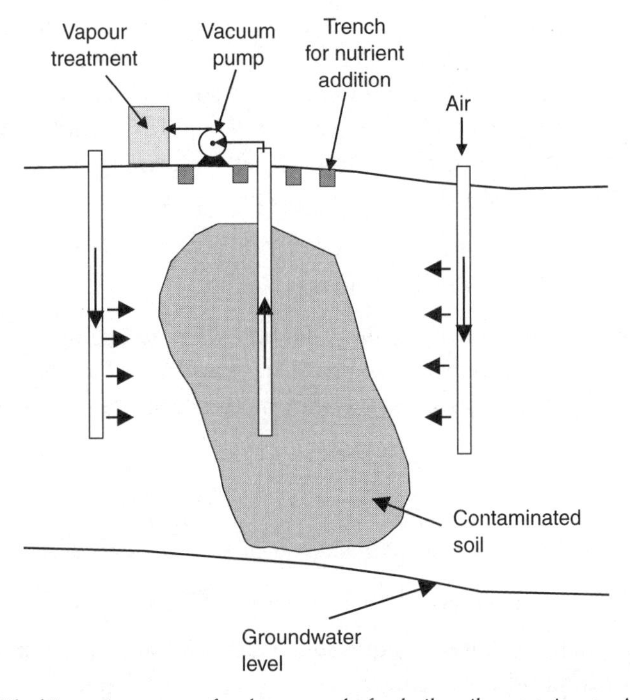

Figure 5.13 The bioventing process for the removal of volatile soil contaminants. A vacuum is applied to a bore hole drilled into the contaminated site that will draw out any volatile compounds. Bore holes are also drilled around the site to replace the air removed by the vacuum. In some cases nutrients are applied to assist degradation of the compounds to volatiles.

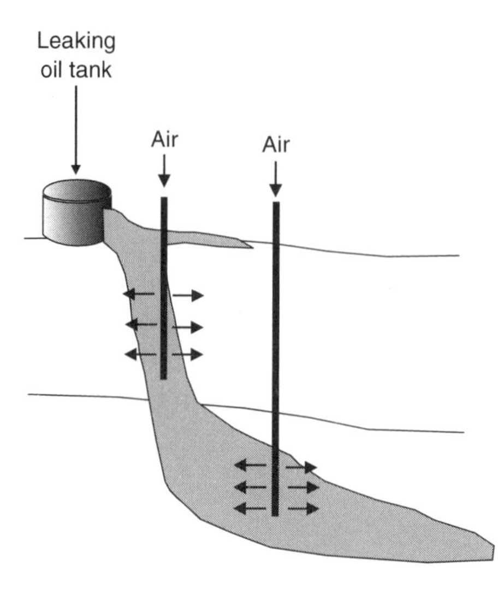

Figure 5.14 The process of biosparging where air is pumped into the contaminated site to encourage the aerobic degradation of the contaminants.

Hydrogen peroxide has been used on a number of sites but it can be toxic at low concentrations to micro-organisms. This process is similar to soil vapour extraction, which can be used for volatile contaminants (Fig. 5.14).

In bioventing and biosparging the structure of the soil can be the predominant factor in its success. The major engineering considerations with *in situ* bioremediation are the delivery of the additions and supply to the contaminants, which are affected by conditions in the soil. The fate of contaminants in soil is affected by channels and pores, in the soil structure, diffusion into closed pores, and soil organic matter. Other features are adsorption on to mineral surfaces and partitioning into organic matter.

5.6.5 Bioaugmentation/stimulation

The addition of nutrients injected into contamination well below the surface can be used to stimulate the indigenous microbial population. This technique can be combined with the addition of specific micro-organisms (**bioaugmentation**) (Fig. 5.15).

5.7 Bioremediation techniques *ex situ*

If the contaminated material is excavated it can be treated on or off site, which is often a more rapid method of de-contaminating the area. The techniques that can be used are given in Fig. 5.12 and can include land farming on or off site in the same way as for shallow contamination.

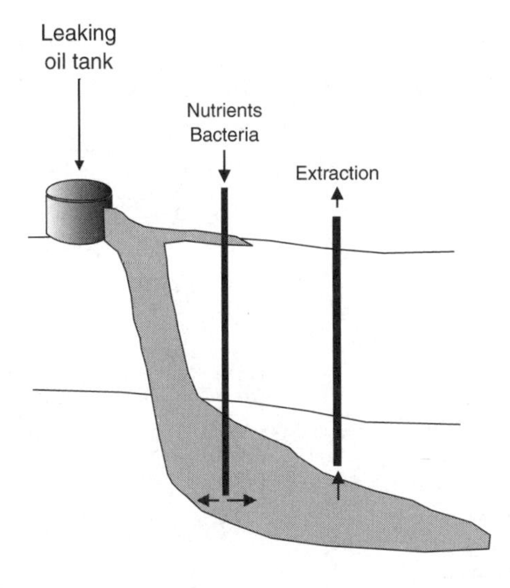

Figure 5.15 Deep contaminated soil can be treated by the addition of a combination of nutrients and micro-organisms in a process known as biostimulation. Another bore hole is used to remove water to make space for the nutrients and micro-organisms.

5.7.1 Composting

The composting process is another solid-phase treatment carried out after extraction. Composting material such as straw, bark, and wood chips is mixed with the contaminated soil and piled into heaps, as for the Windrow process. The process works in the same way as the normal compost system, with a rise of temperature to 60°C and above caused by microbial activity. The higher temperature encourages the growth of thermophilic bacteria. The increased costs of this type of system restrict it to highly contaminated materials, although the process is more rapid than land farming.

There have been a number of studies on the bioremediation of diesel-contaminated soil using composing. The high temperature achieved enhances degradation and diesel availability. The organic materials added were vegetables, fruit, and garden waste and these were added at a concentration of 33–75% (Fig. 5.16). A temperature of above 70°C was achieved after 6–22 days of incubation, with turning every 7 days and 84–86% of the contamination was removed by day 40 compared with 35% in untreated soil (Gestel et al., 2003). Another study was performed with diesel-contaminated soil using either sewage or compost, as these contain a wide range of micro-organisms (Namkoong et al., 2002). Some 80% of the pollution was removed by day 10.

5.7.2 Biopile process

In the **biopile** process the soil is heaped into piles within a lined area to prevent leaching. The piles are covered with polythene and liquid nutrients applied to

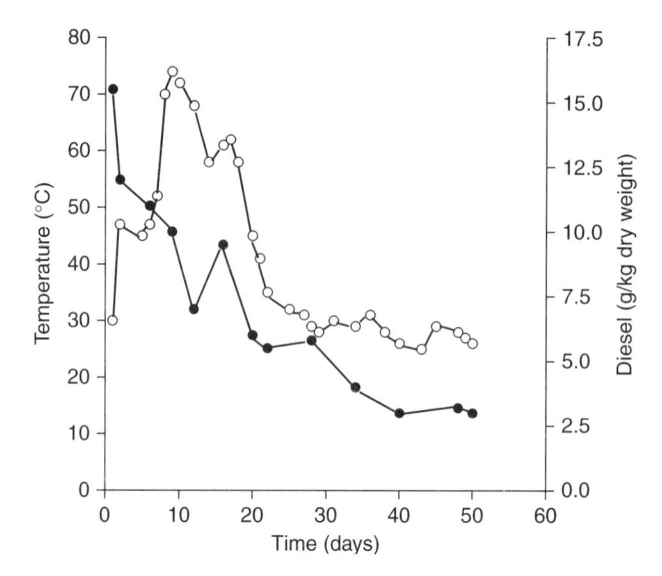

Figure 5.16 The composting of diesel-contaminated soil combined with vegetable, fruit, and garden waste. ○, Temperature (°C); •, concentration of diesel (g/kg of dry weight). Gestel et al. (2003).

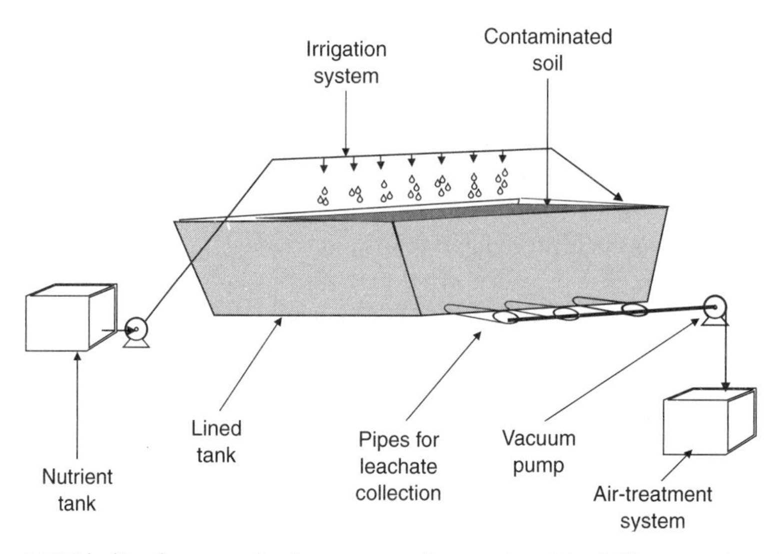

Figure 5.17 The biopile process for the treatment of contaminated land. The contaminated soil is piled on an impermeable layer or tank. The pile is irrigated with water or nutrient and the leachate collected or treated at the base of the pile. In some cases the leachate can be recycled.

the surface (Fig. 5.17). Aeration can be improved by applying suction to the base of the pile as in a composting system (Chapter 4). Any leachate formed is collected by pipes at the base and can be recycled if necessary. This type of system can be used when space is limited, and when vapour emissions need to be restricted some form of biofilter can be added to the system.

5.7.3 Bioreactors

Soil extracted from a contaminated site can also be treated as a solid waste (slurry 10–30%, w/v) or as a liquid leachate in bioreactors of various designs (Fig. 5.18 and Box 5.5). The use of bioreactors gives control of the parameters such as temperature, pH, mixing, and oxygen supply, which can improve degradation rates. The bioreactor designs that can be used for solid-waste slurries can be solid-bed, fluidized-bed, and stirred-tank bioreactors. A list of bioreactor designs is given in Box 5.5 and includes those for the treatment of sewage, contaminated wastewater, and soil slurries. When treating liquid leachates and contaminated groundwater all those reactors that have been used for wastewater treatment can be used (Chapter 4). This includes trickling-filter, rotating-drum-contactor, upflow-fixed-film, and fluidized-bed reactors where the biomass is fixed to some form of solid medium. Freely suspended biomass bioreactors that can be used are stirred-tank, activated-sludge, and upflow anaerobic-sludge-blanket reactors (Langwaldt and Puhakka, 2000). The difference between the treatment of contaminated groundwater and leachates compared with wastewaters is that the contaminants are at low concentrations, have low loads on the system, have slow degradation rates, and have slow biomass growth. Under these conditions an immobilized system, such as an upflow-fixed-film or fluidized-bed bioreactor, is the best option. Activated-sludge systems have also been used to remove xenobiotics such as 2,4-dichlorophenol. In this case the activated sludge was augmented with a special mixed culture (Quan et al., 2004). This type of bioaugmented system has also been used for phenols, chloroanaline, chlorinated solvents, aromatic hydrocarbons, petroleum hydrocarbons (Tellez et al., 2002; see Box 5.6), and coke plant wastewater.

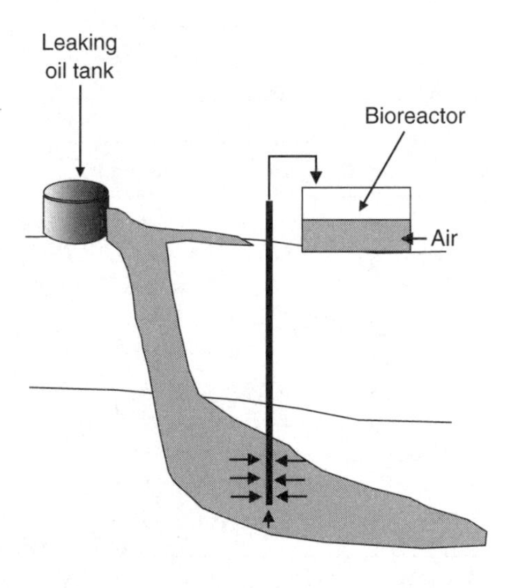

Figure 5.18 Contaminated groundwater can be treated by running into an aerated surface bioreactor. The bioreactor can also be used to treat solid waste as slurries.

BOX 5.5 Some of the bioreactor designs used to treat sewage, to remove metals from wastewater, and to treat solids

The stirred-tank design is outlined in Chapter 2 (Box 2.8) and activated sludge in Chapter 4 (section 4.3.3) and both use suspended microbial cultures. The remaining bioreactors used for sewage treatment (Chapter 4) use microbial cultures immobilized on a number of inert substrates. The immobilization gives a high-surface-area and operates as a continuous system with a high biomass content. In a situation where metals are removed from wastewater, leachates, and soil washes, rotating biological contactors, fluidized-bed, reactors and fixed-bed reactor designs (see figure) have been used (Fig. 5.28). In the case where sulphate-reducing bacteria are used to produce insoluble metal sulphides anaerobic conditions are needed and the upflow anaerobic-sludge-blanket (UASB) bioreactor has been used (Fig. 4.36). When the contaminated material is solid, such as soil, it can be treated in a stirred tank if made into a slurry. Alternatively a fixed-bed, biopile, or rotating-drum design can be used, as can be seen in the figure.

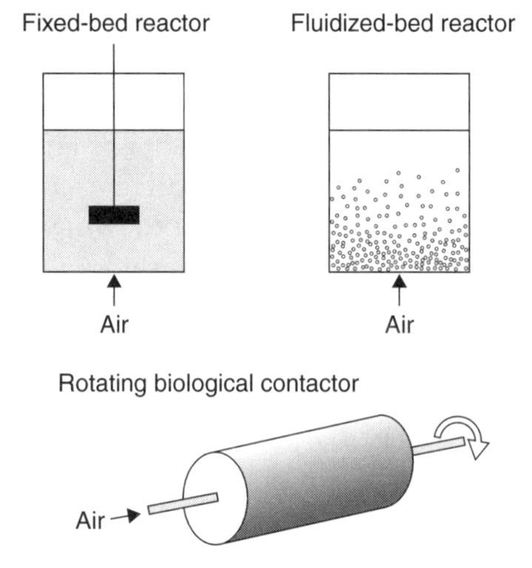

Bioreactors used for sewage treatment, metal removal, and solids treatment

Bioreactor type	Sewage	Metal removal	Solid state
Stirred tank	✔	✔	✔
Activated sludge	✔		
Trickling filter	✔		
Rotating biological contactor	✔	✔	
Fluidized bed	✔	✔	✔
Fixed bed	✔	✔	
Upflow anaerobic-sludge blanket (UASB)	✔	✔	
Biopile			✔
Rotating drum			✔

BOX 5.6 Activated-sludge process for removing petroleum hydrocarbons from oilfield water (data derived from Tellez et al., 2002)

The performance of the activated-sludge system was tested with various solids retention times (SRTs) where SRT is as follows:

$$\mathrm{SRT} = \frac{V_r X}{Q_w X_r}$$

where V_r is volume (l), X is MLSS (mg/l), Q_w is waste-sludge flow rate (l/day), X_r is concentration of sludge in recycled sludge (mg/l), and P_x is waste-activated sludge produced per day (l/day).

$$P_x = Q_w X_r$$

The coefficients were determined to be as follows:

Yield, Y = 0.44 mg of MLSS/mg of TPN
(where TPN is total petroleum hydrocarbon);
K_s, half velocity coefficient = 1.36 mg/l;
k (mg of TPN/mg of MLSS/day) = 1.36;
k_d, endogenous decay coefficient = l/day.

Using these coefficients the results of mixed-liquor suspended solids (MLSS), *f/m* ratio, TPH levels, and cell production are shown in the figure below. In panel (a), as the SRT increases the biomass (MLSS) level also increases, reaching a plateau at 20 days. In panel (b), the removal of TPH reaches a maximum at 20 days with about 90% removal, and the production of cells steadily drops as the SRT increases. Panel (a) shows the effect of SRT on MLSS (m/l; ○) and the *f/m* ratio (•). Panel (b) shows the effect SRT on TPH (mg/l; ○) and cell production (g/day; •).

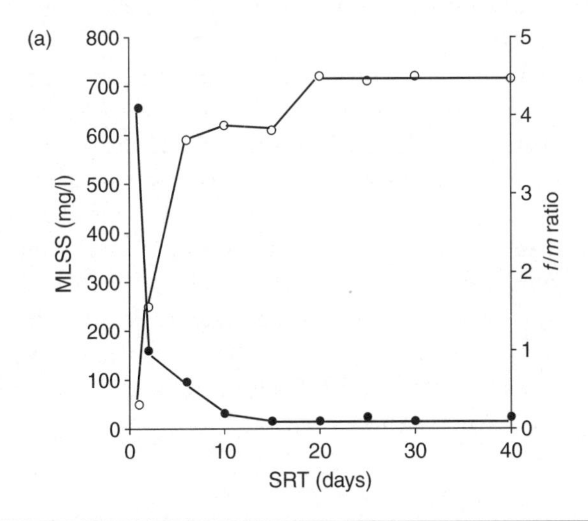

BOX 5.6 **Continued**

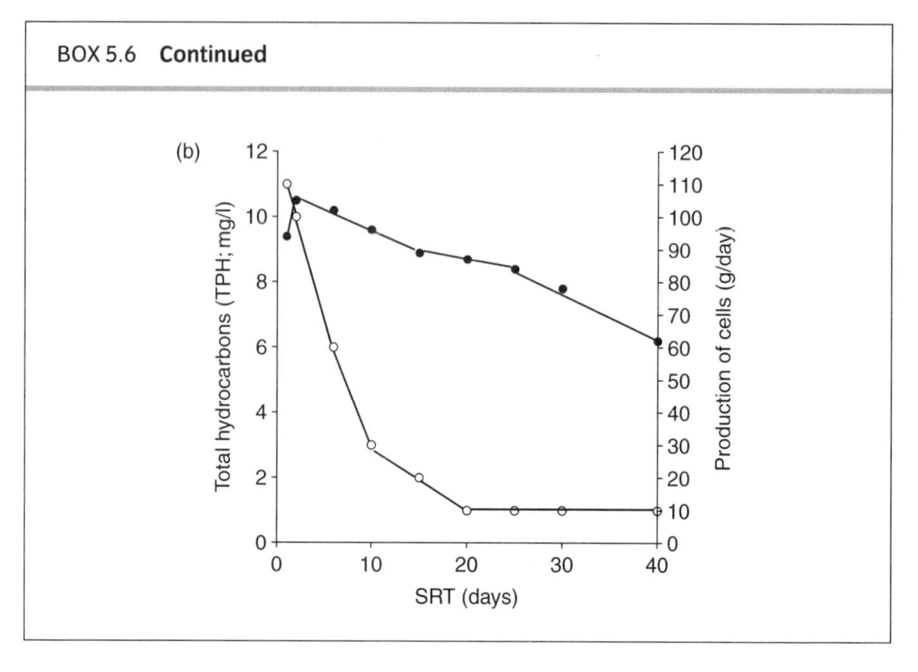

5.7.4 **Novel technologies**

A number of novel techniques for the treatment of contaminated sites have been developed recently. One was to use the specificity of the antibody/ antigen reaction to remove antibody/antigen reaction to remove specific pollutants from contaminated waters (Harris, 1999). At present the use of antibodies will be restricted by the high cost of the technique. A second novel method was to use nanoscale iron particles. Nanoscale iron particles have a very small size (1–100 nm), ten time less than bacteria, are non-toxic, and will transform a wide variety of environmental contaminants (Zhang, 2003). The nanoparticles are so small that they can be injected under pressure to penetrate the spaces between soil particles and reach the contaminants (Fig. 5.19).

A permeable reactive barrier has been used to treat acid mine drainage (Gibert et al., 2002). The active permeability barrier is a trench filled with porous material such as peat placed in the outflow from a mine (Fig. 5.20). The porous material can act as an immobilization matrix for micro-organisms that can degrade or precipitate the contaminants. For the treatment of acid mine drainage sulphate-reducing bacteria have been encouraged to grow in the barrier, converting metal sulphates into sulphides, which precipitate (see equations 5.1 and 5.2 in section 5.9.2).

5.8 **Phytoremediation**

Phytoremediation is the use of plants for the removal of contaminants and metals from soil and water or to render them harmless. The use of plants for

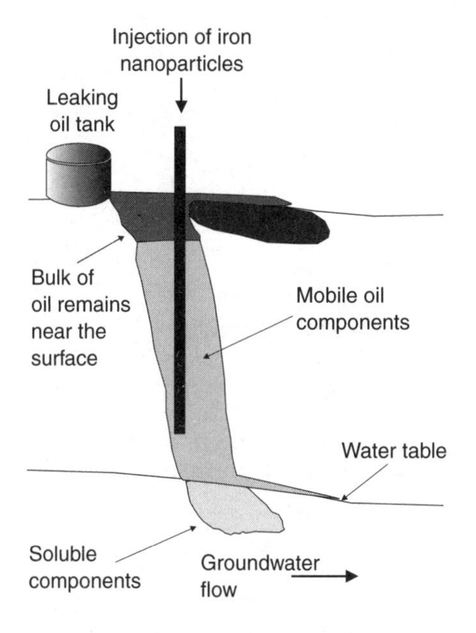

Figure 5.19 A novel treatment for contaminated land is to inject iron nanoparticles into the contaminated site. The particles are small enough to penetrate between the soil particles and degrade the contaminants. Adapted from Zhang (2003).

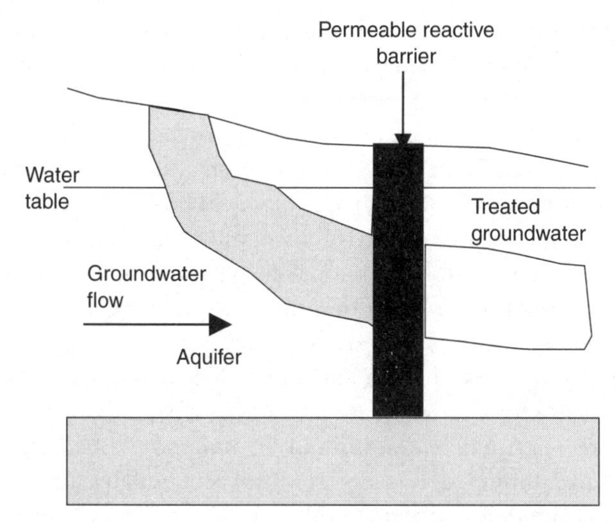

Figure 5.20 The use of a permeable reactive barrier to treat contaminated groundwater. The barrier is placed across the flow and consists of a trench filled with a mixture of stone chips and compost. The stone chips provide a surface for micro-organisms to colonize and the compost provides nutrients.

bioremediation provides an aesthetically pleasing option, has minimal disruption, has no disruption to topsoil, is effective with low levels of mixed contamination, offers the possibility of recovery of metals, and is inexpensive, some 50–80% less than alternatives (Pulford and Watson, 2003). The disadvantages are that the process can be slower than other bioremediation methods, taking a number of growing seasons, the contaminants may reduce the plant growth considerably, and that the plants, which accumulate pollutants,

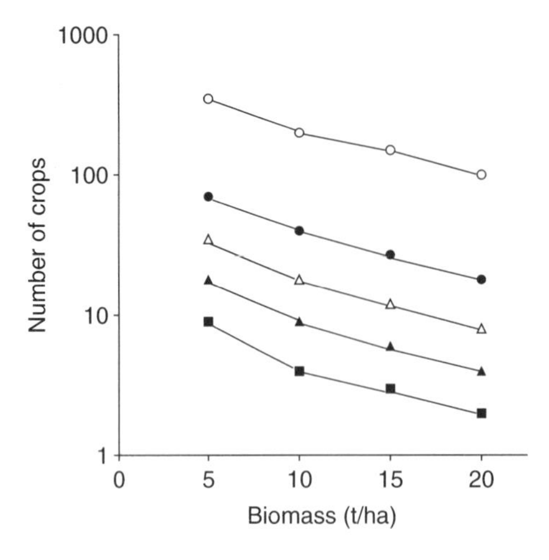

Figure 5.21 The effect of bioconcentration factor on the number of crops required to remove contaminants at various biomass yields (t/ha). Bioconcentration factor: ○, 1; ●, 5; △, 10; ▲, 20; ■, 40. Chaney et al., (1997).

constitute a hazard to wildlife and food chains. A plant used for phytoremediation needs to be tolerant of the pollutant, grow rapidly with a high yield per hectare, accumulate the metal in harvestable parts, have a profuse root system, and have a high bioconversion factor. The bioconversion factor is the concentration of pollutant in the plant compared with that in the environment. Most plants have a bioconversion factor of less than 1. The bioconversion factor needs to be 20 or above for phytoremediation to be able to reduce the contamination by 50% in 10 crops (Fig. 5.21). Clearly the higher the biomass the more rapid the rate of phytoremediation. The plants that accumulate high concentrations of metals are known as hyperaccumulators and can accumulate 50–100 times more metal than normal plants (Chaney et al., 1997). There are about 400 species that are hyperaccumulators and the levels accumulated are 10 000 mg/kg for Zn and Mn, 1000 mg/kg for Co, Cu, Ni, and As, and 100 mg/kg for Cd (McGrath and Zhao, 2002). Examples of hyperaccumulators are *Thlaspi caerulescens* and *Cardaminopsis halleri*, which accumulate zinc and cadmium. *Alyssum lesbiacum* accumulates nickel and the fern *Pteris vittata* accumulates arsenic (Table 5.8).

Unfortunately many hyperaccumulators only produce low biomass levels; for example *Thlaspi caerulescens* only produces 2–5 t/ha. However, there are plants that produce higher biomass levels; *Alyssum bertolonii* produces 9 t/ha and *Berkheya coddii* produces 22 t/ha.

High accumulators must be able to tolerate high levels of metal in their roots and shoots and this is possible by the concentration of the metal in the vacuole or by chelation of the metal. The plant must also be able to take up the metal from the soil at high rates and transfer the metal from the roots to the shoots at a high rate.

Hyperaccumulation involves adsorption, transport, and translocation to areas where large quantities of metal can be stored. One of the most studied

Table 5.8 Some plant hyperaccumulators

Plant	Metal	Concentration (mg/kg)
Dicotyledons		
Cystus ladanifer	Cd	309
	Co	2 667
	Cr	2 667
	Ni	4 164
	Zn	7 695
Thlaspi caerulescens	Cd	10 000–15 000
	Zn	10 000–15 000
Arabidopsis halleri	Cd	5 900–31 000
Alyssum sp.	Ni	4 200–24 400
Brassica junica	Pb	10 000–15 000
	Zn	2 600
Stanleya pinnate	Se	
Sedum alfredii	Zn	
Betula	Zn	528
Grasses		
Vetiveria zizaniodes	Zn	0.03
Paspalum notatum		
Stenotaphrum secundatum		
Pennisetum glaucum		
Ferns		
Pteris vittata	As	22 000
Pityrogramma calomelans	As	
Aquatic plants		
Erchhornia crassipes	Zn	2 008
Azolla pinnata	Zn	4 316
Lemna minor	Zn	3 698
Parrot feather	Cu	3 400
Myriophyum aquaticum	Zn	549
	Cu	3 184
Ludwiigina palustris	Zn	1 243
	Cu	848
Mentha aquatica	Zn	1 498
	Cu	314

mechanisms for metal sequestration is by the peptides metallothioneins and phytochelatins. The metal binds to the organic sulphur in cysteine, which makes up most of these peptides. It has been shown that metallothioneins and phytochelatins are stimulated by exposure to metals (Pawlik-Skowronska, 2001).

The molecular basis of hyperaccumulation in *Arabidopsis halleri* appears to be controlled by a single gene and zinc accumulation appears to be involved with increased uptake in the hyperaccumulator *T. caerulescens*. The transporters belong to the zinc-regulated transporter, ZIP, which is highly expressed in *T. caerulescens*.

Phytoremediation can be divided into a number of processes (Fig. 5.22).

- Phytoextraction (phytoaccumulation): the removal of contaminants and metals from the soil and their storage in the plant.
- Phytodegradation: the uptake and degradation of organic compounds.
- Phytovolatilization: the volatilization of pollutants into the atmosphere.
- Phytostabilization: the transformation of one species of molecule into a less-toxic species (Cr^{6+} to Cr^{3+}) or the reduction of mobility.
- The removal of pollutants from the atmosphere, gaseous contaminants.

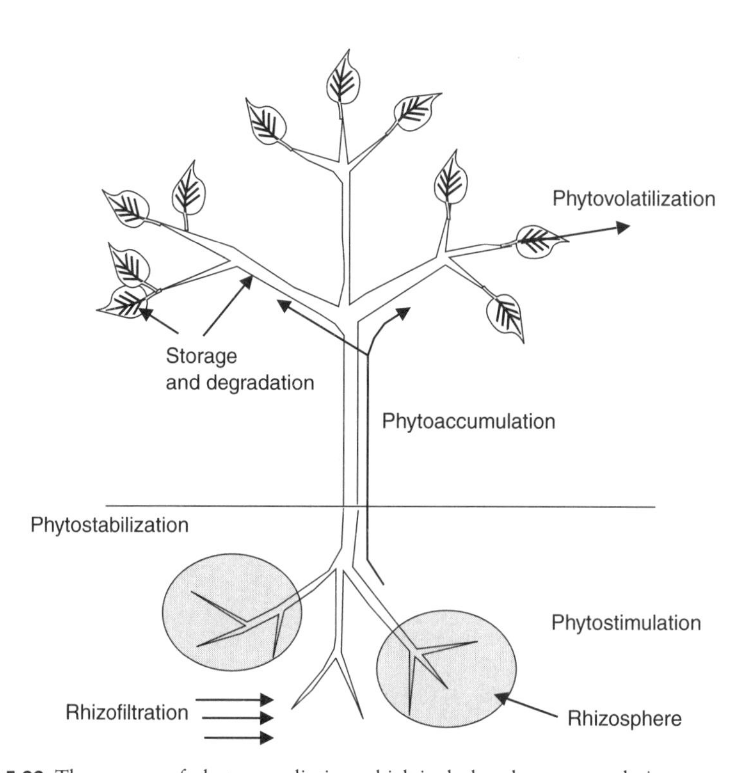

Figure 5.22 The process of phytoremediation which includes phytoaccumulation, phytodegradation, phytovolatilization, and phytostabilization.

There are a number of related processes which involve aquatic plants or associated micro-organisms. The roots of plants are associated with a number of micro-organisms, known as the rhizosphere.

- Rhizofiltration: the uptake of metals or degradation of organic compounds by the micro-organisms making up the rhizosphere.
- Rhizostimulation: the stimulation of plant growth by the rhizosphere by providing better growth conditions or a reduction in toxic compounds.
- Phycoremediation: the use of micro- and macroalgae to remove metals and organic pollutants.

5.8.1 **Phytoextraction**

Recently plant cell cultures have been shown to be capable of the degradation of nitroglycerine and PCBs (Mackova et al., 1997), which suggests that some plants could be isolated that are capable of degrading a number of environmental contaminants. In many respects plants are ideal for the remediation of contaminated soil, as they are a low-cost system, are easy to apply, require the minimum of maintenance, and have an excellent level of public acceptance.

Phytoextraction is the uptake of metals and organic pollutants by the roots of plants and their storage in roots, leaves, and stems. Plant roots constitute a very large area containing high-affinity chemical receptors. Examples of pollutants that can be removed are heavy metals, TNT, TCE, and BTEX. Sunflower roots can concentrate uranium 30 000-fold from contaminated water (Meagher, 2000), and a fern *Dicropteris dichotoma* has been shown to accumulate rare earths La, Ce, Pr, and Nd (Shan et al., 2003).

Herbaceous plants are suitable for phytoremediation as they grow rapidly, have a high biomass, and can stabilize soils and spoil tips. Four grasses, vetiver grass (*Vetiveria zizaniodes*), bahia grass (*Paspalum notatum*), St Augustine grass (*Stenotaphrum secundatum*), and bana grass (*Pennisetum glaucum* × *P. purpureum*) were used to decontaminate an open-cast mine (Xia, 2004). The uptake of cadmium and lead were followed over 6 month's growth and the data are shown in Fig. 5.23. It is clear that vetiver grass was the most efficient followed by bana grass and that the addition of fertilizer had only a small effect in this case. Grasses have been used to remove TNT from soil. The grasses Johnson grass and Canadian wild rye reduced the TNT levels from 10.2 mg/kg to less than 250 μg/kg within 100 days (Sung et al., 2003).

Aquatic plants have also been used to take up metals and three plants, parrot feather (*Myriophyum aquaticum*), creeping primrose (*Ludwiigina palustris*), and water mint (*Mentha aquatica*), were able to remove Fe, Zn, Cu, and Hg (Kamal et al., 2004). The removal rate for zinc was 0.45 mg/l/day with concentrations of zinc in the plants of 291 (*M. aquaticum*), 68 (*L. palustris*) and 209 mg/kg (*M. aquatica*) after a starting concentration of 3.56 mg/l. Water velvet (*Azolla pinnata*) and duckweed (*Lemna minor*) have also been

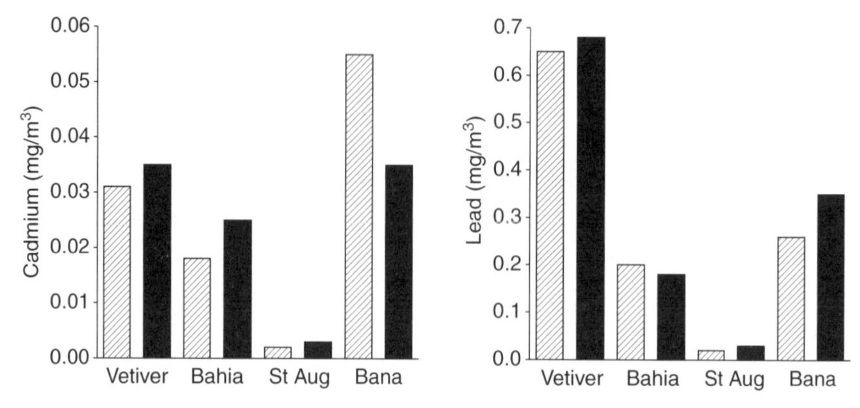

Figure 5.23 The phytoremediation of cadmium and lead with four grasses: vetiver grass, bahia grass, St Augustine grass, and bana grass. The amounts of metal taken up over 6 months' cultivation. Hatched bars, control; solid bars, plus fertilizer. Data from Xia (2004).

used to remove zinc from polluted water. Copper-polluted water has been tested against 12 different plants and water hyacinth (*Eichhornia crassipes*) and duckweed have also been used. Water hyacinth has also been used to treat tannery and dairy waste, and TNT has been shown to be taken up and degraded by the aquatic plant *Myriophyllum spicatum*. Plants have been shown to degrade a number of synthetic organic compounds although the mechanisms involved are still under investigation. Poplar trees have been used to remove trichloroethylene from contaminated water which was applied to their roots and other plants have been used to remove contaminating explosives from various sites (Boyajian and Carreira, 1997).

There has been considerable interest in using trees for phytoremediation as they are high-biomass producers and have been the subject of considerable plant breeding. Trees are a low-cost sustainable and ecologically viable solution for contaminated land. The availability of metals and their accumulation varies considerably with trees and many of the investigations have been carried with metal-containing sewage sludge. The physical and hydraulic conditions of the site are important in the cultivation of trees and sites such as mine spoils suffer from compaction, lack of nutrients, acidity, salinity, and rapid loss of water. Trees need the ability to grow on poor soil, to have a deep root system, fast rate of growth, and metal resistance (Pulford and Watson, 2003). The trees suggested were willow (*Salix*), birch (*Betula*), poplar (*Populus*), alder (*Alnus*), and sycamore (*Acer*). There are over 400 species of willow and many of these grow in wet lowland environments. Willow can be harvested frequently by coppicing, yielding 10–15 t/ha/year. The coppiced willow can be used in a number of ways including as fuel, in paper, in basket weaving, in ethanol production, and as charcoal. A number of studies have shown that willow will take up cadmium in sufficient quantities to clear moderately contaminated soil in a few years.

With the planting of a hyperaccumulating crop on a low-grade ore or mineralized soil the metals can be extracted. Phytomining using *Alyssum* spp. has been put into commercial operation. It has been found that inactivated alfalfa biomass is capable of accumulating appreciable quantities of gold which could be used to extract gold from mine tailing and low-grade ores. A pH value of 2–6 allows the biomass to convert Au^{3+} to Au nanoparticles.

5.8.2 Phytodegradation

Organic compounds can be degraded by plants or sequestered in the vacuole for degradation later. The best-known process is the glutathione-S-conjugate transfer system where this conjugates with organic pollutants. In general organic compounds can undergo a number of changes: partial transformation into a less-toxic compound, partial degradation and subsequent sequestration, and complete degradation. The plant degradation of herbicide and pesticides has been studied extensively and the metabolism of TCE, TNT, PAHs, PCBs, and other chlorinated compounds have been studied. Plants contain aliphatic dehalogenases capable of degrading TCE. Plants can degrade TNT, RDX, and nitroglycerine to carbon dioxide, ammonium, and nitrate as they contain nitroreductases, dehalogenaseses, and laccases. Oxygenation is a common process in pesticide and herbicide degradation which makes the molecules more water soluble and therefore more suitable for attack. Xenobiotics can be oxidized by cytochrome P450 and peroxidases. Other enzymes of potential use in phytoremediation are nitroreductase, dehalogenase, laccase, peroxidase, and nitrilase (Morikawa and Erkin, 2003).

5.8.3 Phytovolatilization

Some plants can convert metal ions to more volatile species in a process known as phytovolatilization, which can reduce toxicity and aid disposal through the stomata. TCE can be volatilized by poplar, methyl t-butyl ether (MTBE) by eucalyptus, selenium converted to dimethylselenide in Indian mustard, and methyl mercury converted to mercury vapour by tobacco. In addition, micro-organisms associated with plant roots can convert Hg^{2+} into volatile Hg.

5.8.4 Phytostabilization

Green plants have been used to stabilize soils, prevent dispersion of metal-contaminated soil, and reduce metal mobility by rhizosphere adsorption and precipitation. In addition, heavy metals cannot be degraded but they can be made more water soluble, less toxic, or insoluble so that they precipitate. Certain metals and organic contaminants can be concentrated on or in the root zone without degradation. For example the toxic Cr^{6+} can be converted by bacteria into the less toxic Cr^{3+} (Garbisu and Alkarta, 2001). *Lolium perenne* has been used to stabilize the soil from an iron-treatment plant and produced vegetative cover and no loss of metals (Arienzo et al., 2004).

5.8.5 Removal of pollutants from the atmosphere

Nitrogen dioxide is an atmospheric pollutant and is one component of the nitrogen oxides, collectively known as NO_x. The Compositae, Mytaceae, Solanaceae, and Salicaceae are the taxa that appear to take up nitrogen dioxide.

5.8.6 Rhizofiltration

Rhizofiltration is the removal of contaminants from flowing water by plant roots, which can be performed by the roots or the micro-organisms associated with the roots (rhizosphere), or the two combined. The contaminants removed can include organic compounds as well as metals. The use of constructed wetlands has been investigated for nutrient removal, pathogen reduction, and metal uptake and stabilization. Wetland plants have a large population of micro-organisms associated with their roots and it is this microbial population that is responsible for the sequestration of heavy metals and breakdown of the organic compounds.

The most common artificial wetland system uses the common reed *Phragmites* sp., which can grow in fresh or slightly brackish water (Moshiri, 1993). The reed bed is capped with clay and/or polypropylene in order to stop leakage into the subsoil and is surrounded by a wall (Fig. 5.24). The reeds, which can grow to 1.5 m in height, have the ability to pass oxygen from the leaves to the roots. The oxygen transfer encourages the development of a large aerobic microbial population on the plant roots (rhizosphere). As a result the roots act like a large microbial film reactor and it is this microbial population which sequesters metals and degrades organic materials.

Reed beds have been used to clean up mine leakage containing heavy metals, such as waste from the Wheal Jane mine in Cornwall, UK. ICI have installed reed beds to remove organic waste from a methylmethacrylate plant and they have been successfully used to reduce dairy waste with a BOD_5 value of 1006 mg/l down to 56 mg/l using a two-stage reed-bed system (Biddlestone et al., 1991).

Municipal wastewaters have also been treated using constructed wetlands. Figure 5.25 gives the construction of a three-stage system for the treatment of wastewater including a water-stabilization pond, an artificial wetland with *Typha latifolia*, and terrestrial section with *Salix atrocinerea* (willow; Ansola et al., 2003). The pilot-plant-scale system removes 60% of organics, 30% of nutrients, and 90% of fecal contamination. There are number of systems that can be used in the construction of artificial wetlands and these are illustrated in Fig. 5.26. Wetlands have been used in the Czech Republic as an alternative to conventional treatment (Vymazal, 2002). The size of these wetlands was

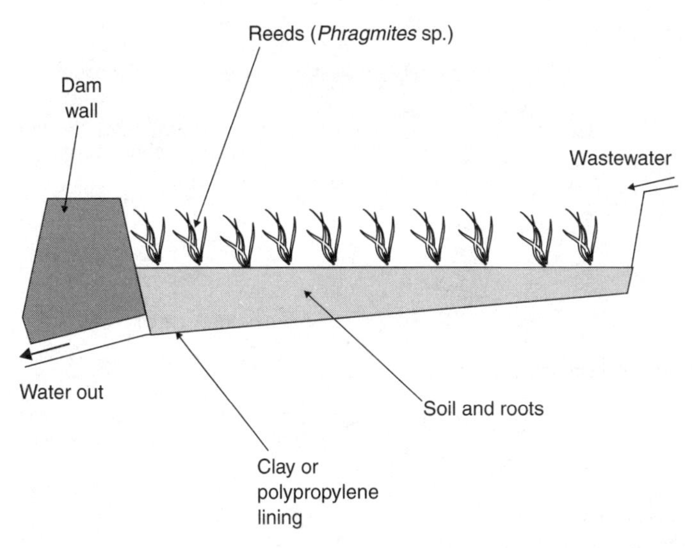

Figure 5.24 The construction of an artificial wetland (reed bed) used to treat contaminated and waste waters. The wetland has an impermeable base and the shallow soil is planted with the common reed (*Phragmites*). The waste is run through the wetland and the clean water removed from the bottom of the wetland.

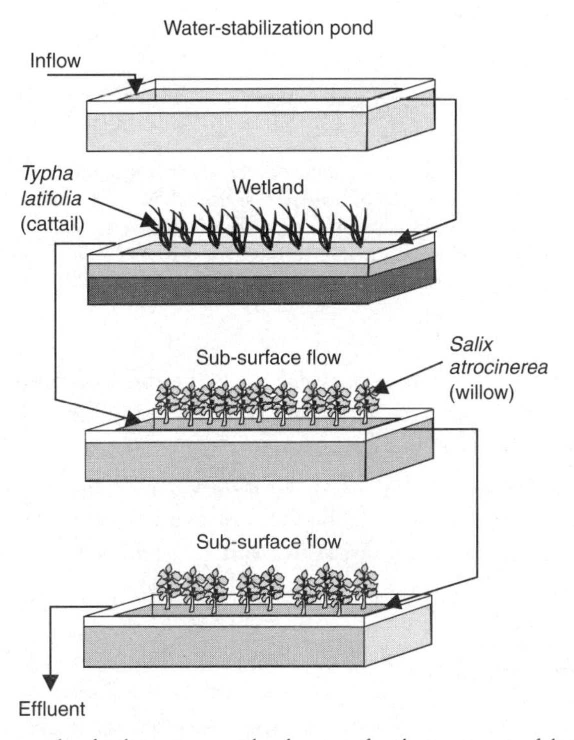

Figure 5.25 An example of a three-stage wetland system for the treatment of domestic wastewater. The first stage was planted with the cattail (*Typha latifolia*) and the next two with willow (*Salix atrocinerea*). The first wetland was operated as a surface flow and the other two a sub-surface flow through the root systems.

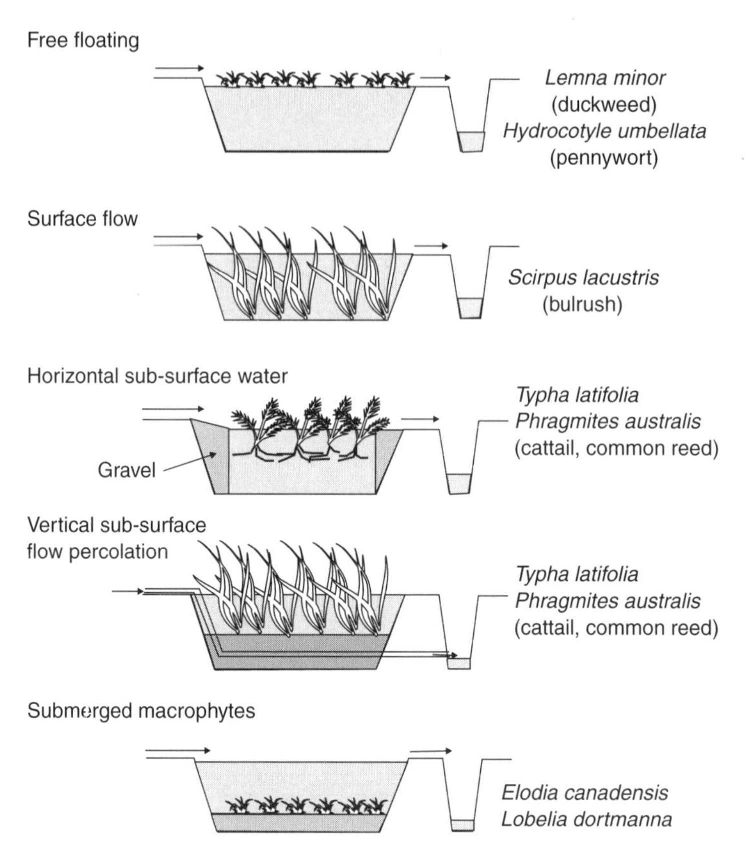

Figure 5.26 The variation in the use of plants to treat wastewaters. The first is free floating and the second surface flow as found in reed beds. Sub-surface flow involves the passage of the wastewater through the substratum in which the plants are growing. The type of flow can be altered by applying the liquid through a sub-surface pipe. The last option is fully submerged plants such as *Elodia canadensis*.

1800–4500 m^2 and *Phragmites australis* was the most commonly used plant. The treatment efficiency was high with an 88% reduction in BOD_5 and 84% for suspended solids. In tropical climates *Cyperus papyrus* (papyrus) and *Miscanthidium violaceum* have been compared and papyrus was more efficient at removing nitrogen and phosphorus from wastewater as it provided more microbial attachment sites, better trapping and settlement of particles, and a higher surface area (Kyambadde et al., 2004).

5.8.7 Rhizostimulation

Beneficial free-living soil bacteria are often referred to as growth-promoting rhizobacteria (Glick, 2003). The high level of micro-organisms associated with plant roots is due to high levels of nutrients that are exuded from the roots, which can be up to 20% of the photosynthetic yield. These bacteria can decrease or prevent the effects of pathogens, provide the plant with compounds that stimulate growth, and help with the uptake of nutrients. These bacteria can fix nitrogen, sequester iron, produce phytohormones, and solublize minerals such as phosphate. Thus they stimulate the growth of plants. Ethylene is important for normal plant growth but high levels inhibit root growth. Some bacteria contain the enzyme ACC (1-aminocyclopropane-1-carboxylic acid) synthase, which converts ACC to ethylene, carbon dioxide, and cyanide. Mycorrhizal fungi also play a role in the carbon flux between the plants, soil, and atmosphere.

Under experimental conditions Indian mustard and *Brassica campestris* hyperaccumulated nickel but under field conditions growth of these plants was inhibited by moderate levels of nickel.

5.8.8 Phycoremediation

Phycoremediation is the ultization of macro- and microalgae, which can be used for nutrient removal, bioabsorbants, acid and metal wastewaters, carbon dioxide sequestration, biodegradation of xenobiotics, and as part of biosensors. Microalgal wastewater-treatment systems are based on good mixing and a good supply of light. Various designs have been proposed to supply these requirements including shallow raceways, rectangular ponds, and closed plastic tubing. Another problem is the removal of the algae at the end of the process, as the cells do not normally aggregate. However, algae do frequently stick to surfaces and this form of immobilization avoids harvesting and mixing. Two microalgal isolates immobilized in such a manner have been shown to remove ammonium, nitrite, and orthophosphate from wastewater using a corrugated raceway bioreactor (Craggs et al., 1997). A number of other immobilization systems have been used for algae for nutrient removal (Oliguin, 2003). Microalgae such as *Chlorella vulgaris* have been shown to accumulate metals principally by adsorption to the cells surface (Kratchovil and Voleski, 1998) (see section 5.9.1). Tributyltin has been shown to be removed from wastewater by algal cells (*Chlorella minuta, Chlorella sorokiniana, Scenedesmus dimorphus,* and *Scenedesmus platydiscus*) and again by adsorption to the cell surface (Tam et al., 2002).

5.8.9 Applications of genetic engineering to phytoremediation

Phytochelatin synthase has been expressed in *Nicotiana glauca* R (shrub tobacco), increasing its tolerance to lead and cadmium (Gisbert et al., 2003). YCF1 is a gene that codes for an MgATP-activated glutathione-S-conjugate transporter which sequesters xenobiotics after conjugation with glutathione.

Overexpression of YCF1 in *Arabidopsis thaliana* and isolated vacuoles gave four times the uptake of cadmium and chromium.

An altered mercuric ion reductase gene (*mer*A) was introduced into *A. thaliana*, producing a plant which was mercury-resistant and which volatilized mercury into the atmosphere (Garbisu and Alkorta, 2001). Another study on mercury metabolism *A. thaliana* plants overexpressing (*mer*A) mercuric reductase and (*mer*B) organomercurial lyase detoxified organomercury converting it to volatile mercury (Bizily et al., 2000). Yellow poplar (*Liriodendron tulipifera*) tissue culture and plantlets have been transformed with the (*mer*A) gene and shown to grow in the presence methyl mercury and to release elemental mercury (Rugh et al., 1998).

Transgenic *Arabidopsis*, *Pittosporum tobira*, and *Raphiolepis umbellata* containing the *NiR* (nitrate reductase) gene from spinach have been produced. Overexpression of the gene appears to increase nitrogen dioxide assimilation (Morikawa and Erkin, 2003).

Tobacco plants have been produced containing the mammalian cytochrome P450 2E1, which has been shown to increase significantly the degradation of TCE and ethylene dibromide (EDB) (Doty et al., 2000). In another study tobacco plants expressing the enzyme pentaerythritol teranitrate reductase were capable of degrading glycerol trinitrate, an explosive chemical. The enzyme had been derived from an explosive-degrading bacterium (French et al., 1999).

5.9 Metals bioremediation

Chemical or biological processes cannot degrade metals which means that any process for the treatment of metal must concentrate the metal so that it can be contained or recycled. Metals like cadmium, zinc, copper, lead, and mercury contaminate many wastewaters. The following methods have been used to treat the aqueous extract or other metal-containing wastewaters from industrial processes and landfill leachates.

5.9.1 Biosorption

Biological material can adsorb a variety of metals. The response of microbial cells to high concentrations of metals can be one or more of the following, which can in some cases confer a degree of tolerance to that metal.

- Exclusion of the metal from the cell.
- Energy-dependent efflux of metals taken into the cell.
- Intracellular sequestration by specific proteins, some of which are known as metallothioneins.

- Extracellular sequestration on either cell-wall or extracellular polysaccharides.
- Chemical modification of the metal.

The internal and external sequestration of metals means that biological material can bind metals to high levels of up to 30% of the dry weight (Voleski and Holan, 1995). The use of biological material to remove metals from wastes can take two forms. The first is the detoxification of the wastewater stream, and the second is the recovery of valuable metals such as gold (Vilchez et al., 1997).

The uptake of metals from wastewater by living material can be active or passive, or both. Passive uptake is independent of cellular metabolism and involves the binding of metals to the polyanionic cell wall or by ion exchange with ions in the cell wall. Microbial extracellular polysaccharides are also known to bind metal. Passive uptake is rapid, reaching completion in 5–10 min, and is unaffected by metabolic inhibitors, but is affected by physical conditions such as pH and ionic strength. Passive binding is reversible and can occur with both living and dead material (Fig. 5.27). The active uptake of metals is slower than passive uptake, dependent upon the cellular metabolism, and is affected by metabolic inhibitors, uncouplers, and temperature. In active uptake the metals are complexed with specific proteins, such as metallothioneins, or contained in the vacuole. Both passive and active uptake can occur at the same time and the adsorption of metals is relatively non-specific in terms of the metal taken up.

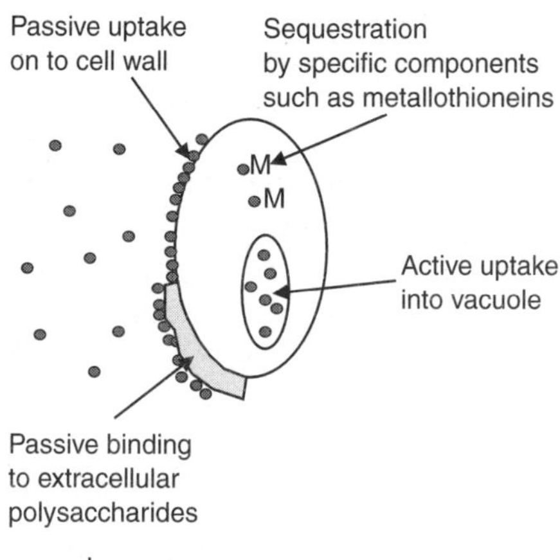

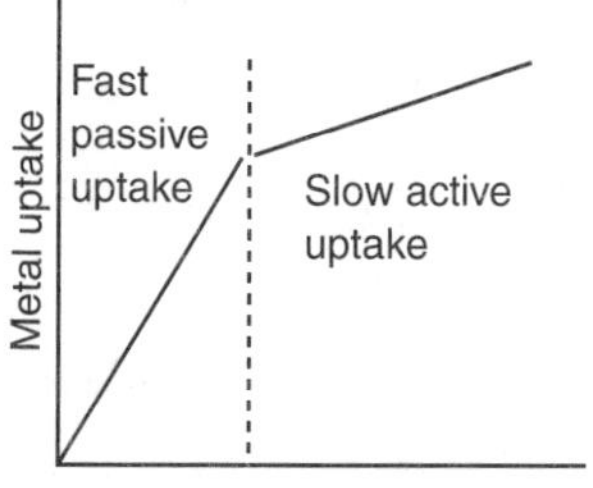

Figure 5.27 The uptake of metals by biological material, which can be both active and passive. The passive uptake is adsorption on to the cell surface, and any extracellular polysaccharides. The active uptake is slower where the metals are taken into the cells and sequestered by reaction with metallothioneins (M) or stored in the vacuole.

Biosorption has been evaluated as a method for the removal of metals from waste streams and in addition a small-scale treatment of mine drainage has been reported (Ledin and Pedersen, 1996). It has been shown that biosorption may be economically competitive with chemical techniques, particularly when the biomass used is inexpensive, such as waste biomass from the fermentation industry, excess sewage sludge, and easily harvested marine algae. The ability of marine macroalgae (seaweed), both living and dead material, to adsorb metals has been well documented (Voleski and Holan, 1995), and has been used in multiple cycles of adsorption and desorption. Immobilized non-living microalgae in a permeable matrix is available commercially as AlgaSORB (Vilchez et al., 1997). This material has properties similar to ion-exchange resins and has been used to remove cadmium. In another case a marine brown macroalga, *Ecklonia radiata*, has been used to bind copper. The response of organisms to high levels of metals is the production of metal-binding proteins such as metallothioneins and various attempts have been made to overexpress metallothioneins in bacteria in order to increase metal binding. In an attempt to increase the specificity of the metal binding of biological material peptides were expressed on the surface of bacteria, which had the ability to bind metals by forming a sphere around the metal ion. *E. coli* cells expressing one or two hexahistidine clusters on the surface were able to bind 11 times as much cadmium as non-engineered cells.

The process formats for the removal of heavy metals using biological materials can take the form of immobilized material in fixed-bed, fluidized-bed, and rotating-disc reactors as shown in Fig. 5.28. The process of regeneration of the biomass is also shown.

5.9.2 Extracellular precipitation

A number of industrial and mining effluents contain not only metals but also sulphates, and processes have been developed for their removal. In the presence of sulphate heavy metals can be removed by the action of anaerobic, sulphate-reducing bacteria such as *Desulfovibrio* and *Desulfotomaculum* strains. Under the anaerobic conditions the bacteria use simple carbon sources, such as lactic acid, to generate hydrogen sulphide from sulphate.

$$3SO_4^{2-} + 2\text{ lactic acid} \rightarrow 3H_2S + 6HCO_3^- \qquad (5.1)$$

The hydrogen sulphide reacts with any metals present, forming insoluble metal sulphides.

$$H_2S + Cu^{2+} \rightarrow CuS + 2H^+ \qquad (5.2)$$

The bicarbonate formed in the first reaction breaks down to carbon dioxide and water, increasing the pH and further encouraging the precipitation of sulphides. The production of excess hydrogen sulphide is a problem as this is

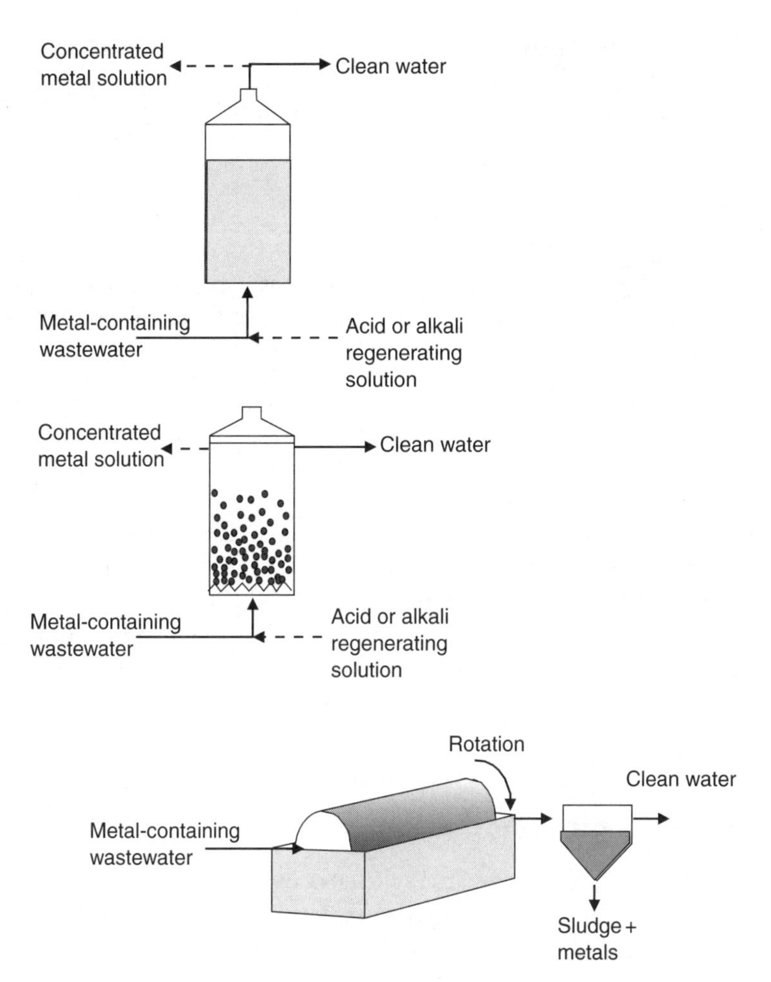

Figure 5.28 Some of the processes for the removal of heavy metals. (Top) A packed-bed bioreactor containing immobilized biomass, where the operation is continuous, with *in situ* regeneration using acid or alkali; (middle) a fluidized-bed reactor where the biomass is immobilized in or on an inert substrate and operated as the fixed-bed reactor. Removal of the metals uses the same process as for the fixed-bed reactor; (bottom) rotating-disc bioreactor where the biomass forms a film on the rotating discs and aeration occurs when the discs are out of the liquid. Any excess biomass sloughs off to be collected in a filter or settling tank for disposal.

poisonous and corrosive, but it can be burnt off or controlled by limiting the supply of organic carbon, although it is not always possible to balance the two. In some cases the excess hydrogen sulphide can be oxidized to sulphur by oxygen or by colourless, green, or purple sulphur bacteria (Kolmert et al., 1997).

The production of insoluble metal sulphides has been used in a number of bioreactor formats including an upflow sludge system as seen in Fig. 5.29 (Gadd and White, 1993), or with the biomass immobilized in a reactor filled with spent mushroom compost which acts as a support. A comparison of

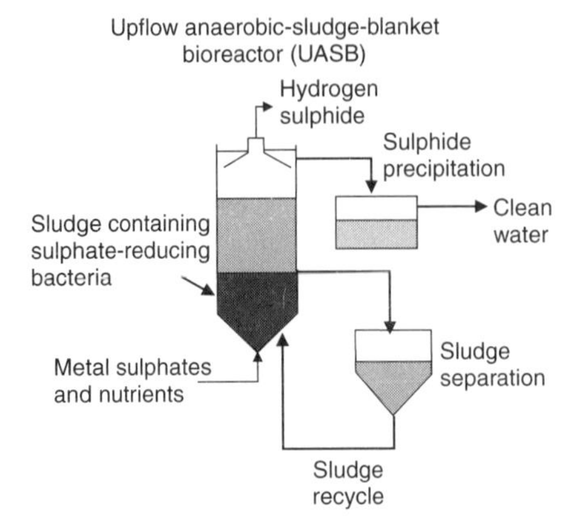

Figure 5.29 An upflow sludge-blanket bioreactor for the anaerobic precipitation of heavy metals such as sulphides. Hydrogen sulphide is produced along with the metal sulphides and the excess hydrogen sulphide can be vented off and burnt.

bioreactor formats suggested that a packed-bed system was better than suspended carrier systems (Kolmert et al., 1997). Another system has been developed using sand filters inoculated with bacteria which adsorb and precipitate metals (Diels et al., 2003). In this system the sand is retained in a moving bed and later the metal-loaded biomass is removed from the sand and treated.

5.10 Gaseous bioremediation

Gaseous contaminants released into the environment can be volatile organic compounds, sulphur dioxide, nitrous oxides, CFCs, and greenhouse gases such as carbon dioxide and methane. These contaminants can originate from a number of sources. Volatile organic compounds come from a number of industrial processes, and the treatment of contaminated sites. Sulphur dioxide and nitrous oxides are derived from the combustion of sulphur-containing oils and coal, and carbon dioxide from the combustion of fossil fuels.

A number of designs have been proposed for the removal of gaseous pollutants (Burgess et al., 2001; Schroeder, 2002). The principles are to pass the polluted gas through a vessel in which the pollutants can be transferred to a water medium where they can be degraded by micro-organisms. There are a number of bioreactor designs for the removal of gaseous pollutants and these are as follows: biofilters, trickling biofilters, bioscrubbers, membrane bioreactors, and activated sludge.

The simplest design is the bioreactor, where the gas is passed through a bed of porous material, which contains micro-organisms, immobilized in the bed. The bed needs to be humidified in order to keep the micro-organisms viable. The bed needs to have a uniform porous matrix that allows the gases to pass through while retaining the micro-organisms. The advantages of biofiltration are the simple flexible design, treatment of high volumes of gas, good removal (up to 99%), and low cost of operation. The disadvantages are that the designs are still developing, large areas and large bed volumes are required, and the solubility of the gas in water is a limiting parameter.

5.10.1 Biofilters

The simplest biofilter design is an open soil or bed bioreactor where the volatile organic compounds are piped into the base of the bed about 1 m down, and as they pass through the soil to the surface the organisms present degrade them (Fig. 5.30a). The soil can be replaced with compost and both have the advantages of containing a very mixed microbial population. A more sophisticated biofilter (Fig. 5.30b) is a bioreactor containing packing with a high surface area, which supports an active microbial biofilm. The support

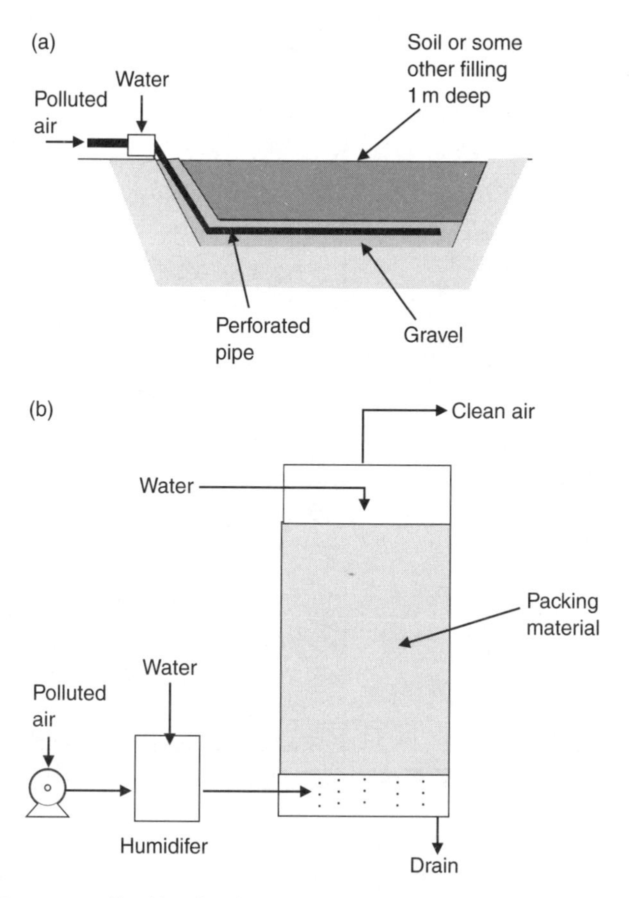

Figure 5.30 Two types of biofilter for the treatment of gaseous pollutants. (a) The simplest design is where the gas is piped into the bottom of a trench about 1 m deep and this is covered with soil or compost. As the gas percolates through the covering material the micro-organisms present degrade the pollutants. (b) In the second design the packing material is peat, compost, wood bark, or soil and this material is kept moist by the continual addition of water.

material can be simple, such as peat, wood bark, compost, leaves, and soil, or a more complex high-porosity plastic material. The biofilm is maintained by a continual supply of nutrients and high humidity (10–50%) is maintained by either humidifying the inlet air or by adding water and nutrient at the top of the vessel.

5.10.2 Trickling biofilters

In the trickling biofilter humidity is maintained by recirculating the medium through the reactor bed (Fig. 5.31). In some cases the top of the filter is closed and the gas can also be recycled through the filter bed. The advantages are that the recycled water can be supplemented with nutrients, maintaining the viability of the microbial population.

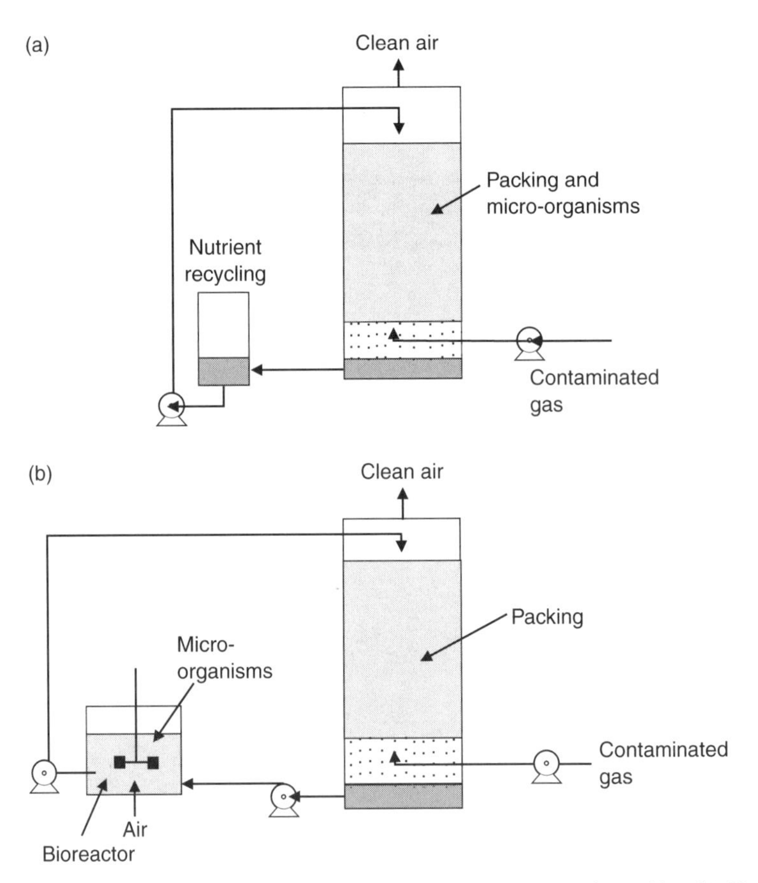

Figure 5.31 Two other systems for the removal of gaseous wastes are (a) the trickling biofilter and (b) the bioscrubber. (a) The packing material in the trickling biofilter can be peat, wood bark, compost, leaves, and soil. The micro-organisms in this packing are maintained by a supply of nutrients to the top of the vessel that is recycled. (b) In the bioscrubber the packing material is inert and just acts as a high surface area for the gas to be dissolved in the descending liquid. The liquid containing the pollutant is treated separately in a bioreactor containing the micro-organisms.

5.10.3 Bioscrubbers

In the bioscrubber design the contaminants in the gas are dissolved in the water trickling down the column and degraded in a separate bioreactor (Fig. 5.31). The use of a separate bioreactor in which the biodegradation is carried out has the advantage that a number of parameters can be controlled, such as temperature and pH. The bioreactor can contain a suspended microbial culture that can be well mixed and operated anaerobically.

5.10.4 Membrane bioreactors

In a membrane bioreactor the gaseous pollutant passes through a membrane into the liquid where it is degraded. The membrane can have a specified pore size that can be selective, and can be formed from silicone rubber or hydrophobic micropore polysulphone. Behind the membrane the microbial culture can be suspended or immobilized (Fig. 5.32).

5.10.5 Activated sludge

Activated-sludge tanks can be used to treat gaseous wastes by mixing the waste gas stream with the air supplied to the base of the tanks but adsorption is limited by the bubble size and residence time. The smaller the bubble the higher the area/volume ratio and the longer the time spent in the liquid the more gas can be transferred. Activated-sludge systems have been used to treat hydrogen

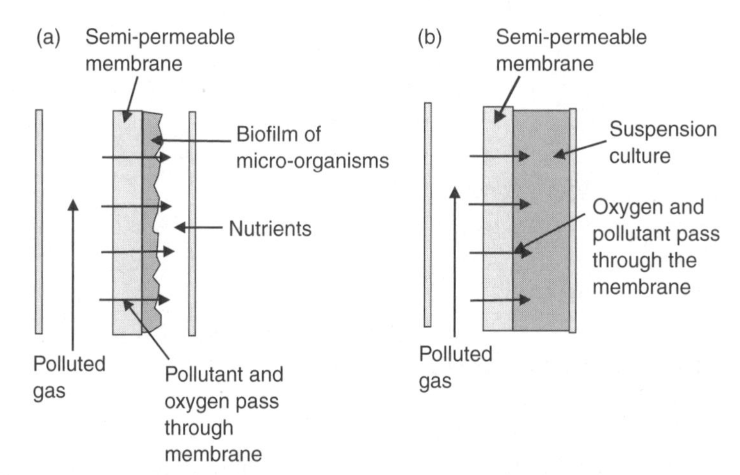

Figure 5.32 Membrane reactors for the removal of gaseous pollutants. (a) the semi-permeable membrane allows the gas to pass through into a biofilm of micro-organisms where the pollutants are degraded. The microbial biofilm is supplied with nutrients by a flow of liquid over the biofilm. In the second case (b) a cell-suspension culture is retained behind the membrane which will need a nutrient supply at intervals.

sulphide, which gave up to 99% removal (Burgess et al., 2001). Other problems are corrosion of the pipework and toxic effects on the microbial population, which would reduce activity.

These filters are beginning to be installed to treat waste gases from industrial processes and have been shown to be capable of removing volatiles such as BTEX, dichloromethane, styrene, toluene, hydrogen sulphide, ammonia, and odorous contaminants (Cox et al., 1997; McNevin and Barford, 2000).

The removal of ammonia and hydrogen sulphide using biofilters has been reported in a number of cases and although the removal rates vary greatly the removal efficiency is at least 99%. The hydrogen sulphide-removal biofilters employ *Thiobacillus* species, and heterotrophic organisms including *Xanthomonas* sp., *Hyphomicrobium* sp., and *P. putida*. Ammonia removal involves immobilized *Arthrobacter oxydans*.

A trickling biofilter containing *Burkholderia cepacia* G4 has been used to treat trichloroethylene and for inlet loading of 8.6 mg of TCE/l/day (Lee et al., 2003). A novel design consists of a two-phase partitioning bioreactor where toluene is trapped by the organic phase. The water phase contains a culture of *Alcaligenes xylosoxidans* which is capable of degrading toluene (Daugulis and Boudreau, 2003). The system was able to treat 748 mg/l/h at a conversion of 98%.

5.11 Biochemical pathways of biodegradation

The biodegradation of complex hydrocarbons, pesticides, herbicides, and xenobiotics generally requires the concerted effort of a number of enzymes and in many cases more than one micro-organism. Hydrocarbons are stable reduced compounds and therefore degradation generally proceeds by oxidation under either aerobic or anaerobic conditions. Microbial degradation of monocyclic and polycyclic aromatic hydrocarbons has been studied extensively (Cerniglia, 1993). Degradation of non-halogen xenobiotics is carried out by a range of enzymes to convert them into catechol or protocatechuate (Fig. 5.33).

The subsequent metabolism of catechol can take one of two pathways; ortho-cleavage yields *cis,cis*-muconate, whereas meta-cleavage yields 2-hydroxymuconic semialdehyde. Both pathways lead to the compounds pyruvate, acetaldehyde, succinate, and acetyl-CoA, which can enter the Krebs cycle (Fig. 5.34). In general aromatic ring hydroxylation is followed by ring cleavage, and both of these reactions are carried out by oxygenases (Fukuda, 1993). The incorporation of two oxygen molecules by dioxygenases introduces two hydroxyl groups that can undergo either meta- or ortho-cleavage. This process is found in bacteria and algae. Incorporation of a single oxygen molecule is catalysed by cytochrome P450 monooxygenases and is found in

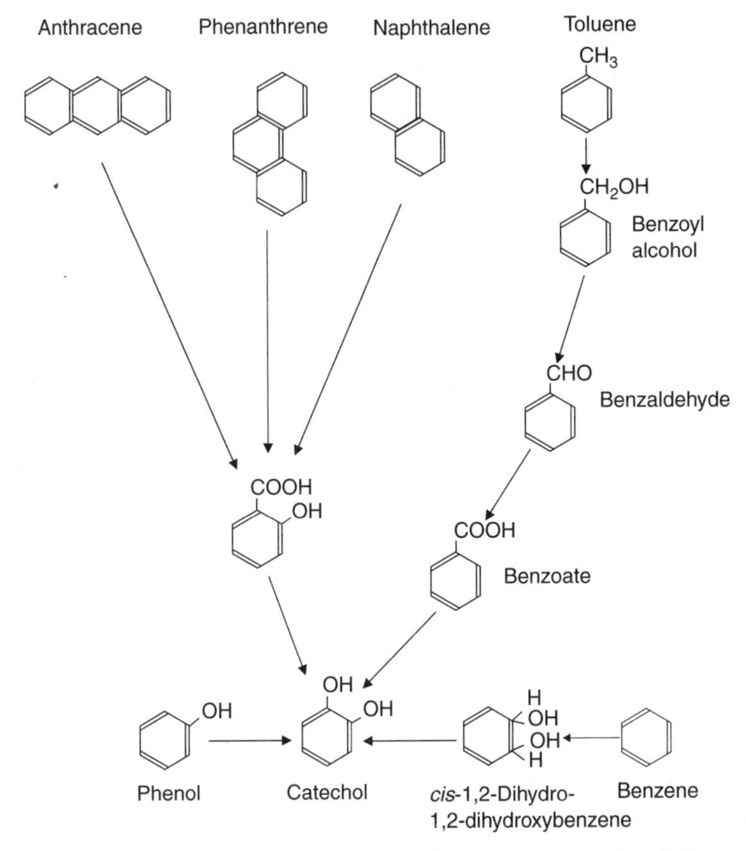

Figure 5.33 The breakdown of toluene and PAHs. The compounds are degraded to catechol.

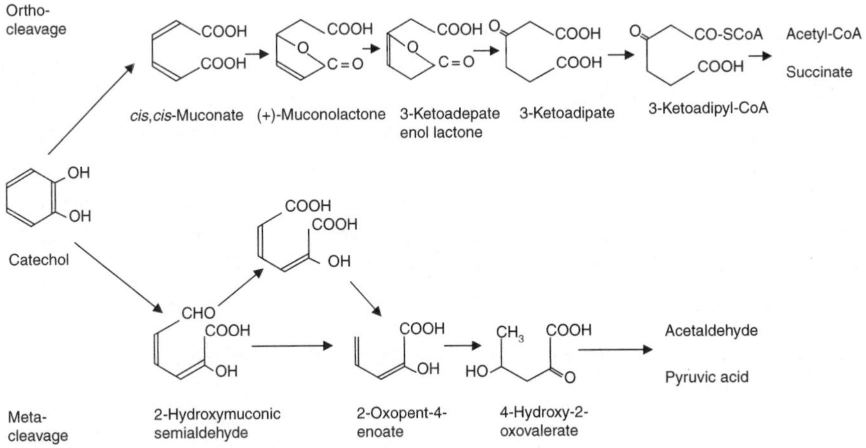

Figure 5.34 The pathway for the degradation of catechol showing both ortho- and meta-cleavage pathways.

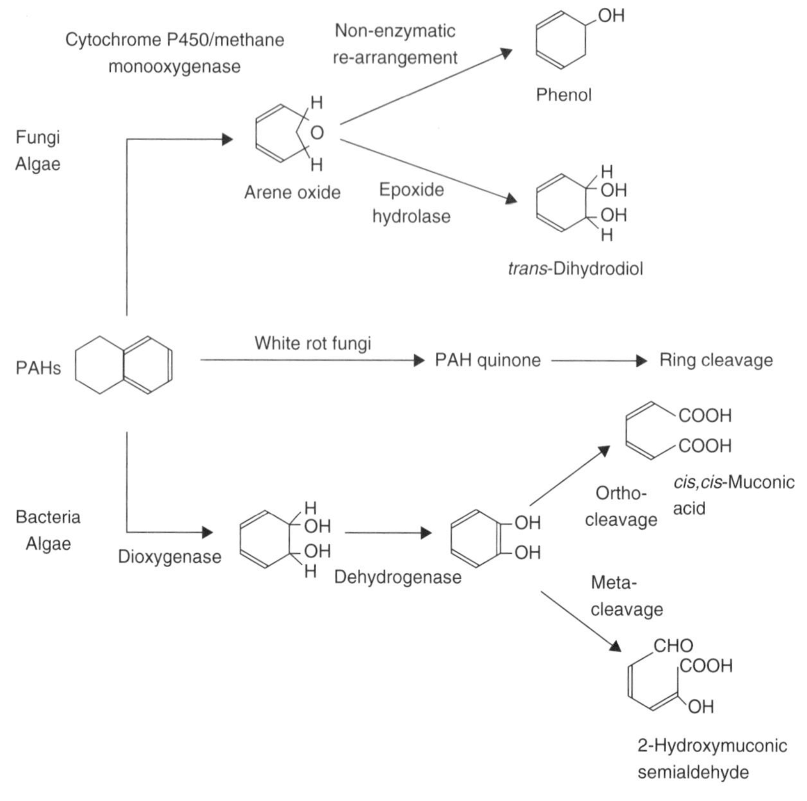

Figure 5.35 The initial steps in the degradation of polycyclic aromatic hydrocarbons by fungi, bacteria, and algae. Adapted from Cerniglia (1993).

fungi and algae (Fig. 5.35). The white rot fungi have a different pathway, forming quinones before ring cleavage.

Halogenated xenobiotics are the main ingredients of herbicides and pesticides. Organisms capable of degrading these xenobiotics have been found in soil and sediment, particularly from contaminated sites, and include bacteria, fungi, and algae. The degradation pathways for some of the chloroaromatics have been determined (Fukuda, 1993) under both aerobic and anaerobic conditions. The chloroaromatics are normally cleaved by monooxygenases and dioxygenases similar to those found with the degradation of PAHs. Superimposed on the degradation is the process of dehalogenation which can be by one of four mechanisms.

- Oxidative dehalogenation; here the halogen is removed and replaced by two hydroxyl ions.
- Eliminative dehalogenation; the simultaneous removal of the halogen and an adjacent hydrogen ion.

- Hydrolytic (substitutive) dehalogenation; the substitution of the halogen with a hydroxyl ion.
- Reductive dehalogenation; the halogen is replaced by a hydrogen ion.

The following are examples of the degradation of organochlorines illustrating individual pathways and general processes. Pentachlorophenol is a herbicide and fungicide used for the preservation of wood and is a priority pollutant (Table 3.2). Because of its toxicity its manufacture has all but ceased in Europe, although treated wood is still imported. A number of micro-organisms have been isolated which can degrade PCP under aerobic and anaerobic conditions and include *Flavobacterium*, *Arthrobacter*, *Rhodococcus*, and the white rot fungus *Phanerochaete chryososporium*. Most pathways for the breakdown of chlorophenols consist of the dechlorination and hydroxylation of the aromatic ring followed by ring cleavage. Both hydroxylation and ring cleavage are catalysed by oxygenases similar in nature to those found in PAH metabolism. The first step appears to be the rate-limiting step in PCP degradation where aerobic degradation starts with reductive dehalogenation.

Atrazine is the most widely used triazine herbicide and is effective against broad-leaf weeds. It had been employed for some 40 years and was considered to be recalcitrant. However, pure cultures of bacteria have been isolated which can degrade atrazine (De Sousa et al., 1998) although bacterial consortia have also been reported to degrade atrazine. Atrazine is converted to cyanuric acid in three steps and the cyanuric acid can be converted to CO_2 and NH_3. Cyanuric acid can also be metabolized by soil bacteria that cannot degrade atrazine. Three genes are involved in coding for the enzymes that convert atrazine to cyanuric acid and are located on a large plasmid. In *Pseudomonas* spp. these genes are found on a plasmid. The isolation of these three enzymes has allowed determination of the form of metabolism sharing that occurs in a bacterial consortium that can degrade atrazine. Figure 5.36 shows the contribution made by *Clavibacter* and *Pseudomonas* where *Clavibacter* are responsible for the first two steps and *Pseudomonas* for the remaining steps.

The bioremediation of hydrocarbon-contaminated soil cannot always be maintained under aerobic conditions due to waterlogging, the fine particle structure of the soil, and blocking of the soil pores with the biomass itself. However, aliphatic, monocyclic, and polycyclic aromatic hydrocarbons can be degraded anaerobically provided oxygen can be obtained from water under methanogenic conditions, from nitrate under nitrifying conditions, and sulphate under sulphur-reducing conditions. The hydrocarbons are converted to central metabolic intermediates by hydration, dehydration, reductive dehydroxylation, nitroreduction, and carboxylation. The central intermediates are benzoyl-CoA and sometimes resorcinol, which are reduced and hydrolysed and finally transformed to compounds which can enter the Krebs

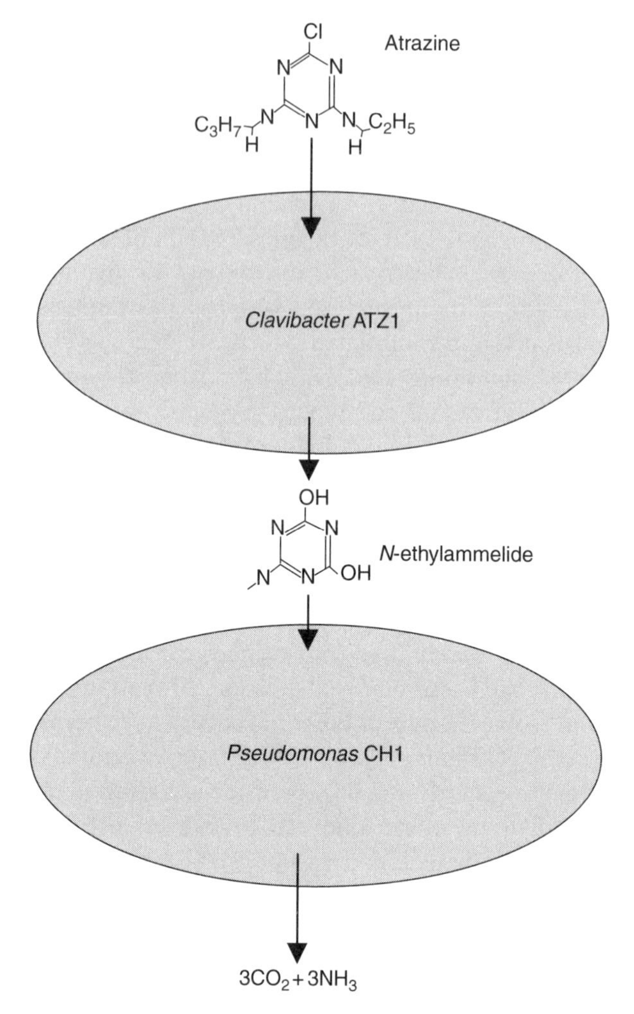

Figure 5.36 The degradation of atrazine by a consortium of bacteria, *Clavibacter* ATZ1 and *Pseudomonas* CN1. From De Souza et al. (1998).

cycle (Holliger and Zehnder, 1997). The only disadvantage with anaerobic degradation is that the process is much slower than the aerobic pathway.

5.12 Conclusions

It is clear that the use of bioremediation for the removal of pollutants from contaminated sites or from process wastes will continue to increase. Increasing legislation will help with the use of bioremediation as a viable alternative to chemical treatment. There are biological processes that can be used for all forms of contamination including soil, water, and gaseous. The processes offer *in situ* treatment or after-excavation treatment on or off site. Whatever process is used the main aim is to encourage the aerobic growth of the indigenous micro-organisms by providing air, nutrients, and the correct pH. The addition of specific micro-organisms tailored for the contaminant in a process known as bioaugmentation has given mixed results. One positive aspect of bioaugmentation is that it probably supplies the plasmids containing the relevant genes, which can be spread through the population even if the culture itself does not survive. Micro-organisms are not the only organism used for remediation. Plants are being used to degrade xenobiotics, sequester metals, and treat wastewaters. Plants offer a slow but cheap method of remediation that is well regarded by the public.

5.13 Further reading

Glazer, A.N. and Nikado, H. (1995) *Microbial Biotechnology*. Freeman and Co, New York.

Head, I.M., Singleton, I., and Milner, M. (2003) *Bioremediation: a Critical Review*. Horizon Scientific Press, Norfolk.

Marschner, H. (1995) *Mineral Nutrition of Higher Plants*. Academic Press, London.

Mason, C.F. (1996) *Biology of Freshwater Pollution*, 3rd edn. Longman Scientific & Technical, Harlow.

McEldowney, S., Hardman, D.J., and Waite, S. (1993) *Pollution: Ecology and Biotreatment*. Addison Wesley Longman, Harlow.

Moshiri, G.A. (1993) *Constructed Wetlands for Water Quality Improvement*. CRC Press, Boca Raton, FL.

6 Biotechnology and sustainable technology

6.1	Introduction	230
6.2	Provision of bulk and fine chemicals	231
6.3	Microbial polymers	245
6.4	Microbial plastics	248
6.5	Industrial processes and clean technology	253
6.6	Conclusions	264
6.7	Further reading	265

6.1 Introduction

It has been clear for some time that the exploitation of fossil fuels, metals, and other non-renewable resources cannot continue unchecked and that some replacement is required. Worldwide interest in sustainable industrialization or development was probably launched in 1987 with a report to the World Commission on Environment and Development (the Brundtland Commission). The report was entitled *Our Common Future*, where sustainable was defined as "Development that meets the needs of the present without compromising the ability of future generations to meet their needs" (World Commission on Environment and Development, 1987). The report was confirmed at the UN Earth Summit in Rio de Janeiro in 1992. The objective was to achieve industrial production and energy generation where environmental and economic systems are in balance. Another definition for sustainable development was "to prolong the productive use of our natural resources over time, while at the same time retaining the integrity of their bases, thereby enabling their continuity" (de Paula and Cavalcanti, 2000).

At a time when the demand for energy continues to increase the International Energy Agency (IEA, 2003) has predicted that the supply of crude oil will peak around 2014 and then decline and that coal will last until 2200 (Evans, 1999). The decline in available coal and crude oil should cause the prices of these fuels to rise, which should limit their use. Other factors in the use of fossil fuels are the environmental impacts of their use. Fossil fuels produce carbon dioxide when they are burnt and this is in part responsible for global warming. The production of carbon dioxide was of concern to a number of governments and at the Kyoto conference in 1997 reductions in carbon dioxide emissions were agreed. The Kyoto agreement was to reduce carbon dioxide emission to 12.5% below the 1990 levels by 2008–12 and a 20% reduction by 2020. The use of renewable energy was to be doubled from its present value of 2.5% to 5% by 2003 and should reach 10% by 2010. In 1990 the UK Government published its first White Paper on sustainability and in 1998 a consultation paper was produced which lead to a revised paper, *A Better Quality of Life*, in 1999. More recently the UK Department of Trade and Industry has produced a White Paper, *Our Energy Future – Creating a Low Carbon Economy* (Department of Trade and Industry, 2003), which indicates what is needed for a sustainable future.

Biotechnology can have a major impact on the development and implementation of sustainable technology and its influence will be mainly in the following areas.

- Provision of bulk and fine chemicals.
- Industrial processes and 'clean technology'.
- Energy supply.
- Waste treatment and bioremediation.
- Environmental monitoring.
- Agriculture.

6.2 Provision of bulk and fine chemicals

Prior to the development of the petrochemical industry many bulk and fine chemicals were extracted from plants. For example, soybeans were used to produce adhesives, linoleum, printing inks, and plastics before they were replaced by synthesis from petrochemicals. The development of the petrochemical industry made available a wide range of chemicals including natural gas (methane), aliphatic alkanes, and aromatic hydrocarbons (benzene, toluene). Fig. 6.1 shows some of the products made from the components of crude oil and natural gas. The supply of the components is finite and these will need to be replaced by renewable sources, as fossil fuels become limited. The best sources of these chemicals will be either plants or micro-organisms.

6.2.1 Plants as a source of chemicals

Plants use sunlight as a supply of energy, and fix carbon dioxide, and therefore are a sustainable source. Plants can be used to supply a very wide range of non-food materials and some examples are shown in Table 6.1. The use of

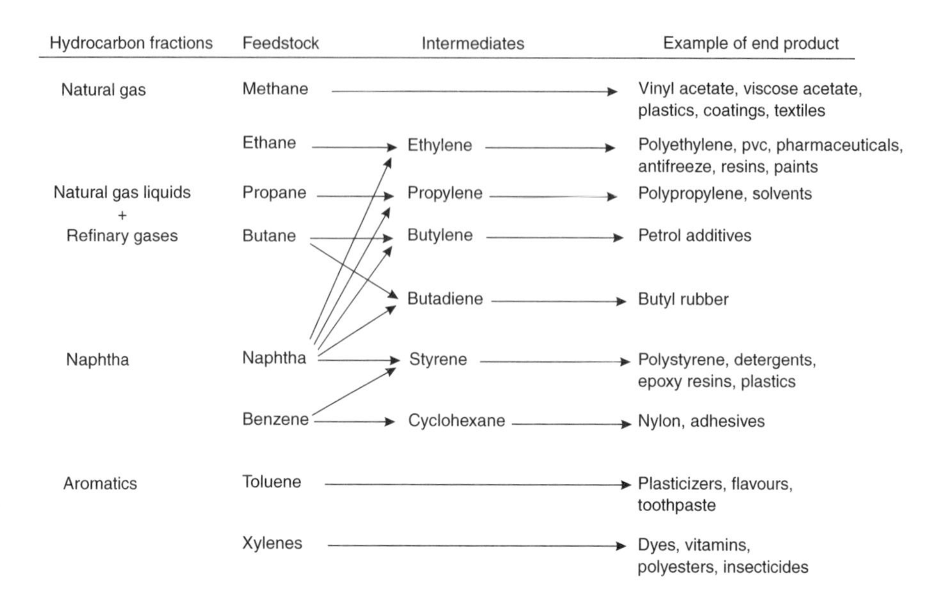

Figure 6.1 The feedstocks and end products produced from crude oil. PVC, polyvinyl chloride. Adapted from Evans (1999).

non-food crops, excluding timber, as a supply of commodity chemicals was 51 million tonnes in 1999 and is set to rise to 71 million tonnes.

The European Foundation produced a report on crops which should help towards the development of a sustainable environment (European Foundation for the Improvement of Living & Working Conditions, 2000). An example of crops are those that can supply fibre such as: *Agave*, cassava, *Crambe*, *Cupea*, elephant grass, fibre hemp, flax, bamboo, *Guar*, *Guayule*, *Jojoba*, *Knaf*, maize, meadow foam, oil palm, peas, *Plantago*, potato, *Pyrethrum*, rapeseed, safflower, soybean, Stokes aster, sugar beet, sunflower, *Vernonia*, and wheat. However, under European conditions the following are the most suitable.

- Flax or linseed (*Linum usitatissimum*). This crop can be grown for oil or as flax for fibre. The flax is harvested and retted by immersing in water, which partially rots the stems. This allows the long fibres to be extracted and used to produce linen, which has a number of uses. Linseed oil can be used in paints, varnishes, enamels, soap, and synthetic resins. The yield of oil is about 2 t/ha and that of fibre is 1 t/ha.
- Hemp (*Cannabis sativa*). Like flax, hemp is an annual and is treated like flax to release the fibre. The plant produces long fibres that can be used in textile mixtures and building materials. The yield is about 3 t/ha.

Table 6.1 Production of bulk and fine chemicals from non-food crops

Chemicals	Plants
Dyes	Woad, madder, safflower
Fuels	Rapeseed, sunflower, soybean, oil palm, *Miscanthus*, willow, poplar, maize, sugarcane
Industrial raw materials	Rapeseed, sunflower, soybean, maize, potato, *Sorghum*, manioc, sugarcane, sugarbeet
Lubricants, waxes	Rapeseed, linseed, rain daisy, honesty, meadow foam
Paints, coatings, and varnishes	Linseed, pot marigold, rain daisy, Stokes aster
Plastics and polymers	Rapeseed, linseed, castor bean, meadow foam
Resins and adhesives	Rain daisy, Stokes aster
Soaps, detergents, solvents, and emulsifiers	Rapeseed, hemp, *Cupea*, coriander, castor bean, poppy
Textiles, fibres	Flax, hemp, nettle, bamboo, grass, maize

Source: Data from Evans (1999) and European Foundation for the Improvement of Living & Working Conditions (2000).

- Bamboo (Bambusoideae). Bamboo is used extensively in Asia and species of bamboo have been established in many parts of the world. Bamboo can produce 10–15 t/ha but how it is going to be used in Europe is under investigation.
- Nettle (*Urtica dioica*). Nettle is a weed, but has an interesting fibre and cellulose content. The use of nettles is under investigation.
- Grass. Grass is used in agriculture directly, or as hay or silage, but it can be converted into protein, sugars, and fibre by a process known as 'biorefining'. These three products can be used to produce paper, alcohol, and a feedstock.
- Maize (corn, *Zea maize*). Maize is grown as food, animal feed, and a source of starch but the plants can also be used for their fibres (stems and husks).

In addition to textile materials plants contain sugars, starch, cellulose, and oils. All these can be used to replace chemicals or products from petrochemicals by conversion using chemical or biological methods (Fig. 6.2).

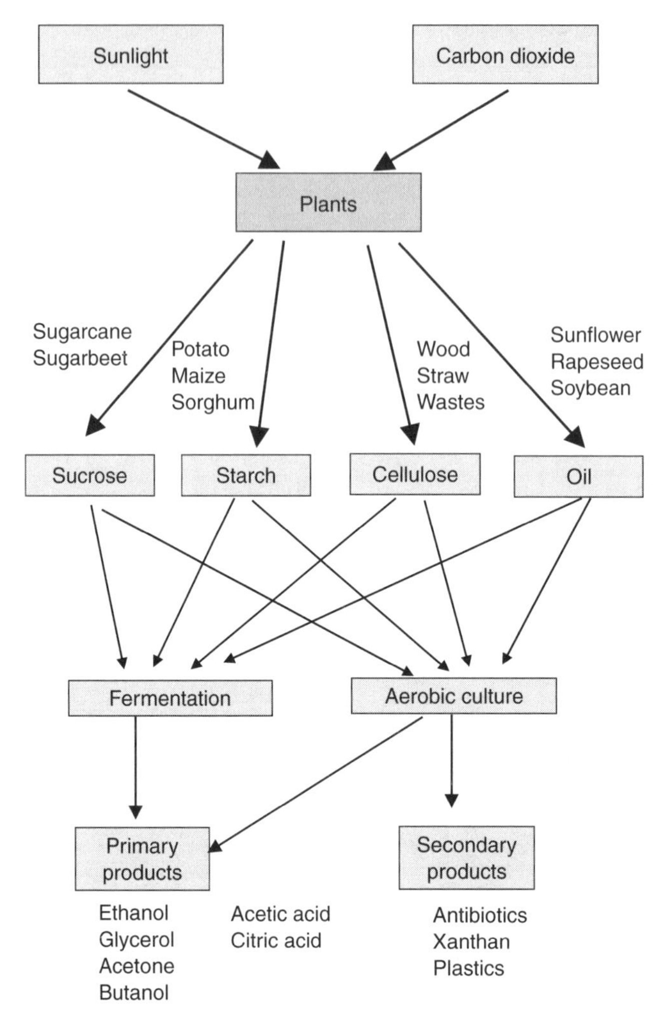

Figure 6.2 The range of feedstocks and chemicals that can be produced from plants. These are renewable as plants derive their energy from sunlight and carbon from atmospheric carbon dioxide.

6.2.2 Microbial production of chemicals

Starch, cellulose, and sugars can also be used a substrates by micro-organisms to form a wide range of chemicals (Fig. 6.2). Micro-organisms have the ability to produce large quantities of both simple and complex compounds, many of which are of commercial value. The compounds that micro-organisms produce are often divided into two types; **primary products** and **secondary products**. Primary products are those products that are essential for growth and division, whereas secondary products are only produced by specific organisms and are not essential for growth. Examples of primary products are RNA, DNA, amino acids, and vitamins, which are clearly essential, but often also included in this group are the simple metabolic end products such as ethanol, acetone, and citric acid (Box 6.1). Secondary products are linked to the primary products via central metabolism and in general they are formed after growth has ceased. Examples of secondary products are antibiotics, biopesticides, and more recently cloned products in both microbial cultures and plants. The chemicals obtained from micro-organisms by fermentation are shown in Table 6.2, which also gives their major uses and value. In the 1920–30s soybeans were used to produce adhesives, linoleum, printing inks, and plastic before they were replaced by petrochemical-derived products. At present in the USA 3.8 million tonnes of ethanol, 0.5 million

BOX 6.1 Primary and secondary microbial products

Primary products in the strictest sense are those components of the cell that are essential for life and reproduction whereas secondary products are more complex and not essential for growth. Primary products often include simple end products formed by fermentation using pyruvate generated by glycolysis. The figure below shows some of the primary products that can be produced from pyruvate. No one organism will produce all these products and the types of fermentation possible is discussed in Chapter 2.

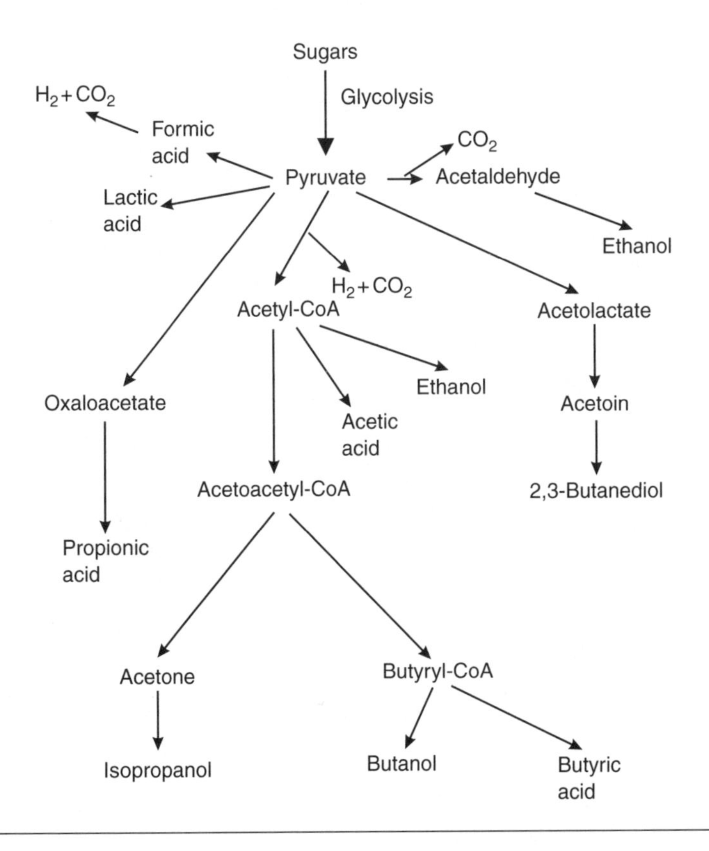

BOX 6.1 ***(Continued)***

The charactereristics of the products are outlined below.

Primary products:

- essential for growth, e.g. DNA, RNA, proteins, amino acids, and vitamins;
- simple end products of fermentation, e.g. ethanol, acetic acid, acetone, butanol, and lactic acid.

Secondary products:

- not essential for growth;
- formation dependent on environmental conditions;
- regulation of secondary products is different from primary products and they are often produced after growth has ceased, triggered by some form of nutrient depletion;
- products are only made by a few organisms and some produce a number of variations;
- the product can infer a competitive advantage to the producing organism but at times this advantage is not obvious;
- examples are antibiotics, medicinals such as paclitaxel, colours, and dyes.

The figure below shows some of the primary and secondary microbial products.

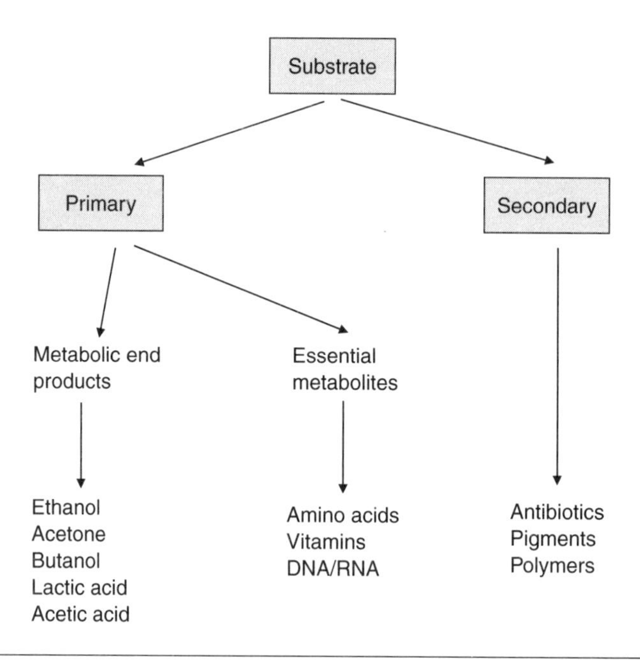

Table 6.2 Chemicals obtained from fermentation

Chemical	Micro-organism	Use
Acetic acid	*Acetobacter* spp.	Food acid, solvent and intermediate for many organic syntheses
Acetone	*Clostridium acetobutylicum*	Solvent, intermediate for many chemicals
Butanol	*Clostridium acetobutylicum*	Solvent, intermediate in chemical synthesis
Citric acid	*Aspergillus niger*	Food acidulant, electroplating
Ethanol	*Saccharomyces cerevisiae*	Petrol enhancer, solvent, used in the production of ethylene, butadiene, esters, and ethers
Fumaric acid	*Rhizopus oryzae*	Polyester resins
Glycerol	*Saccharomyces cerevisiae*	Solvent, plasticizer, soaps, antifreeze
Isopropanol	*Clostridium acetobutylicum*	Solvent, antifreeze, inks
Lactic acid	*Lactobacillus* spp.	Textiles, ester manufacture

tonnes of cellulose esters, 0.19 million tonnes of sorbitol, and 0.16 million tonnes of citric acid are produced per year as commodity chemicals from biological sources. However, this only represents less than 5% of the total compounds produced.

In the case of fine chemicals, plants and micro-organisms have retained their importance and some 25% of prescribed pharmaceuticals are extracted from plants, and most antibiotics are microbially produced. The micro-organisms are also grown on renewable sources such as sugars and starch, which can be extracted from plants.

Some of the products that can be produced by micro-organisms and have potential to replace chemically synthesized products include ethanol, glycerol, acetone and butanol, 1,3-propanediol, lactic acid, citric acid, acetic acid, sorbitol, and fumaric acid. Other products, which are of commercial use but are not primary products, include extracellular polysaccharides, plastics, and polyalkanoates.

6.2.3 Ethanol

Ethanol has been produced by yeasts fermenting sugars for at least 3000 years in the production of wines and beer. More recently ethanol has been produced as a fuel to supplement or replace petrol. More details of the production of ethanol are given in Chapter 7.

6.2.4 Glycerol

Glycerol is a simple alcohol and was observed as a by-product of ethanol fermentation by Pasteur in 1860. Glycerol at 78–85% is used in the manufacture of colours and cosmetics and 100% glycerol is used in antifreeze, ointments, and dynamite and in the oil industry. The chemical synthesis of glycerol was developed in 1872 and at present chemical synthesis starts with allyl chloride and during the process large quantities of chlorinated products are formed. However, there are biotechnological processes for the production of glycerol using a number of micro-organisms, including *Saccharomyces cerevisiae*, *Bacillus subtilis*, and the halotolerant green alga *Dunaliella tertiolecta*.

Glycerol is synthesized from fructose-1,6-diphosphate via dihydroxyacetone phosphate (see the pathway in Box 6.2.). At fructose-1,6-diphosphate glycolysis divides and the other branch leads to ethanol in fermentation. In a

BOX 6.2 Production of glycerol

Yeast (*Saccharomyces cerevisiae*) forms a small amount of glycerol during fermentation. During fermentation pyruvate is converted to acetaldehyde and then to ethanol (Box 6.1), the main product. However, if hydrogen sulphite (H_2SO_3) is added to the culture it binds to acetaldehyde, thus stopping its conversion to ethanol. This forces fructose-1,6-diphosphate down the other pathway to dihydroxyacetone phosphate and the NADH that is not used to convert acetaldehyde to ethanol converts this to glycerol (see below).

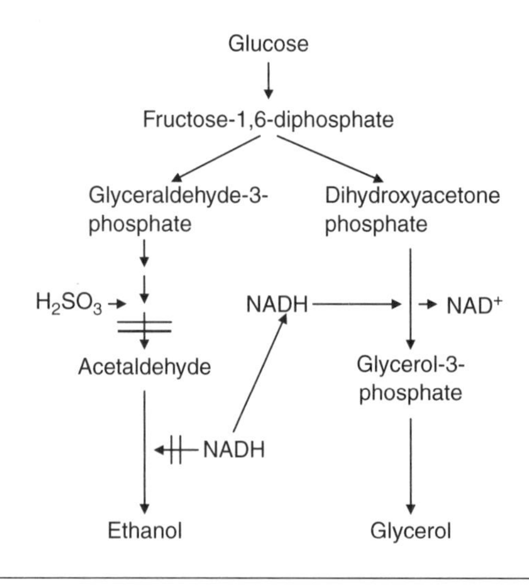

normal ethanolic fermentation the major proportion of the products will be ethanol but there will be a small amount of glycerol. The process for the production of glycerol aims to force glycolysis to make glycerol at the expense of ethanol. There are three strategies for increasing glycerol production, addition of sulphite, alkaline growing conditions between a pH of 7 and 8, and osmotic stress (Taherzadeh et al., 2002). The best-known strategy is the addition of sulphite, a technique that was used for the first time in World War I. The addition of hydrogen sulphite to the culture causes it to bind to acetaldehyde, forming a complex that cannot react with $NADH_2$, which stops the formation of ethanol. The spare $NADH_2$ is therefore free to react with dihydroxyacetone phosphate forming glycerol-3-phosphate and then glycerol. The final concentration of glycerol is about 3%, with 2% ethanol and 1% acetaldehyde. Glycerol recovery is difficult as it has a high boiling point and a number of methods have been described. It is the cost of recovery that is critical in the commercial development of a biological glycerol-production process. In 1980 most glycerol was produced chemically but by 1993 85% was derived from biological sources.

6.2.5 Acetone and butanol

The biological production of butanol was first observed in 1861 when Pasteur isolated a butyric acid-producing bacterium. Studies showed that butanol was also formed and the organism was not able to grow in the presence of air. Beijernick isolated two butanol-forming bacteria, *Granulobacter butylicus* and *Granulobacter saccharobutyricum*, in 1893 (Durre and Bahl, 1995).

Acetone and butanol are examples of primary products or end products of simple metabolic pathways (Fig. 6.3). The industrial production of these two solvents by fermentation has a long history that started in 1914. Acetone and butanol were some of the first biotechnological products and the process that developed was one of the largest. Before 1914 acetone was produced by heating (dry distillation) calcium acetate (Fig. 6.4). Calcium acetate was produced by the dry distillation or pyrolysis of wood. The wood distillate contained about 10% acetic acid that was either distilled off into calcium hydroxide to form calcium acetate or neutralized directly with lime. Between 80 and 100 tonnes of wood was required to produce 1 tonne of acetone. In 1910 Chaim Weizmann had been working in Manchester as part of a group working for Strange and Graham Ltd. trying to produce butanol by fermentation. Butanol was needed as it could be used to form butadiene, a precursor of synthetic rubber. At the time natural rubber was in short supply, as Brazil was the only source and did not allow the export of rubber trees. By 1914 Weizmann and coworkers had isolated an anaerobic organism which was later named as *Clostridium acetobutylicum* that produced both acetone and butanol when grown on starch. In 1914 at the start of World War I the demand for acetone increased rapidly as it was used

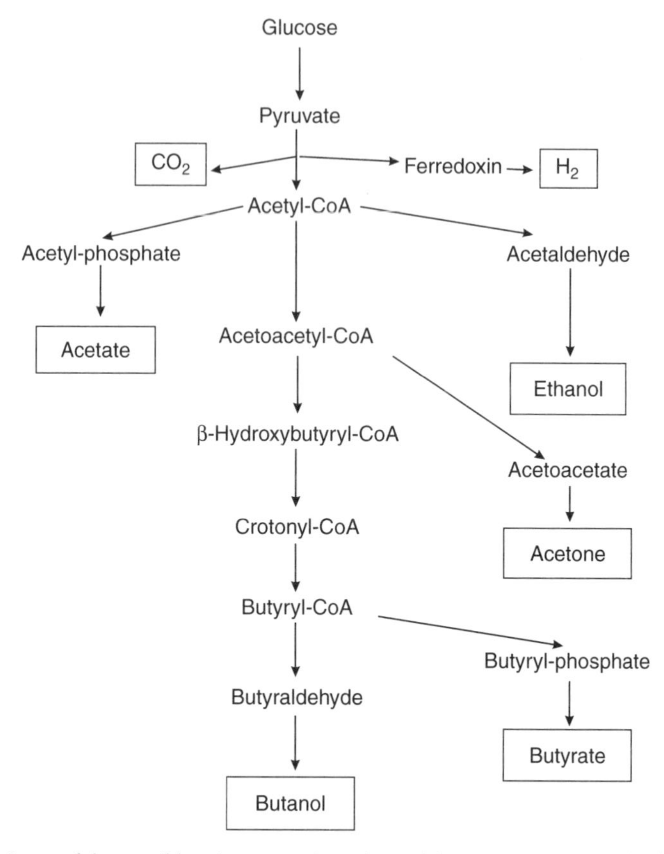

Figure 6.3 Some of the possible primary products formed from pyruvate. Not all of these pathways are carried out by one micro-organism but in *Clostridium* sp. butanol, acetone, ethanol, and butyrate can be formed.

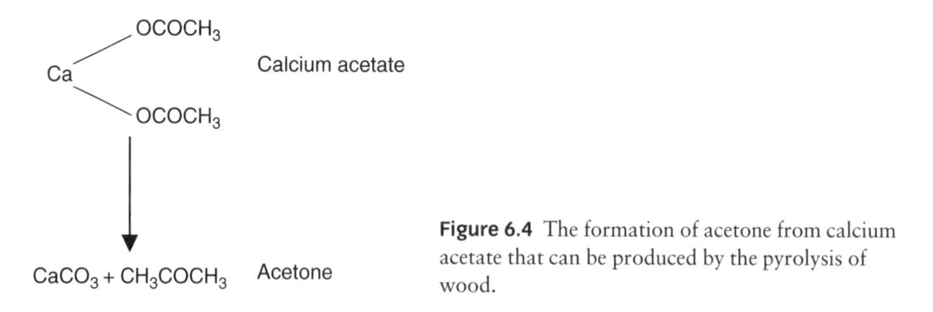

Figure 6.4 The formation of acetone from calcium acetate that can be produced by the pyrolysis of wood.

as a solvent for nitrocellulose in the manufacture of cordite, a smokeless explosive. By 1915 the demand had exceeded supply and the Nobel Company approached Weizmann and the process of biological production of acetone was adopted rapidly. Brewing capacity was commandeered and by 1916 the bioreactor capacity had reached 700 m^3. At the end of

World War I the demand for acetone declined but butanol was still in demand as a solvent for nitrocellulose paints used in the motor industry. Acetone was also being used as a solvent in the production of aircraft dopes and for the production of textiles and isoprene. Therefore the fermentation process continued until the 1950s.

Only certain clostridia are capable of producing reasonable levels of acetone and butanol and *C. acetobutylicum* has been the one most studied and used in industrial processes. *C. acetobutylicum* is a Gram-positive anaerobic spore-forming rod 0.6–0.9 μm wide and 2.4–4.7 μm long. It is motile and will ferment arabinose, galactinol, fructose, galactose, glucose, glycogen, lactose, maltose, mannose, salicin, starch, sucrose, trehalose, and xylose. The optimum growth temperature is 37°C. As the bacterium will form spores readily when the nutrients are exhausted it can be maintained easily as spores mixed with sterile soil. However, loss of solvent-forming potential is a common problem with *C. acetobutylicum* cultures. However, heat treatment restores solvent-forming ability. The spores are placed in boiling water for 60–90 s which removes weak spores and vegetative cells and the remaining culture maintains its potential to produce solvents. The concentration of substrate normally used was 6.0–6.5% and the maximum yield of solvent formed was 37% of the substrate used. However, in practice the yields are around 30% with a ratio of butanol/acetone/ethanol of 6:3:1. Thus 100 tonnes of substrate will yield about 12 tonnes of acetone. The yields depend upon a number of factors including the strain of micro-organism, temperature, pH, and substrate. In the 1930s a bacterium *Clostridium saccharobutylicum* was isolated which when grown on sucrose formed acetone and butanol only.

The progress of the fermentation to produce acetone and butanol can be seen in Fig. 6.5. During the exponential phase little solvent is produced but butyric and acetic acids are formed, causing the pH of the medium to drop from 6.0 to below 5.5. In the stationary phase the accumulation of acetone, butanol, and ethanol proceeds rapidly at the expense of the acids and therefore the pH rises. The pathway for the formation of butanol and acetone in clostridia is shown in Fig. 6.3 and shows the formation of acetic acid, ethanol, hydrogen, and carbon dioxide. In the metabolism of *C. acetobutylicum* sugars are converted to pyruvate with the formation of NADH and ATP via glycolysis. The pyruvate is oxidatively decarboxylated to acetyl-CoA by a pyruvate-ferredoxin oxidoreductase. The reducing equivalents generated are converted to hydrogen by an iron hydrogenase. Acetyl-CoA can be converted to a number of compounds including acetic acid, ethanol, acetone, and butanol. In the main pathway a thiolase converts acetyl-CoA to acetoacetyl-CoA, and subsequently to acetoacetate and then acetone. Acetoacetyl-CoA is also converted to β-hydroxybutyryl-CoA, then crotonyl-CoA, butyryl-CoA, butyraldehyde, and finally butanol.

In culture *C. acetobutylicum* can be in three states; acidogenic where acetic and butyric acids are formed at neutral pH, solventogenic where acetone,

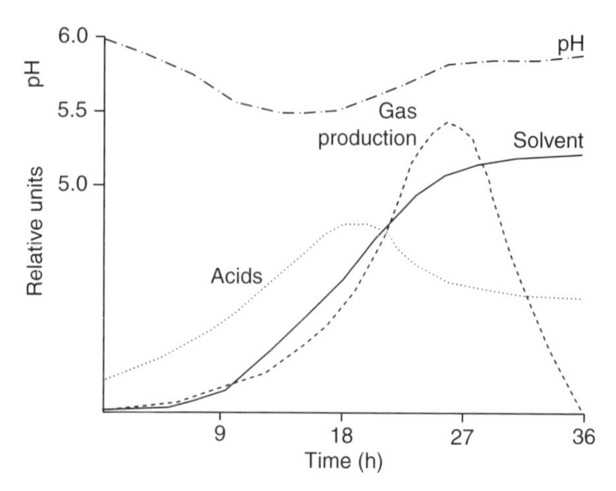

Figure 6.5 The time course of *Clostridium* sp. fermentation. Acid (butyric acid) formation occurs first followed by gas production; a mixture of carbon dioxide and hydrogen. The production of acid is reflected by a change in the medium pH. Solvent production—acetone, butanol, and ethanol—start at 9 h reaching a maximum at 27 h.

butanol, and ethanol are formed at low pH, and alcohologenic where butanol and ethanol are formed, but no acetone, at neutral pH. Thus it is important to monitor or maintain pH.

Although molasses-based fermentations were more economical than the original starch substrate the expansion of the petrochemical industry from 1945 onwards meant that by the 1960s the process had ceased to be used in the USA and UK. The process was used in South Africa until the 1980s and is still used in China today. The reasons for the decline of the acetone/butanol process were:

- low yield of solvents (30–35% of substrate);
- low solvent concentration in medium due to the toxicity of butanol and ethanol at 20–25 g/l;
- phage sensitivity;
- autolysin-induced autolysis in the stationary phase;
- cost of distillation;
- production of considerable amounts of waste;
- high cost of molasses; and
- petrochemical production was cheaper (Fig. 6.6)

However, in the last 10 years the process has been re-evaluated in light of the modern development of genetic manipulation and waste treatment, and the sudden increase in oil prices in 1973 (Durre et al., 1995; Durre, 1998). The reasons for the possible re-introduction are:

- the process uses renewable substrates;
- the newer strains can grow on waste starch and whey and metabolic engineering is being attempted so that it can be grown on cellulose;

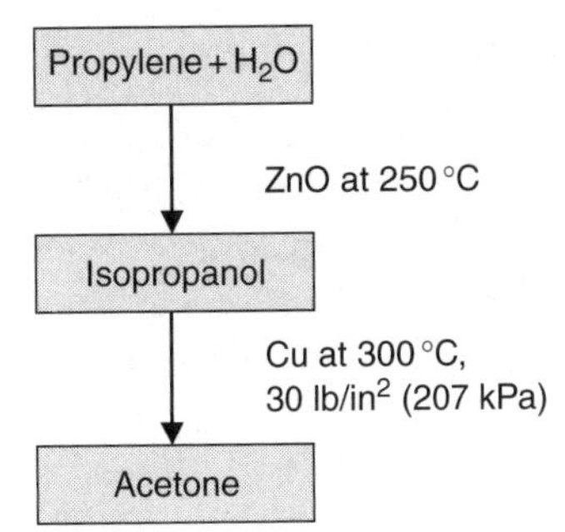

Figure 6.6 The chemical production of acetone from propylene.

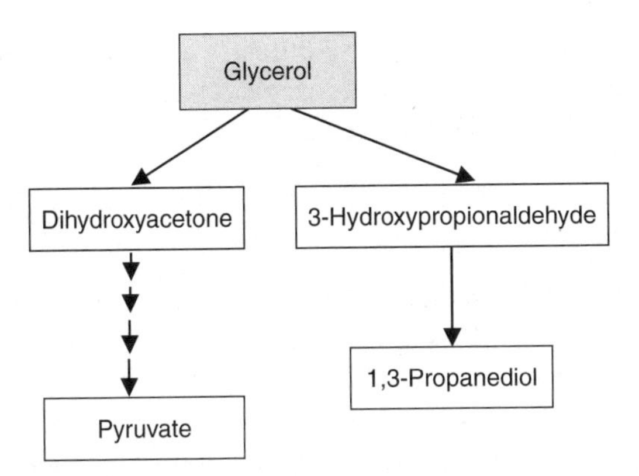

Figure 6.7 The pathway to 1,3-propanediol from glycerol under anaerobic conditions carried out by *Clostridium butyricum*.

- the waste can now be treated anaerobically, forming biogas;
- the process may be able to operate at 60°C so that the solvents can be removed as they are formed; and
- solvent may be recovered during fermentation using reverse osmosis, perstraction, pervaporation, membrane evaporation, liquid/liquid extraction, adsorption, and gas stripping, (Durre, 1998). Any process that avoids distillation will be considerably cheaper.

6.2.6 1,3-Propanediol

1,3-Propanediol can be used to produce polyesters and polyurethanes, as a solvent and lubricant, and is therefore of considerable commercial interest. Much of the compound is produced by chemical synthesis but it can be produced by micro-organisms. The biological production uses *Clostridium* spp. under anaerobic conditions to convert glycerol to 1,3-propanediol (Fig. 6.7). A pilot plant culture of 2000 l with a productivity of 2.8 g/l/h has been run using *Clostridium butyricum* (Gunzel et al., 1991). However, to be economically productive, product concentration and extraction have to be improved, although recently a higher productivity of 5.5 g/l/h has been reported (Papanikolaou et al., 2000).

6.2.7 Lactic acid

The production of lactic acid by micro-organisms has been known since Pasteur. Lactic acid is a product of fermentation (Box 6.1) and is used in the food industry, and in the manufacture of varnishes, lacquers, and plastics. Industrial production is carried out both chemically (oxidation of propene) and by biological processes; fermentation of sucrose, maltose, glucose, and lactose by *Lactobacillus delbrueckii*.

6.2.8 Citric acid

Citric acid is used mainly by the food industry and although some is used in industry no details are given here. Citric acid was originally extracted from citrus fruits but now over 99% is produced by cultures of *Aspergillus* species.

6.2.9 Acetic acid

The production of acetic acid from alcohol has been known as long as the production of wine. Acetic acid bacteria represented by two genera, *Gluconobacter* and *Acetobacter*, are found widely in nature as any person who has tried home brewing will attest. The acetic acid bacteria convert ethanol to acetic acid (Box 6.1), forming vinegar and as this is a food product no further details will be given.

6.2.10 Fumaric acid

Fumaric acid is used in the food industry and in the manufacture of polyester resins. Fumaric acid is mainly produced chemically from benzene but the micro-organisms *Rhizopus* spp. and *Candida* spp. can produce commercial levels of fumaric acid. Glucose or starch are used as the carbon source and the product is easy to recover as it poorly soluble in water.

6.2.11 Fine chemicals

6.2.11.1 *Acrylamide*

Acrylamide is used to form polymers used in oil recovery and as a flocculant in other products. The chemical production of acrylamide involves the hydration of acetonitrile using a copper catalyst. The process produces by-products, which are toxic, and excess acetonitrile also needs to be treated. There are a number of micro-organisms that contain the enzyme nitrile hydratase, which can convert acrylonitrile to acrylamide without the by-product formation. Acrylonitrile added to *Pseudomonas* spp. cultures is converted to acrylamide by up to 40% if the substrate is added in small quantities.

6.2.11.2 *Biosurfactants*

Surfactants are surface-active compounds that can alter the surface properties in processes such as emulsification, foaming, and wetting, and have wide uses in both industry and environmental remediation. Most surfactants in current use are made chemically but there are a very large number that can be produced by micro-organisms. The biological production of surfactants would have the advantage of being produced from renewable sources and the product would be biodegradable, of low toxicity, and effective in extreme conditions. Both bacteria and yeast produce biosurfactants, which are glycolipids where the sugar can be rhamnose, trehalose, sucrose, and glucose (Box 6.3). The organisms can be grown on sugars or hydrocarbons and the product formation can be both growth- and non-growth-associated. Biosurfactants could be used to replace all the surfactants currently in use and may have use in bioremediation and enhanced oil extraction (see Chapter 8). One of the problems in the bioremediation of petrochemical and chlorinated contaminants is the solubilization of these insoluble compounds. Biosurfactants can be added directly, produced by indigenous micro-organisms stimulated to grow and by augmentation with biosurfactant-producing organisms.

6.3 **Microbial polymers**

Many micro-organisms are known to produce extracellular polysaccharides as part of a capsule or slime layer, or free in the medium. The production of a capsule and slime layer is related to resistance to drying and pathogens, and attachment to surfaces and formation of biofilms. Polysaccharides are used extensively in industry as adhesives, gums, thickeners, gelling agents, stabilizers, and binding agents. The polysaccharides used are extracted from plants and large marine algae and include starch, alginate, carrageenan, and agar. The only microbial polysaccharide currently produced on a large scale is xanthan but other microbial products have the potential to replace the various polymers.

Research into microbial polysaccharides started in the 1940s with the development of dextran as a blood plasma extender. Several polysaccharides are now used widely in a number of industries, of which the best known is xanthan (Table 6.3).

Xanthan is a polymer produced by *Xanthomonas campestris* and consists of an alternating glucose backbone carrying side chains of D-mannose and D-glucuronic acid. *X. campestris* is a Gram-negative rod, a plant pathogen causing black rot in brassicas. Production at present exceeds 20 000 tonnes per year (Sutherland, 1998) and it is used extensively in the food industry. Solutions of xanthan are pseudoplastic, being able to regain their viscosity after shearing. The pseudoplastic property is of particular value in drilling

BOX 6.3 **Biosurfactants**

Surfactants are used in a very wide range of industries, which include the metal, paper, petroleum, pigment, textile, agricultural, building, food, cleaning, and leather industries. Biosurfactants could replace many of these. Biosurfactants are generally high-molecular-weight lipid complexes and can be classified as follows: glycolipids, hydroxylated and cross-linked fatty acids, polysaccharide–lipid complexes, lipoproteins, phospholipids, and cell surface molecules.

Both yeast and bacteria produce biosurfactants and some of the organisms that are known to produce them are given in the table below.

Organism	Type	Surfactant type
Torulopsis bombicola	Yeast	Glycolipid (sophorous)
Torulopsis petrophilum	Yeast	Glycolipid
Candida lipolytica	Yeast	Polysaccharide
Candida petrophilum	Yeast	Peptidolipid
Candida tropicalis	Yeast	Polysaccharide
Pseudomonas aeruginosa	Bacterium	Glycolipid (rhamnose)
Bacillus subtilis	Bacterium	Lipoprotein, surfactin
Rhodococcus erythropolis	Bacterium	Trehalose cross-linked fatty acid
Pseudomonas spp.	Bacterium	Glycolipid (rhamnose)*
Pseudomonas paraffineus	Bacterium	Glycolipid (sucrose, fructose)

* The structure of a rhamnolipid produced by *Pseudomonas* spp. is shown below.

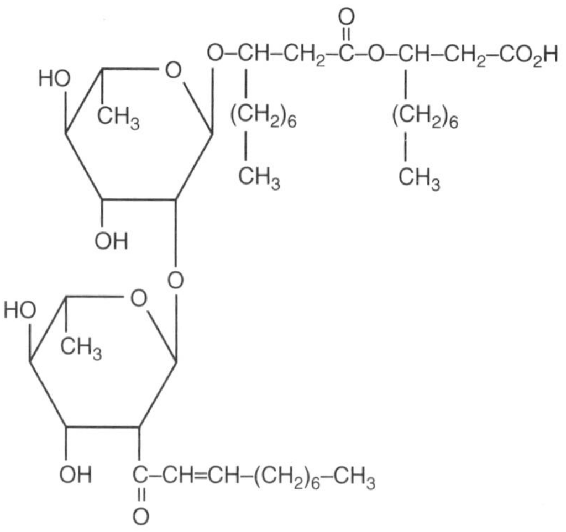

O–CH–CH$_2$–C(=O)–O–CH–CH$_2$–CO$_2$H
(CH$_2$)$_6$ (CH$_2$)$_6$
CH$_3$ CH$_3$
HO CH$_3$ O OH
O
HO CH$_3$ O OH
C(=O)–CH=CH–(CH$_2$)$_6$–CH$_3$

Table 6.3 Microbial polysaccharides

Product	Micro-organism	Molecular weight (Da)
Curdlan	*Agrobacterium* spp. *Rhizobium* spp.	7.4×10^4
Dextran	*Leuconostoc mesenteroides*	$(4-5) \times 10^7$
Gellan	*Spingomonas paucimoblis*	5×10^5
Pullulan	*Aureobasidium pullulans*	1×10^3–3×10^6
Schleroglucan	*Sclerotium rolfsii, Sclerotium glucanium*	1.3×10^5–6×10^6
Xanthan	*Xanthomonas campestris*	2×10^6–1.5×10^7

muds that act as a seal and lubricant during the drilling of oil wells. The mud needs to be viscous to form a good seal but flow when the drill bit is rotated.

Curdlan is a neutral gel-forming 1,3-β-D-glucan (74 000 Da) which can be formed by a number of bacteria such as *Agrobacterium* and *Rhizobium*. It forms a gel, which is elastic and does not melt when heated. The property of forming a gel upon acidification has also suggested its use in oil wells. Scheroglucan is a glucose homopolymer produced by a number of micro-organisms including *Sclerotium glucanium*. Scleroglucan is produced commercially and has been developed as an alternative to xanthan for enhanced oil recovery (EOR) by Elf Aquitane (McNeil and Harvey, 1993). However, the yields are lower than of xanthan, the process takes longer with a lower final concentration. Dextrans are different from the other microbial polysaccharides in that they are produced outside the cell. The substrate sucrose is converted by a extracellular enzyme dextransucrase into the branched α-D-glucan dextran.

The processes for the production of microbial polymers are similar to those for the production of antibiotics apart from the viscosity of the culture, which means that the impeller design has to change. In large stirred-tank bioreactors viscous pseudoplastic materials mean that the culture thins when sheared and thus the viscosity will be lowest near the impeller. This leads to the division of the bioreactor into well-mixed and aerated areas near the impeller and the remainder being poorly aerated and mixed. Poor mixing and low aeration will reduce growth and yield. A number of modifications to the impeller in the normal stirred-tank bioreactor have been proposed in a number of cases and one, the helical ribbon-screw design, did give a significant improvement (McNeil and Harvey, 1993). Microbial polysaccharides have not replaced plant or algal polysaccharides but have established a market of their own, as they possess a number of advantages over conventional rivals.

6.4 Microbial plastics

The world annual production of polymeric material, plastics, was around 150 million tones in 1996 and has shown no reduction (Ren, 2003). Twenty percent of plastics are used for packaging and this is dominated by expanded polystyrene made from naphtha, a fraction of non-renewable crude oil. Although plastics are being used increasingly because of their durability, ease of moulding, and resistance to biodegradation it is this last property that is causing concern. Plastic wastes in rivers, in lakes, and on land do not degrade and threaten the environment, accumulating at a rate of 25 million tonnes per year (Lee, 1996a). Much of the plastic waste that is collected is disposed of in landfill sites—20% in the USA—which is another problem due to its high volume/weight ratio and resistance to degradation. The possibilities of recycling plastics are limited and incineration yields toxic compounds. To regulate the disposal of plastics at sea the Maritime Pollution Treaty (MARPOL) was enacted in 1994 and in the US the Plastic and Pollution Research and Control Act was introduced in 1994. Despite this legislation the demand for biodegradable plastic has not increased.

Biodegradable plastic would reduce pressure on landfill sites and littering and also contribute to a more sustainable society as they would be produced from renewable resources. Degradable plastics can be biodegradable or photodegradable. Photodegrable plastics will break down into smaller fragments and therefore lose their structure, but the smaller fragments are not normally degradable. In contrast biodegradable plastic will be metabolized by microorganisms. A compromise is semidegradable plastics which contain starch, cellulose, and polyethylene, but to achieve complete degradation a 50% mix is required, which compromises the structural properties of the plastic.

However, development is under way of a number of biodegradable plastics including polyalkanoates (PHAs), polyactides, aliphatic polyesters, polysaccharides, and blends of these. One of the most promising is the PHAs, one of the best known of which is polyhydroxybutyrate (PHB). PHB is an intracellular microbial plastic, which is produced by a number of bacteria and was first discovered in 1926 as a component of *Bacillus megaterium* (Poirier et al., 1995). The PHAs have the general structural formula shown in Fig. 6.8. There are over 80 different types of PHA but they are mainly formed from 3-hydroxyalkanoate acid monomers of 3–14 carbons in length and the polymer has a molecular weight of between 2×10^5 and 3×10^6 Da, containing 100–3000 monomers depending on the growth conditions and microorganism. There are over 90 genera of bacteria (300 species) that have been shown to produce PHA (Table 6.4) and it appears to be produced as an energy store under conditions of limited nutrients. When the limitation is lifted the PHA is broken down. PHA appears in the microbial cells as granules of 0.2–0.5 μm in diameter, which are refractile.

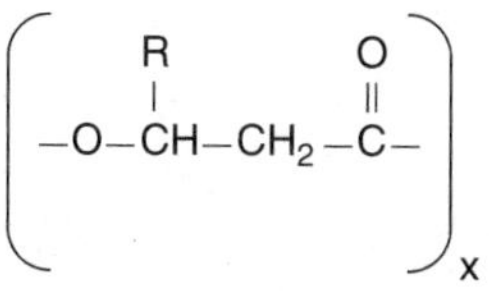

$$\left(-O-\overset{\overset{\displaystyle R}{|}}{CH}-CH_2-\overset{\overset{\displaystyle O}{||}}{C}- \right)_x$$

R	Name	Initial
Hydrogen	3-Hydroxypropionate	3HP
Methyl	3-Hydroxybutyrate	3HB
Ethyl	3-Hydroxyvalerate	3HV
Propyl	3-Hydroxycaproate	3HC
Butyl	3-Hydroxyheptanoate	3HH
Pentyl	3-Hydroxyoctanoate	3HO
Hexyl	3-Hydroxynonanoate	3HN
Heptyl	3-Hydroxydecanoate	3HD
Octyl	3-Hydroxyundecanoate	3HUD
Nonyl	3-Hydroxydodecanoate	3HDD

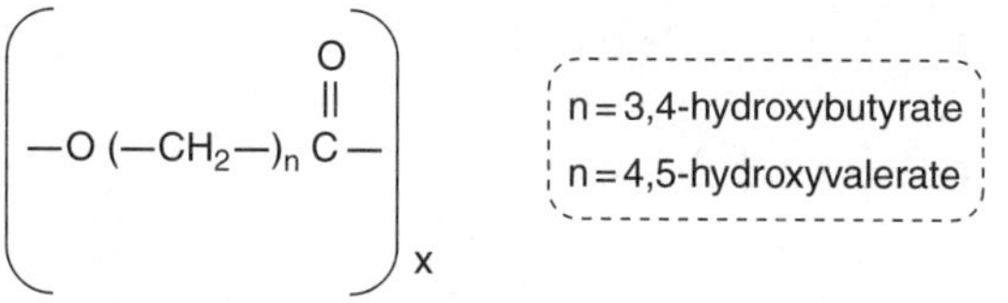

$$\left(-O(-CH_2-)_n\overset{\overset{\displaystyle O}{||}}{C}- \right)_x$$

n = 3,4-hydroxybutyrate
n = 4,5-hydroxyvalerate

Figure 6.8 The general structural formula for polyalkanoates (PHAs).

The commercial value of a biodegradable plastic was clear and ICI used *Alcaligenes eutrophus* (now known as *Ralstonia eutropha*) to produce PHB (poly(3-hydroxybuyrate)). However, the application of PHB was limited as the polymer had a low thermal stability and was brittle. Later a copolymer, P(3HB-co-3HV) or poly(3-hydroxybutyrate-co-3-hydroxyvalerate), was produced by adding propionate to the culture and the polymer was more flexible and tougher. The properties of various polymers are given in Table 6.5. P(3HB-co-3HV) is produced by Zeneca (formerly ICI) and marketed under the name of Biopol^{TM}, and has been used to make films, coated paper, compost bags, disposable food ware, bottles, and razors. Since then Zeneca and Monsanto have concentrated on P(3HB-co-3HV) but despite the good thermal properties the cost is still higher than the chemically synthesized polymers. Some of the companies developing PHA products are Bioscience Ltd. (Finland), BioVentures Alberta Inc. (Canada), Metabolix Inc. (USA), and Polyferm. Inc.(Canada) but Zeneca is the only company producing PHAs on a commercial scale. Polypropylene costs less than 1$/kg whereas PHB or poly hydroxy valerate butyrate (PHVB) cost 3–5$/kg so that considerable effort has been made to reduce the price of biodegradable polymers.

The biosynthesis of PHB is the best-studied pathway although there have been four different pathways found to date (Lee, 1996b). PHB in *R. eutropha* is formed by three enzymatic reactions, starting from acetyl-CoA that can be supplied by the Krebs cycle (Fig. 6.9). In the first stage two molecules of acetyl-CoA are condensed to acetoacetyl-CoA by the enzyme 3-ketothiolase. In

Table 6.4 Production of poly(3-hydroxybutyrate) by various micro-organisms

Organism	Carbon source	PHB content (%)	Productivity (g/l/h)
*Ralstonia eutropha**	Glucose	76	2.42
*Ralstonia eutropha**	CO_2	68	1.55
*Ralstonia eutropha**	Tabioca hydrolyate	58	1.03
Alcaligenes latus	Sucrose	50	3.97
Azotobacter vinelandii	Glucose	80	0.68
Azotobactes chroococcum	Starch	74	0.01
Haloferax mediterrenei	Starch	60	–
Methylobacterium organophilum	Methanol	52	1.86
Methylobacterium sp. ZP24	Whey	60	0.12
Protomonas extorquens	Methanol	64	0.88
Pseudomonas cepacia	Lactose	56	0.02
Pseudomonas cepacia	Xylose	60	0.03
Recombinant *E. coli*	Glucose	80	2.08
Recombinant *Klebsiella aerogenes*	Molasses	65	0.75

Source: Data from Lee (1996a) and Kim (2000).

*New name for *Alcaligenes eutrophus*.

the second stage acetoacetyl-CoA is reduced by an NADPH-dependent acetoacetyl-CoA reductase to 3-hydroxybutryl-CoA. This is then polymerized by the enzyme PHB polymerase. The other pathway differs only in minor ways.

PHB is broken down by a PHB depolymerase to 3-hydroxybutyrate which is in turn converted into acetoacetate by 3-hydroxybutryl-CoA dehydrogenase. Finally the acetoacetate is converted to acetoacetyl-CoA by acetoacetyl-CoA synthetase (Fig. 6.10).

A large number of bacteria and fungi have been shown to degrade PHB and a 1-mm section of P(3HB-co-3HV) was completely degraded in 6, 75, and 350 weeks in anaerobic sewage, soil, and sea water respectively (Lee, 1996b).

In most bacteria PHA is synthesized and accumulated inside the cells under conditions of growth where some nutrient is limiting, for example nitrogen, phosphate, oxygen, and magnesium. Therefore, to maximize PHA production the culture strategy has to provide these conditions while having as high a cell density as possible.

Table 6.5 Physical and thermal properties of various polyhydroxyalkanoates

Polymer	Melting temperature (°C)	Tensile strength (MPa)	Elongation to breakage (%)
Poly(3-hydroxybutyrate)	175–179	40	6
Poly(3-hydroxybutyrate-co-3-hydroxyvalerate) 20%	145	32	–
Poly(3-hydroxybutyrate-co-4-hydroxybutyrate) 10%	159	24	242
Poly(4-hydroxybutyrate)	53	104	1 000
Poly(3-hydroxyhexanoate-co-3-hydroxyoctanoate)	61	10	300
Polypropylene	170–176	34–38	400
Polystyrene	110	50	–

Source: Data from Poirier et al. (1995), Lee (1996a), and Lee (1996b).

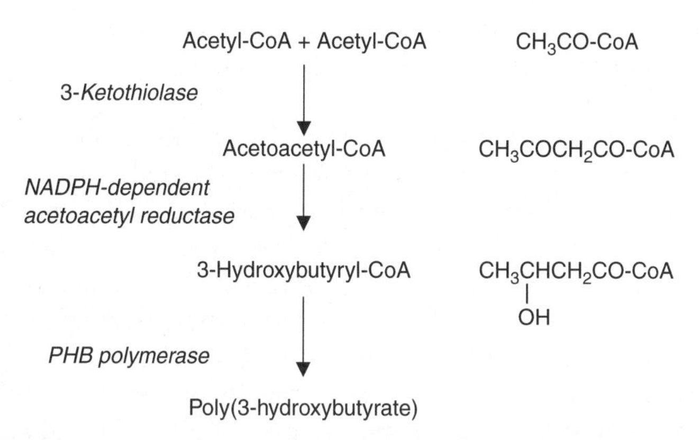

Figure 6.9 The pathway for poly(3-hydroxybutyrate) synthesis from acetyl-CoA found in *Ralstonia eutropha*.

6.4.1 Growth strategy

The bacteria that can accumulate PHA can be divided into two groups; those that require some form of nutrient depletion to trigger accumulation and others that can accumulate PHA during growth. The first group is represented by *R. eutropha*, *Protomonas extorquens*, and *Pseudomonas oleovorans* and the second by *Alcaligenes latus* and *Azotobacter vinelandii*. In the cultivation of *R. eutropha* a glucose-salts medium is used containing a limited amount of phosphate which runs out after 60 h and the eventual PHA concentration

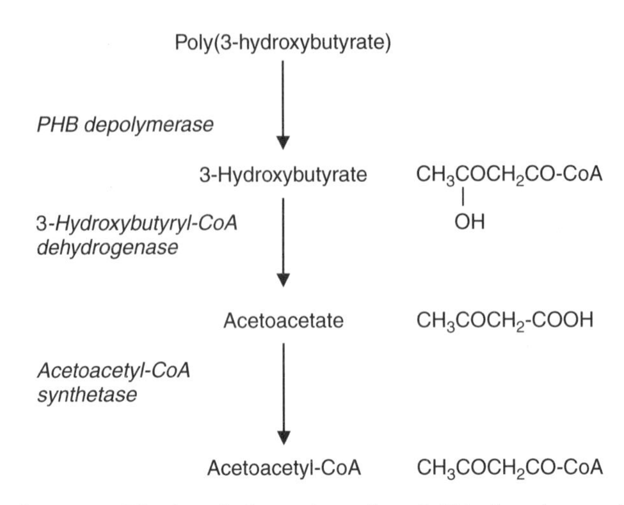

Figure 6.10 The degradation pathway for poly(3-hydroxybutyrate).

reaches 45–80% dry weight. If propionic acid is fed during the accumulation phase P(3HB-co-3HV) is produced. In this group a fed-batch process can also be used where the cells are grown to a high cell concentration in the first stage and in the second stage nutrient limitation is applied. An example of the second group, *Al. latus*, can be grown on glucose in fed-batch and continuous cultures.

Despite the high yields of PHA the cost of the product still requires reducing and a number of strategies have been adopted to reduce the cost. One of the main costs of production is the substrate: a number of carbon sources have been proposed including sugars, oils, alcohols, acids, starch, whey, organic wastewater, and waste from food manufacture. Genetic manipulation of *E. coli* has been used to produce PHA by transferring the genes from *R. eutropha* (Madison and Huisman, 1999). Accumulation of PHA reached 80–90% of the cell dry weight but when a cheap substrate, molasses, was used the yield was 45% and so clearly the cultivation strategy needs optimization.

Another approach was to engineer plants to produce PHA, as crop plants are capable of producing large quantities of material at low cost. Initial research used the genes from *R. eutropha* in the plant *Arabidopsis thaliana*, the '*E. coli*' of plant genetic engineering (Poirier et al., 1995; Slater et al., 1999). In the second generation of engineered plants the genes were targeted to the plastid as the levels of acetyl-CoA were much higher than in the cytoplasm. Oilseed crops were regarded as the best since both oil and PHA are synthesized from acetyl-CoA. PHA has been produced in *Brassica napus*, cotton, and maize but problems of phenotype changes, low product yield, and transgene stability need to be solved before the production is economical (Poirier et al., 1995; Snell and Peoples, 2002). The problem of the growth of transgenic plants containing an antibiotic-selection gene has been solved by the development of a non-antibiotic selection method.

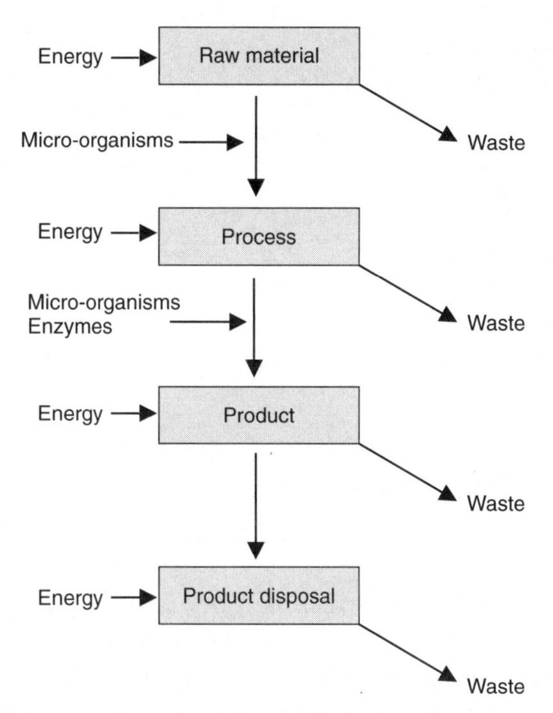

Figure 6.11 The basic stages of an industrial process indicating where enzymes and micro-organisms can be used to replace chemical methods.

6.5 Industrial processes and clean technology

An industrial process can be divided into a number of basic stages (Fig. 6.11):

1. extraction and supply of raw materials;
2. processing of the raw materials, and production of product;
3. use of the product; and
4. disposal of the product.

The objectives of a sustainable industrial process are as follows:

- low consumption of energy;
- low consumption of non-renewable raw materials; and
- reduction or elimination of waste, including energy and material recycling.

At all stages energy may be required and waste produced. The objectives of sustainability in the process would be the use of renewable raw materials and also use of energy from renewable resources. In addition, changes in the process by applying alternative biotechnological technologies also seek to reduce or eliminate waste. This approach is known as **clean technology**. The Organization for Economic Cooperation and Development (OECD) defines

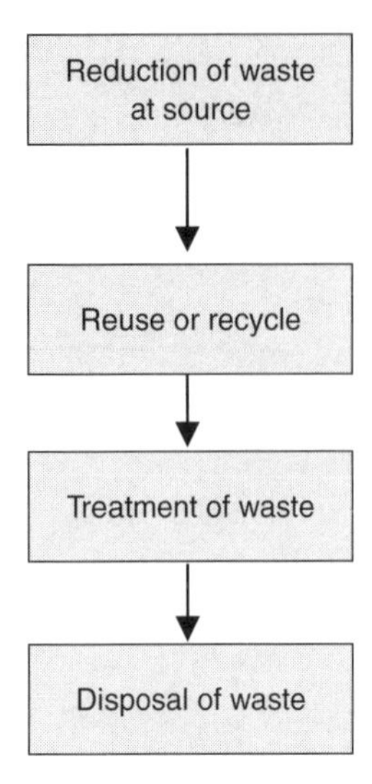

Figure 6.12 The hierarchy of waste production, treatment, and disposal.

clean technology as "any technical measures taken at various industries to reduce or even eliminate at source the production of any nuisance, pollution or waste and to help save raw materials, natural resources and energy"(Hall and Crowther, 1998). Clean technology hopes to reduce waste at source rather than deal with the waste once it has been formed. Fig. 6.12 shows the waste hierarchy where waste treatment has concentrated on the disposal rather than the other stages, particularly the reduction of waste at source. Sustainable technology aims to eliminate the disposal of waste, replacing this with treatment, conversion to another product, reuse, and recycling.

6.5.1 Extraction and supply of raw materials

A process will need a supply of raw materials that will have to be extracted or synthesized, although some should come from recycling. If the raw materials are derived from crude oil it is possible that biotechnology can provide bulk and fine chemicals extracted from plants or by microbial cultivation (section 6.2). Although these biological processes may produce waste, as they are biological they will be easier to convert or treat. If the raw materials are metals then biotechnology can also be used to extract metals from low-grade ores or waste tips, and even used to extract metals *in situ* (Chapter 8). Microbially enhanced oil recovery can also be used to extract crude oil from oil wells where the flow has become slow.

6.5.1.1 *Desulphurization of oils*

The burning of fossil fuels, petrol, diesel, and coal, releases sulphur oxides into the atmosphere which are the principal cause of acid rain. Because of this regulation has been put forward to reduce the sulphur content of petrol

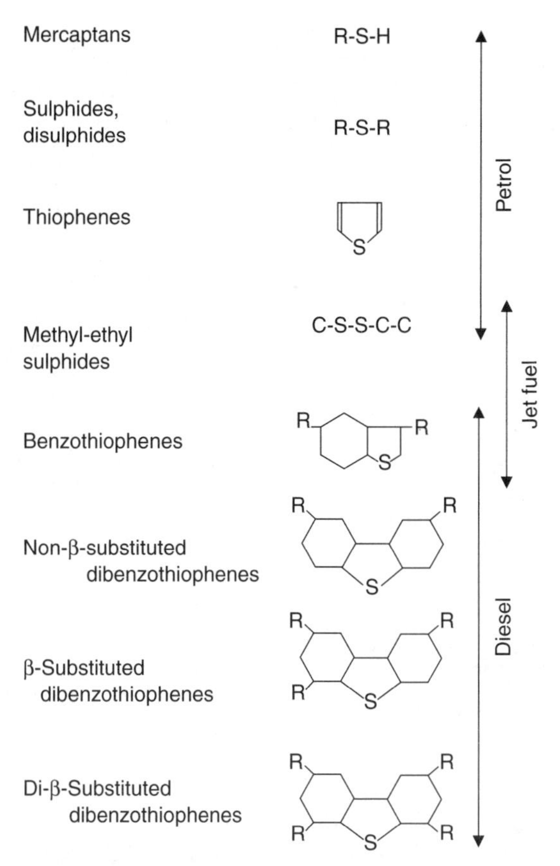

Figure 6.13 The sulphur-containing components found in crude oil and an indication of which fuels these can be found in.

and diesel to 30–50 ppm from the present 500 ppm by 1 January 2005 (Directorate of European Parliament, 2001; Environment Protection Agency, 1999; see Babich and Moulijn, 2003).

Crude oil contains between 0.05 and 5.0% sulphur compounds depending on the source, and these components are given in Fig. 6.13. The majority (70%) of the sulphur in crude oil is found as dibenzothiophene (DBT) or substituted DBT. Reduction in sulphur content is carried out for the distillate stream which make up petrol, diesel, and jet fuel by chemical methods. Three methods can be used: in the first the sulphur compounds are decomposed by treatment with hydrogen at 150–3000 lb/in^2 (1035–20 700 kPa) and 230–455°C; in the second the sulphur-containing compounds are separated and then decomposed; and in the third the suphur compounds are just separated. A decomposition process, which does not involve hydrogen, is also used in some cases. However, biological desulphurization is possible as a number of micro-organisms have been isolated which can degrade DBT and include *Rhodococcus erythropolis*, *Rhodococcus* ECRD-1, *Agrobacterium*, *Gordona*, *Klebsiella*, *Nocardia*

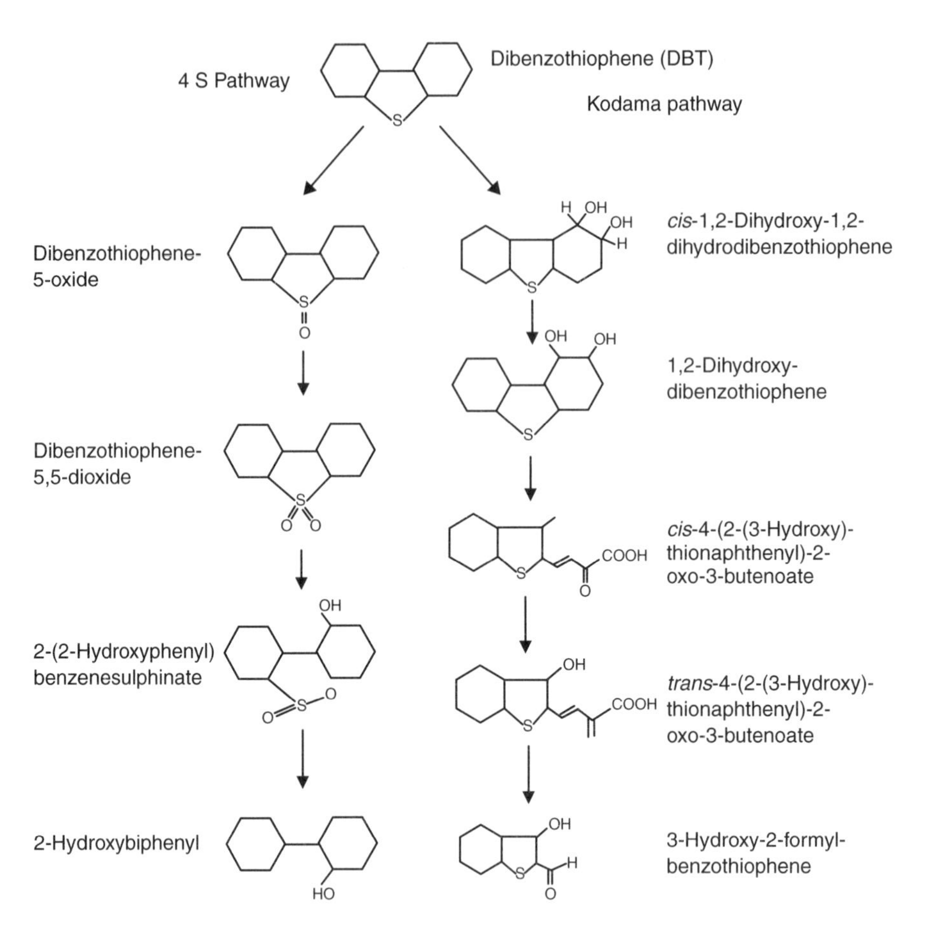

Figure 6.14 The two pathways for the degradation of DBT; the sulphur-specific and the non-sulphur routes.

globelula, Paenibacillus, Pseudonmonas, Bacillus subtilis, and *Mycobacterium* (McFarland, 1999; Monticello, 2000).

There are two pathways in the degradation of DBT; sulphur-specific degradation and non-sulphur degradation. The two pathways are shown in Fig. 6.14. The sulphur-specific pathway or 4 S pathway is found in *Rhodococcus* spp. where the sulphur is removed as sulphate and the degradation stops at 2-hydroxybiphenyl so that no carbon is lost in the process. The other pathway involves ring cleavage (the Kodama pathway) and loss of the whole molecule. The *Rhodococcus* spp. are hydrophobic and therefore bind to the drops of oil in an oil–water system in contrast to other cells such as *Pseudomonas*. Table 6.6 shows the total sulphur removal using biodesulphurization on various oil fractions.

6.5.1.2 *Desulphurization of coal*

Coal is a heterogeneous solid that contains a variety of organic and inorganic compounds that will vary with source. Sulphur occurs as both inorganic and

Table 6.6 Sulphur removal from crude-oil fractions

Crude-oil fraction	Total sulphur removal(%)
Mid-distillates	30–79
Diesel	40–90
Hydrotreated diesel	65–70
Light gas oils	20–60
Crude oil	25–60

Source: Data from McFarland (1999).

organic compounds. The inorganic compounds are mainly iron sulphides (pyrites) and sulphates, which can be up to 6% of the total weight in bituminous coals. The dominant organic constituents are the thiophenes such as DBT.

The inorganic sulphur in coal can be reduced by physical treatments such washing pulverized coal. The main users of coal are the power stations and here they have concentrated on removal of the sulphur oxides from the stack gases. Changes in combustion can reduce the emission of sulphur dioxide by fluidized-bed combustion where pulverized coal is fluidized by an airstream, which ensures that the coal is completely burnt. Limestone is added to trap the sulphur as a molten slag, which can be run off. This process requires advanced combustion designs and is not applicable to all power stations.

There are micro-organisms that will metabolize the components of coal; we know this because coal mines and coal storage are known to produce acidic drainage water, which is due to the formation of sulphuric acid by microbial activity. It is clear that coal can act as a microbial substrate but as coal is a solid, intimate contact between the organisms and the substrate is essential. Thus the structure and state of the substrate is of particular importance. Coal has a porous structure separated by fissures, where the pores can range from 5 mm to 2 nm, which at the lower range can restrict microbial access.

The inorganic sulphur in coal can be oxidized by chemolithotrophs like *Thiobacillus ferrooxidans*, *Thiobacillus thiooxidans*, and *Sulfolobus acidocaldarius*, which are responsible for acidic mine drainage. These organisms are aerobic and can remove inorganic sulphur compounds by both direct and indirect processes. Linked to the oxidation of reduced iron, *T. ferrooxidans* generates energy by the direct oxidation of ferric sulphide to ferrous sulphate.

$$2FeS_2 + 7O_2 + 2H_2O = 2FeSO_4 + 2H_2SO_4 \tag{6.1}$$

The ferrous sulphate is further oxidized to ferric sulphate.

$$4FeSO_4 + O_2 + 2H_2SO_4 = 2Fe_2(SO_4)_3 + 2H_2O \tag{6.2}$$

Microbial action can also directly convert elemental sulphur to sulphuric acid.

$$2S + 3O_2 + 2H_2O = 2H_2SO_4 \tag{6.3}$$

Thermophilic *Sulfolobus* species, which can grow at 65–80°C, are being investigated with a view to improving the removal rate.

The main organic sulphur compounds in coal are the DBTs and these have been used as a model compound in the biological desulphurization of coal. The organic sulphur compounds are an integral part of the coal matrix and therefore are much more difficult to remove than the pyrites. The enzyme needed to remove these organic compounds will need to cleave a C—S bond rather than a C—C bond. Various micro-organisms have been found which are capable of degrading DBTs, including *Rhodococcus* sp. (Gray et al., 1996), *Pseudomonas* spp. TG232 (Kilbane, 1989), *Brevibacterium* sp. (McEldowney et al., 1993), and *Aspergillus niger*. The potential of particular species to cleave the C—S and C—C bonds differs from species to species, with the C—S bond cleavage being the best option as it does not degrade the coal and therefore retains its calorific value. The oxidation of DBTs by *Rhodococcus* sp. follows the same pathway as for the removal of DBT from oils. The sulphite formed at the end of the pathway is converted non-biologically to sulphate, which can be washed out of the coal. An example of the biological removal of inorganic and organic sulphur from coal is shown in Table 6.7. Another pathway has been identified in *Brevibacterium* sp. where DBT is broken down to sulphite and benzoic acid. It would appear that the benzoic acid is also metabolized and therefore this breakdown will result in the loss of carbon from the coal. The enzymes involved in the cleavage to benzoic acid are under investigation.

6.5.1.3 *Denitrogenation of oils*

Crude oil contains up to 0.35% nitrogen-containing compounds that are basic and non-basic, where the non-basic ones contribute 70–75% of the nitrogen. The removal of these compounds from crude oil would reduce the nitrogen oxide produced on burning, although these are not the main source, and reduce poisoning of the refinery catalysts. High-temperature and high-pressure hydrogen treatment will remove some of the nitrogen but a biological

Table 6.7 Biodesulphurization of coal

Sulphur type	Coal (%)	Treated coal (%)	% Reduction
Sulphate	0.38	0.1	74
Inorganic sulphur pyrites	0.22	0.12	45
Organic	2.25	0.2	91
Total	2.85	0.42	85

Source: Kilbane (1989).

method of removal has been investigated. *Pseudomonas aeruginosa* and other micro-organisms have been shown to degrade the nitrogen compounds in oils (Benedik et al., 1998; Le Borhne and Quintero, 2003). The problem with the removal is that it involves a pathway similar to the Kodama pathway for DBT where carbon is lost from the fuel. The process is still under investigation and may be of use with shale oils, which contain higher concentrations (0.5–2.1%) of nitrogen compounds.

6.5.2 Processing the raw material

6.5.2.1 *Enzymes*

In the processing of the product biotechnology can replace the inorganic catalysts normally used with micro-organisms and enzymes.The use of micro-organisms or biocatalysts (**enzymes**) in a process produces a cleaner system because

- the reaction conditions are mild, using less energy;
- the reactions can be very specific;
- the reactions are fast and effective;
- in general they use renewable resources;
- recombinant technology is now available to improve the organism or enzyme;
- enzymes are now available from extremophiles (Chapter 2); and
- they can function in non-aqueous media.

The disadvantages of a biologically based system are that

- in some cases a mixed product can be produced;
- the product is produced in dilute aqueous medium;
- microbial contamination can occur;
- biological systems can be variable;
- short working life (half-life); and
- the enzyme and product are difficult to separate. However, immobilization of the enzyme in or on a surface will keep the enzyme from mixing with the product, but it does increase the half-life (Box 3.1).

The production of chemicals from plants and micro-organisms has been described in section 6.2 but micro-organisms and enzymes can also be used to transform chemicals (Fig. 6.15). Single enzymes are excellent catalysts that can perform regio- and sterio-specific reactions. Micro-organisms can also be regarded as multiple-enzyme containers that can carry out one or multiple reactions. One of the best examples of the very specific nature of enzymes is the production of steroids, which has been used for some time and which illustrates some of the advantages of enzymes such as specificity and operation under mild conditions with few side products (Box 6.4).

The chemical industry is keen to use biocatalysts (enzymes) and micro-organisms for the production of a number of chemicals. Table 6.8 lists some of the industries that use enzymes. A few examples of the use of enzymes and micro-organisms in various industries, with advantages and disadvantages, are given below.

Detergents The first example of the use of enzymes in detergents was by Roehm and Haas, who added tryptic enzymes to the laundry process. The full application had to wait until the 1960s when large quantities of enzymes were

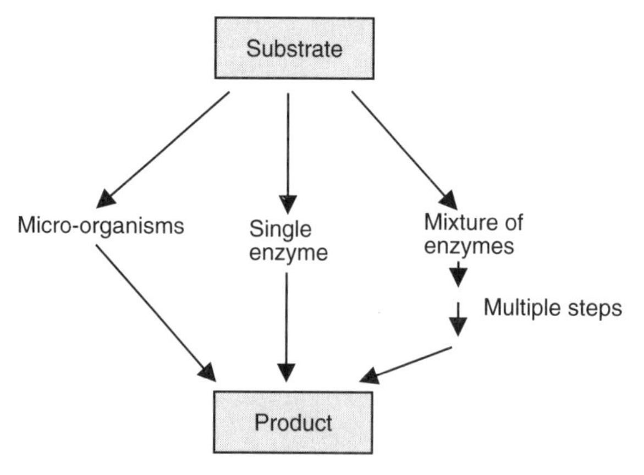

Figure 6.15 The use of enzymes and whole micro-organisms to transform one compound into another, which can involve a single step or multiple steps.

BOX 6.4 The biotransformation of steroids

The specificity of enzymes is illustrated by their use in steroid transformation. The figure shows the hydroxylation of progesterone at one specific position, which would be very difficult to carry out chemically. This biotransformation can be carried out using a culture of *Rhizopus nigricans*.

Progesterone → (*Rhizopus nigricans*) → 11α-Hydroxyprogesterone

Table 6.8 Some applications of enzymes in industrial sectors

Sector	Enzyme(s)	Use
Adhesives	Phenol oxidase	Adhesive production
Agriculture	Phytase	Phosphate release
	Xylanase	Digestibility
Cleaning	Protease, amylase, lipase	Washing powder, drain cleaner
	Cellulase	Restoring colour to cotton
Leather	Protease	Removing hair
	Lipase	Removing fat
Paper	Protease	Biofilm removal
	Amylase	De-inking, starch coating
	Xylanase	Pulp bleaching
	Cellulase	De-inking, drainage
	Laccase	Bleaching pulp
Textile	Cellulase	Cotton treatment
	Laccase	Bleaching
	Amylase	Starch removal
	Peroxidase	Excess dye removal

available (Gupta et al., 2003). The enzymes produced were extracellular so that processing was minimal, which made the enzyme cheap enough to use in an industrial process. The addition of lipase, protease, and amylase to detergents allowed the use of lower temperatures, and the better removal of specific stains. The first detergents containing enzymes did have some problems with allergenic reactions but encapsulation stopped this. There has been continual development of enzyme detergents with increased activity at lower temperatures and at alkaline pH values.

Textiles The textile industry has seen the application of alkaline proteases for degumming silk, removing a protein from the outside of the silk fibres (Gupta et al., 2002). Two pectinases have been used to remove cell-wall components from cotton fibres and glucose oxidase used to bleach the fibres (Tzanov et al., 2001). These processes consume large quantities of water and energy and are expensive so the application of enzymes has made savings in both processes. Cellulases have been used to treat denim garments as an alternative to stonewashing in a process known as biostoning (Belghith et al., 2001).

Leather The treatment of leather produces considerable pollution so that the replacement of the chemical processes by enzymes should have considerable advantages. Proteases have been used to de-hair hides and lipase to remove fat.

Paper The paper industry has seen the application of a number of enzymes including lipase, protease, amylase, xylanase, and cellulase. Cellulases have been used to assist pulping, and to de-ink paper. Laccase enzymes that attack lignin have been used to bleach pulp as an alternative to chlorine bleaching (Kirk et al., 2002).

Phytase The enzyme phytase releases phosphate from phytate, which is *myo*-inositol hexakisphosphate. This is the major phosphorus form in animal feed and its release would improve the nutritional value of animal feeds (Lei and Stahl, 2001).

The application of enzymes to these industries gives processes that use less energy, produce less toxic waste, have reduced process times, and are less expensive. Enzymes are susceptible to higher temperatures and harsher conditions that speed up the process. However, developments in molecular biology and microbiology may solve some of the problems of enzyme life and stability. Molecular biological techniques are available to alter the characteristics of enzymes such as thermostabilty and include site-directed mutagenesis, directed evolution, computational design strategies, and protein engineering (Bull et al., 1999). These techniques are not pertinent to this text but if further information is required refer to the reading list at the end of this chapter. The microbiological techniques available are the isolation of new organisms from the environment, in particular from extreme environmental conditions, and include the Archaea.

6.5.2.2 *Extremophiles*

The micro-organisms found in extreme conditions of temperature, salt, and pressure belong to the domains of Bacteria and Archaea, but the majority are the Archaea. Clearly these types of micro-organism will contain enzymes that can function under extreme conditions (**extremozymes**) (Eichler, 2001). One of the best-known examples of temperature-tolerant enzymes is the DNA polymerases such as that extracted from the thermophile *Thermus aquaticus* (*Taq* polymerase). The temperature tolerance of this enzyme allows temperature cycling to 90–95°C and therefore facilitates the use of the polymerase chain reaction (PCR). Extremozymes could replace the *Taq* polymerase enzyme if the extension rates can be improved. Table 6.9 gives a brief survey of some of the enzymes from hypothermophiles, psychrophiles, and halophiles (Eichler, 2001; Haki and Rakshit, 2003; Huber and Stetter, 1998).

The energy required for the processing stage could be provided from sustainable sources (Chapter 7). The more-efficient use of the energy supply can be achieved by using a combined heat and power (CHP) system where the heat remaining after electricity has been generated is used for other purposes such as the supply of hot water. In addition a CHP system could also be combined with energy extraction from any waste produced from the process. An example of this type of system as applied to a brewery is shown in Box 6.5.

Table 6.9 Enzymes from extremophiles

Organism	Enzyme	Enzyme optimum	Maximum growth temperature (°C)
Thermophiles			
Pyrococcus furiosus	Amylase	100°C	103
Pyrococcus furiosus	Glucosidase	115°C	103
Sulfobolus solfataricus	Glucosidase	105°C	87
Thermococcus profundus	Amylase	80°C	90
Desulfurococcus mucosus	Amylase	100°C	–
Pyrococcus furiosus	Serine protease	85°C	103
Thermococcus aggregans	Serine protease	90°C	98
Sulfobolus acidocaldarius	Acid protease	90°C	85
Thermococcus maritina	Xylanase	95°C	90
Pyrobaculum californica	Lipase	90°C	103
Rhodothermus marinus	Cellulose	95°C	–
Pyrococcus furiosus	DNA processing	90–95°C	103
Thermococcus litoralis	DNA processing	90–95°C	98
Alkaliphiles/acidophiles			
Thermococcus alcaliphilus	Under investigation		90, pH 10.5
Pyrococcus furiosus	Under investigation		103, pH 5–9
Thermotoga maritima	Under investigation		90, pH 9.5
Sulpholobus acidocaldarius	Under investigation		85, pH 1–5
Acidianus infernus	Under investigation		95, pH 1.5–5
Halophiles			
Natronomonas pharonis	Protease	pH 10	
Halobacterium halobium	Serine protease	4 M NaCl	

6.5.3 Use and disposal of the product

Biotechnology can do little to influence the use of a product but can have an influence on its recycling and disposal. With some products micro-organisms or enzymes can be used to recycle or convert the product into another or reclaim some components such as metals. If disposal of the product is required and it is capable of being broken down there are a number of methods available such as composting and landfill, which are discussed in Chapter 4.

BOX 6.5 A combined heat and power system applied to waste treatment from a brewery

The figure illustrates how fuel can be used more efficiently with a combined heat and power system and how waste can be used to generate fuel. In this case low-concentration waste is treated by an aerobic process but the higher-concentration waste is passed through a two-stage anaerobic digester. The digester will produce methane, which after treatment is burnt in a fuel cell. Waste from the anaerobic digester is combined with the low-concentration waste in the aerobic treatment process. The fuel cell produces electricity and the cooling water is heated to about 60°C, which can be used for washing or with a small input of energy can be converted to hot water.

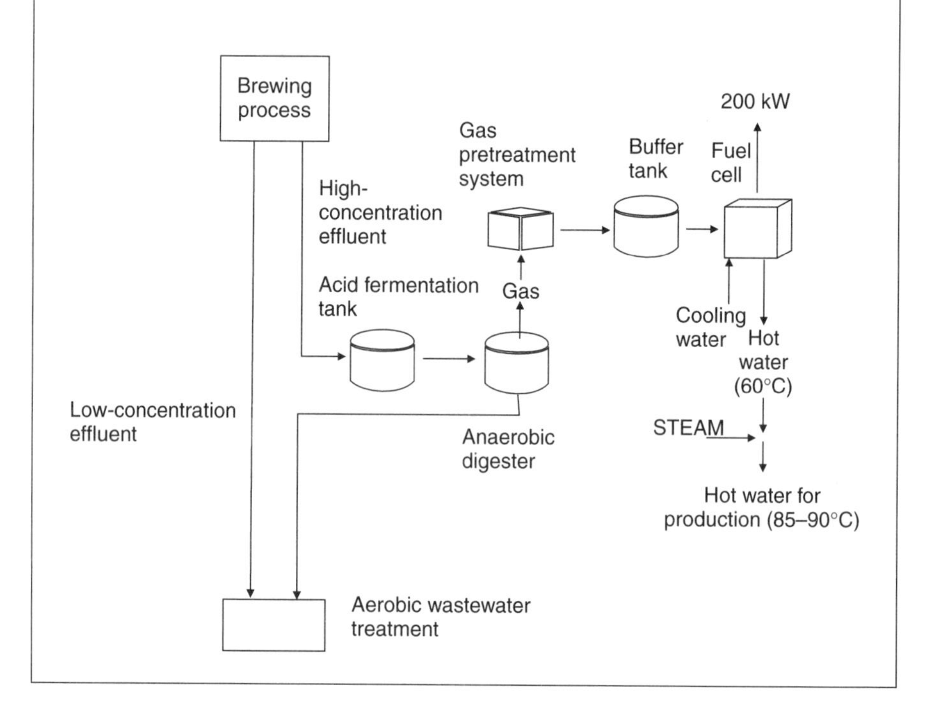

6.6 Conclusions

- It is clear that the potential for bioremediation is immense as legislation and the need for sites for housing and factories increases.
- The process and pathways involved in bioremediation are still under investigation and recombinant DNA technology will help greatly.
- The effect of environmental conditions on the rates of bioremediation requires further investigation in addition to methods of bioremediation.

6.7 Further reading

Glazer, A.N. and Nikaido, H. (1994) *Microbial Biotechnology*. W.H.Freeman & Co., New York.

Rehm, H.-J. and Reed, G. (eds) (1996) *Biotechnology*, 2nd edn, vol. 6, Products of Primary Metabolism. VCH, Weinheim.

7 Biofuels

7.1	Introduction	266
7.2	Finite supply of fossil fuels	268
7.3	Emissions from fossil fuels	271
7.4	Greenhouse gases	274
7.5	Natural sources of greenhouse gases	276
7.6	Ozone	282
7.7	Sulphur dioxide	285
7.8	The effects of industrial (anthropogenic) activity	285
7.9	Remediation of the emissions from fossil fuels	290
7.10	Alternative non-fossil energy sources	292
7.11	Biological energy sources	293
7.12	Combustion of biomass	295
7.13	Biogas	298
7.14	Biodiesel	301
7.15	Ethanol	307
7.16	Hydrogen	316
7.17	Conclusions	319
7.18	Further reading	319

7.1 Introduction

This chapter outlines the world's continual demand for energy and the problems that fossil-fuel use causes. Alternatives to fossil fuels are available and the chapter will concentrate on the biologically produced fuels such as biogas, biomass, biodiesel, and ethanol.

The global use of energy has been increasing steadily since the Industrial Revolution. In the USA it is projected that energy consumption will increase from 97.3 (102.6×10^{18} J) to 130.1 quadrillion Btu (137×10^{18} J; where Btu means British thermal units; 1 Btu = 1055 J; quad = 10^{15}) from 2001 to 2020 an annual increase of 1.5% (Energy Information Administration, EIA, 2003; see **www.eid.doc.gov**). Predicted further increases in energy consumption are 9960 Mtoe (megatonnes of oil equivalents) in 2000 rising to 15200 Mtoe in 2020, an annual increase of 1.75% (Fig. 7.1; International Energy Agency, IEA, 2002). Another predicted global increase in energy demand was 1.85% per annum in 1995; but if the former USSR is excluded the increase in energy demand was 2.9% due to the increased demands of the developing

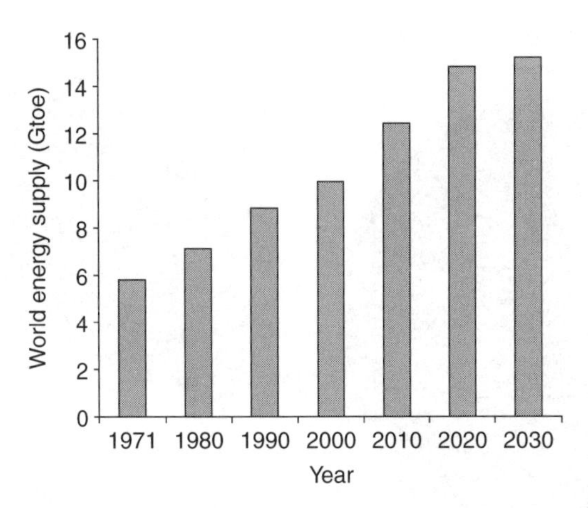

Figure 7.1 The increase in the supply of total energy to the world at present and in the future in Gtoe (Gt of oil equivalents). Data from IEA (2002).

Asian countries (Anon., 1997). It is clear that whatever figures are taken energy demand will continue to increase. The demand does vary from country to country depending on their state of development, as there is a close correlation between energy consumption and improvements in living conditions measured by life expectancy and infant mortality.

The main supplies of energy are the fossil fuels, coal, gas, and oil which supply from 75 to 85% of the total (Fulkerson et al., 1990; EIA 1998; see **www.eia.doc.gov**). The contribution of fossil fuel has changed over the years as can be seen in Fig. 7.2. The contribution by coal has dropped from 74% in 1937 to 23.5% in 2000 (Boyle, 1996; IEA, 2002), which is reflected by the shift from deep coal mining to open-cast extraction. Nuclear power has been introduced during this time and natural gas has seen increasing use. The bulk of coal is now used for the generation of electricity (Fig. 7.3). In the USA coal was used to produce 51% of the electricity and this figure is predicted to be 49% by 2020 (EIA, 1995; see **www.eia.doc.gov**). However, as open-cast coal stocks are exhausted, deep mining may have to be restarted if coal is still required. It has been estimated that the coal stocks will last until 2180 (Fulkerson et al., 1990). Natural gas is replacing coal for the generation of electricity as it is a cheaper and cleaner fuel, and produces more energy on combustion. However, natural gas is less abundant, not evenly distributed, and stocks are said to only last until 2047.

The reasons that fossil fuels continue to be used are that they are found throughout the world, they require only simple technology to extract the energy, and can be transported easily. Another important feature is that crude oil and oil products are liquids, which makes their transport convenient and this property is required for their use as automotive fuels. However, despite their continued use there are problems associated with the use of fossil fuels:

- there is a finite supply of fossil fuels;
- combustion of fossil fuels produces greenhouse gases (**global warming**); and
- combustion of fossil fuels produces other pollutants (air pollution, **acid rain**).

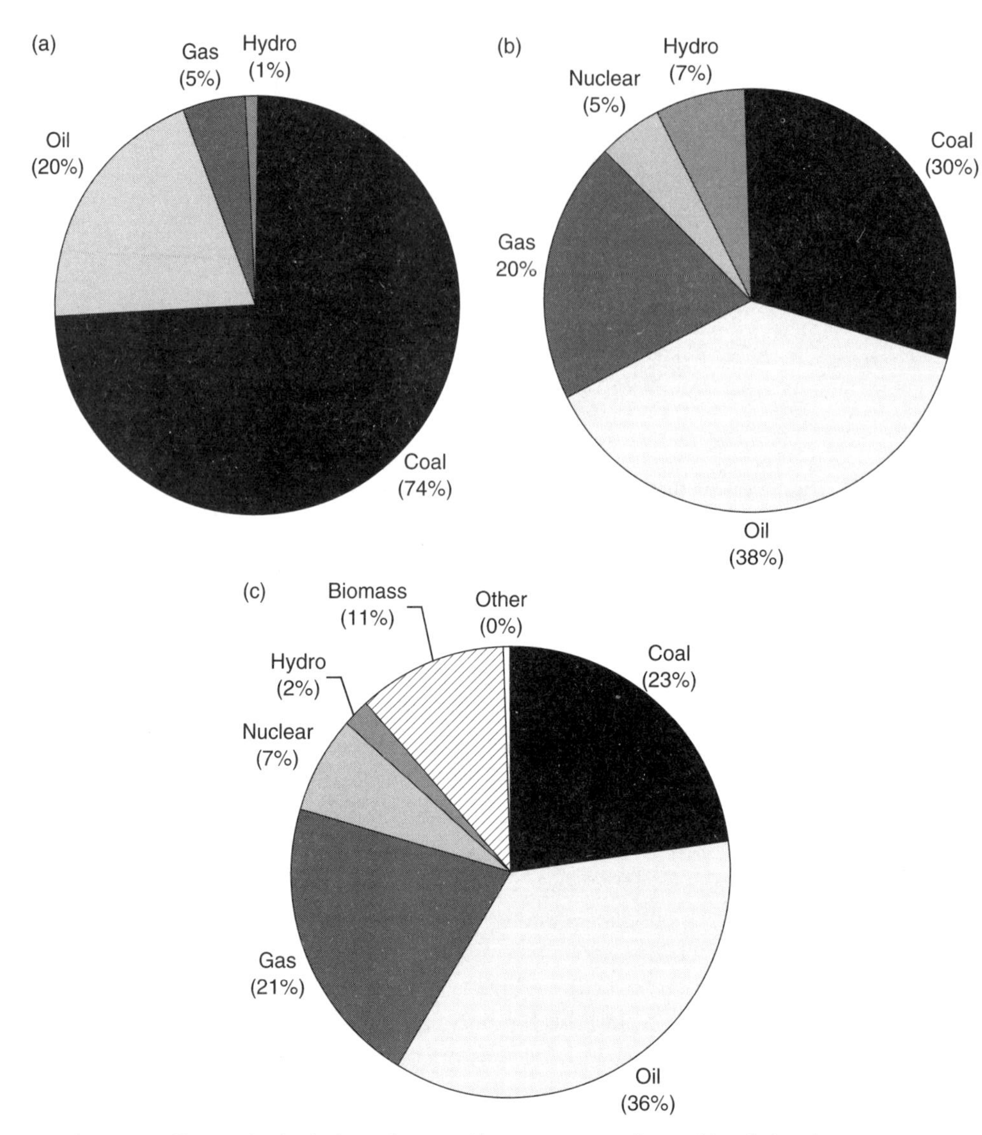

Figure 7.2 Changes in the fuels used to provide energy across the world including biomass. (a) 1937; (b) 1988; (c) 2000. Data from Fulkerson et al. (1990) and IEA (2002).

7.2 Finite supply of fossil fuels

Fossil-fuel stocks, such as oil, are finite and these should run out in the near future, but the precise date varies considerably. Evans (2000) suggests that oil supplies should peak in 2014, gas supplies should last another 70 years, and coal another 200 years. Another estimate is that coal will last until 2180 (Fulkerson et al., 1990). Natural gas is predicted to last until 2047 but its

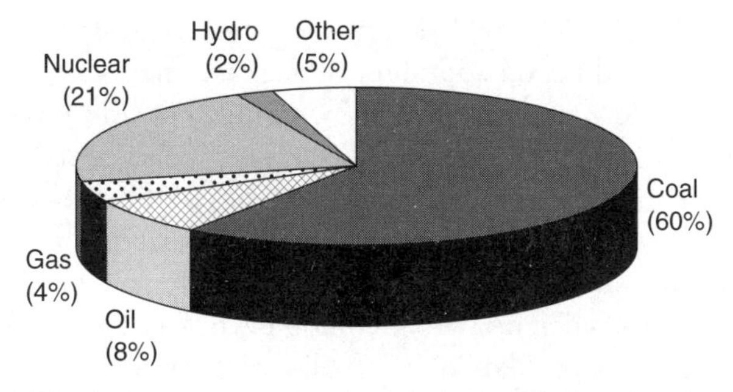

Figure 7.3 The fuels used to generate electricity in the UK. Data from Boyle (1996).

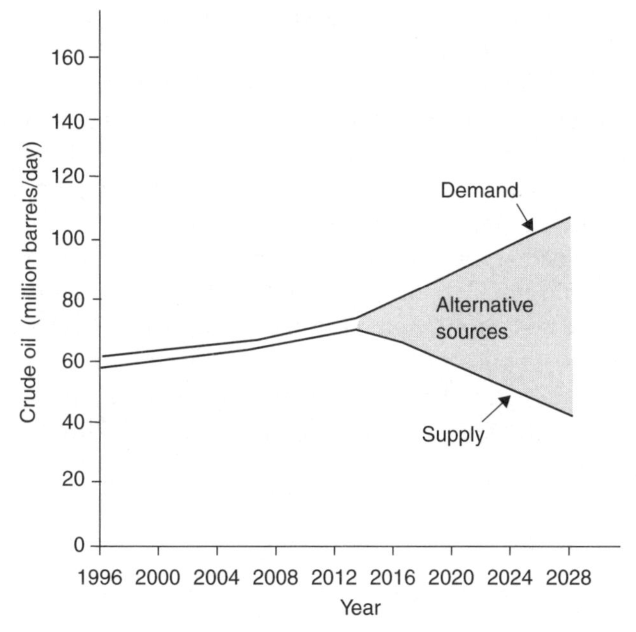

Figure 7.4 The predicted demand and supply of crude oil in millions of barrels per day (1 barrel = 159 l). Derived from Evans (2000).

distribution is uneven and it is less abundant than coal. It is not just the stocks of oil that is the most pressing problem but the rate of oil supply to developed nations may also fail to keep up with demand. The International Energy Agency (IEA) suggest that the demand for oil will soon outstrip supply and that the gap will have to be filled with oil from unconventional sources such as oil shales and tar sands (Fig. 7.4). However, this somewhat pessimistic view is not shared by all. The development of new technology such as horizontal drilling and improved seismic analysis has resulted in the discovery of new reserves of oil. Improved combustion efficiencies and continued exploration and detection of oil sources have extended the estimates of how long oil stocks will last. Despite this it has been estimated that oil supplies will only last until

2080 but the price of crude oil has remained stable and has even decreased. However, the demand for oil continues to increase and as a consequence the extraction of oil has to be attempted in increasingly hostile and difficult conditions, such as in Alaska and the North Sea. This will eventually increase the costs and reduce the supply of crude oil. Estimates of the recoverable reserves of oil are as low as 2 000 000 million barrels (one barrel = 158.97 l) and as high as 3 000 000 million barrels with the IEA giving a figure of 2 300 000 million barrels. With a demand of 72 million barrels a day in 1996 increasing to 112 million barrels per day in 2020, oil supply should peak in 2014 and decline after this time. The sources of crude oil will also change with time. The OPEC Middle Eastern countries currently supply 24% of the world's oil, which will rise to 48% by 2014 and given the unsettled nature of this area, oil supplies could be interrupted.

Unconventional sources of oil do exist and will have to be exploited. Oil shales contain complex organic materials which on heating yield a synthetic crude oil. Tar sands are sandstone or porous rock where crude oil has come to the surface. The volatile fractions have evaporated leaving the heavier fractions such as bitumen mixed with sand and sandstone.

Another view is that forecasters of oil supply believe that alternative crude oil supplies would not develop and therefore they predicted that oil prices would rise above $40/barrel. As a result they underestimated the development of alternative oil supplies and improving oil-recovery technology. When the predictions of production levels were compared with actual production the actual production was considerably better (Fig. 7.5; Piel, 2001).

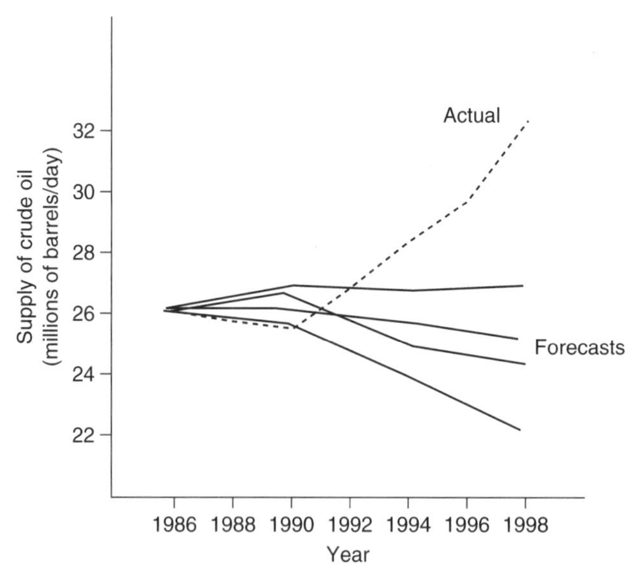

Figure 7.5 Non-OPEC oil production forecasts of the supply of crude oil in millions of barrels per day. Data from Piel (2001).

It appears that the demand for fossil fuel will continue to rise and it is clear that stocks can only last a finite time. It is the length of this time that is the subject of considerable speculation. At present many of the alternatives to oil-based fuel are not competitive on cost, but eventually this will change and alternative power sources will be required.

7.3 Emissions from fossil fuels

Combustion of fossil fuels produces gases that can affect the global climate. **Weather** and **climate** have a profound effect on all forms of life on Earth and one of major influences on climate is the energy derived from sunlight. Weather has been defined as the fluctuating state of the atmosphere in terms of temperature, wind, precipitation (rain, hail, snow), clouds, fog, and other conditions. Climate refers to the mean values of the weather and its variation over time and position. Perhaps in simple terms weather is the conditions experienced by the individual and climate is the conditions experienced by countries and other land areas. The atmospheric circulation and its interaction with large-scale ocean currents and land masses determine the climate. The climate is also affected, often called **forcing**, by these various parameters but the most important is the sun. The climate system consists of five domains (Fig. 7.6):

- the land surface, the soil and vegetation,
- the biosphere, the marine and terrestrial biota,
- atmosphere, the most unstable and rapidly changing domain,
- hydrosphere, consisting of all liquid surfaces, marine and terrestrial, and
- cryosphere, all ice and snow areas.

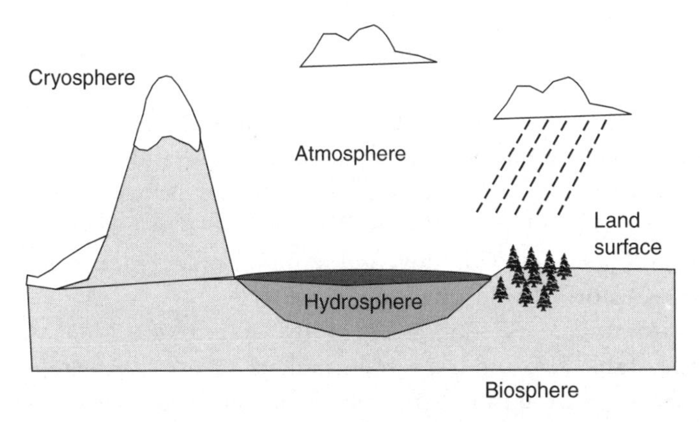

Figure 7.6 The five domains making up the climate system.

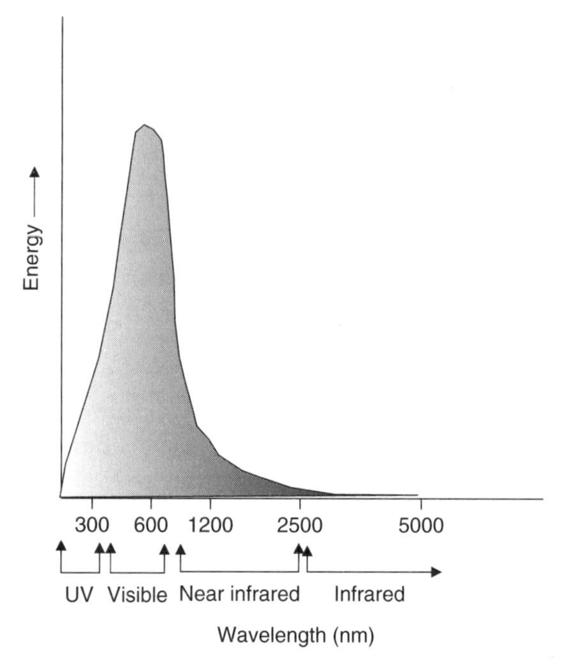

Figure 7.7 The spread of radiation in the range of ultraviolet (UV) to infrared that arrives from the sun.

The land surface, consisting of soil and vegetation, influences how much of the sunlight energy absorbed by the surface is returned to the atmosphere. About half the radiation which arrives from the sun is in the visible range (shortwave, 400–700 nm) and the other half is made up of near infrared (1200–2500 nm) and ultraviolet (UV; 290–400 nm) light (Fig. 7.7). The biosphere, both marine and terrestrial, has a considerable influence on the climate as respiration in the biosphere releases carbon dioxide. Photosynthesis in terrestrial and aquatic plants fixes carbon dioxide and releases oxygen. The fixation of carbon dioxide stores a significant amount of carbon as part of the carbon cycle (Fig. 7.8). Other biosphere products are methane, nitrous oxide, and volatile organic compounds (VOCs), which can also affect the climate.

The hydrosphere consists of bodies of both fresh and salt water, and 70% of the Earth is covered with water. The oceans store and transport large quantities of energy and can store large quantities of carbon dioxide. The cryosphere consists of all forms of ice and snow, permafrost, glaciers, snow fields, sea ice, and polar ice. The snow and ice areas reflect radiation from the sun and act as a store of large quantities of water.

The atmosphere is composed mainly of nitrogen (78%) and oxygen (20.95%; Table 7.1) and has recognizable layers; troposphere, stratosphere, mesosphere, and thermosphere (Fig. 7.9). The troposphere is the layer; in

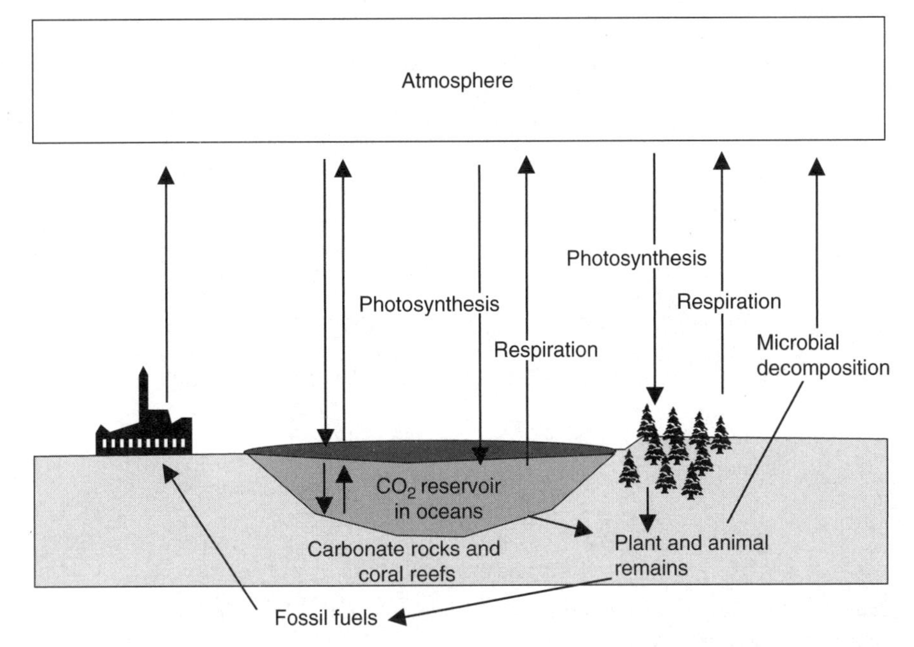

Figure 7.8 The cycle of carbon dioxide to and from the atmosphere.

Table 7.1 The main components of air

Component	Composition (%; v/v)
Nitrogen	78.09
Oxygen	20.95
Argon	0.93
Carbon dioxide	0.03
Neon	0.0018
Helium	0.0005
Methane	0.00017

Note: The figures are for dry air in the trophosphere.

which weather occurs and contains 90% of the gases that make up the atmosphere. In this layer there are a number of gases other than nitrogen and oxygen present in trace amounts, which include carbon dioxide, methane, nitrous oxide, and ozone. These gases absorb and emit infrared radiation and are known as the **greenhouse gases** (Table 7.2).

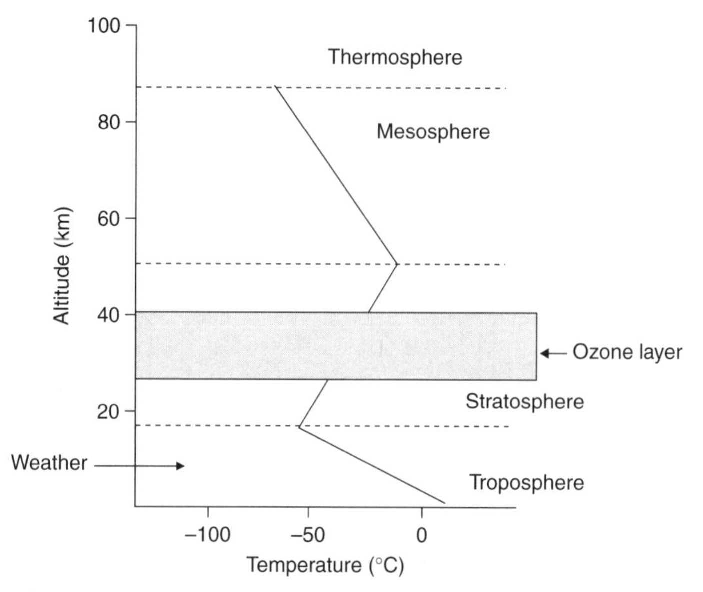

Figure 7.9 The layers that constitute the Earth's atmosphere, their depth in km, and the temperature found in the layers.

Table 7.2 Contribution of greenhouse gases to global warming (1980–1990)

Gas	Contribution (%)
Carbon dioxide	55
Chloroflurocarbons (CFCs)	24
Methane	15
Ozone	NA*
Nitrous oxide (N_2O)	6

Note: Water vapour is a major greenhouse gas but is not included as its contribution is difficult to ascertain.
*NA, not available.

7.4 Greenhouse gases

Because these gases absorb the infrared radiation from the surface they can either radiate the infrared radiation (energy) back to the surface or out into space. These all effect the energy movement both in and out of the atmosphere. In addition to these gases the atmosphere contains solid and liquid particles and clouds (water vapour). Water vapour, which is also a greenhouse gas, is the most variable. Ozone in the atmosphere is found in trace amounts at all levels but is at a maximum (8–10 ppm) in the stratosphere, in a region known as the ozone layer. This ozone absorbs harmful ultraviolet radiation (230–320 nm).

Prior to the Industrial Revolution the world's atmosphere had a stable composition and the climate had been relatively stable since 1000. To be stable the input of energy from the sun must be balanced by outgoing radiation. Figs 7.10 and 7.11 show the input and export of energy from the sun to the global climate. The sun's radiation arrives as short-wavelength radiation and some 50% is absorbed by the Earth's surface. The other 50% is absorbed by the atmosphere, and reflected back by clouds and the surface of the Earth, particularly glaciers and ice sheets. The surface of the Earth radiates the shortwave energy absorbed at a longer wavelength in the infrared. This is due to the lower temperature of the Earth compared with the sun. The higher the temperature the shorter the wavelength of the radiation emitted. The Earth's infrared radiation is affected by a number of factors, principally by the gases in the atmosphere, known as greenhouse gases (Table 7.2). The greenhouse gases derived from natural sources are water vapour, carbon dioxide, methane, ozone, and nitrous oxide, and the man-made gases are the **chlorofluorocarbons** (CFCs). All these gases absorb the longwave radiation emitted by the Earth's surface and then radiate the energy in all directions, including back to the surface increasing the surface temperature. Some gases are better at absorbing radiation than others and therefore their effects are not directly linked to their concentrations (Table 7.2). Some infrared radiation escapes directly through the atmosphere through what is known as an atmospheric window. This effect can be felt on a clear night when the temperature drops quickly and a ground frost can form

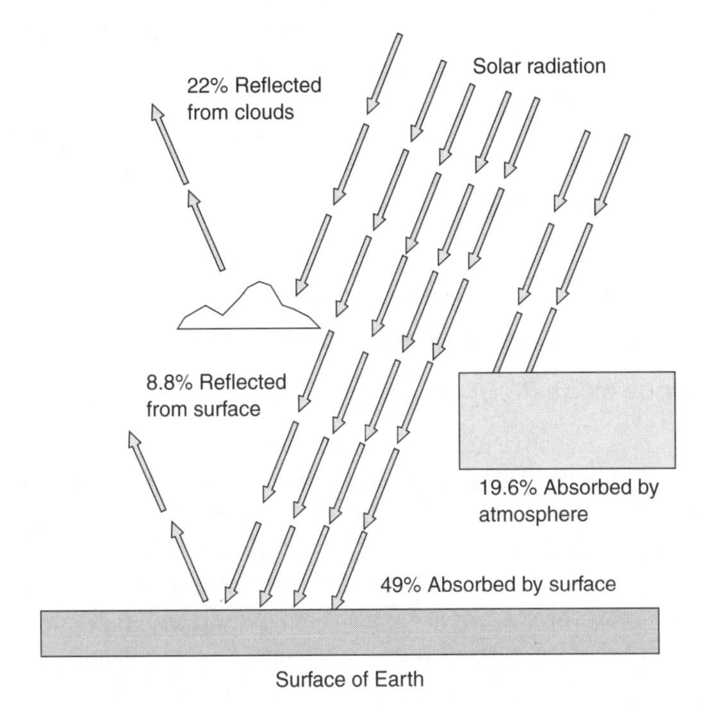

Figure 7.10 The fate of the radiation reaching the Earth from the sun.

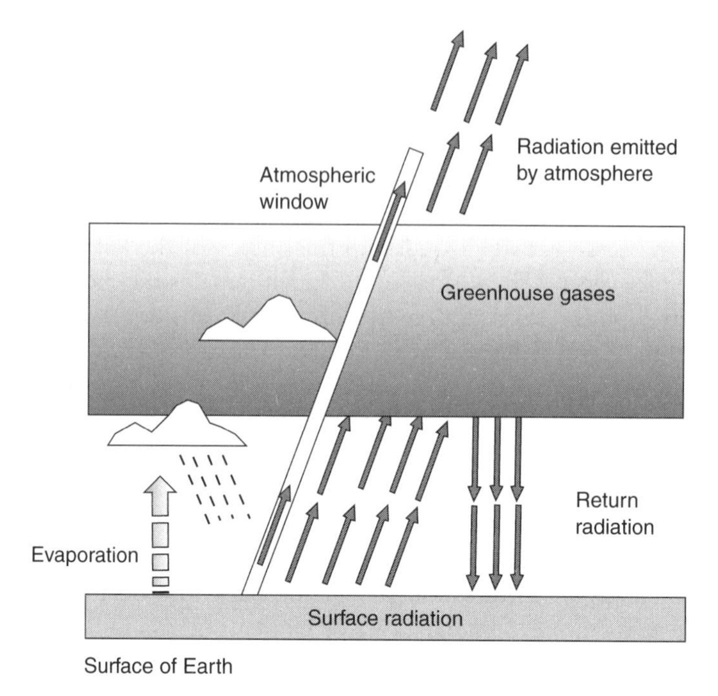

Figure 7.11 The fate of the energy from the sun and adsorbed by the Earth as it is radiated from the Earth.

rapidly, and in deserts when after very hot daytime temperatures this can drop to near zero. The outcome of the return of energy to the surface is that the average temperature of the Earth's surface is 14°C, rather than the −19°C it would be if all the energy was lost. On Venus, where over 90% of the atmosphere is carbon dioxide at a higher pressure than Earth's, the surface temperature is 523°C. Thus the greenhouse gases trap the heat within the atmosphere and some of this is transferred upwards to the colder layers at around 5 km.

7.5 Natural sources of greenhouse gases

7.5.1 Nitrous oxide (N_2O)

Nitrous oxide is an active greenhouse gas found at a very low concentration of 310 ppbv (parts per billion by value) in the atmosphere. On a molecule-to-molecule basis nitrous oxide is 200 times more effective than carbon dioxide at absorbing infrared radiation and is involved in the destruction of ozone. The gas is produced naturally by the denitrification of nitrate by microbial activity in soil and the sea. Table 7.3 gives the natural and man-made input of nitrous oxide per year. Nitrous oxide is produced in a sequence of reactions leading from nitrate to nitrogen gas and is shown below.

$$NO_3 > NO_2 > NO > N_2O > N_2 \qquad (7.1)$$

Table 7.3 Sources of nitrous oxide (N_2O)

Source	Nitrous oxide (Mt* of N/year)	Proportion (%)
Natural		
Oceans	1.4–3.0	17.2
Tropical soil	2.7–5.7	32.8
Temperate soil	0.5–2.0	11.5
Anthropogenic		
Biomass burning	0.2–1.0	5.8
Cultivated soils	0.03–3.5	20.1
Industry	1.3–1.8	10.3
Cattle and feed	0.2–0.4	2.3

*Mt, megatonne, 10^6 tonnes.

Table 7.4 Bacteria capable of denitrification or nitrate breakdown

Organism	Substrate	Product
Thiobacillus thiparus	NO_3^-	NO_2^-
Hysobacter antibioticum	NO_3^-	NO_2^-
Pseudomonas	NO_3^-	N_2O
Achromobacter	NO_3^-	N_2O
Paracoccus denitrificans	NO_3^-	N_2
Thiobacillus denitrificans	NO_3^-	N_2
Hyphomicrobium	NO_3^-	N_2
Halobacterium	NO_3^-	N_2
Pseudomonas	NO_3^-	N_2
*Ralstonia eutropha**	NO_3^-	N_2
Klebsiella aerogenes	NO_3^-	NH_3
Escherichia coli	NO_3^-	NH_3
Flavobacterium sp.	NO_2^-	NO_2
Vibrio succinogenes	NO_2^-	N_2

*New name for *Alcaligenes eutrophus*.

In many cases not all of the steps are carried out by a single micro-organism but rather a consortium of micro-organisms. Table 7.4 shows some of the micro-organisms involved in denitrification and that the end products are not always nitrogen gas. The addition of nitrogen-based fertilizer to soils will increase the rate of denitrification. Nitrous oxide is lost mainly in the stratosphere by photodegradation.

$$2N_2O + h\nu\ (\text{light}) = 2N_2 + O_2 \quad (7.2)$$
$$O_3 + h\nu\ (\text{light}) = O^* + O_2 \quad (7.3)$$
$$O^* + N_2O = 2NO \quad (7.4)$$

where O^* is a reactive oxygen molecule.

7.5.2 Nitrogen oxides (NO_x)

The natural sources of nitric oxide and nitrogen oxide are soils, ammonia oxidation, and lightning. High temperatures such as those generated by lightning can produce these nitrogen oxides. At high temperatures nitrogen reacts with oxygen by a number of mechanisms including the Zeldovitch mechanism (Box 7.1). The reaction yields a mixture of NO and N_2O with NO dominating at approximately 90%. The high-temperature reaction occurs in vehicle engines, aircraft engines, and during biomass burning. If the fuel contains nitrogen the two gases may also be produced. Table 7.5 gives the rates of production from both natural and anthropogenic sources. It is clear that a major contribution is from combustion of fuels in engines of various types. Table 7.6 shows the road emissions from petrol and diesel vehicles. The overall emissions of nitrogenous compounds in the UK can be seen in Table 7.7, where the main output is NO.

7.5.3 Methane (CH_4)

Methane levels are twice what they were in pre-industrial times and have been increased by human activities such as rice cultivation, coal mining, waste disposal, biomass burning, landfills, and cattle farms. Cows can produce up to 40 l of methane per day. Methane is also released from natural sources such as wetlands, termites, ruminants, oceans, and hydrates (Table 7.8). Yields are 138–464 Mt of methane per year for biological systems and 48–100 Mt of methane per year for fossil fuels. Methane is mainly removed from the atmosphere through reaction with hydroxyl radicals where it is a significant source of stratospheric water vapour.

$$CH_4 + OH = CH_3 + H_2O \quad (7.5)$$

The remainder is removed through reactions with the soil and loss into the stratosphere. Although the concentration of methane is over 200 times lower, it is 50 times more effective at absorbing infrared radiation than carbon dioxide (Table 7.2).

7.5.4 Carbon dioxide (CO_2)

The carbon dioxide concentration in the atmosphere is low (353 ppmv; parts per million by value) compared with oxygen and nitrogen, but it is a greenhouse

BOX 7.1 Pollutants produced by combustion

Combustion produces three main pollutants; unburnt hydrocarbons, carbon monoxide, and nitrogen oxides (NO and NO_2). Unburnt hydrocarbons and carbon dioxide are products of incomplete combustion whereas nitrogen oxides are products of the high combustion temperatures. Complete combustion can be achieved by a longer residence time in the burning zone and higher temperatures. The main pathway for the production of nitrogen oxides is shown below.

Zeldovitch mechanism

$$O_2 + N_2 \rightleftharpoons NO + N$$
$$N + O_2 \rightleftharpoons NO + O$$
$$N + OH \rightleftharpoons NO + H$$

Nitrous oxide

$$N_2 + O + M \rightleftharpoons N_2O + M$$
$$N_2O + O \rightleftharpoons NO + NO$$
$$N_2O + H \rightleftharpoons NO + NH$$

Prompt

$$N_2 + CH \rightleftharpoons HCN + N \rightleftharpoons NO$$

The Zeldovitch mechanism functions above 1800 K so that gases like methane will produce less nitrogen oxides than those from synthesis gas (carbon dioxide/hydrogen mix) as it burns at a lower temperature (see table).

Gas	Combustion temperature (°C)
Methane	1950
Propane	1988
Carbon monoxide	2108
Hydrogen	2097

gas. There is a continual flow between the atmosphere and organic and inorganic carbon in the soils and oceans. The primitive atmosphere contained nitrogen, carbon dioxide, and water vapour with possibly ammonia and methane at lower concentrations. Oxygen was only present in trace amounts. Primitive plants developed the process of photosynthesis in which carbon dioxide is fixed in order to make the cellular components and oxygen was released. Now the oxygen concentration is about 20%. Plants on land and sea carry out photosynthesis and this is balanced by carbon dioxide produced by respiration of animals and plants and microbial decomposition (Fig. 7.8). Carbon dioxide is also locked up in plant and animal debris and the oceans are a very large sink. Carbonate rocks and reefs also store carbon. Over many millennia some of the plant and animal debris has been converted by high pressure and temperature

Table 7.5 Global emissions of NO_x

Source	Output (kt* of NO_2/year)	Proportion (%)
Natural		
Soil release	18 100	11.8
Ammonia oxidation	10 200	6.6
Lightning	16 400	10.7
Anthropogenic		
Fuel combustion	65 100	42.3
Aircraft and stratosphere	3 000	2.1
Industry	4 000	2.6
Biomass burning	36 800	23.9

* kt, kilotonne, 10^3 tonnes.

Table 7.6 Emission from vehicles

	Mean on-road emissions (g km^{-1})			
Fuel type	CO	HCs*	NO_x	Particles
Unleaded petrol	27.0	2.8	1.7	0.02
Petrol, catalyst	2.0	0.2	0.4	0.05
Diesel	0.9	0.3	0.8	0.15

Source: Data derived from Colls (1997).
* HC, hydrocarbons.

Table 7.7 Nitrogen emissions in the UK

Nitrogen type	Emission (kt of N/year)	Proportion (%)
NO_x (NO and NO_2)	800	62
Ammonia (NH_3)	380	29.4
Nitrous oxide (N_2O)	110	8.6
Total	1290	

into fossil fuels. However, since the Industrial Revolution the consumption of fossil fuels has increased the atmospheric concentration of carbon dioxide (Fig. 7.8). The increases in carbon dioxide, methane, nitrous oxides, and CFCs are shown in Table 7.9.

Annual carbon dioxide emissions from the use of coal, gas, and oil were above 23 Gt in 2000 having risen from 15.7 Gt in 1973 (Gt, gigatonne = 10^9 tonnes of

Table 7.8 Sources and sinks for methane (CH_4)

Source	Annual release (Mt)
Natural	
Wetlands	115
Rice paddies	110
Ruminants	80
Biomass burning	40
Termites	40
Oceans	10
Freshwaters	5
Anthropogenic	
Gas drilling, venting	45
Coal mining	40
Hydrate distillation	5
Sinks	
Removal by soil	30
Reaction with hydroxyl ions in the atmosphere	400
Atmospheric increase	60

Source: Houghton *et al.* (1990).

Table 7.9 The effect of human activities on greenhouse gas levels

Gas	Pre-industrial levels (1750–1800)	Post-industrial levels (1990)	Rate of increase per annum (%)
Carbon dioxide (ppmv)	280	353	0.5
Methane (ppmv)	0.8	1.72	0.9
Nitrous oxide (ppbv)	288	310	0.25
Chloroflurocarbon (CFC11 and 12; pptv*)	0	764	4

Source: Houghton et al. (1990).
*pptv, parts per trillion by volume.

carbon dioxide; IEA, 2002). Carbon dioxide emissions depend on energy and carbon content of the fuel, which ranges from 13.6 to 14.0 Mt of C/EJ (megatonnes of carbon per exajoule = 10^6 tons of carbon per 10^{18} J of energy) for natural gas, 19.0–20.3 Mt of C/EJ for oil and 23.0–24.5 Mt of C/EJ for coal (Wuebbles et al., 1999). The past and present carbon dioxide concentrations are given in Fig. 7.12 (Houghton et al., 2001). With the increase in greenhouse

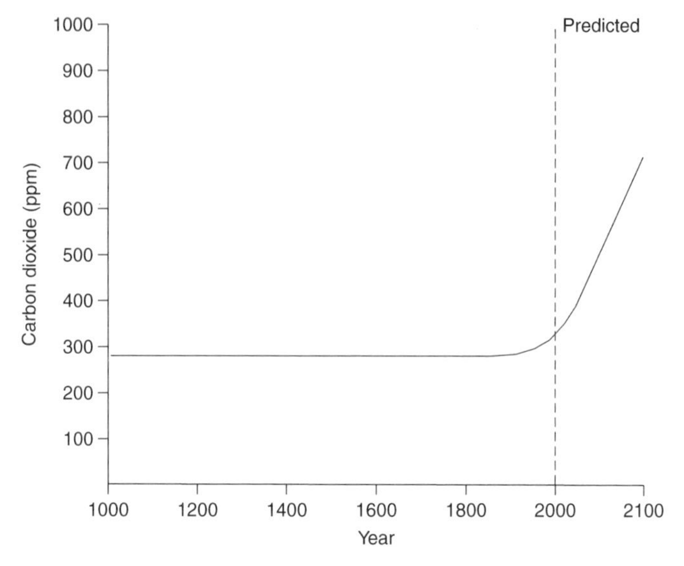

Figure 7.12 The history of carbon dioxide levels in the Earth's atmosphere and the predicted rise in the future. Data from Intergovernmental Panel on Climate Change (IPCC, 2001; www.ipcc.ch).

gases, in particular carbon dioxide, and the addition of new ones such as CFCs, human activity is raising the mean global temperature. The predicted increase in temperature is given in Fig. 7.13. At the predicted rate a global increase in temperature of 2.5°C is expected by 2100 (Houghton, 1996).

7.6 Ozone

Another important gas in the stratosphere and troposphere is ozone (O_3), which is only a trace component (8–10 ppm) but has the ability to absorb UV light (230–310 nm). Ozone is generated from oxygen by a photochemical reaction.

$$O_2 + h\nu \text{ (light)}(<240 \text{ nm}) = 2O \tag{7.6}$$
$$O + O_2 + M = O_3 + M \tag{7.7}$$

where M is any molecule such as N_2 or O_2.

A small quantity of ozone can be formed by the action of UV on naturally produced NO_2 (nitrite).

$$NO_2 + h\nu \text{ (light)}(280\text{–}430 \text{ nm}) = NO + O \tag{7.8}$$
$$O + O_2 + M = O_3 + M \tag{7.9}$$

The nitric oxide (NO) can also degrade ozone.

$$NO + O_3 = NO_2 + O_2 \tag{7.10}$$

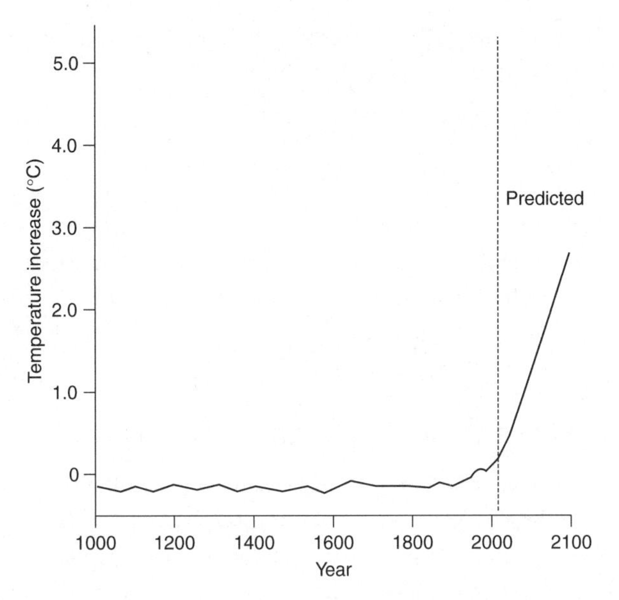

Figure 7.13 The predicted temperature rise until the year 2100. Data from IPCC (2001; www.ipcc.ch).

Ozone can also be broken down.

$$O_3 + h\nu \text{ (light) } (<325 \text{ nm}) = O_2 + O^*$$
$$O^* + O_3 = 2O_2 \quad (7.11)$$

In unpolluted atmosphere the equilibrium between the reactions above produces a low (8–10 ppm) but stable ozone concentration.

The O* reactive oxygen can also react with methane and water forming a reactive hydroxyl radical.

$$O^* + CH_4 = OH^{\cdot} + CH_3 \quad (7.12)$$

The oxygen can also react with water, forming hydroxyl radicals.

$$O^* + H_2O = 2OH^{\cdot} \quad (7.13)$$

In clean air the highly reactive hydroxyl radicals can react with carbon monoxide and methane.

$$CO + OH^{\cdot} = H + CO_2 \quad (7.14)$$
$$CH_4 + OH^{\cdot} = CH_3 + H_2O \quad (7.15)$$

The oxidation of methane is perhaps more complex, particularly in the presence of NO.

The following radicals (represented by X below) have been shown to catalyse the breakdown of ozone: Cl, NO, H, and OH$^{\cdot}$.

$$O_3 + X = O_2 + XO \quad (7.16)$$

$$O + XO = O_2 + X \quad (7.17)$$

The concentration of nitrate radical is minimal in the troposphere in the light because it is broken down by a photochemical reaction. However, at night it can accumulate by the reaction of nitrite with ozone.

$$O_3 + NO_2 = NO_3 + O_2 \quad (7.18)$$

7.6.1 The influence of industry on the ozone layer

Human activity is responsible for an increase in nitric oxide (NO) and nitrogen dioxide (NO_2), which increase the destruction of ozone as X in reaction 7.16. The most important atmospheric constituents in ozone depletion are the halogenated hydrocarbons, particularly the CFCs. CFCs were used in refrigerators, as propellants in aerosols, in packaging and insulation, and in electronic cleansing fluids. This use caused the release into the atmosphere of these compounds, which are very stable and unreactive (Table 7.10). However, when the CFCs reach the stratosphere they are broken down by UV light, which releases chlorine or bromine ions, which then act as X in reaction 7.16, breaking down ozone.

The consequence of ozone depletion is that more UV light in the region of 230–320 nm reaches the surface. The increase in surface UV light causes sunburn, pigmentation, skin aging and skin cancer, eye damage, immune-system damage, and long-term effects on plants. The effect of human activities on the ozone layer was highlighted first in 1985 when the British Antarctic Survey showed that there was a significant decline in the ozone over the Antarctic, called the ozone hole. Since then both national and international agreements, including the Montreal Protocol, have been implemented to reduce or replace CFC use.

Table 7.10 CFCs involved in ozone depletion

Halocarbon	Formula	Concentration (ppt)	Lifetime (years)
CFC11	$CFCl_3$	3 346	60
CFC12	CF_2Cl_2	6 349	120
CFC113	$C_2F_3Cl_3$	1 683	90
CFC114	$C_2Cl_2F_4$		200
CFC115	C_2ClF_5		400
Methyl chloroform	CH_3CCl_3	623	6.3
Carbon tetrachloride	CCl_4	728	50

Table 7.11 Global emissions of sulphur compounds into the atmosphere

Source	Annual flux (Tg of S*)
Anthropogenic (fossil fuel burning)	80
Biomass burning	7
Oceans	40
Soils and plants	10
Volcanos	10
Total	147

*Tg, teragrams, 1×10^{12} g.
Source: Houghton et al. (1990).

7.7 Sulphur dioxide

The concentration of sulphur dioxide is less than 1 ppb in clean air, rising to 2 ppm in highly polluted areas, with levels typically at 0.1–0.5 ppm. Sulphur dioxide is a respiratory irritant, which can affect human health and damage plants. There are a number of natural and anthropogenic sources of sulphur dioxide as shown in Table 7.11, but the latter is by far the largest source. Marine phytoplankton produce dimethyl sulphide, which is converted to sulphur dioxide in the atmosphere; hydrogen sulphide is formed by anaerobic decay; and volcanoes emit sulphur dioxide. Most of the sulphur dioxide produced by human activity is from the burning of fossil fuels.

- Oil products contain between 0.1 and 3% sulphides.
- Natural gas can contain hydrogen sulphide which is removed before use.
- Coal contains between 0.1 and 4% sulphur as inorganic iron pyrites and organic thiophenes (Chapter 6).

Of these three sources, burning coal in power stations is the main source of sulphur dioxide as can be seen from the typical emissions from a coal-fired power station (Table 7.12). The main emissions include carbon dioxide, sulphur dioxide from the sulphur compounds, and nitrous oxides from the nitrogenous compounds in the coal.

7.8 The effects of industrial (anthropogenic) activity

The climate system remained stable for at least 1000 years prior to the Industrial Revolution, as measured by ice-core determinations. Since the start

Table 7.12 Composition of emissions from a coal-fired power station

Chemical	Concentration
Air (oxygen-depleted)	80%
Water	4.5%
Carbon dioxide (CO_2)	12%
Carbon monoxide (CO)	40 ppm
Sulphur dioxide (SO_2)	1000–1700 ppm
Sulphur trioxide (SO_3)	1–5 ppm
Nitric oxide (NO)	400–600 ppm
Nitrogen dioxide (NO_2)	20 ppm
Nitrous oxide (N_2O)	40 ppm
Hydrochloric acid (HCl)	250 ppm
Hydrofluric acid (HF)	<20 ppm
Particulates	<115 mg m^{-3}
Mercury	3 ppb

Source: Data from Roberts et al. (1990).

of the Industrial Revolution atmospheric concentrations of carbon dioxide, methane, nitrogen oxides, and carbon monoxide have all increased. The reasons for these increases are the burning of fossil fuels, deforestation, and agricultural activities (Fig. 7.14).

7.8.1 Consequences of emissions

This is the most rapid change in global temperature for the last 10 000 years and will have a number of consequences; a number of scenarios have been put forward. The trends in observed climate changes are given in Table 7.13 (Weubbles et al., 1999; Houghton et al., 2001). The melting of the sea ice and glaciers will increase the sea level by 0.5 m, which will directly affect people living in low-lying areas such as the delta regions of Egypt, China, and Bangladesh, where 6 million people live below the 1-m contour. A summary of the impact of climate change is given by Weubbles et al. (1999), who concluded that "the balance of evidence suggests that there is discernable human influence on global climate".

7.8.2 Acid rain

In the atmosphere sulphur dioxide is rapidly oxidized to sulphuric acid in the following reactions.

$$2SO_2 + O_2 = 2SO_3 \quad (7.19)$$

$$SO_3 + H_2O = H_2SO_4 \quad (7.20)$$

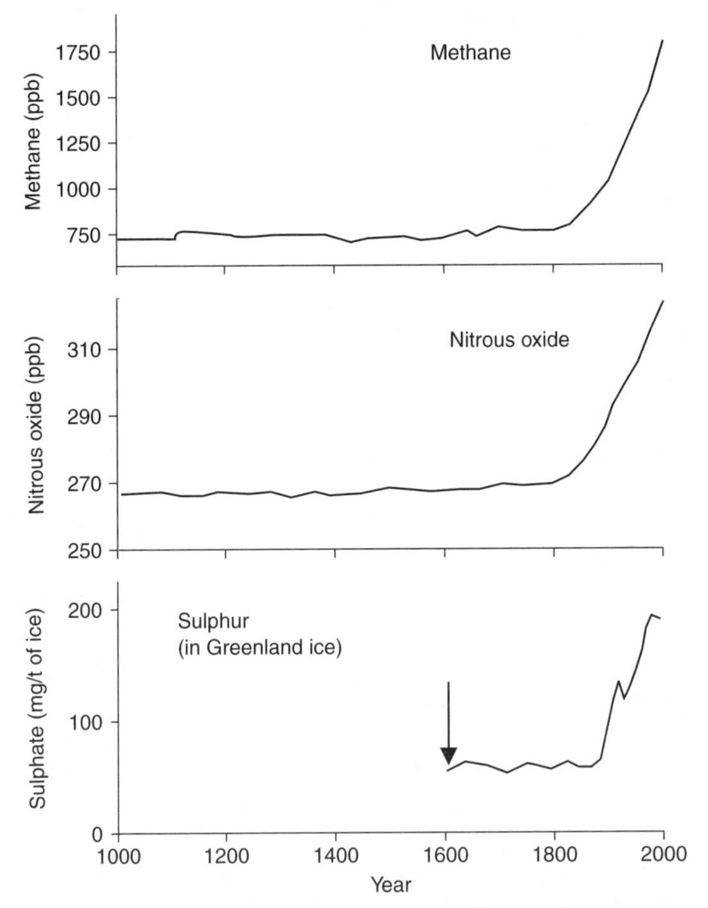

Figure 7.14 The increases in methane (ppb), nitrous oxide (ppb), and sulphate (mg/t of ice) as detected from ice cores from 1000 to 2000. In the lower panel the arrow indicates the time limitation of sulphur detection in ice samples. Data from IPCC (2001; **www.ipcc.ch**).

This oxidation is rapid in the presence of water and particulates and the sulphur dioxide will also react with hydroxyl radicals.

$$SO_2 + OH^{\cdot} + M = HOSO_2 + M \quad (7.21)$$

where M is a particle or molecule acting as a catalyst.

$$HOSO_3 + H_2O + O_2 = H_2SO_4 + HO_2 \quad (7.22)$$

$$HO_2 + NO = OH + NO_2 \quad (7.23)$$

Sulphur dioxide is also very soluble in water, forming sulphurous acid.

$$SO_2 + H_2O = H_2SO_3 \quad (7.24)$$

$$H_2SO_3 \rightleftharpoons HSO_3^- + H^+ \quad (7.25)$$

$$HSO_3^- \rightleftharpoons SO_3^{2-} + H^+ \quad (7.26)$$

Table 7.13 Trends in the climate that indicate global warming

Variable	Analysis period	Change
Surface air temperature and sea surface temperature	1851–1995	0.65 ± 0.15°C
Alpine glaciers	20th century	Indicates warming of 0.6–1.0°C
Extent of snowcover in northern hemisphere	1972–1992	10% decrease in annual mean
Extent of sea ice in the northern hemisphere	1973–1994	Downward trend since 1977
Extent of sea ice in southern hemisphere	1973–1994	No change, possible decrease 1950–1970
Length of northern hemisphere growing season	1981–1991	12 ± 4 days longer
Precipitation	1900–1994	Generally increasing outside tropics
Heavy precipitation	1910–1990	Growing in importance
Antarctic snowfall	Recent decades	5–20% increase
Global mean sea level	20th century	1.8 ± 0.7 mm/year

Sources: Based on Harvey (1999), Houghton (1996), and Houghton et al. (2001).

The pH of clean rainwater is about 5.6 (see Box 7.2) due to dissolved carbon dioxide. However, in the presence of pollutants such as sulphur dioxide and nitrous oxides, sulphuric, sulphurous, and nitric acid are formed which reduce the pH down to 1. These acids have a short residence time in the atmosphere due to their returning to the surface as rain. The acid rain has an effect on water bodies, vegetation, and buildings. The problem of changes in water and soil pH has been of concern in the last 40 years. Acidification of waters causes an increase in the leaching of toxic metal ions into the water and also changes the flora and fauna. Acid rain has been blamed for the death of trees in a number of forests in Scandinavia and the USA. The effect is perhaps indirect as the change in pH of the soil may leach out toxic metals and change the uptake of ions. Acid rain has an effect on buildings, particularly those made from limestone.

7.8.3 Smog

The burning of fossil fuels also produces particulates, soot, and black smoke from vehicles and power stations. Power stations now control these types of emission and there are regulations on the particulate emissions from vehicles.

BOX 7.2 **Acid rain**

In pure water the concentration of hydrogen (H^+) and hydroxyl (OH^-) ions is 10^{-14} M at 25°C. Therefore, the concentration of the hydrogen ions is 10^{-7}M.

The definition of pH is pH $= -\log_{10}[H^+]$ and therefore the pH of pure water is 7. However, carbon dioxide dissolves in water as follows.

$$CO_2 + H_2O \rightleftharpoons H_2CO_3 \rightleftharpoons H^+ + HCO_3^-$$

$$HCO_3^- \rightleftharpoons CO_3^{2-} + H^+$$

These give water a pH of about 5.6.

But air does contain other gases such as sulphur dioxide and nitrogen oxides that can also affect the pH of water.

Sulphur dioxide will dissolve in water.

$$SO_2 + 2H_2O \rightarrow HSO_3^- + H_3O$$

The equilibrium constant (K) for the reaction is 2.1×10^{-2}M.

$$K = \frac{[HSO_3^-][H_3O^+]}{P_{SO_2}} = 2.1 \times 10^{-2}\ M$$

If the concentration of sulphur dioxide is 0.2 ppm, this is equivalent to a partial pressure (P) of 0.2×10^{-6} atm. If $[H_3O^+] = [HSO_3^-]$ therefore the total is $[H_3O^+]^2$. Thus

$$K = \frac{[H_3O^+]}{P_{SO_2}}$$

$$[H_3O^+]^2 = K \times Pso_2$$

$$[H_3O^+] = \sqrt{K \times P_{SO_2}}$$

$$= \sqrt{(2.1 \times 10^{-2}) \times (0.2 \times 10^{-6})}$$

$$= 6.48 \times 10^{-3} \approx H_3O^+$$

Therefore

$$pH + -\log_{10}[H_3O^+] = 4.19$$

Before the Clean Air Act in the UK coal fires were responsible for the creation of smog, a mixture of fog and smoke, in large cities, but these do not occur now. However, photochemical smogs do occur at the present time where NO_x, mainly NO, and unburnt hydrocarbons build up due to high traffic density in cities such as Mexico City, Bangkok, and Los Angeles. Nitric oxide is converted to nitrogen dioxide and nitrogen dioxide is photochemically used to produce ozone.

$$NO_2 + h\nu\ (\text{light}) = NO + O \qquad (7.27)$$

$$O + O_2 + M = O_3 + M \qquad (7.28)$$

In addition, aldehydes are produced which are the precursors of peroxyacyl nitrates (PANs). The unhealthy parts of the smog are ozone, PANs, NO_2, and particles which can cause respiratory problems.

7.9 Remediation of the emissions from fossil fuels

7.9.1 Reduction in sulphur dioxide

The methods of reducing sulphur dioxide emissions are as follows.

- Burning less fuel is an obvious remedy but the demand for energy is still increasing as countries become more developed.
- A reduction in fuel use can be achieved by reducing energy loss and with more efficient combustion.
- Use low-sulphur fuels. Inorganic sulphur compounds, such as pyrites, in coal can be removed by catalytic hydrodesulphuration. The organic compounds in coal, mainly thiophenes, can be removed by microbial action (Chapter 6). Natural gas contains little sulphur and this is one reason for its increasing use in electricity generation.
- Improved combustion in power stations can reduce emissions such as pressurized fluidized-bed combustion.
- Sulphur compounds can be removed from flue gas by a number of methods. The most common is to spray calcium carbonate into the gas where the sulphur dioxide combines to form calcium sulphate or gypsum, which can be used to make plasterboard.

$$SO_2 + CaCO_3 = CaSO_4 \cdot 2H_2O \tag{7.29}$$

- Replace fossil fuels with cleaner renewable energy sources, which contain low levels of sulphur.

7.9.2 Reduction in nitrogen oxides

- Burn less fuel (see above).
- The higher the combustion temperature the more nitric oxide is formed. Reduction in combustion temperature will reduce the formation of nitric oxide. The production of nitric oxide can be reduced by changing the air/fuel ratio, making it fuel-rich, but this reduces the thermal efficiency. Low-NO_x burners have also been developed which also reduce thermal hot spots.
- Similar to flue-gas treatment for sulphur compounds, nitric oxide can be removed from flue gas by injecting ammonia. In the presence of a catalyst the nitric oxide is converted to nitrogen gas.

$$4NH_3 + 4NO + O_2 = 4N_2 + 6H_2O \tag{7.30}$$

- Catalytic converters can be fitted to vehicles, which can convert nitric oxide to nitrogen gas.
- Replace fossil fuels with cleaner, low-nitrogen, renewable sources.

7.9.3 Reduction in carbon dioxide

- Burn less fuel (as above).
- The global carbon flow is shown in Fig. 7.15, which shows the large input of fossil fuels to atmospheric carbon dioxide (Kirschbaum, 2003). The removal of carbon dioxide from the atmosphere is carried out by the oceans and the biosphere. The ocean's uptake potential is very large but the process is slow and the system will take many centuries to equilibrate. However, plants and soil can sequester carbon dioxide rapidly compared with the oceans and enhanced cropping and growth may offset some 2000–5000 Mt of C/year, but a more realistic figure is 200–1000 Mt of C/year (Cannell, 2003). Changes in the carbon dioxide uptake by the biosphere through deforestation, reforestation, increased vegetation, and soil populations will alter the atmospheric carbon dioxide levels. Biotechnology may only affect these alternatives by the use of genetic engineering to improve photosynthesis or increase lignin concentrations in plants. Lignin is the most difficult constituent to break down so it would lock up carbon dioxide for longer. The use of land to produce energy crops is the best option as it both removes carbon dioxide and replaces fossil fuels (Cannell, 2003).
- The deep water in the oceans is not saturated with carbon dioxide and there are schemes to inject carbon dioxide into the deep sea. An alternative is to inject carbon dioxide into old gas and oil reservoirs. This has the advantage that it may help to recover additional oil from these reservoirs. These techniques are under investigation.
- The last possibility is to replace fossil fuels with renewable carbon dioxide-zero or -neutral energy sources.

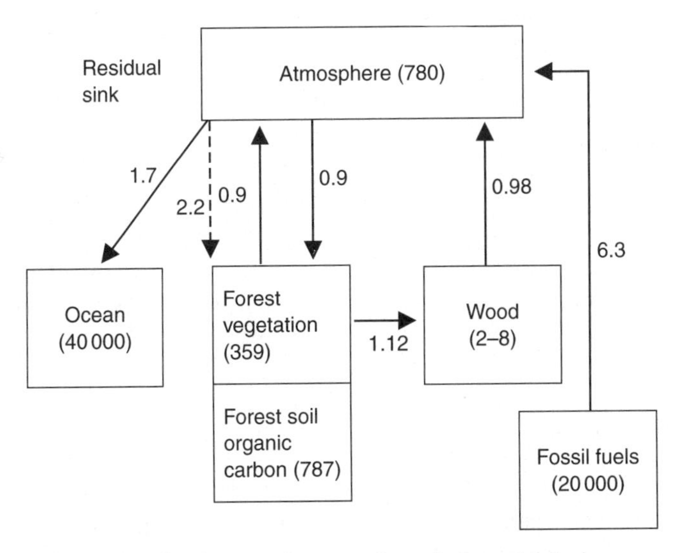

Figure 7.15 Global carbon flow between the atmosphere, the large sink in the oceans, and the contribution from fossil fuels. Values are in Gt of carbon per year. Data from Kirschbaum (2003).

7.10 Alternative non-fossil energy sources

Alternative systems for the supply of all forms of energy are being sought and those now being used or under development are detailed below.

7.10.1 Nuclear power

The fission process releases large amounts of energy, about 50 million times that of coal on a weight basis, which means that very little uranium fuel is required. No combustion is involved so that there are no emissions but fission generates radioactive materials, some of which have very long half-lives. There are also considerable problems in the reprocessing and disposal of spent fuel, the possibility of leaks or accidents, and the decommissioning of the power stations at the end of their working life. The accidents at the nuclear generating plants at Three Mile Island and Chernobyl have shown that despite very stringent safety arrangements accidents can occur. This has made the public wary of nuclear power and more likely to accept alternatives sources of power.

7.10.2 Hydroelectric power

Hydroelectric power is a clean, non-polluting, long-lasting, renewable source, which does not produce carbon dioxide. Large-scale hydroelectric plants are responsible for about 17% of the electricity supply in developed countries and 31% in developing countries. However, hydroelectric systems have environmental impacts and can only be sited in certain areas, thus restricting their application.

7.10.3 Tidal power

The rise and fall of water levels due to tides can be harnessed to generate electricity and, like hydropower, it is clean, reliable, long-lasting, renewable, and does not produce carbon dioxide. Sites with a sufficient tidal range and area are limited and this power source represents only 10% of the energy that is available from hydroelectricity.

7.10.4 Wave power

Schemes for the harnessing the rise and fall of waves are under investigation in a number of countries. Devices for the conversion of wave energy to shaft power or compression have been proposed and a number have been tested.

7.10.5 Wind power

Harnessing the power of the wind is one of the most promising alternative methods of electricity generation as it has the potential to generate substantial amounts of energy without pollution. Wind can also be used to drive water pumps in order to store energy, to charge batteries in remote regions, or as off-grid power sources. The potential for wind power has been recognized and wind farms have been installed in 15 countries including Brazil, China, Denmark, Spain, USA, India, and the UK.

7.10.6 Geothermal power

The centre of the Earth is very hot at about 4000°C and most heat which reaches the surface cannot be utilized, but in areas of volcanic activity high-grade heat is retained in molten or hot rocks at a depth of 2–10 km. The heat from these hot or molten rocks can be extracted from hot springs and used to run steam turbines directly for the generation of electricity. If the water is below 150°C it can be used as a supply of hot water for industrial or domestic heating.

7.10.7 Solar energy

Sunlight can be used either directly or indirectly for solar panels for hot-water generation, solar collectors for steam generation, used to create electricity, solar architecture for heating buildings, photovoltaic generation of electricity, and solar hydrogen generation.

7.10.8 Biological power

It is in this area that biotechnology will make a major contribution, as discussed in the next section.

7.11 Biological energy sources

Biological materials, such as wood, have always been used by humans as a source of energy but recently the use of biological materials to provide a source of energy that is renewable and carbon dioxide-mitigating has attracted considerable attention. The use of renewable biological materials to replace fossil fuels has a number of advantages, which are the reduction in coal, gas, and oil use, a reduction in the emission of greenhouse gases, and the reduction in imported energy sources. One tonne of biomass used to generate electricity prevents 0.5 t of carbon from being emitted as carbon dioxide from coal, 0.44 t of carbon from oil, and 0.28 t of carbon from natural gas. Biomass used to produce liquid fuels prevents 0.2–2.0 t of carbon/hectare

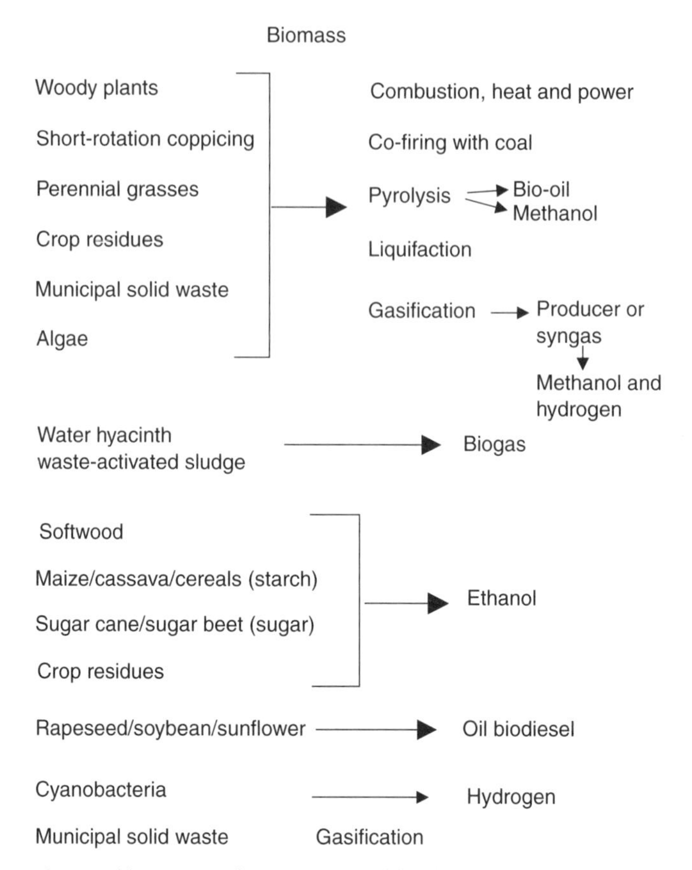

Figure 7.16 The possible sources of energy, gas, and liquid fuels from a range of biological materials.

from being emitted into the atmosphere (Cannell, 2003). In the UK the former Ministry of Agriculture, Food and Fisheries (MAFF) encouraged the planting of energy crops with the Arable Area Payments Scheme and the UK also has the Non-Fossil Fuel Obligation (NFFO), where regional electricity companies must purchase some of their electricity derived from a non-fossil source and this supply commands a premium price. One estimate states that biomass is supplying some 15% of the world's energy, or 55 EJ (55×10^{18} J) (25 million barrels oil/day) (Scurlock et al., 1993). Another estimate gives renewables as 13.8% of the world's total energy supply and it is predicted that this will decline to 12.5% by 2030 (IEA, 2002).

Biomass energy can be used to produce heat, electricity, gas, or liquid fuels. Figure 7.16 outlines the possible sources of biomass energy. The energy can be extracted as follows:

- bio-oils from pyrolysis,
- production of methanol from gasification,
- production of biogas (methane),
- plant-derived oils; biodiesel,

- production of ethanol; gasohol,
- production of hydrogen from gasification of biomass and direct biological production.

7.12 Combustion of biomass

Biomass is the term for organic matter, both living and dead, such as trees, crops, grasses, roots, micro-organisms, algae, and plant process wastes. The types of biomass are therefore very diverse and a number of methods of extracting energy from biomass are currently being investigated in a number of countries. The extraction of energy from biomass can take a number of routes, as shown in Fig. 7.17.

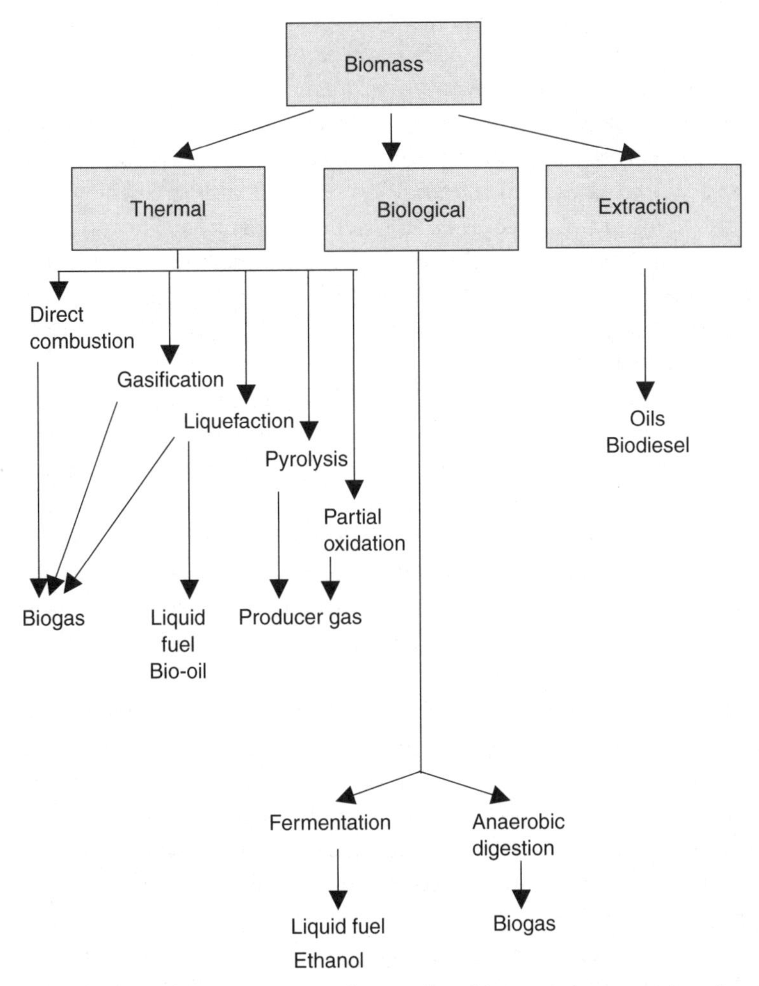

Figure 7.17 The routes for the extraction of energy from biomass using extraction, thermal, and biological methods.

The direct combustion of biomass in the form of wood has been with us for thousands of years as a source of heat. However, wood chips, grasses, and waste can be burnt in order to produce electricity. Some 90% of the biomass energy is held in trees and in developing counties timber is used not only for energy but also for a number of other industries. At the same time the trees are not being replaced at the same rate as they are being harvested so that the resource is being depleted and carbon dioxide added to the atmosphere. A problem with the use of wood in boilers is that although combustion technology is advanced for coal and oil, the combustion of wood is not well understood.

Gasification is a process where the biomass is heated in an oxygen-limited atmosphere to produce a gaseous fuel. The resulting gas is a mixture of carbon monoxide, hydrogen, methane, carbon dioxide, and nitrogen. There is a number of gasification methods operating at various temperatures. The value of the gasification process is that it produces a cleaner fuel, which is more versatile, and can be burnt in boilers, in internal combustion engines, and gas turbines. Under some conditions gasification can produce synthesis gas (syngas), which is a mixture of methane and hydrogen. The methane in the syngas can be removed by reacting it with steam over a catalyst at around 900°C, which converts it to carbon monoxide and hydrogen. A further reaction with water converts any excess carbon monoxide to carbon dioxide, which can be removed by a solvent process or pressure-swing adsorption. The remaining mixture of carbon monoxide and hydrogen can be either converted to methanol by reaction over a catalyst at 450°C or if hydrogen is required the two gases can be separated (Larson et al., 1996).

Pyrolysis is the heating of the biomass in the absence of air at temperatures of 300–500°C. Under these conditions the solids which remain are charcoal, and the volatiles if they are collected can, after treatment, be used as fuel oil (bio-oil). The production of charcoal has been known for centuries and as a fuel it has twice the energy density of wood and burns at a higher temperature. Liquifaction is similar to pyrolysis but with a lower temperature under high pressure.

Co-firing is the firing of a renewable fuel such as biomass with coal, natural gas, and oil mainly for the generation of electricity. Burning of biomass has been shown to reduce emissions of sulphur dioxide and NO_x and will clearly reduce carbon dioxide emissions. This is due to the biomass containing little nitrogen and sulphur and during firing the nitrogen is converted to ammonia, which reduces nitric oxide to nitrogen. Some of the biomass types which have been co-fired are cattle manure, sawdust, sewage sludge, straw, and municipal solid waste (Sami et al., 2001) so that it can be part of a waste treatment system. However, there are some technological problems due to the corrosive alkaline of the ash, the biomass needs to be particulate and combustion studies are required (Table 7.14).

The sources of biomass for the above processes can be agricultural, domestic, and industrial wastes, and purpose-grown crops. The energy contents of the wastes or crops are given in Table 7.15 and compared with petrol. The

Table 7.14 Estimated carbon dioxide reduction by replacing fossil fuels with plant-derived fuels.

Type of biomass	Carbon content (dry weight)	Change in soil organic content during growth	Fossil energy input during growth	Estimated total CO_2 emission reduction
Maize	400	–	−20/−40	300 ± 80
Perennial herbaceous crops	400	55–150	−12	400 ± 140
Short-rotation woody crops				
3-year rotation	540	0–200	−8	550 ± 210
10-year rotation	540	0–200	−8	600 ± 220
Forests				
100-year rotation	540	–	–	140 ± 30
400-year rotation	540	–	–	30 ± 10

Note: All values are kg of carbon/Mg of biomass.
Source: Cook and Beyea (2000).

Table 7.15 Mean energy contents of various fuels

Fuel	Energy (GJ/t)
Natural gas	55
Coal	28
Petrol	47.3
Diesel	43.0
Wood (20% moisture)	15
Paper	17
Dung	16
Straw	14
Sugarcane	14
Domestic waste	9
Commercial waste	16
Grass	4

residues from forestry and timber processing are obvious sources but are often discarded. Straw burning has been stopped in the UK and this straw represents a considerable source of energy. At present some 200 000 tonnes of straw, less than 1% of the total straw produced in the UK, are burnt in boilers. In tropical countries wastes like bagasse (sugar cane), rice husks, and old cotton plants are under investigation as boiler fuels. Domestic and industrial wastes also contain combustible material and this can be used as fuel. Most of the municipal solid waste in the UK goes to landfill sites, but there are a few solid-waste burners operating in the UK and these sell non-fossil-fuel-generated electricity. Municipal solid waste has been used to generate either methanol or hydrogen as a possible fuel for metropolitan buses (Larson et al., 1996). The disadvantage of wastes is that they have to be treated to remove non-combustible materials prior to combustion.

The growth of crops specifically for the production of energy has attracted attention, in particular in the EU, with the provision of renewable material for bioenergy generation. The bioenergy systems have the advantages of being renewable, carbon dioxide-neutral, and utilizing the excess land in the EU, but the main limitations are the costs of harvesting, delivery, drying, storage, and the combustion technology. If these crops are to replace coal or gas as a fuel they need to be able to compete on a cost basis. One of the main parameters influencing cost is productivity as measured in terms of dry weight of biomass/ha/year. Three main crops have been considered as possible candidates; forestry (wood chips), short-rotation coppicing, and perennial grasses. Forestry has long been recognized as an economic crop with yields up to 20 t/ha/year, and can be used directly as wood chips or for the production of charcoal and oil via pyrolysis. Short-rotation coppicing of willow (*Salix*) and poplar (*Populus*) appear to be most promising crops with yields of 9–20 t/ha/year (Table 7.16). The third group are the perennial grasses such as *Miscanthus*, napier grass (*Pennisetum*), and limpograss (*Hemmthria*). These are not native to the UK or Europe but have a fast growth rate and give high yields (Table 7.16). *Miscanthus* can be grown in Europe and is under trial at present within the EU. When considering a large-scale bioenergy program the following need to be considered;

- land availability,
- productivity of species,
- environmental sustainability,
- social factors, and
- economic feasibility.

7.13 Biogas

Biogas is a mixture of gases produced when organic material is broken down anaerobically. It contains 50–75% methane, carbon dioxide, hydrogen suphide, and hydrogen. It contains sufficient energy (20–25 MJ/kg) to be used as a fuel in boilers and dual-fuel engines. The anaerobic digestion of all sorts of organic matter follows the same pattern as that found in the anaerobic

Table 7.16 Possible biological sources of energy and fuel

Source	Fuel type	Plant/organism	Yield (dry weight/ha/year)
Woodland/forests	Wood chips	Many species	10–35
Short-rotation	Wood chips	Willow (*Salix dasyclodo*)	6–15
coppicing		Poplar (*Populus* sp.)	10–17
Perennial grasses	Straw etc.	*Miscanthus* sp.	20
		Limpograss (*Hemmthria*)	7–22
		Napier grass (*Pennisetum*)	34–55
Crop residues	Waste	Cane fibre (from sugar cane, *Saccharum officinarum)*	
		Maize straw (*Zea mays*)	20
		Straw from rice, wheat, barley, millet	
Municipal waste	Waste	–	
Cultivated crops	Sugar	Sugar cane	36–70
		Sugar beet	7.8–15.4
	Starch	Maize	26
		Cassava (*Manihot*)	6.1–13.2
		Potato	5–21
	Oil	Rapeseed (*Brassica napus*)	2–3
		Sunflower (*Helianthus annuus*)	
		Soybean (*Glycine max*)	
Aquatic plants	Plant material	Water hyacinth	
		(*Eichornia* sp.)	52–100
		Cattails (*Typha* sp.)	8–34
Microbial cells	Hydrogen	Cyanobacteria, microalgae	
	Oil	*Botryococcus braunii*	

digestion of excess sewage sludge in sewage works. Methane is produced in this system by the strict anaerobic methanogens. The anaerobic digestion systems that can produce large quantities of biogas are sludge digestion, landfill sites, agricultural and industrial wastes, and specifically constructed digesters.

Activated-sludge digestion has a dual use; it reduces the amount of sludge needed to be disposed of and it produces gas, which can be used to power pumps and heaters (Chapter 4). A digester of 200–400 m^3 can produce gas at a rate of 1.0 m^3/m^3/day that represents between 4.2 and 10.4 GJ

(Table 7.17). In some cases biogas has been used to generate electricity as in some cases it commands a non-fossil-fuel premium as part of the Non-Fossil Fuel origin (NFFO) agreement. Landfill sites when capped can produce methane, which can be collected if pipes or channels are incorporated into the construction. In some cases landfill sites have been designed to produce gas for domestic heating (Chapter 4). Agricultural and some industrial wastes are sufficiently strong to be suitable for anaerobic digestion and small anaerobic digesters have been installed on farms. A number of industrial wastes are treated anaerobically and the biogas used to run the pumps and heaters. In one case, a brewery, the biogas produced is used in a combined heat and power system which produces electricity and hot water (see Box 6.5).

In a number of developing countries anaerobic digestion has been used for on-site treatment of waste and the production of low-pressure gas for domestic use. These systems have been used widely in India and China and a typical design is shown in Fig. 7.18. The digester is run at ambient temperature,

Table 7.17 Biogas production rates

Substrate	Biogas m^3/m^3/day	Hydraulic retention time (hours)
Primary sewage sludge	0.9–3.0	5–22
Secondary sewage sludge	0.7–2.4	5–22
Municipal waste	2.4–3.6	19–30
Non-recyclable municipal waste	0.3–0.6	20
Primary and secondary sewage sludge	1.0	10
Cattle manure	1.0	10
Pig manure	1.0	10

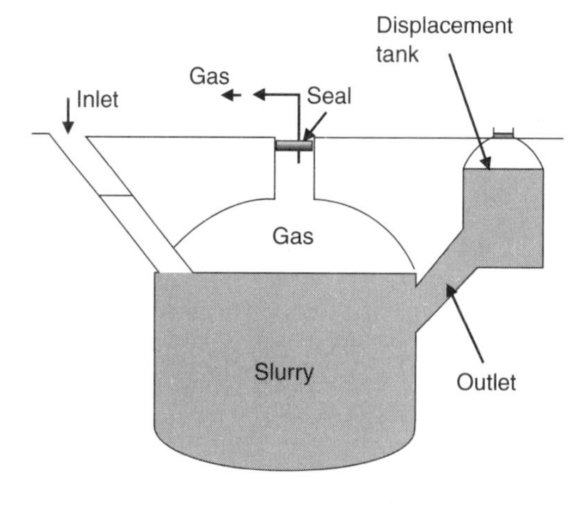

Figure 7.18 One of the designs for the production of biogas from human and animal waste. The gas provides a low-pressure source which can be used for cooking and heating.

which must reduce gas production in winter when the temperature is low. The loading rate in these systems is about 10 kg of cattle dung/m^3/day, producing gas at 0.15 m^3/m^3/day that is used mainly for cooking and lighting.

7.14 Biodiesel

Both transport and industry rely heavily on the diesel engine that is widely used to power tractors, pumps, and generators. The USA uses 50 billion (50×10^9) gallons annually (Louwrier, 1998) and the UK consumption was 16.42×10^6 tonnes in 2001. The engine designed by Diesel ran for the first time on the 10th August 1893 and the patent, when filed, proposed that the fuel could be powdered coal, groundnut oil, castor oil, or a petroleum-based fuel (Shay, 1993; Machacon et al., 2001). At the time the growing petrochemical industry provided the best fuel, a crude oil fraction now called diesel, which has been the fuel of choice ever since this time. Conventional diesel is produced by the distillation of crude oil collecting middle distillate fractions in the range of 175–370°C. The fuel contains hydrocarbons such as paraffins, naphthenes, olefins, and aromatics containing 15–20-carbon molecules. To replace diesel the substitute will have to be similar in the following properties:

- the calorific value, a measure of the energy available in the fuel;
- the cetane number, a measure of the ignition quality of the fuel;
- the viscosity of the fuel, which is important as it affects the flow of the fuel through pipelines and injector nozzles where a high viscosity can cause poor atomization in the engine cylinder;
- the flash point, a measure of the volatile content of the fuel and gives a measure of the safety of the fuel;
- obtained from renewable resources; and
- available in large quantities.

Plant-derived oils have been used to replace diesel in emergency situations but recently interest has been revived in the use of oils as a renewable and carbon-neutral replacement for diesel. It is clear from Diesel's patent that the use of plant-derived oils in the internal combustion engine is not new. Oils are produced from plants throughout the world in considerable quantities (Shay, 1993). Plant oils are normally extracted from oil-containing seeds where the plant uses oil rather than starch as an energy store for the seed. Seed oil can be extracted from a wide range of crops such as soybean, sunflower, rapeseed (canola), and oil palm and these can be grown in most climates and locations. A list of oil-producing plants is given in Table 7.18 where it is clear that perennial crops have a higher yield of oil per hectare. Despite the higher yields

Table 7.18 Oil yields from annual and perennial crops

Plant	Yield (kg/ha/year)
Perennials	
Tung oil tree (*Aleurites fordii*)	790
Cocoa (*Theobroma cacao*)	863
Olive (*Olea europea*)	1 019
Brazil nut (*Bertholietia excelsa*)	2 010
Avocado (*Persea americana*)	2 217
Coconut (*Cocos nucifera*)	2 260
Macauba palm (*Acrocomia aculeata*)	3 775
Oil palm (*Elaeis guineensis*)	5 000
Annuals	
Maize (*Zea mays*)	145
Cotton (*Gossypium hirsutum*)	273
Soybean (*Glycine max*)	375
Linseed (*Linum usitatissimum*)	402
Sunflower (*Helianthus annuus*)	800
Peanut (*Arachis hypogaea*)	890
Rapeseed (*Brassica napus*)	1 000
Gopher plant (*Euphorbia lathyris*)	1 119
Castor bean (*Ricinus communis*)	1 188
Jojoba (*Simmondsia chinensis*)	1 528

Sources: Data from Tickell (2000) and Shay (1993).

from the perennial plants, annual crops like rapeseed and sunflower have commanded most interest, probably because there is already a market for their oil and annuals are a more flexible crop.

However, there is other plant sources of oils or compounds suitable as fuels such as herbaceous plants, trees, and algae. A number of herbaceous plants produce hydrocarbons (terpenes), particularly those in the Euphorbiaceae such as *Hevea brasiliensis, Euphorbia lathyris* (3–10 t of dry weight/ha/year) and *Calotropis procera* (10.8–21.9 t of dry weight/ha/year). The hydrocarbons are produced as latex, which consists largely of a C_{30} triterpenoid which can be cracked (by pyrolysis) to form petrol. The herbaceous plants can be grown in various parts of the world and give quite good yields in terms of dry weight per hectare. The annual plants do have the problems of soil erosion and annual planting which are not found for the trees. Trees like *Eucalyptus globus, Pittosporum resiniferum*, and *Copaifera multijuga* also produce oils; often in the fruit as in *P. resiniferum* but with

Table 7.19 Lipid contents of some microalgae

Species	Maximum lipid content (%; w/w)
Monalanthus salina	72
Botrycoccus braunii	53–75
Dunaliella primolecta	54
Dunaliella bardawil (salina)	47
Navicula pelliculsa	45
Radipsphaera negevensis	43
Biddulphia aurita	40
Chlorella vulgaris	40–58
Nitzschia palea	40
Ochromonas dannica	39–71
Chlorella pyrenoidosa	36
Peridinium cinctum	36
Neochloris oleabundans	35–54
Oocystis polymorpha	35
Chrysochromulina sp.	33–48
Phaeodactylum tricornutum	31
Stichococcus bacillaris	32

Sources: Data from Kosaric and Velikonja (1995) and Scragg et al. (2002).

the Brazilian tree *C. multijuga* the trunk can be tapped and the oil used directly as a diesel replacement.

A number of algae are capable of producing terpenoid oils, one of which is *Botrycoccus braunii* which is reported to accumulate up to 86% dry weight as oil (Calvin, 1985). Hydrocracking of the oil yields 62% petroleum, 15% aviation fuel, 15% diesel, and 3% heavy oil. Low-nitrogen conditions can stimulate lipid accumulation, some of which can be very high (50–60%) (Table 7.19). This has been the subject of a comprehensive National Renewable Energy Laboratory study.

The advantages of plant-derived oils are,

- liquid form,
- calorific content is 80% of diesel (Table 7.20),
- readily available,
- renewable/sustainable,
- non-toxic and biodegradable,
- carbon dioxide-neutral, and
- low-sulphur.

Table 7.20 Energy content of plant oils, microalgae, petrol, and diesel

Fuel	Energy content (MJ/kg)
Petrol	47.3
Diesel	43.0
Ethanol	29.4
Methanol	22.4
Rapeseed oil	39.5
Castor oil	37.0
Sunflower oil	36.9
Euphorbia oil	39.3
Chlorella vulgaris	28.0

The disadvantages are

- High viscosity,
- low volatility and high flash point, and
- contain reactive unsaturated hydrocarbon chains.

To function correctly in a diesel engine the fuel must form a fine mist, which should burn rapidly and evenly. It would appear that plant-derived oils could not be used directly for long periods in diesel engines for a number of reasons (Ma and Hanna, 1999).

Untreated plant-derived oil contains residual components such as waxes, gum, and high-molecular-weight fatty components, which clog the fuel lines and filters. High oil viscosity causes poor atomization, affecting ignition and combustion, which gives carbon deposits on injectors, combustion chamber walls, and pistons. The polymerization of unsaturated fatty acids in the combustion chamber also causes deposits on the walls and some components mix with the lubricating oils, increasing their viscosity (Peterson et al., 1996; Ma and Hanna, 1999). The presence of water in the oils can allow microbial growth that can block the fuel filters.

A comparison of plant-derived oils with diesel is shown in Table 7.21. Other concerns were the cost of the oil, the effects on engine performance, and fuel preparation (Ma and Hanna, 1999). Therefore, plant-derived oils have been modified and four methods have been used to modify plant oils; blending, microemulsification, pyrolysis, and transesterification.

7.14.1 **Blending**

In order to reduce their viscosity plant oils can be blended with diesel and alcohols to form a mixture that can be used in unmodified engines. For example

Table 7.21 A comparison of the properties of diesel, plant oils, and altered plant oils

Property	Diesel	Rapeseed oil	Sunflower oil	Soybean oil	Cracked soybean oil	Catalytic heated coconut oil
Density (kg/l)	0.84–0.848	0.778–0.91	0.86–0.92	0.885–0.91	0.88	0.81
Viscosity (cSt)	2.8–3.51	37–47	33.9–46	32.6–45	7.74	2.58
Flash point (°C)	64–80	246–273	183–274	178–254	ND*	ND
Cetane number	47.8–51	37.6–50	37.1–49	37.9–45	43	60.5
Calorific value (MJ/kg)	38.5–45.6	36.9–40.2	33.5–39.6	33.6–39.6	40.6	47.5

*ND not determined.
Sources: Data from Srivastava and Prasad (2000), Williamson and Badr (1998), Antolin et al. (2002), and Laforgia and Ardito (1995).

the viscosity of canola (rapeseed) at 10°C was 100 cSt (centistokes) and diesel 4 cSt but a 75:25 mixture of canola/diesel was 40 cSt, and a 50:50 mix 19 cSt. A mixture of diesel and vegetable oil (80:20) has been tested successfully in a diesel engine (Louwrier, 1998). However, the blend separated into two phases after some time. A dilution of sunflower oil with diesel (1:3) reduced the viscosity to 4.88 cSt but long-term use in a direct-injection engine indicated severe coking of the injectors (Srivastava and Prasad, 2000) and other blends also showed carbon deposits and engine wear. A blend of coconut oil and diesel has been tested in a diesel engine (Machacon et al., 2001). Increasing the coconut oil content reduced the emission of NO_x, and smoke, but the fuel consumption decreased by 16%. Thus blending is a simple and cheap method, which requires no modification to the oil but has problems with its long-term use and storage.

7.14.2 **Microemulsification**

Microemulsions can be formed by the dispersion of a mixture of oil, diesel, water, surfactant, and short-chain alcohols such as methanol, ethanol, and butanol. Microemulsions can be made with plant oil and an ester and dispersant, or with oil, an alcohol, and surfactant with or without diesel and contains drops of 0.1–1 μm. Microemulsions can improve the spray characteristics of the fuel as they have lower viscosities. Mixtures of methanol and triolein have given good results but have a reduced calorific value although the alcohol content increases the latent heat of vaporization, which cools the combustion.

7.14.3 Pyrolysis

Plant oils contain a high proportion of triglycerides where three fatty acids are linked to glycerol. A reduction in triglyceride content will reduce the viscosity of the oil. Pyrolysis is the heating of oils to 300–500°C in the presence of a catalyst which will break up the triglycerides. Soybean oil has been treated with an aluminum oxide catalyst and has yielded oil with properties close to those of diesel (Table 7.21). However, the yields of pyrolysis are poor (80%), and the process is expensive so that it may not be adopted.

7.14.4 Transesterification

Transesterification of plant oils is the conversion of the triglycerides making up the oils to fatty acid esters and glycerol. This can be achieved by treating the oil with methanol or ethanol in acid or alkaline conditions (Fig. 7.19). The methyl or ethyl ester mixture is known as **biodiesel**. The fatty acid methyl esters have properties similar to those of diesel and oils such as soybean, sunflower, and particularly rapeseed oil have been used to form biodiesel (Table 7.22).

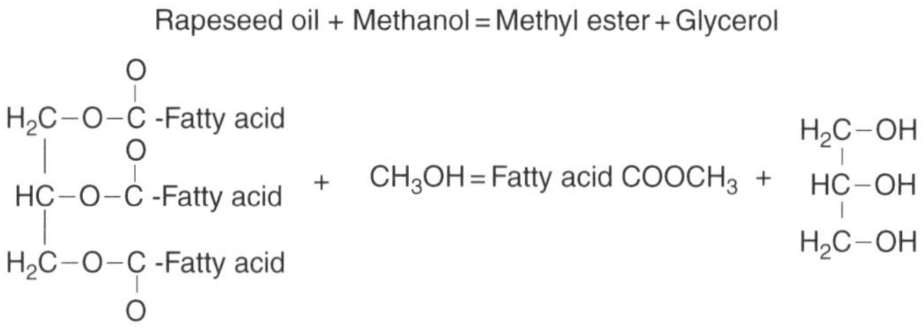

Rapeseed oil + Methanol = Methyl ester + Glycerol

$$\begin{array}{l} H_2C-O-\overset{O}{\overset{|}{C}}\text{-Fatty acid} \\ \quad | \\ HC-O-\overset{O}{\overset{|}{C}}\text{-Fatty acid} \\ \quad | \\ H_2C-O-\underset{O}{\underset{|}{C}}\text{-Fatty acid} \end{array} + CH_3OH = \text{Fatty acid } COOCH_3 + \begin{array}{c} H_2C-OH \\ | \\ HC-OH \\ | \\ H_2C-OH \end{array}$$

Figure 7.19 The transesterification process for the production of biodiesel from rapeseed oil.

Table 7.22 The properties of diesel and esters of plant oils

Property	Diesel	Rapeseed methyl esters	Rapeseed ethyl esters	Sunflower methyl esters	Biodiesel specification EN 14214
Density (kg/l)	0.84–0.848	0.768–0.88	0.876	0.886	0.86–0.9
Viscosity (cSt)	2.8–3.51	6.1–7.2	6.17	4.3	3.5–5.0
Flash point (°C)	64–80	170–185	124	110	>101
Cetane number	47.8–51	51.8–54.4	59.7	46.9	>51
Calorific value (MJ/kg)	38.5–45.6	35.3–40.5	40.5	40.0	NA*

Sources: Data from Srivastava and Prasad (2000), Williamson and Badr (1998), and Peterson et al. (1996).
*NA, not available.

Methanol and ethanol are most frequently used and although a ratio of 3:1 alcohol/oil should be sufficient to force the reaction towards ester formation in practice a higher 6:1 ratio is used, and NaOH and KOH are used as the catalysts. Alkali transesterification is the fastest process but if water is present soap formation can occur. In this case an acid-catalysed (hydrochloric and sulphuric acid) process should be used. After transesterification the products are a mixture of esters, glycerol, alcohol, catalysts, and tri-, di-, and monoglycerides. The conversion is over 90% in most cases so the amount of glycerides is limited. The glycerol is removed by settling, as it is valuable as an industrial intermediate. In some cases a lipase has been used to carry out the conversion as these will function in the presence of water and do not need the removal of catalyst and salts at the end of the reaction. Although the enzyme is more expensive, immobilized enzyme or enzyme-containing cells allow the development of a continuous process (Ban et al., 2001). Biodiesel compared with diesel is non-toxic and biodegradable (Ma and Hanna, 1999), with a very low sulphur content, which on combustion produces less carbon monoxide, soot, and hydrocarbons (Peterson et al., 1996; Altin et al., 2001; Dorado et al., 2003) and can be used in diesel engines without modification (Table 7.22).

Biodiesel has been used by a number of organizations, bus companies in particular, to power diesel engines successfully (Louwrier, 1998). For example a truck with a 5.9-l diesel engine was run for 8742 miles consuming 1771 l of biodiesel at 7.76 km/l. There were no problems and the emissions showed a reduction in hydrocarbons (55.6%), CO (50.6%), and NO_x (11.8%) but an increase in carbon dioxide (1.1%) and particulate matter (10.3%) (Peterson et al., 1996).

A number of trials have been carried out with blends of biodiesel and diesel that have characteristics similar to diesel and which warranted further investigation (Ali et al., 1995; Romig and Spataru, 1996). A 5% blend of biodiesel and diesel is used in France, Austria, Germany, and Italy and a 20% blend in the USA.

7.15 Ethanol

Ethanol can be used as a fuel to replace petrol and ethanol-fuelled cars were planned as early as 1880 by Henry Ford. The use of ethanol as a fuel started in the 1930s in the USA where ethanol produced from maize was used at a concentration of 20% to produce gasohol called Agrol. In the UK gasohol was marketed by the Cleveland Oil Company under the name of Discol in the 1930s and continued until the 1960s. In the USA gasohol was dropped by 1945 due to the availability of cheaper petrol.

The large-scale production of ethanol as a fuel started in Brazil in 1975 followed by the USA in 1978, probably initiated by the increases in crude oil

Table 7.23 Ethanol-producing bacteria and yeasts

Organism	Substrate utilized
Yeasts	
Saccharomyces cerevisiae	Glucose, fructose, galactose, maltose, maltotriose, and xylulose
S. carlsbergensis	Glucose, fructose, galactose, maltose, maltotriose, and xylulose
Kluyeromyces fragilis	Glucose, galactose, and lactose
Candida tropicalis	Glucose, xylose, and xylulose
Bacteria	
Zymomonas mobilis	Glucose, fructose, and sucrose
Clostridium thermocellum	Glucose, cellobiose, and cellulose

prices due to the oil crisis in the 1970s. At the present the replacement of petrol with ethanol is also driven by the need to reduce carbon dioxide emissions, as ethanol is a carbon dioxide-neutral, sustainable product.

The ability of micro-organisms to produce alcohol from sugars has been known since Egyptian times and could be regarded as one of the first uses of biotechnology. The principal organism involved with the production of ethanol is the yeast *Saccharomyces cerevisiae* that in the absence of oxygen will ferment sugars to ethanol and carbon dioxide. There are other micro-organisms capable of producing ethanol (Table 7.23) which can ferment different sugars from *S. cerevisiae*. The pathway involved in the production of ethanol in *S. cerevisiae* is initially glycolysis (Chapter 2) and then pyruvate is converted to acetaldehyde by the enzyme pyruvate decarboxylase with the release of carbon dioxide (Fig. 7.20). Acetaldehyde is converted by the enzyme alcohol dehydrogenase to ethanol. The overall equation is given below.

$$C_6H_{12}O_6 = 2C_2H_5OH + 2CO_2 \tag{7.31}$$

The theoretical yield of ethanol from this equation is 51% of the substrate added but some of the energy is required to maintain the cells so that the yield is about 95% with pure substrates. However, with industrial systems the best yields are around 91%. The concentration of ethanol obtained by fermentation is normally from 5 to 10% as ethanol becomes inhibitory above 5%. Concentrations of 10% ethanol can be obtained with pure substrates or high concentrations of substrates and ethanol-tolerant (10–18%) strains have been isolated. The reason for the loss of viability as the ethanol concentration increases is that ethanol is a solvent and disrupts the cells lipid–protein membrane making it increasingly leaky. Yeast strains with a higher tolerance to ethanol have membranes containing a higher proportion of longer-chain unsaturated fatty acids.

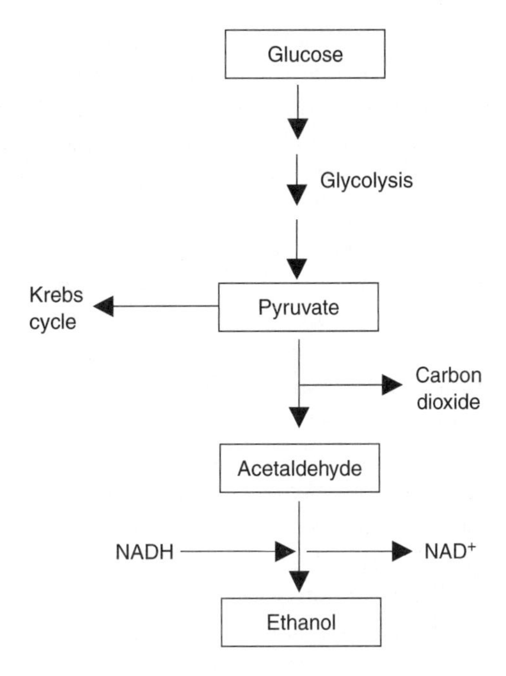

Figure 7.20 The pathway from glucose to ethanol triggered under anaerobic conditions in alcohol-producing micro-organisms.

Table 7.24 The characteristics of petrol and ethanol

Characteristics	Ethanol	Petrol
Boiling point (°C)	78	35–200
Density (kg/l)	0.79	0.74
Gross energy (MJ/kg)	27.2	44.0
Latent heat of vaporization (MJ/kg)	855	293
Flash point (°C)	45	13
Octane number	99	90–100

Other organisms have been investigated to improve the yield and concentration of ethanol formed. One example, the bacterium *Zymomonas mobilis,* is under investigation. This organism is more ethanol-tolerant, has a higher growth rate, and the fermentation uses the Entner–Doudoroff pathway (Chapter 2) which consumes less ATP than glycolysis.

As can be seen from Table 7.23 none of the yeasts are capable of metabolizing either starch or cellulose, which are two of the most abundant substrates available for biofuel production. To make starch and cellulose suitable for fermentation both need to be converted into sugars and the technology for starch conversion exists but that for lignocellulose is under development.

Petrol engines will run on ethanol, as the properties of ethanol are similar to petrol in many areas (Table 7.24). The octane rating of ethanol is higher than petrol as is the latent heat of vaporization. The octane rating is a measure of the resistance of the fuel to pre-ignition when compressed in the cylinder of the engine. A low-octane fuel will pre-ignite, causing a condition known as 'pinking' and this will result in a loss of power. The higher heat of vaporization of ethanol means that as the fuel is vaporized in the carburetor the mixture is cooled to a lower temperature than that for petrol. This means that more fuel enters the engine, in part compensating for the lower energy content. To avoid separation of an aqueous layer in cold weather the ethanol needs to be anhydrous as ethanol normally contains 4.5% water. The heat of combustion (or gross energy) is lower than petrol, which leads to some reduction in performance and a 15–25% increase in fuel consumption. Ethanol has some disadvantages in that it mixes with water and this type of mixture will corrode steel tanks. Table 7.25 gives the variation of ethanol/petrol blends used in a number of countries. Hydrous ethanol which contains 4.5% water (Alcool) has been used in all-ethanol vehicles in Brazil, but sales of these vehicles ceased in 1990s to be replaced with a blend containing 24% ethanol. In the USA the initial blend contained 10% ethanol (Gasohol; E10) but more recently a blend containing 85% ethanol (E85) has been introduced. Ethanol is also used in the USA to increase the oxygen levels in petrol to 7.6% and as a replacement for methyl tertiary butyl ether (MTBE) in reformulated petrol.

Two countries, Brazil and the USA, initiated the production of biomass-derived ethanol and each had different approaches. The initial reasons for the development of an ethanol-production system was the dependence of both countries on imported oil. This importation of oil cost the USA $106 billion in 2000. The price of chemically produced ethanol increased, which made biologically produced ethanol more economical. In the 1970s chemically produced ethanol was selling at $0.145/l but in the 1980s the increase in the feedstock increased ethanol prices to $0.53/l, which was the same price as biologically produced ethanol. An additional reason for the production of alcohol as a fuel was the low prices that the farmers were getting for their

Table 7.25 Ethanol/petrol combinations

Fuels	Country	Ethanol content (%)
Hydrous ethanol (Alcool)	Brazil	95.5
Gasoline	Brazil	24
E85	USA	85
E10 (Gasohol)	USA	10
Oxygenated fuel	USA	7.6
Reformulated gasoline	USA	5.7

Note: Hydrous ethanol contains 4.5% water
Source: Wheals et al. (1999).

Table 7.26 Ethanol production for fuel

Country	Substrate	Production (10^9 l)
Brazil (1999)	Sugar cane juice or molasses	10.5 (hydrous) 6.5 (anhydrous)
USA (1998)	Maize starch (95%) and other starch crops	5.3 (anhydrous)
Canada (1998)	Maize starch and 15% wheat starch	0.24 (anhydrous)

Source: Wheals et al. (1999).

maize. At present fuel ethanol accounts for 7% of the maize crop, boosting farm incomes by $4.5 billion, and is responsible for 200 000 jobs. Production was initially stimulated by the removal of a 4 ¢/gallon tax from ethanol by the Carter Administration's Energy Act in 1979. More recently the effect of the burning of fossil fuels on the environment has been of concern. Transport fuels are responsible for a considerable amount of air pollution and less polluting replacements are being introduced.

The reasons for the development of ethanol as a fuel in Brazil were similar to those of the USA, to reduce the imports of petrol as Brazil had few oil fields, to open up areas of the country for cultivation, to provide employment, to increase the industrial base, and to develop ethanol exports of plant and expertise. In addition, Brazil is one of the largest producers of sugar from sugar cane so that a good substrate was readily available which did not require processing. At present only three countries are major fuel ethanol producers, Brazil, USA, and Canada (Table 7.26).

7.15.1 **Production in Brazil**

The production of ethanol in Brazil started in 1975 with a National Alcohol Programme to produce 95.5% hydrous ethanol (Alcool) and by 1980s the majority of the cars used this fuel. However, in the 1990 this fuel was replaced by a mixture containing 24% ethanol. Initially ethanol was produced by the fermentation of sugar from sugar cane (*Saccharum officinarum*) in simple batch fermenters with capacities of up to 1.5 million l although continuous processes are also being run. At present Brazil produces 2.6×10^8 tons of sugar cane which is processed by 324 mills to produce sugar and ethanol or in some cases only ethanol. Sugar cane can be grown easily in Brazil and contains 12–17% sugar on a wet-weight basis. Sugar cane can be crushed to yield sugar (90% sucrose) and cane fibre, known as bagasse. Sugar cane also has one of the highest yields of potential biofuel crops (Table 7.26).

In the processes that combine sugar production and ethanol formation, after concentration sugar crystals will form. These are removed, leaving molasses, which contains up to 65% sugars. In other systems the cane juice is heated to 110°C to reduce microbial contamination. The fermentation is carried out at 33–35°C at high cell densities to yield 8–12% ethanol with a short fermentation time of 6–10 hours. The bagasse, the fiber and cellulose remaining after the juice is removed is often burnt to provide heat for the distillation.

7.15.2 Production in the USA

The main renewable substrate for fermentation in the USA is starch extracted from maize. Other starchy crops (sorghum, cassava, and barley), and wastes such as cane or citrus molasses, cheese whey, sulphite liquor, and potato wastes were also considered. It was thought at first that there would be both small and large producers of ethanol in the USA. In 1988 51 plants were operating, of which 12 used waste material and one situated in Hawaii used molasses. However, some 78% of the ethanol was produced by eight plants, with maize was the main source of starch, and by the 1990s only one waste-utilizing plant remained. Maize is normally wet milled to produce starch, oil, gluten, and high-protein material (Fig. 7.21). The organism used in fermentation

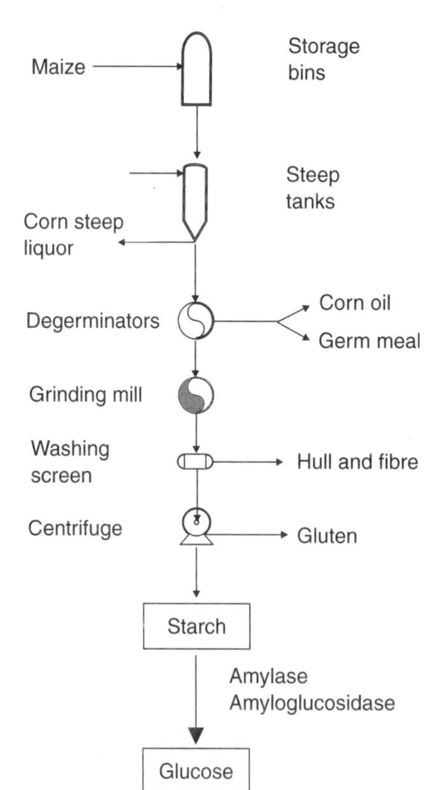

Figure 7.21 The processing of maize in the USA for the production of starch and an outline of the enzymatic production of glucose from starch.

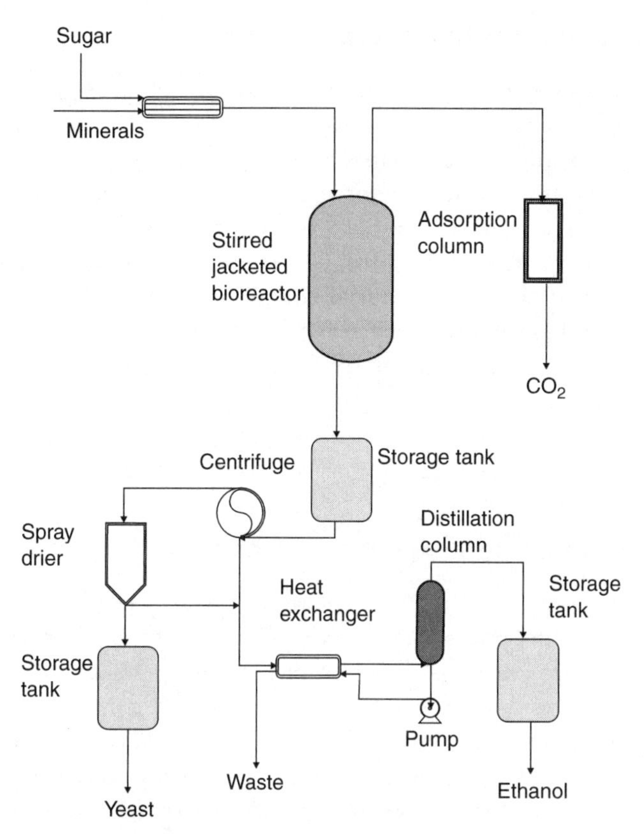

Figure 7.22 A process for the production of ethanol from glucose. The products are yeast, carbon dioxide, and ethanol, all of which can be collected and sold.

is almost always the yeast *S. cerevisiae* but it cannot metabolize starch so that the starch needs to be converted into sugar before fermentation can proceed. Starch is converted to sugar by heating the starch to 150–169°C to gelatinize the starch granules, followed by hydrolysis by amylase enzymes or acid. Often an α-amylase is added before heating to reduce the viscosity of the gelatinized starch. The α-amylase enzyme is inactivated by the high temperature but remains active during the heating process and thins the starch slurry. The gelatinized starch, called the mash, is cooled to 90°C and more α-amylase added and incubated for 30–60 min. This is the liquefaction stage where the starch is converted to long-chain dextrins. The dextrins are converted to glucose by the addition of an *Aspergillus niger* amyloglucosidase and incubating at 50–60°C for a further 60–120 min. The sugar mix is then cooled to 30°C and the yeast added to start the fermentation (Fig. 7.22). The fermenter (bioreactor) is run in a batch mode at 30–35°C for 42–72 h until a final concentration of 8–12% ethanol is obtained. The yeast cells are removed by centrifugation and the ethanol recovered from the medium by distillation.

7.15.3 Economics of ethanol production

It has been calculated that with maize the energy input to grow the crop is about 30% of the energy contained within the crop. If the crop is used to produce ethanol the output is less than that required to grow and process the crop (Table 7.15) and this is also true for sugar cane. However, if the fossil fuel is replaced by bagasse or the coal input ignored for the distillation stage then the input/output energy rises to 1.29 for maize and 2.03 for sugar cane.

There are fixed costs in biological energy production. In the case of the Brazilian ethanol programme there is some disagreement about the economics of ethanol production. Some authors have suggested that the cost of ethanol was as low as $0.185/ l in the 1980s at which price it could compete with fossil fuels. However, the continued low prices for oil appear to make ethanol in Brazil an expensive alternative at the moment.

7.15.4 Improvements in alcohol production

A number of process changes have been investigated in order to improve the economics of ethanol production. The traditional method of fermentation has been batch culture but other forms of bioreactor operation exist which can improve ethanol productivity and these are fed-batch, continuous, and multi-vessel continuous cultures. The productivity of these various systems is given in Table 7.27. The explanation of the types of fermentation can be see in Chapter 2. It is clear from the table that continuous culture with cell recycling is far superior to both batch and continuous culture and is further improved by vacuum removal of ethanol as this inhibits growth at high concentrations.

Cellulose and lignocellulose represent a cheap, abundant substrate for the production of ethanol. Cellulose makes up much of the biomass available from crop residues, fine feed, corn stover, bagasse, sugar beet pulp, softwood, wheat straw, rice straw, pulp and paper-mill residue, sawdust, forest thinings, municipal solid waste, winter cereals, and recycled paper. All of these sources have been used to produce ethanol. The process of forming sugar from cellulose can be either chemical or biological. Alkali and either dilute or

Table 7.27 Ethanol productivity using various bioreactor operations

System	Ethanol (g/l/h)
Batch	2
Continuous	5
Continuous, multi-stage	12
Batch with cell recycle	15
Continuous with cell recycle	40
Continuous with vacuum removal of ethanol and cell recycle	80

concentrated acid treatment will produce sugars. In the case of dilute acid hydrolysis the material is pyrolysed prior to treatment to improve breakdown. Ethanol production from cellulose is often carried out in two stages, the hydrolysis of cellulosic material to sugars by cellulase enzymes, and the fermentation of the sugars. Because of the presence of hemicellulose and lignin and the crystalline natures of cellulose some form of pretreatment is required before enzyme hydrolysis. These pretreatments are listed in Fig. 7.23 and include carbon dioxide, steam, and ammonia explosion, mechanical grinding, acid, white rot fungi treatment, and ozonolysis. Once the structure of the cellulose has been opened up by the pretreatment then enzymatic hydrolysis can proceed. The crude cellulase enzyme is a consortium of enzymes, which operate under mild conditions; pH 4.8 and 45–50°C. Although cellulase is commercially available they are usually obtained from fungi such as *Trichoderma reesei*, although bacteria also produce cellulase. The pathway of cellulose breakdown

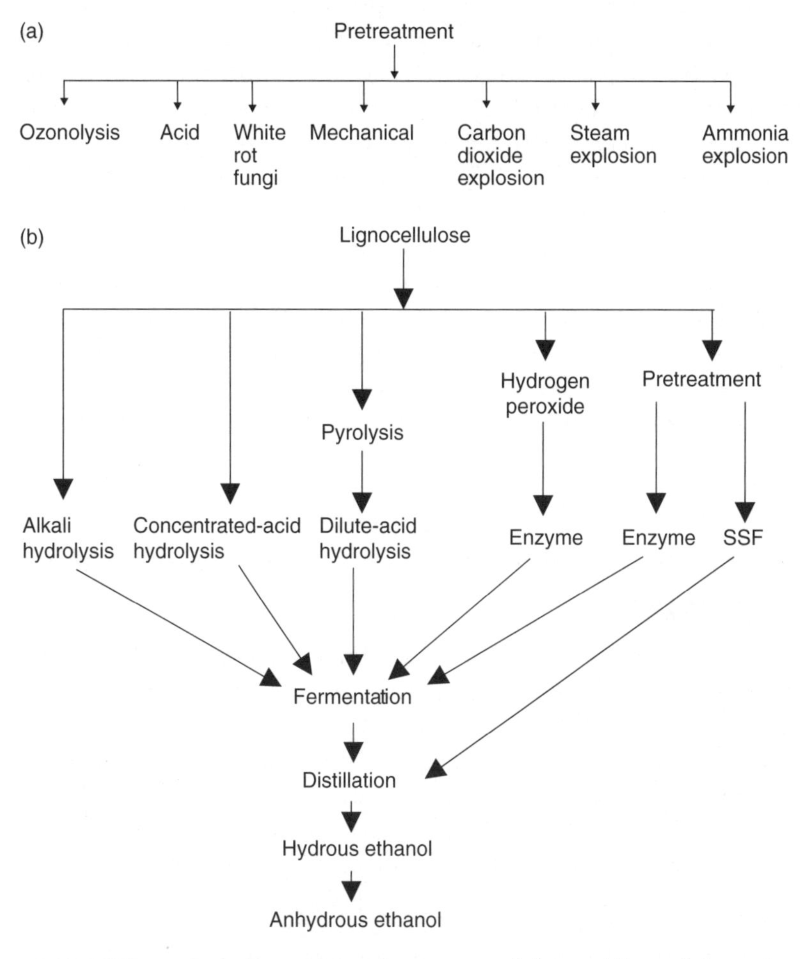

Figure 7.23 (a) The methods that can be used to pretreat cellulose and lignocellulose prior to its degradation to glucose. (b) The chemical and enzymatic methods available for the conversion of cellulose/lignocellulose to glucose which is then fermented to ethanol.

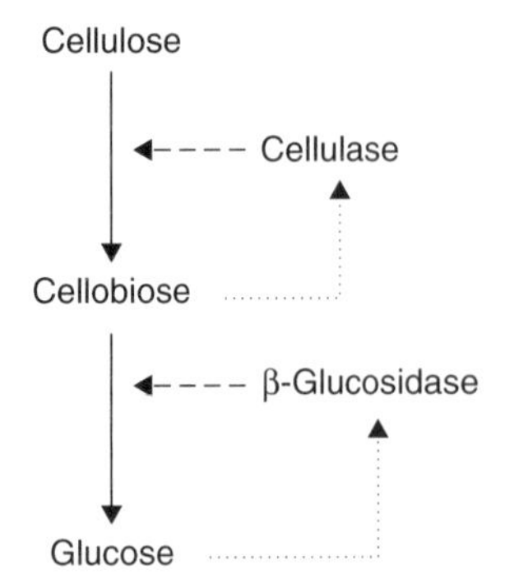

Figure 7.24 The pathway of cellulose breakdown to glucose, indicating the feedback inhibition that can occur.

is shown in Fig. 7.24 where cellulose on hydrolysis liberates cellobiose which is cleaved into two molecules of glucose by the enzyme β-glucosidase. The disadvantage of the enzyme process is that both products, glucose and cellobiose, act as inhibitors of cellulase and β-glucosidase. One method of circumventing this inhibition is to combine cellulose breakdown (saccharification) and fermentation (SSF). In the SSF process as the cellulase forms glucose it is removed by the yeast to form ethanol. The SSF process was introduced to reduce the inhibition of the enzyme found with cellobiose and glucose. The microorganisms used in SSF are mainly *T. reesei* and *S. cerevisiae* at 38°C, which is a compromise between the optimum for yeast growth of 30°C and for enzyme hydrolysis of 45–50°C. More-recent developments have introduced thermotolerant yeasts *Kluyvermyces marxianus* and *Kluyvermyces fragilis* which can grow at 42°C. The SSF has the other advantages that it increases the hydrolysis rate and lowers the amount of enzyme required, and has higher product yield, fewer problems with sterility, and smaller bioreactor volumes.

Genetic manipulation has produced yeasts containing an amylase, which can metabolize starch.

7.16 Hydrogen

Hydrogen is an abundant, invisible gas that is considered to be an ideal fuel as on combustion it yields only water, although NO_x can be formed during the combustion. Hydrogen can be carried as a liquid (cryogenic; −253°C), a compressed gas, or solid state, or produced *in situ*. Hydrogen can be used directly in an internal combustion engine (Pehr et al., 2001), in fuel cells, and in turbines. A hydrogen filling station has been built at Munich airport as a trial. There are a number of possibilities for the delivery of hydrogen:

- centralized reforming truck or pipeline delivery,
- on-site reforming from natural gas or methanol, and
- on-site electrolysis.

Hydrogen can be produced from a number of sources using various different technologies, some of which are renewable while others are not. These are as follows. Non-renewable processes:

- steam methane reforming,
- coal gasification,
- partial oxidation of hydrocarbons, and
- thermocatalytic treatment of water.

Renewable processes:

- electrolysis,
- biomass gasification,
- biomass pyrolysis, and
- biological.

Steam reforming of methane or natural gas is the most widely used method of producing hydrogen and the least-expensive method. It is a well-established commercial process. Another well-established process for the production of hydrogen is coal gasification. Hydrocarbons such as oil can form hydrogen with non-catalytic partial oxidation but the process is expensive, as it requires pure oxygen. The thermocatalytic process forms hydrogen from water at high temperatures (550°C) using zeolite catalysts. All these processes use fossil fuels as the feedstock and use energy generated from fossil fuels. However, all the following processes produce hydrogen from renewable sources and use renewable energy.

Electrolysis of water produces a small amount of hydrogen at present and requires electricity. If the electricity can be generated in a sustainable way such as hydroelectric, wind, wave, photovoltaic, or solar power then hydrogen production could be renewable. Biomass or waste gasification produces syngas, a mixture of carbon monoxide and hydrogen, which can contain other hydrocarbons. The hydrocarbons are steam reformed to carbon monoxide and hydrogen and the ratio of $CO:H_2$ may need altering. If the syngas is passed over a Cu/Zn/Al catalyst methanol is formed whereas if hydrogen is required this is separated using pressure swing adsorption and compressed. If the biomass is treated at high temperature (450–550°C) without air a bio-oil is produced. The bio-oil can be steam reformed to produce hydrogen.

7.16.1 Biological production of hydrogen

The biological production of hydrogen would be a renewable path and if combined with the use of a waste material would be ideal. The production of hydrogen by micro-organisms, principally bacteria and microalgae, has been known for some time. Hydrogen production by green algae was first observed in the 1940s.

Hydrogen production is triggered in green algae by anaerobic conditions, which induce the reversible hydrogenase. The reversible hydrogenase is however inhibited by oxygen and therefore this system will only work if the two reactions are separated. This can be achieved by sulphur deprivation which reduces the synthesis of the amino acids cysteine and methionine and the photosystem II (PSII) repair cycle is blocked (Ghirardi et al., 2000). The PSII complex requires frequent repair and is therefore sensitive to a reduction in protein synthesis. The pathway is shown in Fig. 7.25. Direct hydrogen production cannot be sustained and at present the indirect pathway is more efficient. In the indirect process light is used to fix carbon dioxide and the stored carbohydrates can then be used to generate hydrogen when placed in anaerobic conditions (Fig. 7.25). The separation of the two processes occurs in the cyanobacterium *Anabena cylindrica* where the vegetative cells generate oxygen and fix carbon dioxide whereas heterocysts containing nitrogenase produce hydrogen.

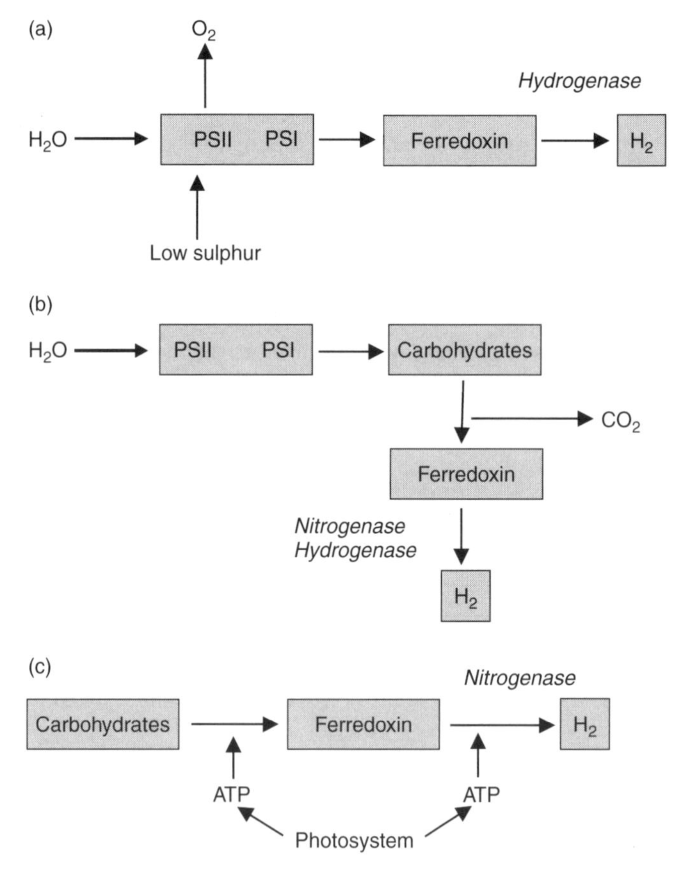

Figure 7.25 Three pathways to the biological production of hydrogen. (a) Direct hydrogen production occurs in low-sulphur conditions in a photosynthetic micro-organism. (b) In the indirect process the photosynthetic organism produces storage carbohydrates which can subsequently be used to form hydrogen. (c) In the third system bacteria can use their photosystem to form hydrogen. PS, photosystem.

Bacterial production of hydrogen can occur with photosynthetic bacteria such as *Rhodospirillum rubrum and Rhodobacter sphaeroides* under anaerobic conditions. Other bacteria such as *Clostridium bifermentans, Enterobacter aeroenes,* and *Clostridium butylicum* use substrates such as whey, starch, and glutamate to produce hydrogen, again under anaerobic conditions, at rate of 4–6 g of H_2/kg of substrate (Wang et al., 2003).

7.17 Conclusions

Environmental biotechnology will have a considerable part to play in the biofuels area.

- It is clear that human activities, particularly the burning of fossil fuel, are having an effect on the global climate. This could produce significant effects in the long term such as rises in sea level and the reduction in ice fields.
- It is also true that fossil fuels are limited and that some form of alternative energy supply will be required either in the short term or long term.
- Energy for the generation of electricity can be provided by biomass and ethanol; biodiesel and hydrogen can provide liquid fuels most suitable for transport. All these fuels can be produced from renewable, sustainable sources and should see much greater use in the future and have the added advantage of being carbon dioxide-neutral.

7.18 Further reading

Jackson, A.R.W. and Jackson, J.M. (1996) *Environmental Science*. Addison Wesley Longman, Harlow.

Johansson, T.B, Kelly, H., Reddy, A.K.N., and Williams, R.H. (1993) *Renewable Energy*. Earthscan Publications, London.

Lee, S. (1996) *Alternative Fuel*. Taylor & Francis, London.

8 Natural resource recovery

8.1	Introduction	320
8.2	Oil recovery	321
8.3	Recovery of metals	330
8.4	Conclusions	343
8.5	Further reading	344

8.1 Introduction

Biological processes are not only harnessed to reduce or eliminate pollution but can also be used to assist in the extraction of oil and metals from low-grade ores and waste dumps. Metals and crude oil are both non-renewable resources, which may have adequate supplies at present, but demand is unlikely to diminish so that new sources or improved extraction methods will eventually be needed. Micro-organisms have the ability to extract metals from ores and have been used industrially since the 1950s for copper extraction (Agate, 1996). At present microbial mining is being used on an industrial scale for copper, uranium, and gold.

The primary recovery of crude oil can be from 0 to 50% depending on the conditions of the source (Farouq Ali and Thomas, 1996). Some form of secondary recovery is normally applied to extract more of the oil and the techniques are collectively known as **enhanced oil recovery** (EOR). One method, which can be included in EOR, is based on the use of micro-organisms to help to extract oil and the method is known as **microbially enhanced oil recovery** (MEOR). MEOR is not used commercially at present because of its high costs, as the present price of crude oil is low. However, crude oil stocks will diminish and the prices may rise, and under these conditions MEOR may be required.

This chapter outlines the possible use of micro-organisms and microbial products for the extraction of crude oil once the primary flow has ceased. The second half of the chapter covers the use of micro-organisms for the recovery of metals from mining wastes, low-grade ores, and worked-out mines.

8.2 Oil recovery

Crude oil is an extremely complex and variable mixture of organic compounds. Crude oil has accumulated underground as a result of the anaerobic degradation of organisms over a very long time. Under the conditions of high temperature and pressure the organic material has been converted to natural gas, liquid crude oil, shale oil, and tars. At underground temperatures shale oils and tars do not flow, but the crude oil is liquid, and unless contained will escape to the surface, where the volatiles evaporate, forming a tar bed. The majority of the compounds in crude oil are hydrocarbons, which can range in molecular weight from the gas methane to high-molecular-weight tars and bitumen (section 5.3.1).

These hydrocarbons can also come in a wide range of molecular structures; straight and branched chains, single or condensed rings, and aromatic rings. The proportion of individual compounds can vary greatly between crude oil sources and this variation in composition affects the properties of the oil. Oils with a high proportion of low-molecular-weight material are known as 'light' oil and flow easily, while 'heavy oils' are the reverse. In addition to the hydrocarbons, crude oil contains 0.05–3.0% heterocyclic compounds, containing sulphur, nitrogen, oxygen, and some heavy metals.

If the oil reaches the surface the volatiles will evaporate and the heavy components will remain, forming tar sands and bitumen lakes. These tar sands can be mined by open-cast methods and the tar stripped from the sand by hot water and alkaline conditions. The oil needs to be dried before it can be shipped or treated. Treatment usually consists of upgrading the tar by cracking with heat and catalysis, to form lower-molecular-weight hydrocarbons. Underground accumulation of crude oil occurs at sites all over the world in porous sandstone, limestone, and chalk. As oil is less dense than water and does not mix with it, oil is often forced upwards through the porous rock by rising water until the oil reaches a layer of impervious rock. If the impervious rock forms some type of dome an oil reservoir will form (Fig. 8.1). Reservoirs occur at all depths and often the reservoir is under considerable pressure, mainly from dissolved gas and the pressure of the overlying rock, and the presence of an aquifer can also increase the pressure. In many cases gas collects at the top of the reservoir. Oil is recovered from these reservoirs by first finding the dome by seismic investigation and then drilling into it. The pressure in the reservoir will force the lighter crude oils to the surface, as they are liquid at the well temperatures, which can be up to 90°C, but any bitumen and asphalt will remain as they are too viscous to flow, even at the elevated temperatures. If the pressure is very high in the reservoir the oil will be forced out of the well in what is known as a 'gusher'. The well is normally capped and the oil runs off into storage and transport facilities.

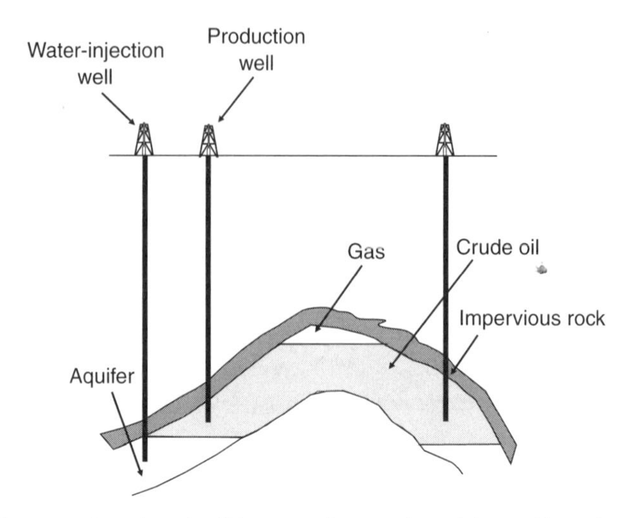

Figure 8.1 The extraction of crude oil from an oil reservoir and the position of water-injection wells.

As the primary removal of the oil proceeds the reservoir pressure drops, although pressure may be maintained by the aquifer. This primary extraction usually accounts for about 10–15% of the original oil in place (Millington et al., 1994). Once the pressure has reduced to low levels the oil needs to be extracted by pumping. Mechanical (nodding donkey) or electrical pumps are placed at the base of the well and used to extract the oil. The suction applied at the base of the well may cause the water in the aquifer to rise if the rock is sufficiently permeable. If the water reaches the pump, water will be pumped out and this is a condition known as 'coning' (Fig. 8.2). This condition is a major problem in oil extraction and the pump may have to be stopped for some time to allow the cone to subside. Eventually the flow of oil is reduced to uneconomic levels, which generally represents some 15–20% of the total oil in the reservoir.

Any further recovery of oil is known as the secondary production and generally involves the injection of water or gas into the well to force out the oil. Water is used normally as the gas from the reservoir is usually piped away and sold. Seawater can be used but the sulphate present will need to be removed, as any sulphur present will encourage the growth of sulphate-reducing bacteria, which will degrade the oil. Water flooding involves the drilling of another well or wells (up to five) some distance from the production well and the water is normally injected below the oil layers to force it out. This pattern can be arranged across an oil field (Fig. 8.1). Water can also be pumped into the oil-bearing rock itself to sweep out the oil. However, this technique is not without its problems. The rock, although porous, is unlikely to be homogeneous in structure and some channels will be considerably larger and allow water to flow much more easily as it has a higher mobility than oil,

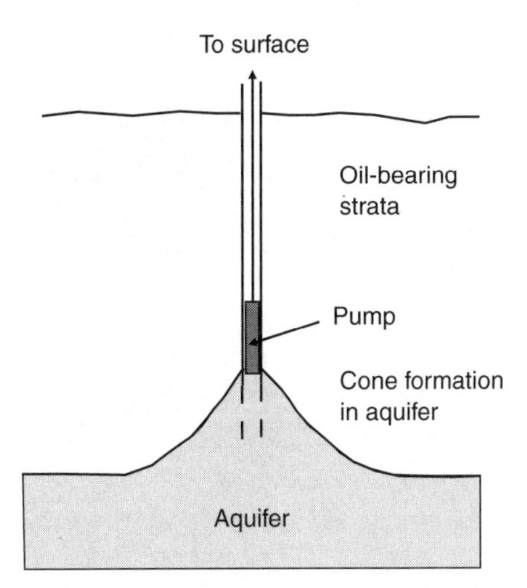

Figure 8.2 As the oil becomes more difficult to extract from the oil well 'coning', which is the extraction of water from the aquifer, can occur.

causing a process called 'fingering' (Fig. 8.3). Another problem is that oil droplets can clog the small pores of low-porosity rocks, stopping any flow of liquid. In addition, fractures and channels will also allow the water to pass through without displacing the oil, forming what are known as thief zones (Fig. 8.3). The yield at the end of the secondary extraction is in the region of 35% of the original oil in place, which means that there is a considerable amount of oil remaining.

8.2.1 Enhanced oil recovery (EOR)

Secondary recovery of oil can be enhanced and the methods involved can be divided into thermal or non-thermal (Fig. 8.4). The objective of any EOR is to influence the flow characteristics of the crude oil and the flow properties can be described by its mobility ratio and the capillary number (Farouq Ali and Thomas, 1996). The mobility ratio (M) is the ratio of the mobility of the displacing fluid (Y displacing) to the mobility of the oil (Y oil).

$$M = \frac{Y\,(\text{displacing})}{Y\,(\text{oil})} \tag{8.1}$$

$$Y = \frac{k}{\mu} \tag{8.2}$$

where k is the effective permeability and μ the viscosity. If M is less than 1 the displacing liquid water moves more easily than the oil and therefore can move past the oil, leaving much of it in place. If M is very much larger than 1 the liquid will be very viscous and flow only in wide channels, leaving oil droplets behind in the smaller pores in what is known as viscous fingering. The ideal situation is $M \geq 1$. The value of M can be reduced by increasing the viscosity of the waterflood by addition of high-molecular-weight polymers. The other

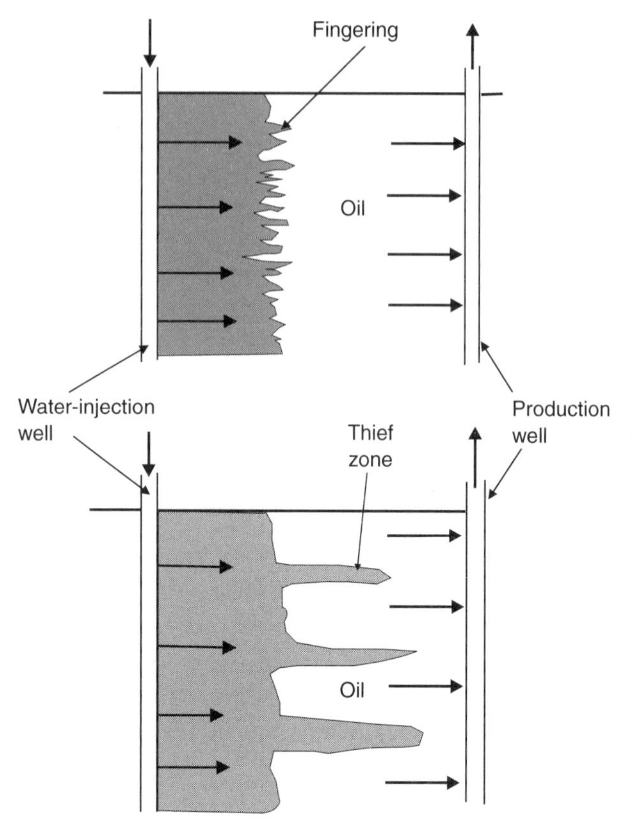

Figure 8.3 In secondary oil extraction water is pumped into the well to force the oil from the porous rocks. Some of the problems associated with this type of extraction are 'fingering' (top) and breakthrough or 'thief' zones (bottom).

property which influences oil flow is the capillary number. The capillary number, N_c, is a measure of the permeability of the oil. It is defined as

$$N_c = \frac{\Phi_p k}{Lq} \tag{8.3}$$

where q = interfacial tension, k = effective permeability of the displaced fluid, and Φ_p/L = pressure gradient. If the capillary number is increased this will decrease the residual oil saturation and release more oil. The capillary number can be increased by reducing oil viscosity, which increases k, increasing the pressure gradient or decreasing interfacial tension. It is these factors that EOR techniques attempt to influence but in reality conditions in oil reservoirs are complex, with the formation of emulsions and rock–fluid interactions.

A number of non-thermal methods of EOR are shown in Fig. 8.4 but in practice these consist mainly of chemical and miscible floods. Chemical floods consist of addition of high-molecular-weight polymers, surfactants, alkalis,

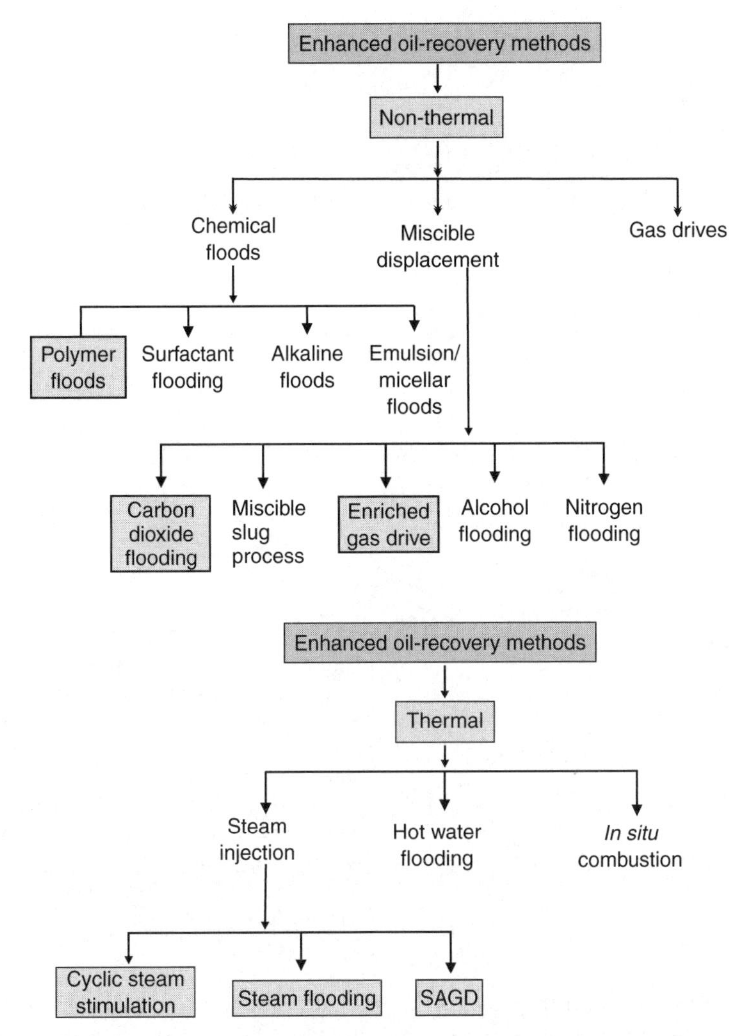

Figure 8.4 Some of the thermal and non-thermal methods used in EOR. Methods in boxes are those which have been used commercially. SAGD, steam-assisted gravity distillation.

and emulsions or combinations of all four to the flood water, but the only commercially used method is the polymer flood. A high-molecular-weight (2–5 million Da) polymer is added to the flood water at a concentration of 200–1000 mg/l, which has the effect of increasing its viscosity and thus lowering the mobility of the flood water. A decrease in water mobility will bring the mobility factor close to 1, which will considerably reduce fingering and enhance oil recovery. The polymers that have been used are the chemically derived polyacrylamides or the biopolymers xanthan, curdlan, and scleroglucan (Fig. 8.5 and Table 8.1). The problems with polymer addition are the high cost of the polymers, the degradation of the polymers under the conditions found in the oil well and their adsorption to rock strata. Miscible

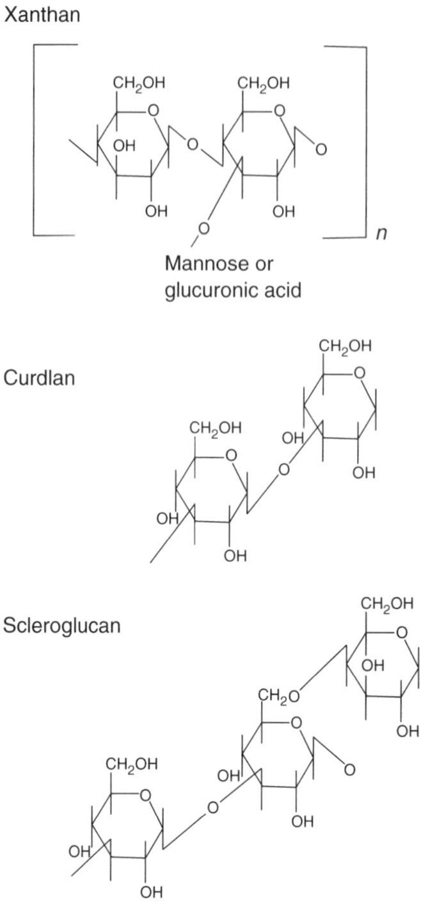

Figure 8.5 The basic structures of microbial polysaccharides, xanthan, curdlan, and scleroglucan, used in enhanced oil extraction and drilling muds. Xanthan has a 1,4-β-D-glucose backbone with side chains of D-glucuronic acid between two D-mannose units. Curdlan is a 1,3-β-glucose chain with no side chains. Scleroglucan has a backbone of 1,3-β-D-glucose with occasional 1,6-β-D-glucosyl units.

oil recovery is the injection of a solvent, such as a hydrocarbon gas (propane), liquid carbon dioxide, liquid hydrocarbons, and alcohols. All these compounds are miscible with the oil and therefore reduce the oil's interfacial tension to zero so that the oil can be swept from the pores in the rock. The commercial methods use liquid carbon dioxide and an enriched gas drive where solvent is added to a gas drive. Chemical flood and miscible displacement are mainly used for light oils, although they have been applied to some heavy oils.

Thermal extraction consists of steam injection, hot-water flooding, and *in situ* combustion. In all cases the heat applied reduces the viscosity of the oil and this can mobilize even heavy, viscous oils (Fig. 8.4). Steam injection can take the form of cyclic steam stimulation where steam is applied for a period of time followed by removal of oil and when this declines another steam cycle is applied. This method is likely to be best with viscous heavy oils and is one of the most successful methods used to date. Steamflooding is the continuous application of steam. **Steam-assisted gravity drainage** (SAGD) has been developed

Table 8.1 Microbial polysaccharides used for oil extraction

Polysaccharide	Micro-organism	Molecular weight (Da)	Structure
Xanthan	*Xanthomonas campestris*	$2 \times 10^6 - 1.5 \times 10^7$	1,4-β-D-Glucose backbone with side chains of mannose, glucuronic acid
Curdlan	*Agrobacterium* sp., *Alcaligenes faecalis*	7.4×10^4	1,3-β-D-Glucose
Scleroglucan	*Sclerotium glucanicum* *S. rolfsii*	$1.3 \times 10^5 - 6 \times 10^6$	1,3-β-D-Linked glucose with some 1,6-β-D-glucosyl residues

for tar sands. In this method steam is injected horizontally into the sands and the mobilized oil collected with another horizontal well at a lower level. *In situ* combustion is unusual in that the heat is generated by burning some 10% of the residual oil. Compressed air or oxygen is pumped into the well until the oil ignites spontaneously or is ignited. The combustion front is sustained by the pumping of air or oxygen and the heat generated lowers the viscosity of the oil and in some cases cracks the oil into lighter fractions. Successful application of the technique has been carried out in Russia, Rumania, California, Texas, and Canada (Castanier and Brigham, 2003). Although all the techniques have been shown to be effective they are expensive due to the high cost of the chemicals and the energy required for steam generation.

8.2.2 Microbially enhanced oil recovery (MEOR)

MEOR is the introduction of micro-organisms or stimulation of existing micro-organisms in the well. Micro-organisms are known to produce gas and polymers and in some cases both. The *in situ* growth and production of microbial products such as polymers will clearly be more economical than the *ex situ* preparation. However, the conditions in the oil well are extreme and cannot be easily manipulated and therefore any micro-organism will have to function in anaerobic conditions at high temperatures, pressures, and salinity (Table 8.2). Some species of *Clostridium* are capable of growth in anaerobic conditions at 45°C but are sensitive to high salt levels (<5%). Some strains of

Table 8.2 Typical conditions found in oil well

Parameter	Value
Depth (m)	600–2 200
Pressure (MPa)	6.5–21.0
Temperature (°C)	40–90
Oil density (g/ml)	0.84–0.90
Salinity (g/l)	170–230
pH	6.25
Oxygen (mg/l)	<0.05

Source: Yakimov et al. (1997).

Bacillus are capable of growth under oil-well conditions, as are some strains of *Bacillus licheniformis* (Yakimov et al., 1997). Bacteria associated with oil wells may belong to the genera *Bacillus, Pseudomonas, Micrococcus*, and *Acinetobacter*, and Archaea, where partially aerobic conditions occur at the base of the well. In the remainder of the well there will be anaerobic species such as *Clostridium, Desulfovibrio*, and *Methanobacterium*. Water injection will clearly decrease the temperature such that the full range of temperatures from 40 to 90°C may well be represented across the well. In general bacteria will grow at temperatures of up to 40°C, but above 40°C thermophiles will dominate.

Recently an increasing number of micro-organisms have been isolated which are capable of growth under extreme conditions. The extremophiles have been found in hot-water springs, deep-sea hydrothermal vents, alkaline lakes, and the Arctic and often belong to the division Archaea (Woese et al., 1990). The pressure at the bottom of the wells will be high but most micro-organisms appear to be pressure-tolerant. Another possible problem is the presence of toxins in the crude oil but data so far do not indicate this.

MEOR can participate in four processes for the enhanced recovery of oil (Moses, 1991):

- well stimulation (production of gas, surfactants, and acids),
- matrix acidizing,
- profile improvement, and
- downhole polymer and surfactant flooding.

One reason for the reduction in oil flow in the well is the pore size and particulate formation of some limestone and sandstone rocks, which can block with asphalt and bitumen. Metallic scale deposits may also be present and add to the problem of oil flow. Acid treatment will remove much of these particulates, particularly with limestone, but the application of acid is very expensive.

Certain anaerobic micro-organisms can grow under extreme conditions and on oil as a substrate, and produce acids, surfactants, and gases. Acid production *in situ* could help to reduce this problem and the production of gas and surfactant would also help to mobilize the oil. If micro-organisms are injected into the well along with a carbon and nitrogen source and the well shut for some time acid, gas, and solvents will be produced *in situ*. The process of fermentation can be followed by monitoring well pressure. Although this type of injection may only have an affect for some 2–5 m around the base of the well, there have been reports of a significant improvement in production, and the biological method is cheaper than the conventional application of acids.

Matrix acidizing is a process where some of the rock structure is dissolved by the *in situ* generation of acid by microbial activity. One of the traditional methods of increasing oil flow is to fracture the rock strata in which the oil is held by applying high-pressure water. This procedure can be expensive and the fractures may close up after the pressure is released. Some of the rocks associated with oil are limestone or sandstone, held together with carbonates, and acid generated by bacterial growth in the well will dissolve these and open up the strata.

Profile improvement is the plugging of the large cracks (thief zones) in the rocks, which allows the injected water to force the oil out of the small pores without passing through the large fissures only. The plugging of the fissures can be achieved by growing polymer-forming bacteria in the well. Although there are a number of anaerobic polymer-forming bacteria, the conditions at the bottom of the well are not easy to predict, particularly in terms of temperature. This will mean that the time scale of polymer formation, whether the polymers are produced at the correct site, and how long the polymer will exist, are also unpredictable. Polymer formation can also be used to reduce 'coning' but in this case the bacteria will need to be injected into the production well.

Polymer and surfactant flooding is the continuous microbial formation of both surfactant and polymers, which is different to the batch growth and production of polymers used in profile improvement. A number of anaerobic bacteria have been isolated which can produce polymers or surfactants and many of the bacteria will become attached to the particles in the well. If nutrients are provided continuously as growth proceeds the growing zone will expand outwards and the polymers will force the oil towards the production well.

Most of the practical work on MEOR has been carried out in Eastern Europe as oil prices are too low in the USA and the West to justify the development of this form of recovery at present. Well stimulation has been carried out and anecdotal evidence suggests that this has been successful. None of the other techniques appear to have been used to date. The real application of

MEOR will probably have to wait until crude oil prices reach values where enhanced extraction becomes economical.

8.2.3 Microbial polymers

Microbial polymers such as xanthan, curdlan, and scleroglucan have been added to oil wells to increase the viscosity of the water flood. Microbial polymers have other uses in oil recovery as their pseudoplasticity has seen their addition to drilling muds. In addition, microbial polymers have found uses as thickeners and stabilizers in the food industries. The majority of the microbial polymers are polysaccharides that accumulate outside the cells or can remain attached to the cell surface. Research into microbial polysaccharides started in the 1940s with the development of dextran as a blood-plasma extender. Several polysaccharides are now used widely in a number of industries, of which the best known is xanthan (Table 8.1 and Fig. 8.5). Xanthan is a polymer produced by *Xanthomonas campestris* and consists of an alternating glucose backbone carrying side chains of D-mannose and D-glucuronic acid. Production at present exceeds 20 000 t/year (Sutherland, 1998). Solutions of xanthan are pseudoplastic, being able to regain their viscosity after shearing. The pseudoplastic property is of particular value in drilling muds that act as a seal and lubricant during the drilling of oil wells. The mud needs to be viscous to form a good seal but flow when the drill bit is rotated. Curdlan is a neutral gel forming 1,3-β-D-glucan (74 000 Da), which can be formed by a number of bacteria such as *Agrobacterium* and *Rhizobium* spp. It forms a gel that is elastic and does not melt when heated. The property of forming a gel upon acidification has also suggested its use in oil wells. Scleroglucan is a glucose homopolymer produced by a number of micro-organisms including *Sclerotium glucanium*. Scleroglucan is produced commercially and has been developed as an alternative to xanthan for EOR by Elf Aquitane (McNeil and Harvey, 1993).

8.3 Recovery of metals

Micro-organisms are active in the formation and degradation of minerals in the Earth's crust and the use of micro-organisms to extract or leach metals from ores has been carried out for hundreds of years without us knowing that micro-organisms were involved. Research on iron- and sulphur-oxidizing bacteria in 1920–1930 laid the foundation of bioleaching. Bryner et al. (1954) described how iron pyrites and copper sulphide could be oxidized by *Thiobacillus* sp. from the Kennecott Bingham Canyon open-cast mine. This work lead to the first patent in 1958 (Zimmerley et al., 1958). The ability of micro-organisms to solubilize metals from insoluble metals is known as '**bioleaching**' and is the basis of a growing biomining and biowinning minerals industry.

This will become important as high-grade surface mineral deposits are worked out and become less viable, and mining companies will be forced to find other mineral sources. These will include the working of low-grade ore deposits, mine tailings, mine dumps, and worked-out mines. The extraction of metals from these low-grade sources using mechanical and chemical methods is difficult and expensive but biological methods can be used as a replacement. Biological extraction can function well at the low concentration of metals in contrast to the mechanical and chemical methods. The advantages of biological extraction of mineral are that the process uses little energy, does not produce harmful emissions and reduces the potential pollution of metal-containing wastes. Successful commercial metal-leaching processes include the extraction of gold, copper, and uranium (Suzuki, 2001).

8.3.1 The processes of bioleaching

The most important mineral-decomposing micro-organisms are the iron- and sulphur-oxidizing chemolithotrophs (section 2.10). Chemolithothrophs obtain energy from inorganic chemicals, use carbon dioxide as their carbon sources, and are represented by hydrogen-, sulphur-, and iron-reducing bacteria and Archaea. The most important metal-leaching organisms are the iron- and sulphur-reducing micro-organisms that use ferrous iron and reduced sulphur compounds as electron donors and fix carbon dioxide as a carbon source. In the metabolism of sulphur these organisms produce sulphuric acid, which means that the organisms grow at low-pH environments of 1.4–1.6. At low pH ferric iron is soluble and is often used as an electron acceptor instead of oxygen so that the organisms can grow anaerobically, an important feature in the working of mines and dumps.

Metal sulphides are the major mineral forms of many metals and iron sulphide (pyrite) is the most abundant sulphide. The role of micro-organisms in the leaching of metals is at present the subject of considerable discussion. Their original role in bioleaching was regarded just to be the re-oxidation of ferric sulphate to ferrous sulphate, after the ferric sulphate had leached the iron from the iron-ore pyrite (FeS_2) in the following reactions.

$$FeS_2 + 8H_2O + 14Fe^{3+} = 15Fe^{2+} + 2SO_4^{2-} + 16H^+ \qquad (8.4)$$

$$14Fe^{2+} + 3.5O_2 + 14H^+ \text{ (bacteria)} = 14Fe^{3+} + 7H_2O \qquad (8.5)$$

However, further study has been carried out, and three mechanisms have been proposed to explain the influence of microbial culture on the leaching of metal sulphides. These are shown in Fig. 8.6 and are

- indirect, where the microbes just recycle the ferric ions;
- indirect contact, where the recycling of the ferric ion occurs within an extracellular polymeric layer; and
- direct, where the mineral is oxidized directly by the microbial culture.

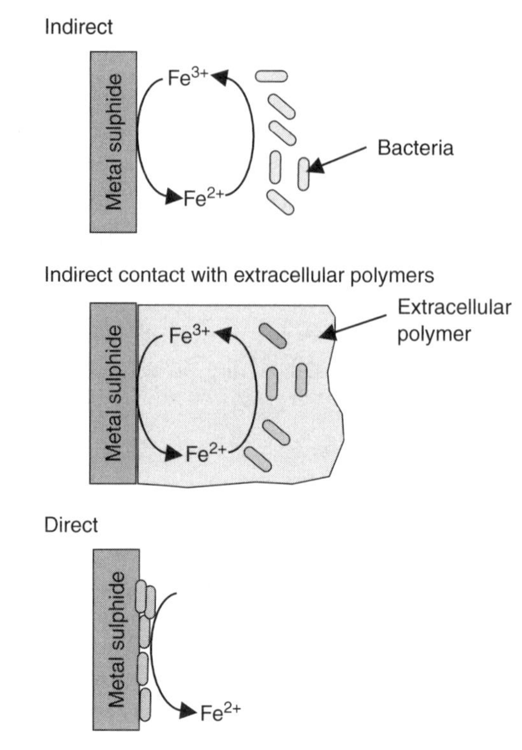

Figure 8.6 The three processes of microbial metal sulphide leaching. Modified from Crundwell (2003).

8.3.2 Indirect leaching

Metals can be leached from the metal sulphide ores by two mechanisms, oxidation with ferric ions and solubilization with acid. The differences between pure chemical and bacterial leaching of copper is shown in Fig. 8.7. This depends on the ability of the chemolithotrophs to generate ferric ion by the oxidation of the soluble ferrous ions. The ferric ions are oxidizing agents that can oxidize the metal sulphite, releasing the metal as a soluble sulphate. The production of ferric ions is as follows:

$$2Fe^{2+} + 0.5O_2 + 2H^+ = 2Fe^{3+} + H_2O \qquad (8.6)$$

The ferric ion formed is an oxidizing agent that can react with other copper sulphides such as chalcopyrite ($CuFeS_2$), chalcocite (Cu_2S), bornite ($CuFeS_4$), and covellite (CuS). In some cases the thiosulphate route is followed in the absence of oxygen. The thiosulphate route is as follows:

$$FeS_2 + 6Fe^{3+} + 3H_2O = 7Fe^{2+} + S_2O_3^{\,2-} + 6H^+ \qquad (8.7)$$

$$S_2O_3^{\,2-} + 8Fe^{3+} + 5H_2O = 8Fe^{2+} + 2SO_4^{\,2-} + 10H^+ \qquad (8.8)$$

In some cases elemental suphur is formed but this soon reacts to form sulphuric acid.

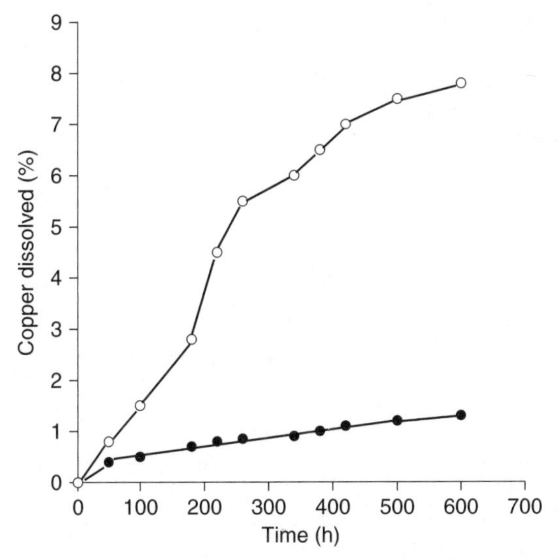

Figure 8.7 The leaching of copper from enargite (Cu_3AsS_4) at pH 1.6 with bacteria (•) and acid leaching (○). Data from Escobar et al. (1997).

$$ZnS + 2Fe^{3+} = Zn^{2+} + S^0 + 2Fe^{2+} \quad (8.9)$$
$$S^0 + 1.5O_2 + H_2O = SO_4^{2-} + 2H^+ \quad (8.10)$$

In acid conditions hydrogen sulphide can also be formed which can react with ferric ions to form elemental suphur.

$$ZnS + 2H^+ = Zn^{2+} + H_2S \quad (8.11)$$
$$H_2S + 2Fe^{3+} = S^0 + 2Fe^{2+} + 2H^+ \quad (8.12)$$

In the case of polysulphide ores the leaching is by a combination of ferric ions and protons with elemental sulphur as an intermediate.

8.3.3 Direct leaching

Direct leaching occurs when the microbial cell is attached to the mineral (Fig. 8.6) and again involves acid and proton production. Figure 8.8 shows the kinetics of iron leaching and the attachment of microbial cells to an iron ore (Rodriguez et al., 2003). The attached microbes can oxidize the sulphide directly and also elemental sulphur, forming acid that can assist with leaching.

$$FeS_2 + H_2O + 3.5O_2 = Fe^{2+} + 2SO_4^{2-} + 2H^+ \quad (8.13)$$
$$S^0 + 1.5O_2 + H_2O = SO_4^{2-} + 2H^+ \quad (8.14)$$

These reactions have been shown to be catalysed by the bacterium *Thiobacillus ferrooxidans* and the full cycle of reactions for both direct and indirect leaching is shown in Fig. 8.9.

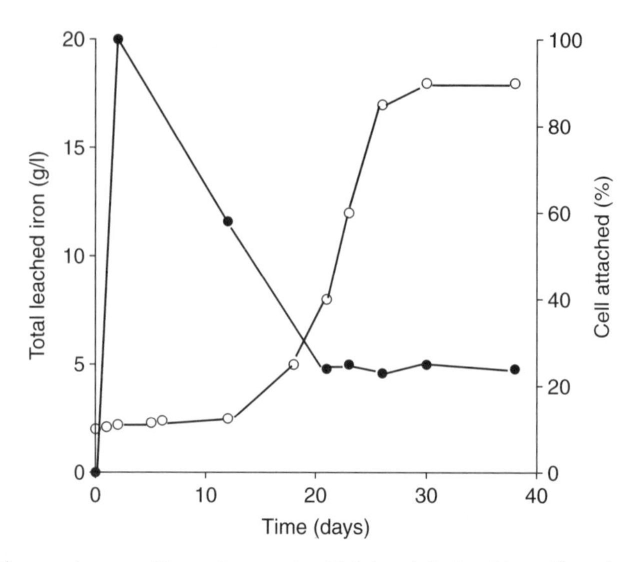

Figure 8.8 The attachment of bacteria to pyrite (FeS_2) and the leaching of iron from pyrite. •, Bacterial numbers (%); ○, iron concentration (g/l). Data from Rodriguez et al. (2003).

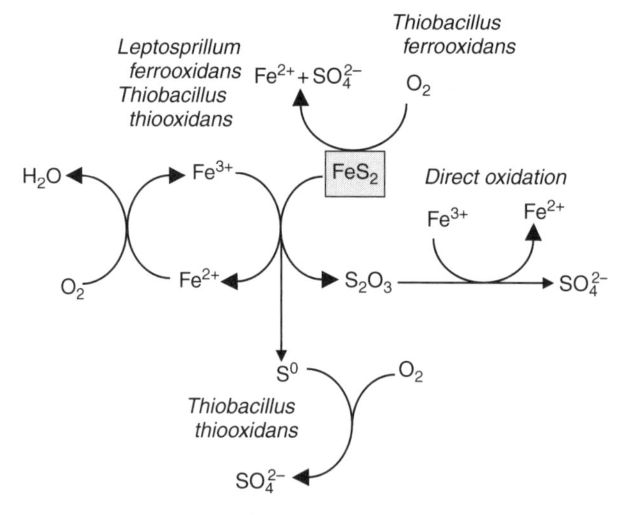

Figure 8.9 The cycle of bioleaching, both direct and indirect.

8.3.4 Bioleaching micro-organisms

Many different forms of micro-organisms have been isolated from natural and commercial bioleaching systems that are capable of degrading metal sulphides. The micro-organisms include fungi, bacteria, and Archaea and these can be divided into groups depending on their temperature optimum. The groups are the mesophiles, 20–40°C, the moderate thermophiles, 40–60°C, and the extreme thermophiles, above 60°C (Table 8.3). These organisms are predominantly chemolithotrophs with a few autotrophs as some have been

Table 8.3 Some micro-organisms that carry out bioleaching

Organism	Type	Metabolism	pH optimum	Temperature range (°C)
Iron oxidizing				
Mesophile (20–40°C)				
Thiobacillus ferrooxidans*	Bacterium	Anaerobe/ Fe/acid	2.4	28–35
T. prosperus	Bacterium	Halotolerant/ Fe/acid	2.5	30
Leptospirillum ferrooxidans	Bacterium	Fe only	2.5–3.0	30
Moderate thermophile				
Sulfobacillus acidophilus	Bacterium	Fe/acid	–	50
S. thermosulfi-dooxidans	Bacterium	Fe/acid	–	50
L. thermoferro-oxidans	Bacterium	Fe	2.5–3.0	40–50
Extreme thermophile				
Acidianus brierleyi	Bacterium	Acid	1.5–3.0	45–75
A. infernus	Bacterium	Acid	1.5–3.0	45–75
A. ambivalens	Bacterium	Acid	1.5–3.0	45–75
Sulfurococcus yellowstonii	Bacterium	Fe/acid	–	60–75
Sulphur oxidizing				
Mesophile				
T. thiooxidans	Bacterium	Acid	–	25–40
T. acidophilus	Bacterium	Acid	3.0	25–30
Moderate thermophile				
T. caldus	Bacterium	Acid	–	40–60
Extreme thermophile				
Sulfolobus solfataricus	Archaean	Fe/acid	–	55–85
S. rivotincti	Archaean	Fe/acid	2.0	69
S. yellowstonii	Archaean	Fe/acid	–	55–85

Sources: Agate (1996), Johnson (1998), and Gomez et al. (1999).
* Now named *Acidithiobacillus*.

shown to assimilate organic compounds such as formic acid. The metabolism of metal sulphides produces sulphuric acid and not surprisingly almost all the micro-organisms are acidophiles growing below pH values of 3.0 (Johnson, 1998). The micro-organisms can also be divided into three metabolic groups; those only oxidizing iron compounds, those only oxidizing sulphur compounds, and those capable of oxidizing both (Table 8.3).

A more extensive list of organisms known to have bioleaching abilities is given in Krebs et al. (1997). The important mesophiles, growing at 25–35°C, are chemolithotropic and highly acidophilic (pH 1.5–2.0); *T. ferrooxidans, Thiobacillus thiooxidans,* and *Leptosprillum ferroxidans* (Rawlings and Silver, 1995). *T. thiooxidans* can only use reduced sulphur compounds and *L. ferroxidans* can only use the ferrous ions. Thus while neither can reduce metal sulphides alone together they rapidly degrade pyrites (FeS_2). Thermophilic bacteria, *Thiobacillus* TH-1 and *Sulfolobus brieleyi,* have been found to grow on chalcopyrite ($CuFeS_2$). Most of these bacteria require some form of organic substrate for vigorous growth. *S. brierleyi* is an extreme thermophile which can grow at 70°C and can metabolize pyrite, chalcopyrite, and pyrrhotite (FeS).

8.3.5 Recovery of metals from mining wastes

Mining leaves behind very large quantities of tailings and low-grade ores, which are too low in metal content to warrant conventional extraction. This represents a considerable loss of metal and any process for the economical extraction of metal from this material will be of considerable value. It has been shown that micro-organisms can extract cobalt, nickel, cadmium, antimony, zinc, lead, gallium, indium, manganese, copper, and tin from sulphur-based ores. The basis of microbial extraction is that the metal sulphides, the principal component in many ores, are not soluble but when oxidized to sulphate become soluble so that the metal salt can be extracted. In the case of gold extraction this is somewhat different as the leaching process removes metal sulphide, allowing the gold to be extracted by cyanide.

There are three main methods of applying bioleaching; *in situ* treatment, heaps or dumps, and bioreactors (Fig. 8.10). *In situ* bioleaching can be used in mines that have come to the end of their useful life but still contain metal ores. The leaching solution containing *T. ferrooxidans* is pumped into the mine where it is injected into the ore (Fig. 8.11). The leachate is recovered from lower down the mine, pumped to the surface where the metal is recovered, and the bacterial suspension aerated before pumping back to the mine.

8.3.5.1 *Heap and dump extraction*

The treatment of low-grade ores or mine tailings in heaps or dumps is the most common method of bioleaching. An example of such a system is given in Fig. 8.12 and examples of operations are given in Table 8.4. The dump is often

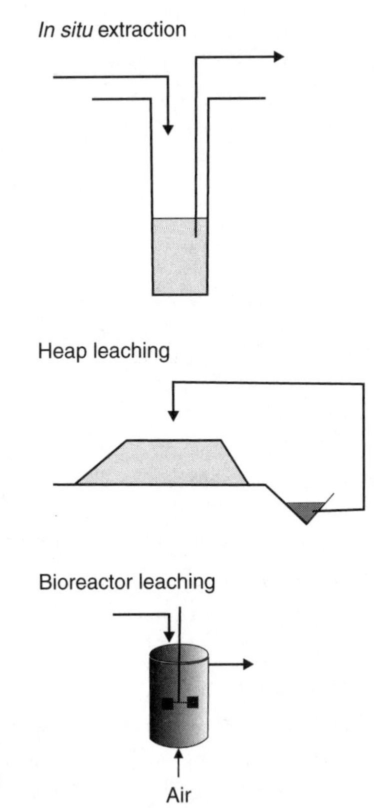

Figure 8.10 The three processes for the microbial extraction of metals from sulphides. *In situ* extraction, heap leaching, and bioreactors.

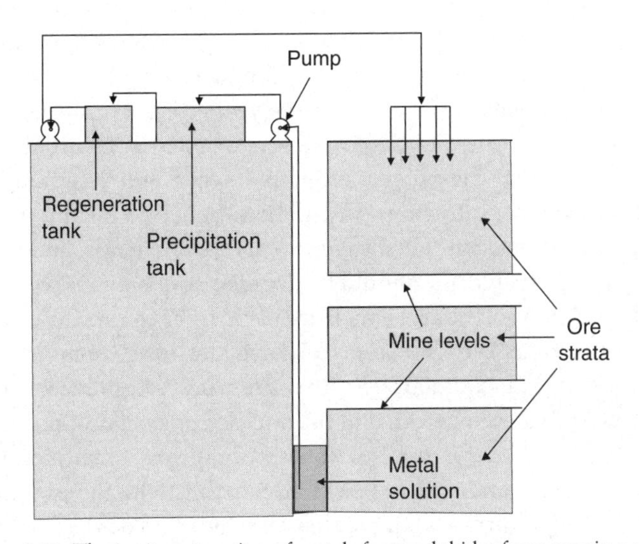

Figure 8.11 The *in situ* extraction of metals from sulphides from ores in a mine.

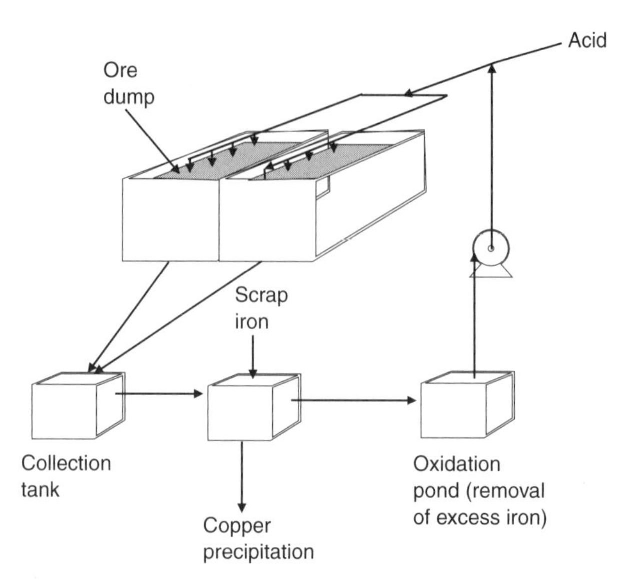

Figure 8.12 The heap or dump extraction of copper from low-grade ores or wastes. The copper leached out is collected on scrap iron and the liquid recycled to the heap.

set on a slope with a depth of 7–20 m of crushed ore or tailings. The dump is sprayed with water acidified with sulphuric acid to ensure that the pH is between 1.5 and 3, low enough to encourage the growth of *Thiobacillus* to increase the leaching rate, which gives bacterial populations of up to 10^6 cells/g of rock. The leaching solution is often collected at the base and recirculated and in this way a large population of *Thiobacillus* spp. develop. The *Thiobacillus* spp. will leach metals such as copper into solution and this can be collected by precipitation from the leachate. At this stage excess iron is removed in an oxidation pond and the liquid returned to the dump. The design of the dumps has improved over the years to give a more-even particle size to avoid anoxic zones. The construction of the dumps often do not take into consideration that the process requires oxygen and it should be as open as possible to allow for diffusion. Air can be supplied to the dump from pipes at the base of the dump but it is difficult to aerate efficiently. Acid-consuming rock such as calcium chloride should be avoided and some rocks may contain compounds such as molybdate, which are toxic to bacteria. Galena (PbS) also presents a problem as the lead sulphate formed during leaching is relatively insoluble and will accumulate on the ore surface, preventing leaching. Anaerobes have also been detected in the anoxic zones of dumps including the sulphate-reducing *Desulfovibrio* sp. These organisms form metal sulphides that coat the mineral and reduce leaching, which is why the process needs to be aerobic. In newly formed dumps, that have not had acid added a group of mesophilic sulphur-oxidizing bacteria might occur. These micro-organisms are useful as they oxidize sulphides at pH values of 4–7 to give sulphuric acid that in turn reduces the pH and allows the other sulphur-oxidizing organisms

Table 8.4 Some of the characteristics of copper-heap leaching processes

Country	Ore used	Amount of ore used (10^6 t)	Surface area (1000 m^2)	Height of dump (m)	Method	pH	Temperature (°C)	Copper extracted (kg/m^2)
USA	Chalcocite, chrysocolia	1 300	525	60	Ponds, trenches	2.4	NA*	1.09
USA	Chalcocite	100	36	60	Perforated plastic pipes	2.2	19.4	0.8
USA	Chalcocite	40 000	2 880	366	Channels	2.5	38–52	1.8
Bulgaria	Chalcopyrite, pyrite	0.6×6	16 500	25	Spray trenches and channels	1.8	11–13	1.8
Mexico	Chalcocpyrite, pyrite, chalcocite	10	NA	NA	Sprays	2.1	NA	2.5
USA	Chalcocite	41	40 000	6	Ponds (60×30 m)	3.5	NA	0.18

Source: Data from Rawlings and Silver (1995).
* NA, not available.

to flourish. Thus the dump represents a very diverse and dynamic microbial community of up to 10^6 cells/g of rock. The microbial population is hard to follow as sampling and enumeration are not easy; as dump sampling is difficult and special media are needed for the culture of many micro-organisms. The use of DNA probes and PCR techniques will allow a better understanding of population dynamics.

Ores such as chalcopyrite ($CuFeS_2$) and energite (Cu_3AsS_4) require temperatures as high as 75–80°C for leaching which cannot be generated in dumps and therefore can only be carried out in bioreactors.

8.3.5.2 *Bioreactors*

The bioreactors used are the highly aerated stirred-tank designs where finely ground ore is treated. Often nutrients such as ammonia and phosphate are added and the bioreactor operated in a continuous manner. The use of bioreactors allows a close control of the process parameters such as temperature, pH, and aeration, which are critical to the process, and these can be used in series. The leaching can take days rather than the weeks required with dump extraction as temperatures of 40–50°C are used, although the ore loading is 20%. The commercial processes using bioreactors are given in Table 8.5 where most of the processes use the commercial BIOX®, and all but one are gold-extraction plants. Stirred-tank technology could also work for zinc, copper, and zinc ores but the high capital and running costs probably preclude these at present.

Table 8.5 Bioreactor extraction

Plant	Date	Yield (per day)	Bioreactor volume (m^3)	Ore used
Sao Bento BIOX®, Brazil	1991	150	580	Pyrite, arsenopyrite, pyrrhotite
Wiluna BIOX®, Australia	1993	115	2 820	Pyrite, arsenopyrite
Youanmi BacTech, Australia	1994	120	3 000	Pyrite, arsenopyrite
Ashanti-Sansu BIOX®, Ghana	1994	720	16 200	Pyrite, arsenopyrite, pyrrhotite
Tamboraque BIOX®, Peru	1998	60	1 570	Pyrite, arsenopyrite
Leizhou BacTech, China	2001	100	4 050	Pyrite, arsenopyrite

Source: Data from Rawlings et al. (2003).

Conventional bioreactors such as stirred-tanks and alternatives such as the airlift and immobilized-bed bioreactors have been investigated up to pilot-plant scale. Initial results have indicated that the airlift design is more efficient than the conventional designs.

8.3.6 Extraction of copper

The estimate in 1991 was that the biological recovery of copper exceeded $1000 million and accounted for 25% of the world's copper production. Bioleaching has considerable advantages as it functions at low temperatures and is a cheap method of utilizing waste. The waste formed is generally that remaining after extraction of rock from a mine where the copper level is too low for it to be extracted economically (0.1–0.5%). In most mining operations the waste material is formed into terraced dumps 100 m wide and 5 m deep with an impermeable base. Dilute sulphuric acid is sprinkled or sprayed on to the dump so that as it percolates through the dump the pH is reduced to 2–3, which promotes the growth of *T. ferrooxidans* and other leaching micro-organisms. The copper, on oxidation to copper sulphate, is dissolved in the dilute acid and is collected at the bottom of the dump (Fig. 8.12). The copper is removed from the solution by precipitation on to scrap iron although newer methods of solvent extraction or electrowinning are being tried. Often, once the copper has been removed the dilute acid can be recycled but the recycling has to be limited as it contains high levels of ferric salts that coat the ore, decreasing the availability of copper (Fig. 8.12). Some examples of copper leaching from dumps are given in Table 8.4.

8.3.7 Extraction of uranium

There are two possible processes in the biological extraction of uranium from ores. Direct leaching by *T. ferrooxidans* has been proposed in the following equation.

$$2UO_2 + O_2 + 2H_2SO_4 = 2UO_2SO_4 + 2H_2O \quad (8.15)$$

However, in conditions where oxygen is limited this cannot operate, and the indirect bioleaching process operates using pyrite, which is often associated with uranium ores. In this case *T. ferrooxidans* produces the ferric ion, which reacts with the uranium ore, and sulphuric acid, which also leaches uranium.

$$UO_2 + Fe_2(SO_4)_3 = UO_2SO_4 + 2FeSO_4 \quad (8.16)$$

$$UO_3 + H_2SO_4 = UO_2SO_4 + H_2O \quad (8.17)$$

One example of uranium leaching *in situ* is the Dennison mine in the Elliot Lake district of Canada which extracted 300 tons of uranium worth $25 million in 1988 (Rawlings and Silver, 1995). The ore was fractured, a bulkhead formed, and the ore behind this flooded with leach liquor. After 1–3 weeks at

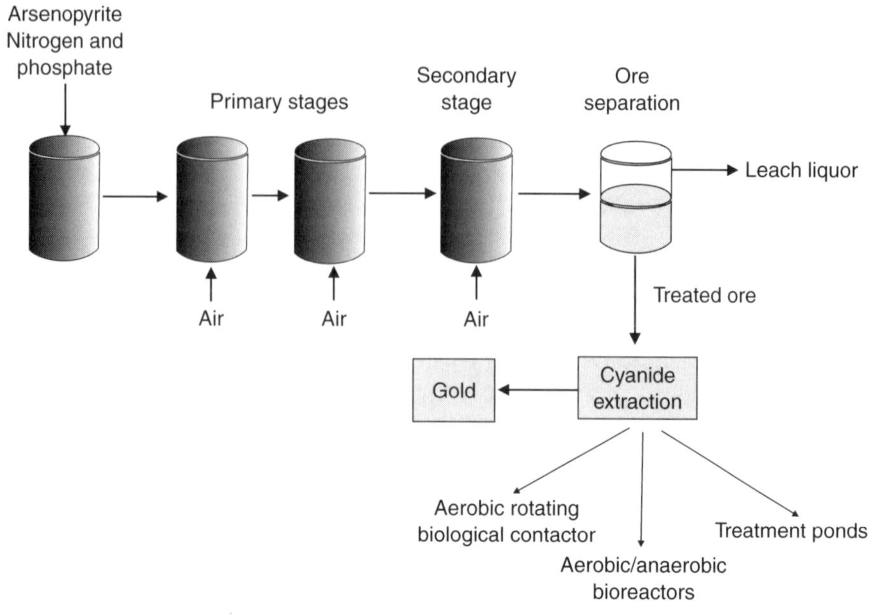

Figure 8.13 The process for the extraction of gold from arsenopyrite-containing ore and the treatment of the cyanide used to extract the gold. The gold is made available for extraction by the removal of arsenopyrite that masks the gold. The gold is removed from the cyanide by the addition of carbon (charcoal) and the waste cyanide can be treated in aerobic rotating biological contactors, aerobic/anaerobic bioreactors, and treatment ponds.

12–15°C the liquor was drained for extraction of uranium. The cycle is repeated until the ore has been exhausted.

8.3.8 Extraction of gold

The normal method of extracting gold (90%) from ores is to treat it with cyanide and then extract the gold from the cyanide extract with carbon. However, some ores are recalcitrant to cyanide treatment as the gold is enmeshed in pyrite (FeS_2) and arsenopyrite (FeAsS) and without treatment only 50% of the gold can be extracted. These ores can be treated by roasting and pressure leaching but biological leaching offers a low-energy alternative. Generally leaching is carried out in open aerated bioreactors, which are more expensive to operate, but with a product as valuable as gold the expense can be justified. The leaching is carried out in a sequence of bioreactors (Fig. 8.13); primary and secondary bioreactors. The treated ore is then treated with cyanide and the gold removed from the cyanide extract with carbon. The cyanide waste from this and direct extraction represent a major pollutant and has to be treated before release into the environment (Akcil and Mudder, 2003). Cyanide can be destroyed by a sulphur dioxide and air mixture or a copper-catalysed hydrogen peroxide mixture. However, there are biological methods, both aerobic and anaerobic, for the treatment of cyanide.

The processes include aerobic rotating biological contactors, bioreactors, and stimulated ponds. The aerobic reactions are as follows:

$$CN^- + 0.5O_2 + 2H_2O = HCO_3^- + NH_3 \quad (8.18)$$

$$NH_3 + 1.5O_2 = NO_2 + H^+ + H_2O = NO_3^- \quad (8.19)$$

Some of the micro-organisms known to oxidize cyanide include species of the genera *Actinomyces, Alcaligenes, Arthobacter, Bacillus, Micrococcus, Neisseria, Paracoccus, Thiobacillus*, and *Pseudomonas*.

8.3.9 Recent developments

All bioleaching has been carried out with indigenous microbial strains whose proliferation has been encouraged by the acid conditions. However, there are methods whereby these micro-organisms could be improved. The first is to isolate new bacterial strains from extreme environments, such as mine-drainage sites, hot springs, and waste sites, and use these to seed bioleaching processes. Secondly, the existing micro-organism could be improved by conventional mutation and selection or by genetic engineering. A number of genes have been isolated and characterized from *Thiobacillus* spp. (Rawlings and Silver, 1995). One possibility would be to introduce arsenic resistance into some bioleaching organisms, which could then be used in gold bioleaching. Also more needs to be known about the population dynamics within the bioleaching dumps and the relative importance of various organisms and mechanisms. Various plating methods are available for the determination of the bacterial population but these are not always accurate, as the conditions in the dump are difficult to reproduce. PCR techniques using small-subunit rRNA sequences can now be used to replace the plating techniques (De Wulf-Durand et al., 1997).

Heterotrophic leaching, the production of organic acids, is a solution for wastes and ores of high pH (5.5) where many of the acidophiles would not grow. Metal-containing waste of high pH need not only include ore but can include metal wastes from all forms of industry. Heterotrophic leaching also includes fungi and *Trichoderma horzianum* has been shown to solubilize MnO_2, Fe_2O_3, Zn, and calcium phosphate minerals.

The development of bioleaching at elevated temperatures of up to 50°C has been shown to be feasible, but temperatures up to 70°C are being developed for use in bioreactors.

8.4 Conclusions

In this chapter it can be seen that:

- micro-organisms are currently being used to produce polysaccharides used in drilling muds;

- micro-organisms could also be used to enhance oil recovery by growth *in situ* (MEOR) if the price of crude oil rose sufficiently to warrant further extraction; in addition the methods involved in MEOR require further research before the process can be fully understood;
- micro-organisms can leach a number of metals either directly or indirectly from low-grade ores, tailings, and from low-yielding mines in an *in situ* process; and
- at present bioleaching is used extensively for copper and increasingly for uranium and gold extraction.

8.5 Further reading

Glazer, A. and Nikaido, H. (1995) *Microbial Biotechnology.* W.H. Freeman, Basingstoke.

Moses, V. and Cape, R.E. (1991) *Biotechnology; The Science and Business.* Harwood Academic, London.

9 Agricultural biotechnology

9.1	Introduction	345
9.2	Detection and diagnostics	346
9.3	Micropropagation	347
9.4	Somatic cell genetics	347
9.5	Production of transgenic plants	352
9.6	Safety of transgenic crops	362
9.7	Transgenic plants	364
9.8	Transgenic animals	373
9.9	Disease control	374
9.10	Germplasm and biodiversity	376
9.11	Conclusions	376
9.12	Further reading	377

9.1 Introduction

Agriculture has developed slowly over thousands of years with the domestication of plants and animals. Biotechnology similarly has a long history starting with the discovery of brewing some 5000 years ago and a timeline for the two is given in Table 9.1. The breeding of plants and animals was boosted when genetic inheritance was more widely understood. Mendel discovered the principles of genetics but it was some time before the data were appreciated fully. Traditional breeding continued up to the 1970s with considerable success, one of which was the high-yield rice in the 'green revolution'. The development of cloning starting in 1974 and the production of transgenic plants in 1980 profoundly changed plant and animal breeding. The ability to transfer genes across breeding barriers can produce novel products and, in addition, genetically engineering and micropropagation can respond more rapidly than conventional breeding in the development of a new plant variety. However, there are some concerns over the release of genetically engineered plants into the environment, but the positive effects of the reduction in the use of pesticides and herbicides by use of these types of plants should not be ignored. The techniques of molecular biology, genetic manipulation, are outlined in Chapters 2 and 3. This chapter will cover some aspects of the application of biotechnology to the agricultural industry and will include the following:

- detection and diagnostics,
- micropropagation,

Table 9.1 Landmarks for agriculture and biotechnology

Date	Agriculture	Biotechnology
10 000 BC	Domestication of plants and animals	
2000–3000 BC		Brewing, cheese, wine
14 AD		Distillation
1680		Description of micro-organisms by Leeuwenhoek
1818		Discovery of fermentation by yeasts
1860		Lactic acid production Pasteur
1865	Mendel, genetics	
1914		Sewage treatment
1920		Citric acid production
1940		Penicillin production
1950	Hybrid crops	
1953	DNA double helix	DNA double helix
1960	Green revolution, rice	
1974		Cloning
1980	Transgenic plants	Transgenic plants
1982		Production of human insulin in *E. coli*

- somatic cell genetics,
- transgenic plants,
- transgenic animals,
- disease control, and
- germplasm and biodiversity.

9.2 Detection and diagnostics

Biotechnology and in particular molecular biology have produced powerful techniques for the detection of micro-organisms and chemicals in the environment. The detection and enumeration of the microbial populations found in a range of environments is now possible where it was not possible previously. Techniques such as PCR and small-subunit rRNA sequencing have revealed a much greater microbial diversity than was expected from previous data based on growth. Thus microbial processes such as composting can be followed with much greater understanding. The use of biomarkers and bioindicators has been described in Chapter 3 and these can be used to follow the effect of any process or chemical used or released in agriculture. Similarly biosensors can provide online, real-time measurements of many processes and environments.

The detection of disease in both animals and plants is important in terms of both health and commerce. New biotechnology-based vaccines against brucellosis, encephalitis, and hepatitis have been developed. The production of pure antigens by genetic manipulation will allow the formation of monoclonal antibodies, which are more specific than the normal antibodies. This allows accurate and specific detection of infections in both plants and animals.

9.3 Micropropagation

The technique of micropropagation has developed over a number of years and can produce whole plants from protoplasts, calluses, suspension cells, and parts of plants. The processes for the production of calluses and suspension cultures and organogenesis are described in Box 9.1. The part of the plant taken, known as an explant, can be encouraged to form shoots and then the shoots can be induced to form roots to regenerate the whole plant (Fig. 9.1). The process of regeneration is controlled by the use of growth regulators and can produce many hundreds of plants from one explant. Thus micropropagation is used to produce commercial quantities of plants difficult to cultivate traditionally, such as orchids, to produce virus-free plants and to multiply elite plants.

The first protoplasts were isolated from plant tissue in the 1960s when plant material was incubated with a mixture of fungal enzymes including cellulases. These enzymes removed the cell wall from a plant cell, resulting in the formation of a spherical cell bounded by only a cell membrane (Box 9.2). The cell minus the wall will normally take up water and eventually burst but it can be stabilized by the addition of mannitol and sorbitol. Protoplasts have been isolated by mechanical or more often enzymatic treatment from many plant species, but protoplasts capable of sustained division and regeneration are still restricted to a limited number of species.

9.4 Somatic cell genetics

This is the improvement of plants using plant cell-culture techniques and is a useful contribution to plant breeding. Its applications include the use of somaclonal

BOX 9.1 Production of callus and suspension cultures

The figure shows the production of callus and suspension cultures. Parts of a plant (explant), such as leaves, stem, or root, and seeds are surface sterilized and then placed on a solid medium containing a carbon source (sucrose) and a mixture of growth regulators. If the explant such as a stem section is placed on solid medium containing auxin and cytokinins in a ratio of 10:1 the cells will begin to proliferate in an unorganized manner forming an amorphous mass known as a callus. If the callus is placed in a small volume of liquid medium a suspension culture will develop consisting a small groups of cells.

By changing the balance of auxin to cytokinin tissues and cells can be induced to go through a process known as organogenesis, forming either roots or shoots. An auxin-to-cytokinin ratio of 4:1 will encourage shoot formation, whereas a ratio of 100:1 will cause roots to form. Thus the shoots formed can be excised, placed on rooting medium where the shoot can be encouraged to root, forming a plantlet which with careful handling will mature into a plant. This appears to be simple but in reality it is far from it as for each species the exact conditions have to determined experimentally, although there now is a body of information related to a large number of species. Under some conditions the plant tissue will undergo a process known as embryogenesis where somatic embryos are formed and these can be cultured to form whole plants.

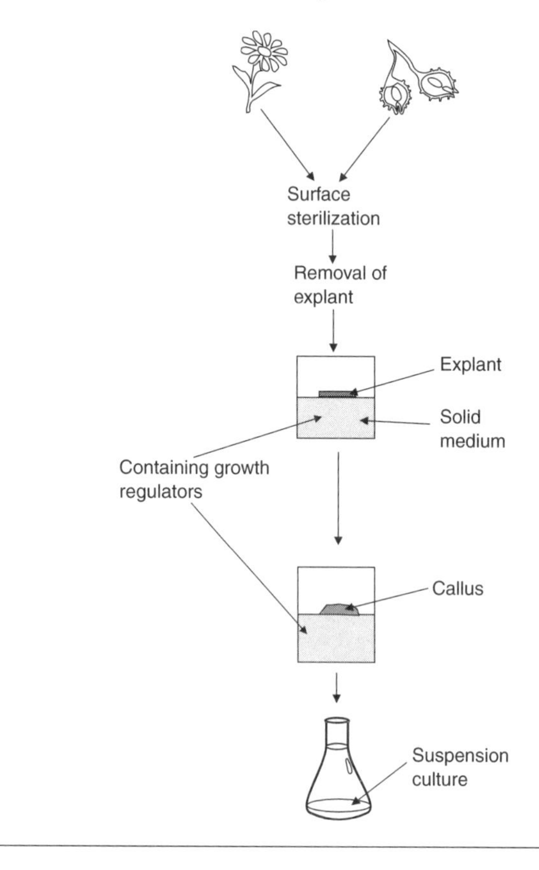

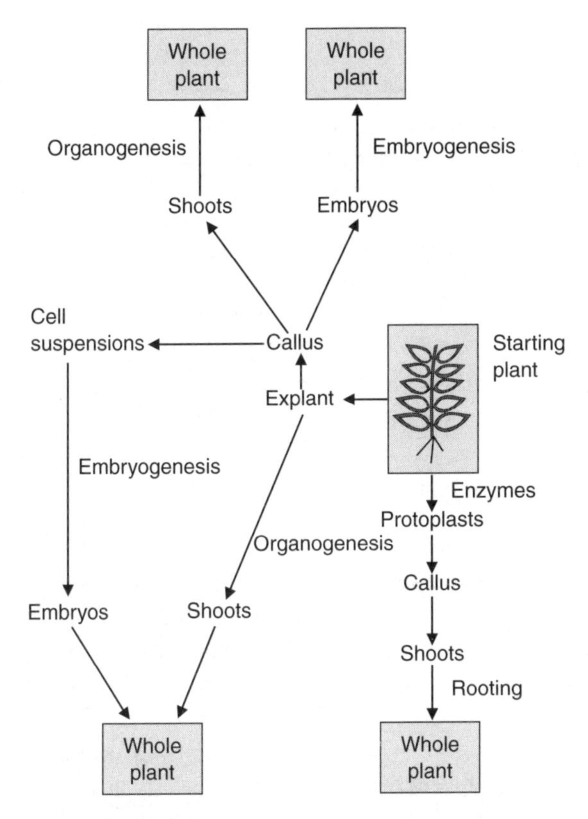

Figure 9.1 The pathways for the regeneration of plants from fragments of plants (explants) and protoplasts. By adding growth regulators to the solid medium the explant can be induced to form shoots in a process known as direct organogenesis. Under a different growth-regulator regime an amorphous mass of cells form, known as a callus. Callus material can be induced to form shoots (organogenesis) or embryos (indirect embryogenesis). The shoots can be induced to form roots and therefore a plant and the embryos will also develop into plants. Calluses, when placed in a liquid medium, will form a suspension culture, which can also form embryos. Plant material can also be used to produce protoplasts that can be manipulated to regenerate plants.

BOX 9.2 **The production of protoplasts**

Protoplasts are normally isolated from soft plant tissue such as young, fully expanded leaves. The leaves are cut into strips and floated on an enzyme mixture (osmoticum) containing cellulase, hemicellulase, pectinase, and 0.4 M sorbitol (or mannitol). The sorbitol is added to stop the protoplasts taking up water and rupturing. The strips are incubated for 8–24 h at 25°C, typically overnight. After incubation the leaf fragments are removed by filtration through a mesh (50 μm). The protoplasts are harvested by low-speed centrifugation (100 g for 5 min) but they still need to be separated from cells, still having some cell wall present. This can be achieved by centrifugation of the mixture through a 23% sucrose layer (100 g, 5 min). The cells with some cell wall present will pass through the sucrose layer but the protoplasts will form a layer at the interface. These can be removed and retained.

BOX 9.2 **(Continued)**

The process of regeneration can be carried out in multiwell plates or on solid medium for 3–7 days in the dark at 25 °C. The protoplasts are observed at intervals to see if the cell wall has begun to form and the cells are dividing. If division has occurred the cells are transferred to solid medium, repeating every 7 days for up to 3 weeks, by which time small colonies will have formed. These colonies can be induced to form shoots by applying growth regulators.

variation to produce variant plants, protoplast fusion to overcome interspecific crossing barriers, and the regeneration of haploid cell lines.

9.4.1 **Somaclonal variation**

Micropropagation can not only be used to produce many identical plants, it can also be used to produce variant plants. The procedures used in some of the plant cell-culture techniques can introduce a high level of variation in the regenerated plants, known as somaclonal variation. Somaclonal variation depends strongly on the method used for regeneration with the highest levels of variation found in regeneration from protoplasts. The genetic changes associated with somaclonal variation are point mutations, chromosome changes in number and structure, cryptic changes in chromosome rearrangements, altered sequence copy number, transposable elements, somatic crossover, sister-chromatid exchange, and DNA amplification and deletion (Karp, 1995). This variation is a problem in the production of large numbers of identical plants by micropropagation, but it is of considerable advantage in plant breeding. The use of somaclonal variation involves the following steps:

- induction and growth of calluses or cell suspensions (Box 9.1),
- regeneration of a large number of plants from these cultures,
- screening for desirable traits,
- testing the selected variants for the stability of the trait, and
- the multiplication of the selected variant.

The variation is unpredictable, which is a disadvantage, but variants have been produced in a number of plants including maize, tomato, spruce, and ornamentals (Jain and DeKlerk, 1998). A particular tomato line has been isolated via somaclonal variation, which has a higher solids content and has been used commercially for soup preparation.

9.4.2 **Protoplast fusion**

Protoplasts can not only be used in micropropagation but under certain conditions two different protoplasts can be fused together (Box 9.3). Since any

BOX 9.3 Fusion of protoplasts

Protoplast fusion can be mediated by either chemical or electrical techniques. In the case of chemical fusion high concentrations of polyethylene glycol (PEG), dextran, or polyvinyl alcohol (PVA) are used in combination with high pH in calcium-containing buffers. When protoplasts are in contact and subjected to an electrical field transient reversible pores form in the plasma membrane and the protoplasts fuse together. Two steps are usually involved; an alternating-current (AC) needs to be applied to align the protoplasts and bring them into direct contact. Subsequently a short direct-current (DC) pulse is employed to induce membrane breakdown (see figure). This form of alteration of the membrane fluidity can also be used to insert DNA into protoplasts.

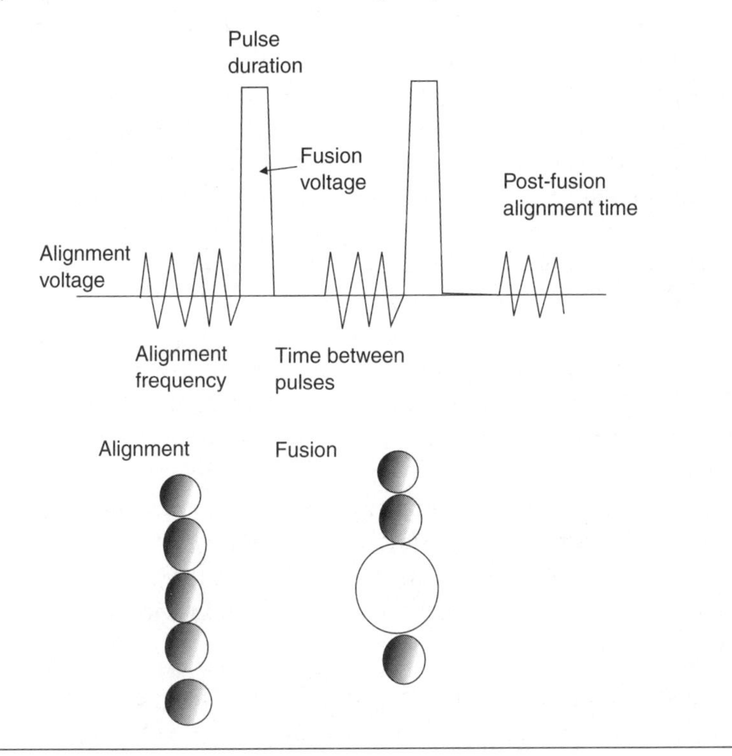

protoplasts can be fused this provides a means to circumvent sexual barriers in plant breeding. Somatic hybrid cells are totipotent and therefore capable of developing into whole plants, although complex nuclear and cytoplasmic combinations may follow protoplast fusion. Although considerable research has been carried out the commercial products have been limited.

9.4.3 Haploid production

Plant breeders traditionally need the establishment of true-breeding lines which are produced by self-fertilization and backcrossing, which is time-consuming and laborious. The production of haploid plants would be of great use in plant breeding. A method of producing haploids is to culture immature pollen, which will be haploid. Anther culture is used widely for the production of homozygous plants and involves the culturing of immature anthers.

9.5 Production of transgenic plants

The term transgenic plant refers to 'plants with unique gene combinations that do not occur naturally and are produced by using recombinant DNA technology'. The techniques for genetic engineering of micro-organisms (recombinant DNA technology) were developed in the 1970s and the rapid development of these techniques has enabled the genes from a wide range of organisms to be identified, isolated, and transferred to other species. Genetic engineering has been one of the fastest growing areas of biotechnology and has seen major advances over recent years. The techniques have allowed genes to be transferred between unrelated species forming new combinations not possible by conventional techniques. The production of transgenic plants has advanced rapidly and saw commercialization in 1994; products include the Flavr Savr™ tomato, herbicide-resistant soya plants, and insect-resistant maize (Kok and Kuiper, 2003). At present there are many field trials and 58.7 million hectares have been planted with transgenic plants. The particular traits that have been transferred are those that confer single characteristics like herbicide resistance and the traits have been restricted mainly to those which involve single genes. Complex characteristics such as drought tolerance are more difficult to transfer as more than one gene is involved and the exact biochemical basis of the traits are not understood fully.

In order to transfer a gene into a plant the following basic requirements need to be satisfied.

- The gene for a particular trait needs to be identified, which requires information on the pathway and the controls involved.
- The gene needs to be isolated (Box 9.4). The source of the gene for a particular trait can be a related plant species or it can be a totally unrelated species such as a bacterium and animal, as in the case of the transfer of the flounder antifreeze gene to strawberries in order to reduce frost damage.
- The gene needs to be transferred to the target plant (transformation) and be placed under the correct control so that the gene functions (expression).
- The transformed plant material needs to be regenerated into a whole plant, tested for the gene activity and, if successful, the line multiplied and field-tested.

BOX 9.4 The isolation of plant genes

The first step is to isolate the DNA from the donor cells or tissue and the method used will depend on the tissue and one's knowledge of the system. There are a number of manuals covering the techniques involved in DNA extraction (Chung et al., 1998; Kobayashi et al., 1998; Sul and Korban, 1996).

When total DNA is extracted the next step is to produce a collection or library of DNA fragments by cutting the DNA into a large number of fragments with restriction enzymes. The fragments are inserted into the appropriate vector. The vector can be used to transform bacteria and the colonies probed for the specific gene (see figure).

If the whole gene is required including the sequences removed during DNA processing the gene will need to be isolated from the total DNA. However, if only the gene is required complementary DNA (cDNA) can be used. cDNA is synthesized from the mRNA which can be extracted from the organism or tissue expressing the product in question. The isolated mRNA will contain the information for the protein in question and can be copied as DNA by using an enzyme, reverse transcriptase. More recently the PCR technique has been used to select and amplify the gene required from either the isolated DNA or cDNA. PCR requires primers that can recognize the start of the gene and these can be prepared if the gene sequence or that of a related gene is known. This method is being used increasingly as it needs only a very small amount of DNA and combined with a variety of either specific or more general primers genes can be amplified, pathogens detected, and the relationship between genomes assessed.

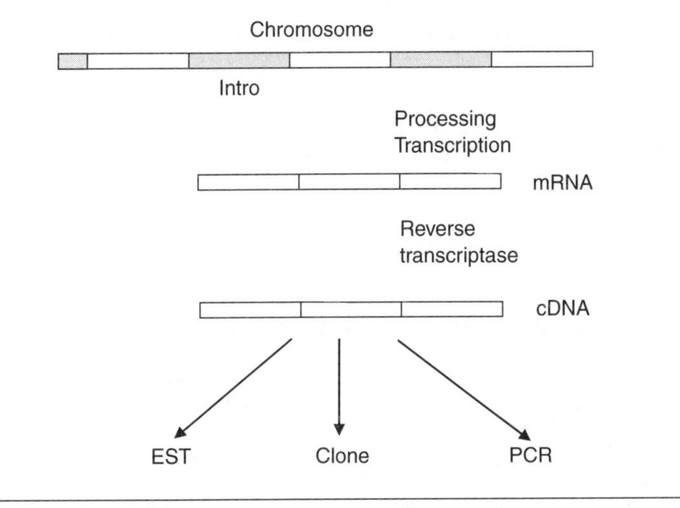

9.5.1 Gene isolation

The isolation of a gene, from whatever source, can be carried out using a number of methods (Fig. 9.2, Box 9.4):

- from total DNA, the cloning of restriction fragments,
- by chemical synthesis, if the protein sequence is known,

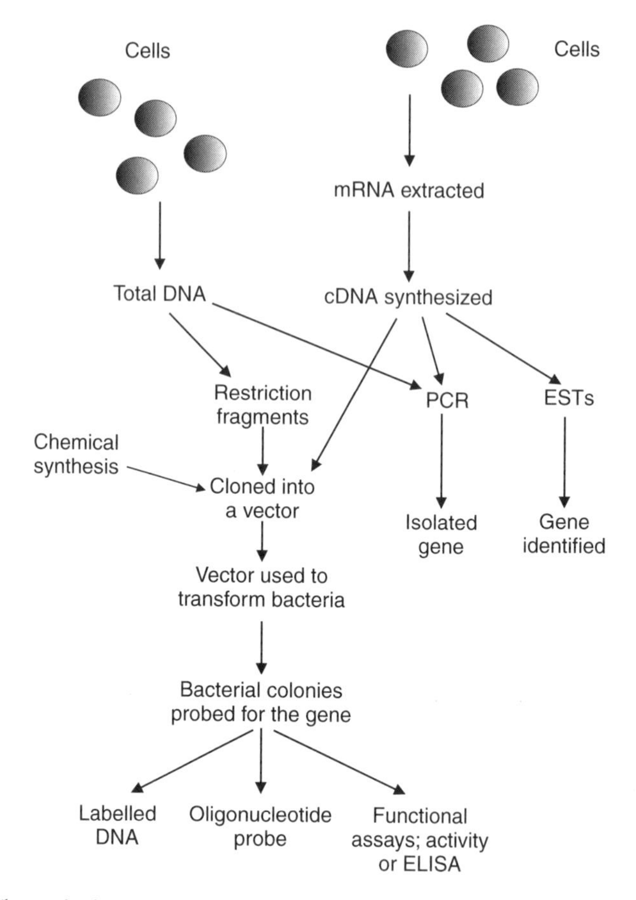

Figure 9.2 The methods involved in the isolation of a specific gene by isolating total DNA or mRNA. The mRNA is used to form complementary DNA (cDNA) that can be amplified by PCR or cloned into a vector. The cDNA can be partially sequenced as an expressed sequence tag (EST) and the sequence compared with extensive databases for a match with known genes. ELISA, enzyme-linked immunosorbant assay.

- from mRNA by the production of complementary DNA (cDNA),
- by PCR using primers derived from related genes, and
- using genomics.

Genomics is a recent development, and defines the sequence of the complete set of chromosomes of an organism and identifies the genes within the whole genome. Genomics should aid the identification of novel genes that control or influence traits that are of value to agriculture. The first step in identifying new genes in plants is the rapid sequencing of cDNA. The partial sequences of these cDNA clones are known as expressed sequence tags (ESTs). These sequences can be compared with known gene sequences present in databases, which should allow the inference of its activity. The databases contain very

BOX 9.5 An example of a bacterial vector

An example of one of the best known plasmids pBR322 is given in the figure. This plasmid has two selectable markers, for ampicillin and tetracycline resistance, and a number of restriction sites. If the vector and DNA are cut using the same restriction enzyme they can be annealed together and the join completed with the enzyme DNA ligase. The circular DNA formed can then be taken up by a bacterium, usually *E. coli*. Bacteria can be treated using a simple procedure so that they are capable of taking up DNA. If the vector that is taken up by the bacteria confers antibiotic resistance then those cells which have been transformed will be the only ones capable of growing on medium containing the antibiotic for which resistance is conferred. If a large number of fragments of DNA are cloned into bacteria a large number of transformed colonies can be obtained. This collection is known as a library and if sufficient colonies are collected statistically the whole of the original DNA will be represented in these colonies. In the case of cDNA the number of colonies will be smaller as the mRNA, from which cDNA is made, does not represent the whole genome.

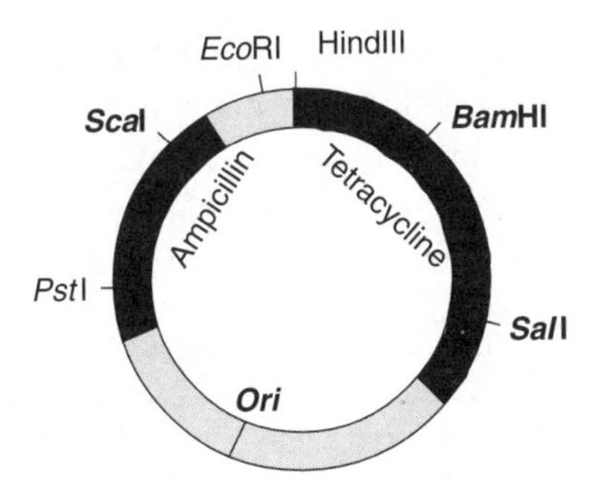

large numbers of ESTs, for example there are 37 000 from the small plant *Arabidopsis* sp. and others from rice, maize, tomato, and soybean.

9.5.2 Vectors

There are, in general, two forms of vector available in bacterial systems for the transfer of DNA, plasmids or bacteriophages. Plasmids are extrachromosomal DNA molecules found in many microbial species in the form of closed circles, which can occur in multiple copies (Box 9.5). The second type of vector is the bacteriophages that infect and are reproduced in bacteria. Plasmids have been modified extensively to give the following characteristics:

- as small a size as possible for ease of isolation and so that the maximum DNA can be added,
- autonomous replication,
- a number of single restriction sites, and
- the addition of a selectable marker such as antibiotic resistance.

9.5.3 Selection of the gene

If a specific gene is to be detected in the library some form of probe is needed to detect the colony or clone containing the gene of interest. A direct method is to use a nucleic acid probe to hybridize to the plasmid DNA extracted from each individual bacterial colony. Hybridization can only be carried out if the gene has already been cloned, the base sequence of the gene is already known, or a related gene has been sequenced. If the sequence of the whole protein is known or at least the N-terminal sequence then an oligonucleotide probe can be synthesized chemically. The conserved protein sequence from a related gene can also be used as the basis of a probe. The probe, once prepared, can be labelled with radioactivity, biotin, or a fluorescent protein and this label can be used to indicate which bacterial clone contains the gene. The gene can also be detected as an enzyme activity if it is capable of producing a functional protein, or by reaction with antibodies.

Once the bacterial clone containing the vector incorporating the gene has been identified, the gene can be isolated and manipulated in a number of ways depending on its final use. The gene can be placed under the control of a strong promoter that can function in a plant such as CaMV, the 35 S RNA cauliflower mosaic virus promoter. The gene can also be transferred to a plant vector, which should also have a marker gene in order to select transformed plants. A list of selectable markers is given in Table 9.2; these are based on

Table 9.2 Some marker genes used in plant transformation

Gene (protein)	Selective agent	Action
*npt*11* (NPT11)	Kanamycin	30 S Ribosome
neo (NPT11)	Neomycin	30 S Ribosome
hpt† (HPT)	Hygromycin	Peptide-chain elongation
bar (PAT), *pat*‡	Phosphinothricin	Phosphinothricin acetyl transferase
gox	Glyphosate	Glyphosate oxidoreductase
EPSPS§	Glyphosate	EPSP synthase
dhfr	Methotrexate	Dihydrofolate reductase
csr-1-1	Sulphonylurea	Acetolactate synthase
uid (GUS)	β-Glucuronidase	Formation of blue colour
luc	Luciferase	Light emission
gfp (GFP)	Fluorescent protein	Light emission

**npt*, Neomycin phosphotransferase.
†*hpt*, Hygromycin phosphotransferase.
‡*pat*, Phosphinothricin acetyltransferase.
§*EPSPS*, 5-Enolpyruvylshikimate-3-phosphate synthase.

positive selection using antibiotic or herbicide resistance such as kanamycin and glyphosate or are non-selectable markers such as luciferase and green fluorescent protein (GFP). At present 90% of the selection has been carried out using the antibiotic kanamycin (neomycin), and the herbicides phosphinothricin and glyphosate (Miki and McHugh, 2004). Recent developments have been to use marker genes that produce products that are non-invasive and easy to use. GUS (from the *uid* gene) and GFP are two examples; GUS codes for the enzyme β-glucuronidase, which will convert the colourless 5-bromo-4-chloro-3-indolyl-glucuronide (X-Gluc) to a blue colour and GFP is a green fluorescent protein derived from the jellyfish *Aequorea victoria* (Chapter 2). The GFP protein is very useful as it does not require cofactors to function and exhibits a green light upon exposure to long-wave UV light (Stewart, 2001). Marker-free transgenic plants are under development but these are more difficult to implement and are less efficient. The most promising method is the use of site-specific recombinases that excise the marker genes.

9.5.4 **Transformation of plants**

Once the gene has been isolated much of the manipulation can be carried out in bacteria, but once this has been achieved the gene needs to be used to transform the target plant. The processes of transforming plant cells are not without their problems. Plants have strong cellulose-based cell walls which makes it difficult to transfer DNA into the cells, and whole plants have bark and waxy cuticles which compound the problem. Gene transfer to whole plants is therefore difficult and to avoid these difficulties various techniques have been developed (Hansen and Wright, 1999) (Fig. 9.3).

9.5.4.1 *Protoplasts*

DNA can be transferred directly into protoplasts using polyethylene glycol (PEG), liposomes, or electroporation (Box 9.6). The exposure of protoplasts to PEG (4000–8000 Da; up to 28%) in a high-magnesium chloride medium can result in DNA uptake. Liposomes containing DNA have been used to transfer DNA to protoplasts, also using PEG. Electroporation is the application of high-voltage direct-current (DC) pulses, which opens up the cell membrane and allows any DNA in the medium to enter the cell (Fig. 9.3). The voltage and pulse length for optimal DNA transference vary according to the source of protoplasts, the electroporation medium, and the type of pulse. Electroporation reduces the viability of the protoplasts. The transformation frequency of electroporation or PEG treatment is approximately 1400 per 3×10^5 treated protoplasts (0.46%) and for some species it is much lower. The key to successful transformation is the ability to regenerate plants directly from the microcalluses.

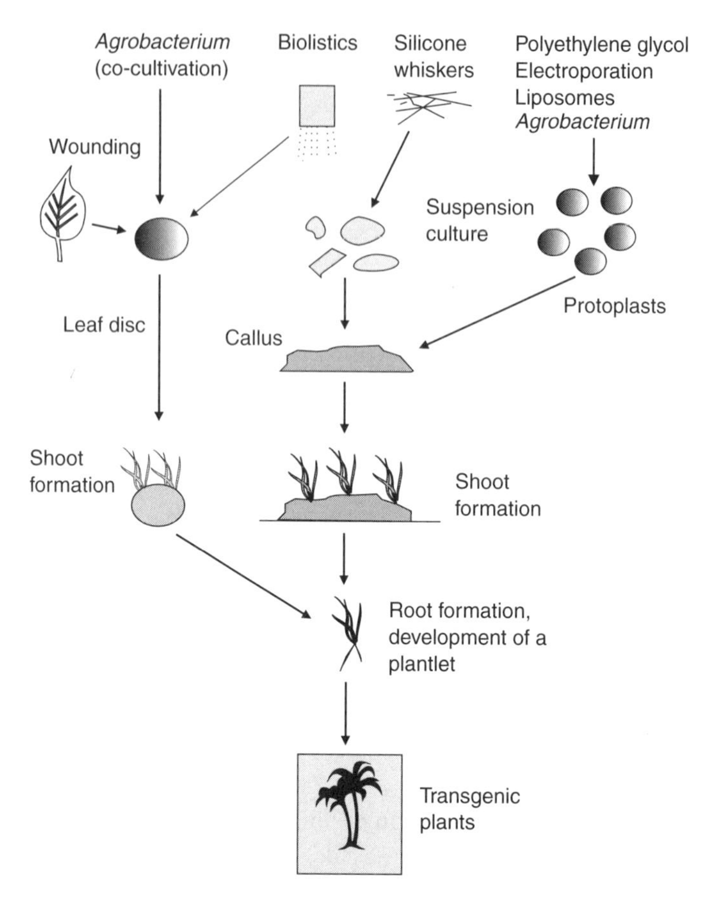

Figure 9.3 The techniques used to produce transgenic plants. The plant material can be protoplasts where the genetic material can be introduced by treatment with polyethylene glycol, electroporation, liposomes, and incubation with *Agrobacterium*. Cell-suspension cultures can be transformed by shaking with thin silicone whiskers. Explants of all types can be transformed by incubation with *Agrobacterium* and microprojectile treatment (Biolistics, Box 9.6). Once transformed the whole plant needs to be regenerated.

BOX 9.6 **Biolistics (microprojectiles)**

Sanford (1988) described the original design where tungsten particles were propelled by a gunpowder charge. This design has been replaced by a number of methods for the acceleration of the particles as the gunpowder system lacked control in the power applied and required a gun licence in some countries. In one design the explosive force of the gunpowder has been replaced by a burst of helium gas which accelerates the macrocarrier (Kikkert, 1993). The macrocarrier carries the particles and when this is stopped by a screen the particles continue onwards to the target (see figure). In another design the gun has been replaced by an airgun which makes the apparatus cheaper and causes less trauma to the target tissues (Oard, 1993). A commercial design, ACCELL™, uses an electrical discharge to accelerate the particles (McCabe and Christou, 1993). A development

BOX 9.6 ***(Continued)***

of the helium accelerator has been low-pressure helium where the macrocarrier has been eliminated and the particle accelerated directly (Gray et al., 1994). The small gold or tungsten particles 0.1–2 μm in diameter are coated with DNA and these are fired into plant tissue. The particles penetrate the tissue and transformation occurs. This method has been used for a wide range of plants particularly those difficult to transform by *Agrobacterium*. The figure shows the components of the Biolistic PDS-100/He device. The gas-acceleration tube is filled with helium at high pressure until the rupture disc breaks. The gas shock is transferred to the macrocarrier. The progress of the macrocarrier is stopped by a stopping ring, launching the particles towards the target. In the particle accelerator (ACCELL™) 15 000 volts (DC) is applied to the electrodes creating an arc and evaporating the water droplet. The shockwave propels the carrier sheet towards the retainer screen which stops its progress but allows the gold particles to continue to the target (see figure).

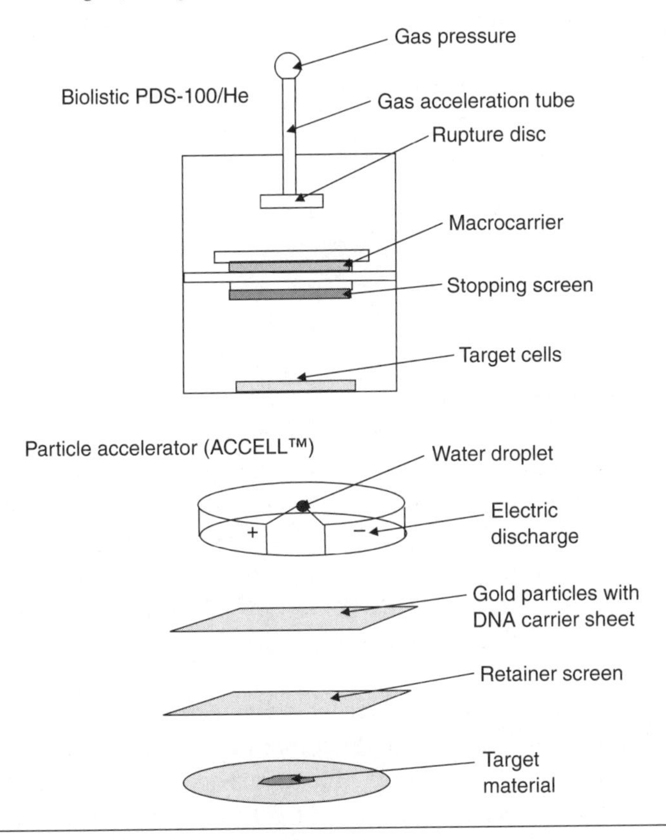

9.5.4.2 *Direct gene transfer*

The direct transfer of DNA into intact plant tissue or cell cultures avoids the problem of the removal of the cell wall and its subsequent regeneration and the technique is not limited by the ability to produce protoplasts. DNA can be transferred directly by microinjection, electroporation, silicone carbide whiskers, and biolistics.

Microinjection, the physical injection of DNA into a cell, is the most difficult and laborious, but it has the advantages that the number of DNA molecules transferred can be controlled and that co-transfer of other DNA, RNA, or even mitochondria is possible. Microinjection can be carried out with individual cells in large explants or whole plants. Despite these advantages the difficulties of the technique mean that the method has only been used for a limited number of plants.

Electroporation is carried out in a similar manner to that used with protoplast fusion and with this technique explants and embryos of maize, rice, sugar cane, and cassava have been transformed (Siemens and Schieder, 1996). A recently developed technique for direct DNA transfer has been to mix silicone carbide whiskers with DNA and cell suspensions. When this mixture is agitated the needle-like whiskers penetrate the cells which results in transfer of DNA (Fig. 9.3), but to date this method has only been applied to sunflower and maize cultures.

The most widely used method of direct transfer of DNA is carried out by particle bombardment (biolistics) using microscopic particles coated with DNA (Box 9.6). The technique has several advantages over other methods:

- it is a mechanical process which is not limited by the host range of *Agrobacterium* (see below) and the restrictions of protoplasts,
- no plasmid manipulation is required, and
- it is a simple process which can be used with large pieces of tissue.

9.5.4.3 *Agrobacterium*

Agrobacterium-mediated DNA transfer (Box 9.7) is normally carried out by co-cultivation or co-cultivation with wounding of the tissue, followed by cultivation on the appropriate medium containing an antibiotic and growth regulators to remove the bacteria and to initiate shoot regeneration. Another variation of the technique involves the sonication of the cells or tissue prior to treatment with *Agrobacterium* (Trick et al., 1997). In addition, *Agrobacterium* has been used to transform seeds and whole plants of *Arabidopsis thaliana* by vacuum infiltration.

Agrobacterium-mediated gene transfer is the method of choice as its host range continues to be extended and the technique is simple. The disadvantage is that the method depends on being able to regenerate the whole plant from the transformed tissue or cell. The size of the DNA transferred (T-DNA) was also thought to be a limiting factor for the amount of DNA which can be transferred is generally 25 kb, but a new binary vector combined with *Agrobacterium* strains with enhanced *vir* genes can transfer up to 150 kb. The Ti plasmid contained in *Agrobacterium* has a number of genes including the *vir* (virulence) gene, the products of which transfer and integrate into the host DNA at a specific point on the plasmid called the T-region. The T-region

BOX 9.7 Agrobacterium

The Gram-negative soil bacterium *Agrobacterium* will form a crown gall if it infects dicotyledonous plants. The agent which is responsible for the gall formation is the presence in *Agrobacterium* of a large plasmid (Ti, tumour inducing), part of which integrates into the genome of the plant. The DNA transferred (T-DNA) is transported to the nucleus where it is integrated into the genome. The T-DNA codes for a number of genes which are responsible for the production of plant-growth regulators auxins and cytokinins (*onc*), and those which code for the enzymes involved in the synthesis of opines, amino acids, and sugar derivatives (see figure, which shows the structure of the *Agrobacterium* plasmid including the T-DNA region). The opines are secreted by the plant cells and consumed by the *Agrobacterium*. The Ti plasmid also contains genes for the overproduction of plant-growth regulators, auxins, that cause rapid growth of the plant cells. There are a number of other components; the *vir* (virulence) region is required for infection. The Ti plasmid has been modified as a vector for the transfer of genes to the plant by removing most of the genes between the border sequences so that the plasmid does not induce the overproduction of plant-growth regulators and therefore does not form calluses. The reduction in size means that there is more room for the added DNA, which includes an antibiotic-resistance gene. The system has proved very successful, giving the integration of a number of copies of the gene into the plant genome. However, the *Agrobacterium* system is restricted by the size of the DNA that the T-DNA can contain and the host range of the bacterium. It was thought that *Agrobacterium* would only infect monocotyledons in the Liliales and Arales orders but important monocotyledons such as wheat and maize have now been transformed.

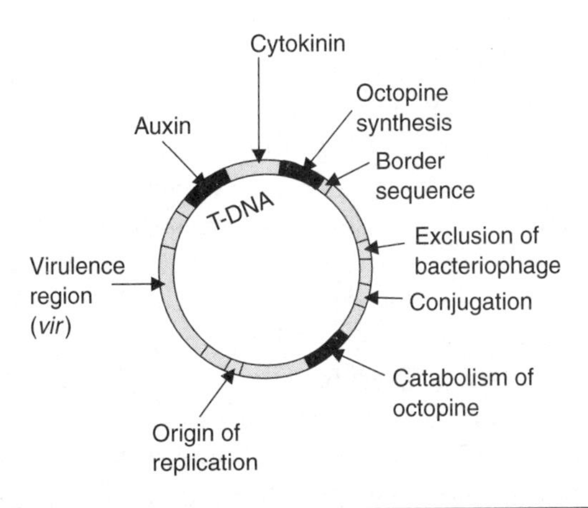

contains genes for growth-regulator synthesis and opine production. To enable this section of DNA to be recognized by the *vir* products the T-region is flanked by 25 bp of DNA known as **border sequences**. The transfer of the genes between two border sequences is far more precise than the random introduction of DNA used in the other techniques, which are suspected of introducing fragmented or multiple copies. Recent data suggest that DNA sequences outside the border sequences of the T-DNA can be transferred to the plant genome (Kononov et al., 1997). The integration of these other sequences may affect the regulatory guideline for the release of *Agrobacterium*-transformed plants as these may require a complete description of the transferred DNA sequences which may now include sequences outside the border regions.

9.6 Safety of transgenic crops

Any novel food, which includes genetically engineered plants grown for food, has to be approved by the Advisory Committee on Novel Foods and Processes (ACNFP) in the UK and the Food and Drug Administration (FDA) in the USA before it can be sold. The food safety issues are as follows and regulations require

- molecular characterization of the gene introduced,
- analysis of the composition of the plant with respect of nutrients, toxins, and allergens,
- an assessment of the potential for the transfer of the gene in the food to human and animal intestinal micro-organisms,
- estimated intake levels of the new proteins,
- toxicological and nutritional evaluation of the new protein, and
- additional toxicity tests of the whole food.

In addition, the deliberate release of genetically engineered plants and micro-organisms also needs approval by the relevant bodies such as the Deliberate Release Directive (90/220/EEC). The consensus holds that the technique of genetic engineering involves refinements or extensions of older techniques and that any risks are only from the genes transferred. The release of genetically modified organisms (GMOs) into the environment is a very different situation from the production of vaccines and hormones in genetically engineered organisms, which is carried out under controlled conditions in production plants. In conventional breeding crop plants have been hybridized with wild lines and other breeding techniques have produced a number of different plant lines, all of which have been released with no problems. However, profound ecological changes have been produced when certain plants have been introduced to different ecological conditions, such as the water hyacinth, which is an invasive weed. Thus, it is the introduction of genetic information from very distant organisms such as fish that causes some concern, as such combinations would not be possible by traditional methods.

The use of antibiotic-resistance genes as markers in transgenic crops has raised concerns about the potential transfer of these genes to soil bacterial, intestinal bacteria, and the cells of animals that eat these plants. One of the most commonly used selectable markers is the gene from *E. coli* coding for aminoglycoside 3-phosphotransferase, commonly known as neomycin phosphotransferase 11 (*npt* 11) which gives resistance to kanamycin (neomycin). The concern is that the *npt* 11 gene product will be toxic or, more worrying, that the antibiotic resistance may be transferred to other micro-organisms in the environment. The consensus from the data available indicates that the transfer of genes from transgenic plants to other organisms would be an extremely rare occurrence. In addition it has been suggested that the transfer of the resistance gene to bacteria in the human gut is not a problem as there are already kanamycin-resistant bacteria in humans and in the soil due to the widespread use of antibiotics.

One other problem, which could affect humans, would be the production of allergenic pollen from transgenic plants that could cause a problem on a large scale. The transfer of the genes from the GMOs to micro-organisms in the environment could also occur. Bacteria can transfer DNA by transformation, conjugation, and transduction and plants by cross-pollination. Transference of the gene to bacteria will only survive in the bacterial population if it constitutes an advantage and in this case antibiotic-resistance genes should be avoided. To date no one has demonstrated that this occurs in nature. The extent of gene transfer from crop plants to wild plant populations may also be a problem and depends on a number of factors. The crop plants and wild species must be compatible—growing in the same location, flowering at the same time—and have a means of pollen transference. Cultivated rapeseed (*Brassica napus*) will cross with wild rapeseed (*Brassica rapa*). The transfer of genes from transgenic rapeseed to the wild variety was tested and this has been confirmed in two commercial fields. The conclusions were that transfer would occur if the two were growing in close proximity but that there was no transfer to other wild relatives (Miki and McHugh, 2004). The genetic strategies to avoid such transference include the introduction of male sterility so that the transgenic plant produces no pollen, linking the gene with a gene that is lethal in pollen, the removal of flowers from the transgenic plant, the removal of compatible species, and the planting of buffer plants. A recent solution was to direct the introduced gene to the chloroplast, which is not transferred to the pollen and therefore cannot be transferred in cross-breeding. A study with rapeseed has shown a very low probability of chloroplast transfer to the wild species (Scott and Wilkinson, 1999). This also has the advantage that the chloroplast has a high copy number or large number of gene copies (Daniell et al., 1998). Other methods of gene-transfer reduction are to treat the plants with growth regulators so that they flower before any compatible plants.

Another problem of transgenic plants may be that the new gene will give the plant a selective advantage, causing it to become a new pest or superweed. A similar problem may be the transfer of the gene from the transgenic plant to a weed, which gives the weed advantages, makes it more of a problem. The transgenic plant may also compete with local beneficial plants and upset the plant communities. Therefore, when considering the release of transgenic plants the risks are considered on a case-by-case basis and the following should not be forgotten.

- Of the antibiotic-resistance genes only kanamycin has been tested for its effect on humans and this type of marker should be replaced by non-antibiotic markers if possible.
- Field trials are needed to determine the best strategies to avoid the transference of genes on a case-by-case basis.
- Field evaluation is needed to determine the level of expression that can vary greatly. Field trials run recently in the UK concluded that transgenic maize could be cultivated commercially, but not rapeseed.

9.7 Transgenic plants

Genetic transformation gives direct access to the vast pool of genes not previously accessible to plant breeders. The first transgenic plants involved single genes for simple traits but improvement in techniques allows more than one gene to be transferred at one time (Sharma et al., 2002). In many plants transformation systems are under development but there has been rapid progress in the development of crops for herbicide resistance, pest resistance, and male sterility. Some of the traits that have been developed and those under investigation are as follows:

- herbicide resistance,
- increased resistance to insects,
- resistance to plant pathogens,
- production of biocontrol agents,
- development of new hybrid crops with male sterility,
- fixed hybrid vigour in inbred crops,
- improvement in nutritional and product quality,
- increase in uptake of phosphorus and nitrogen,
- adaptation to soil salinity and drought,
- oil accumulation,
- increased photosynthetic activity, and
- production of pharmaceuticals, vaccines, and industrial chemicals.

A number of genetically engineered plants and plant products have been approved for sale and consumption in Europe and the USA and at present 58.7 million ha are being cultivated. Table 9.3 gives the number of transgenic

Table 9.3 Genetically engineered crops available worldwide

Crop	Number of strains available	Trait	Approved country
Rapeseed (canola)	16	Herbicide resistance, oil modification, male sterility	USA, Japan, Canada
Carnation	3	Colour, herbicide resistance, extended shelf life	Australia, EU
Chicory	1	Herbicide resistance, male sterility	EU
Cotton	5	Herbicide resistance, insect resistance	Argentina, Australia, China, Japan, Mexico, South Africa, USA
Flax	1	Herbicide resistance	Canada, USA
Maize (corn)	22	Herbicide resistance, insect resistance, male sterility	Argentina, Canada, EU, Japan, South Africa, USA
Melon	1	Delayed ripening	USA
Papaya	1	Virus resistance	USA
Potato	4	Insect resistance, virus resistance	Canada, USA
Rice	2	Herbicide resistance	USA
Soybean	7	Herbicide resistance, oil modification	Argentina, Brazil, Canada, Japan, Mexico, South Africa, USA, Uruguay
Squash	2	Virus resistance	USA
Sugar beet	2	Herbicide resistance	Canada, USA
Tomato	6	Delayed ripening, insect resistance	Japan, Mexico, USA

Source: Data from Phillips (2002).

plants that have been approved for commercial growth, the traits engineered, and the countries where their growth has been approved. Table 9.4 gives a measure of the proportion of exports that transgenic plant products represent. The most frequently engineered trait is herbicide resistance, which is clearly of commercial value. Six crops dominate: soybean, cotton, maize (corn), rapeseed, tomato, and potato.

Table 9.4 The production and trade in genetically manipulated products

Crop	Number of countries producing the crop	Share of global exports (%)	Numbers of countries that import products
Maize (corn)	8	85	168
Soybeans	6	88	114
Rapeseed (canola)	2	50	68
Flax	2	81	74
Potato	4	12	177
Tomato	5	54	140

Source: Data from Phillips (2002).

9.7.1 Tolerance to pesticides and herbicides

The mechanism of action of many herbicides is not well understood but in the case where it is understood the strategy of resistance can take three forms. These are an increase in the affected enzyme, the expression of an altered enzyme not affected by the herbicide, and the expression of an enzyme which detoxifies the herbicide. One of the most frequently studied is the herbicide glyphosate, which inhibits the enzyme 5-enolpyruvylshikimate-3-phosphate synthase (Fig. 9.4), part of the pathway leading to the formation of aromatic amino acids. Overexpression of this gene in a plant gave tolerance to glyphosate, and in another case the gene was cloned from a bacterium in which some of the amino acids had been substituted to make the enzyme less susceptible to glyphosate. A large number of plants have been transformed to glyphosate resistance (Table 9.4) including soybean, maize, and cotton. In soybean an *Agrobacterium* EPSPS (5-enolpyruvylshikimate-3-phosphate synthase) gene, which is insensitive to glyphosate, was used, including a gene from *Petunia hybrida* that transfers the enzyme to the chloroplast.

In 1997 glyphosate-resistant cotton plants were grown commercially for the first time. Some of the growers have found that the cotton bolls drop prematurely or are deformed (Fox, 1997), which indicates some of the problems which may occur with transgenic plants.

9.7.2 Resistance to insects

Plants are often attacked by insects, nematodes, and molluscs that can severely affect the crop yield. An alternative to the use of chemical pesticides is to use genetic engineering to produce plant resistance to these pests. The best-known

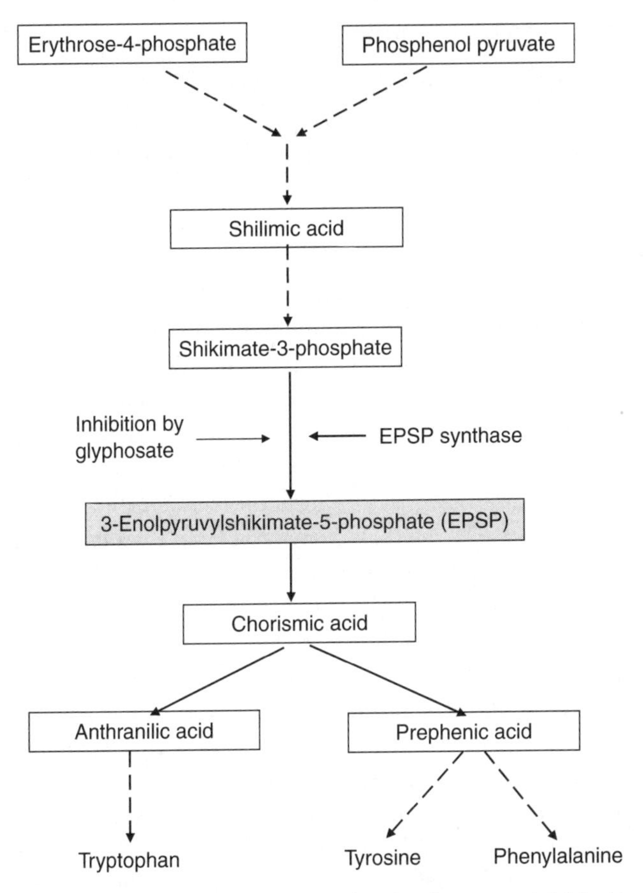

Figure 9.4 An outline of the shikimate pathway indicating the enzyme inhibited by the herbicide glyphosate.

example of a biological form of pest control is the production by *Bacillus thuringiensis* of a protein endotoxin which is toxic to a number of caterpillars but does not harm animals or other insects. The protein is produced as the bacillus forms spores in the form of an inactive protein of 1200 amino acids. If the protein is ingested by a caterpillar the protease activity in the gut of the caterpillar cleaves the protein to form the active 68 000-Da protein. The protein binds to the surface of the midgut cells, killing the caterpillar. Plants transformed by inclusion of the gene responsible for the *B. thuringiensis* toxin include plants like populus, elm, spruce, maize, potato, tobacco, rice, broccoli, lettuce, apple, alfalfa, soybean, and cotton. Maize showed resistance to caterpillar attack, in particular to the important pest the European corn borer. Cotton plants are resistant to cotton bollworm and transformed trees showed resistance to the gypsy moth (Podila and Karnosky, 1996). Other possibilities of introducing pest resistance may be the inclusion of the protein cholesterol oxidase Vip3A, amylase inhibitor, and systemic wound-response proteins (Dempsey et al., 1998).

9.7.3 Resistance to plant pathogens

Viral infections are normally treated by killing the virus vectors with pesticides to stop further transmission. If plant cells are transformed with genes or sequences from the viral genomes the plants become resistant to those viruses (Dempsey et al., 1998). This resistance is known as pathogen-derived resistance (PDR) and provides protection against a variety of viruses. The expression of the coat protein of tobacco mosaic virus (TMV) in tobacco plants made them resistant to TMV. Transgenic lines of squash and papaya exhibiting PDR have been approved for commercial cultivation. Transgenic potato lines have been produced which express a double-stranded RNA-specific ribonuclease *pac*1 from a yeast. These lines are resistant to potato spindle viroid as the ribonuclease digests the double-stranded RNA regions of the viroid (Sano et al., 1997).

Bacterial diseases are of considerable importance to crop plants, causing losses in valuable crops like cereals, vegetables, and fruits. There have been two major approaches to the development of resistance; firstly to introduce antibacterial proteins and secondly to enhance the plants' natural defences (Table 9.5). There are a number of proteins which can be introduced to give bacterial resistance, which include insect-lytic peptides, lysozymes, lactoferrin, and toxins. The enzyme lysozyme is found widely and will degrade peptidoglycans that form part of the bacterial cell wall. Lysozyme gene has been inserted into tobacco plants and the plants have shown partial resistance to *Erwinia carotovora atroseptica*, a plant pathogen. Plants have a number of natural defences to bacterial infection and these can be enhanced to give increased resistance to bacterial infections. One example is the cloning of a pathogen-recognition gene (R) from a resistant line into a susceptible plant. In this way rice has been transformed to resistance to *Xanthomonas orzyae*, which causes bacterial leaf blight in rice.

Phytoalexins are small antimicrobial compounds which are produced in response to the infection of a plant and have been the target for genetic engineering (Dempsey et al., 1998). Attempts have been made to increase phytoalexin levels in order to confer resistance to both bacterial and fungal pathogens.

9.7.4 Hybrid crops with male sterility

Male sterility is related to the lack of pollen production and is found in maize, rice, and sunflower. The approach is to transfer this trait to other plants so that no pollen is produced, stopping gene transfer from cultivated transgenic plants.

9.7.5 Improvement in nutritional and product quality

There are a number of targets that can be altered to improve the quality of the final crop. This is an example of a move away from herbicide resistance, and

Table 9.5 Examples of transgenic plants with improved resistance to bacterial diseases

Protein	Origin	Plant	Resistance
Antibacterial proteins from non-plant origin			
Shiva-1	Giant silk moth	Tobacco	*Ralstonia solanacearum*
MB 39	Giant silk moth	Tobacco	*Pseudomonas syringae* pv. *tabaci*
Attacin E	Giant silk moth	Apple	*Erwinia amylovora*
Lysozyme	T4 Bacteriophage	Potato	*Erwinia carotovora* pv. *tabaci*
Lysozyme	Human	Tobacco	*Pseudomonas syringae*
Lactoferrin	Human	Tobacco	*Ralstonia solanaceearum*
Tachyplesin	Horseshoe crab	Potato	*Erwinia carotovora*
Inhibition of bacterial pathogenicity or virulence factors			
Tabtoxin-resistance	*Pseudomonas syringae*	Tobacco	*Pseudomonas syringae*
Phaseolotoxib-insensitive OCTase*	*Pseudomonas syringae* pv. phaseolicola	Bean	*Pseudonmonas syringae* pv. *phaseolicola*
Enhancement of natural plant defences			
Pectate lyase	*Erwinia carotovora*	Potato	*Erwinia carotovora*
Glucose oxidase	*Aspergillus niger*	Potato	*Erwinia carotovora*
Thionin	Barley	Tobacco	*Pseudomonas syringae* pv. *tabaci*
Artifically induced cell death			
Bacterio-opsin	*Halobacterium halobium*	Tobacco	*Pseudomonas syringae* pv. *tabai*

*OCTase, ornithine carbamoyltransferase.
Source: Mourges et al. (1998).

the products have been called nutrifoods. Examples are the production in rice of provitamin A and iron-binding protein to increase the iron and vitamin A levels. One of the first commercial products in Europe and the USA was the Flavr Savr^{TM} tomato developed by Calgene. The Calgene tomato has an increased shelf life, as antisense technology silencing stopped the production of the enzyme polygalacturonase (Box 9.8). The polygalacturonase enzyme is involved in cell-wall degradation as part of fruit ripening. A similar tomato developed by Zeneca used a truncated sense polygalacturonase gene to reduce the activity of polygalacturonase in order to produce the longer shelf life and

BOX 9.8 Antisense technology

Antisense technology involves the reversal of a coding region of a gene, which results in the transcription of the antisense DNA rather than the normal sense strand (see figure). The gene for polygalacturonase has been inserted in the reverse phase so that when transcribed into mRNA it forms a duplex with the normal mRNA. This double-stranded RNA is then degraded.

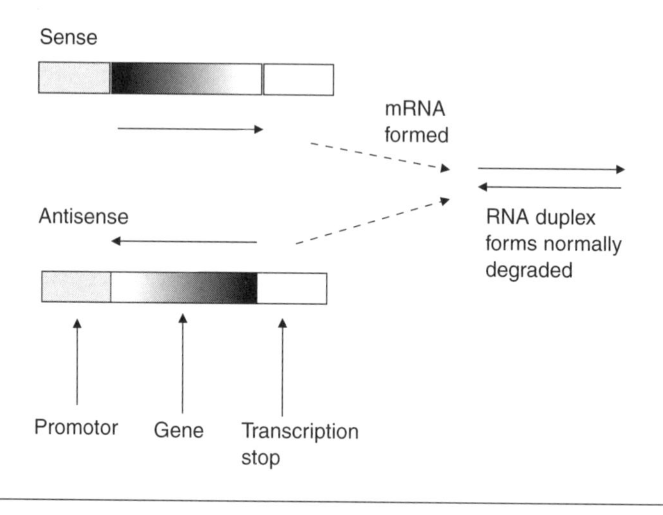

a higher solids content. The Calgene product is sold as fresh tomatoes, whereas Zeneca's was sold processed as a tinned paste.

9.7.6 Adaptation to soil salinity and drought

The ability of plants to withstand a number of stresses would be of great value. The salt-tolerance gene from mangrove plants has been cloned and the introduction of calcineurin provides tolerance to salinity. Genetic engineering can offer considerable potential in the change of the traits of crops such as colour, nitrogen fixation, starch synthesis, increased photosynthetic efficiency, and modification of oils accumulated. Soybean, oil palm, rapeseed, and sunflower account for 72% of the world's vegetable oil, producing 100 Mt at a value of $70 billion (Murphy, 1996). Rapeseed was one of the first targets for genetic manipulation, as there has been a good transformation system in place for some time. Herbicide-resistant lines of rapeseed have been approved for commercial cultivation since 1993. Current interest in rapeseed is in the alteration of the oil composition in the main oil-producing plants. Genetic engineering can be used to change flower colour by the inactivation of the endogenous enzymes or by introducing new genes. Although considerable research has been carried out on nitrogen fixation much more is needed before nitrogen fixation, which is a bacterial process, can be transformed into plants.

Table 9.6 Transgenic plants expressing genes for stress tolerance

Origin	Gene	Host	Stress
E. coli	BetA	Tobacco	Salinity
E. coli	BetA	Potato	Freezing
Arthrobacter globiformis	codA	*Arabidopsis*	Salinity and drought
Vigna aconitifolia	p5cs	Tobacco	Drought
E. coli	Mltd	*Arabidopsis*	Salinity
E. coli	Mltd	Tobacco	Salinity
Saccharomyces cerevisiae	TPS1	Tobacco	Drought
Bacillus subtilis	SacB	Tobacco	Drought
Arabidopsis	fad7	Tobacco	Chilling
Anacystis nidulans	Des9	Tobacco	Chilling
Barley	HVA 1	Rice	Salinity and drought
Winter flounder	Afp and afa3	Tobacco	Freezing
Nicotiana plumbaginifolia	Mn-SOD*	Alfalfa	Drought and freezing
Arabidopsis	Fe-SOD	Tobacco	Oxidative
E. coli; rice	Cr/Cu, Zn-SOD	Tobacco	Oxidative
Vitreoscilla stercoraria	vhb	Tobacco	Hypoxia and anoxia

Source: Holmberg and Bulow (1998).
*SOD, superoxide dismutase.

One of the problems is that nitrogen fixation is very sensitive to oxygen and therefore a low-oxygen environment will be required for the process to function. The engineering of plants to tolerate various stresses such as chilling, heat, drought, salinity, freezing, and flooding has clear advantages, but the tolerance is likely to be a multifaceted characteristic involving a number of genes and is therefore not easy to manipulate. However, a number of methods have been attempted to improve stress tolerance (Holmberg and Bulow, 1998). These are outlined in Table 9.6 and the most successful have been the formation of osmoregulatory products, membrane-modifying enzymes, radical-scavenging enzymes, and stress-induced proteins.

9.7.7 Plants as bioreactors (biopharming)

Most of the commercial transgenic products are produced in bacterial and animal cell cultures. Animal cell cultures are used because if a eukaryote product

Table 9.7 The production of proteins in transgenic plants

Compound	Plant	Application
Vaccines		
Glycoprotein	Tobacco/spinach	Rabies
Spike protein	Maize	Piglet diarrhoea
Avidin	Maize	Avidin
Monoclonal antibody	Tobacco	Colon cancer
Peptides	Cowpea	Pulmonary infections
Toxin	Potato	Cholera
Toxin	Potato	Diarrhoea
Surface antigen	Potato	Hepatitis
Pharmaceuticals		
α-Trichosantin	Tobacco	HIV
Enkephalin	Rapeseed	Opiate activity
Erythropoietin	Tobacco	Erythrocyte regulation
Hirudin	Rapeseed	Thrombin inhibitor
Human serum albumin	Tobacco	Plasma extender
Interferon	Turnip	Anti-viral
Industrial products		
α-Amylase	Tobacco	Starch liquefaction
β-Glucanase	Barley	Brewing
Phytase	Tobacco	Animal feed
β-Glucuronidase	Maize	Molecular biology
Laccase	Maize	Lignin enzyme
Xylanase	Tobacco	Paper pulp

Sources: Data from Goddijn and Pen (1995), Hood (2002), and Hood and Jilka (1999).

is required the cell will process the product. However, the culture of animal cells is expensive, difficult, slow, and requires good sterile control to avoid contamination. Transgenic plants can provide an inexpensive and convenient system for the large-scale production of recombinant proteins. Some examples of the products are given in Table 9.7, and are dominated by pharmaceuticals and vaccines. Whether these plants containing human genes will be given permission to be grown in the field is the subject of some debate. One possible product is poly-3-hydroxybutyrate (PHB), a polyester with thermoplastic properties (see Chapter 6) which has been transferred to flax (Wrobel et al., 2004).

9.8 Transgenic animals

The improvement of animal stocks by breeding has been carried out over many years using conventional techniques, but it is a slow process. The production of transgenic animals has two applications, the use of animals to produce foreign proteins of medical use and as part of a breeding scheme.

The use of animals to produce foreign products is also known as biopharming as most of the products are high-value pharmaceuticals. For example the blood proteins factor VIII and factor IX have been made in sheep, and lactoferrin in cattle. Extraction of the product without sacrificing the animal can be achieved by linking it to protein being produced in milk to ensure that the protein can be extracted from the milk. Examples of this system are the production of human protein C, lactoferrin protein, antithrombin-III, human serum albumin, and monoclonal antibodies. The normal method of producing a transgenic animal is to microinject DNA into a one-cell embryo or into the nuclei of a two-cell embryo, but the success rate is low at 1–4%. The gene transferred is placed under the control of promoters from albumin, prolactin, skeletal actin, transferrin, and phosphoenolpyruvate carboxykinase, and these have been tested mostly in pigs. The low efficiency of transformation has severely restricted the use of transgenic technology.

The use of genetic engineering in the breeding of animals has focused mainly on improvements in growth and both animals and fish have been engineered to increase their levels of growth hormone to ensure faster growth.

The improvement of animals does not always involve genetic manipulation; one example is embryo cloning, which has had considerable publicity recently. In this technique a 16-cell embryo from the artificial insemination of a favoured breeding line of cattle is used as a source of nuclei. These nuclei are transplanted into enucleated eggs from other cows and cultured until they can be used for implantation into a number of other cows. In this way the 16 nuclei will give rise to 16 identical calves. More recently it has been shown that nuclei from an embryo-derived mouse cell line can be transplanted into unfertilized eggs (Willmut et al., 1997) to yield viable offspring. This has been repeated with the transfer of nuclei from cell populations of sheep mammary gland, foetus, and embryo to unfertilized sheep eggs to produce lambs, one of which was named Dolly. This type of advance is perceived as 'cloning' by the public and regarded with some suspicion. Whether this type of technique will be adopted widely will depend on public perception as well as the results.

Another example is bovine somatotrophin (BST) which is a hormone involved with growth promotion and increased lactation. Under normal conditions BST is too expensive to use in the dairy industry, but the gene has been cloned and expressed in *E. coli* and produced in large quantities. The genetically engineered BST, when given to lactating cows, increased milk production by between 10 and 25%. Recombinant BST has been approved for use in milk production but its use has been resisted. Although BST has been approved for injection, the additional milk production will require an increase in feed, and other adverse effects have been reported including increased mastitis, reduced pregnancy rates, and lesions of the knee. Perhaps the best argument against the use of BST is its economic impact on the dairy industry. It has been argued that extra milk is not required at a time when milk production is in surplus, particularly in Europe where milk quotas have been introduced. In addition, at a time when food is marketed as natural it is questioned whether the public will accept milk produced with the help of an injected hormone (Potter, 1994).

9.9 Disease control

Biological disease control can be applied to both animals and plants. The main control applied to animals is immunization and with plants it is the application of micro-organisms.

Traditional vaccines can be either live, containing attenuated viral or bacterial cultures, or killed, containing dead whole cells, viruses, or inactivated toxins. Developments in recombinant DNA technology has enabled antigenic molecules to be identified, cloned, and expressed in bacteria. These antigens are difficult to purify by conventional methods so their expression in bacteria means that there is a source of large quantities of pure antigen. The availability of pure antigens has enabled a new form of vaccine to be developed as these vaccines use a single molecule rather than a mixture or the whole organism. These are known as subunit vaccines and their production has the advantages that they are easy and inexpensive to produce, the antigen is devoid of other material, and the vaccine contains no live organisms. However, the production of subunit vaccines does have some problems of low expression of the cloned gene, and the protein when formed may not fold properly and will therefore not be antigenic. Incorrect folding is often caused by producing an animal protein in a bacterium, where the post-translational processing is different or absent. For example, a protein for fowl plague has been expressed in *E. coli* but the product was not antigenic. To eliminate this problem the protein should be expressed in animal cell cultures where the processing would be compatible. However, animal cell culture is expensive and the extraction process has to ensure that all animal DNA is removed from the antigen. The problems of protein folding and the short life of these vaccines are the reasons that subunit vaccines have only seen restricted use, although a subunit vaccine against human hepatitis B has been produced. Using a modification a vaccine has been produced against rabies using as the antigen a glycoprotein from the virus. The viral vaccine has been shown to eradicate rabies in field trials and a product, Rabora®, is now on sale in France to control rabies in the wild-animal population. Other live recombinant vaccines are available for Newcastle disease and fowlpox in poultry.

It is possible to chemically synthesize the antigen, which eliminates the need for purification. This method has been used to develop a vaccine against the foot and mouth virus. The traditional vaccine against this virus uses killed virus particles but it loses activity rapidly. The foot and mouth virus contains four capsid proteins and one of these (VP1) has been produced in *E. coli* in large amounts, but does not fold properly. The 140–160-amino acid VP1 peptide was linked to a large carrier protein expressed in *E. coli* and the vaccine was successful in guinea pigs but was ineffective in cattle.

Biological substitutes for pesticides have been known for some time and include micro-organisms that attack pests. These micro-organisms consist of bacteria, yeast, and fungi that can reduce a number of plant diseases. Table 9.8 gives some examples of commercial preparations. These micro-organisms affect the disease by a number of mechanisms:

- production of antibiotics,
- production of digesting enzymes,

Table 9.8 Commercially available biological-control micro-organisms

Trade name	Micro-organism	Disease
Bacterial		
Serenade	*Bacillus subtilis*	Fungal
Cedomon	*Pseudomonas chlororaphis*	Cereal disease
Camperico	*Xanthomonas campestris*	Fungal grass disease
Many	*Bacillus thuringiensis*	Caterpillar/insect attack
Fungal		
AQ10	*Ampelomyces quisqualis* M-10	Powdery mildews
Mycostop	*Streptomyces griseoviridis*	Fungi in greenhouses
Binab	*Trichoderma* sp.	Root rot, wilt
Contans WG	*Coniothyrium minitans*	Root rot
Primastop	*Gliociadium catenulatum*	Root rot, wilt
Root pro	*Trichoderma harzianum*	Root rot
Sporodex	*Pseudozyme flocculose*	Powdery mildews
Trieco	*Trichoderma viride*	Root rot, wilt

Sources: Data from Punja and Utkhede (2003), and Gerhardson (2002).

- parasitism,
- competition for nutrients,
- interference with pathogenicity factors, and
- induction of resistance in the plant.

The last mechanism has parallels with animal vaccines in that application can induce host-defence responses.

9.10 Germplasm and biodiversity

The world is losing plant and animal species at an increasing rate and lost genetic material cannot be replaced. In the main this loss is being caused by agricultural practices of land clearance and cultivation.

The Convention on Biological Diversity (United Nations Environment Programme, 1992) during the Rio de Janeiro conference defined biodiversity as "the variability among living organisms from all sources including terrestrial, marine and other aquatic ecosystems and the ecological complexes of which they are part; this includes diversity within species, between species and of ecosystems". It is essential that biodiversity is preserved for the following reasons:

- All species have genuine intrinsic value regardless of any value they have to human welfare.
- Many species do have a value to human society and only a limited number of species have been tested for their potential value. An example is the screening of plants for new drugs, which has come up with medicines such as the anti-leukaemic drug vincristine from the plant *Catharanthus roseus*. The availability of plant and animal species is required for breeders to maintain a high degree of variation.
- Species do make up ecological communities that are involved with processes such as nutrient recycling, provision of oxygen, and carbon dioxide removal. In only a few cases are there sufficient data to evaluate the ecological importance of particular species but this does not diminish their importance.

Plant and animal cell culture biotechnology can rescue endangered species where multiplication is not possible by conventional means and cryopreservation technology can also store germplasm indefinitely.

9.11 Conclusions

Biotechnology through the application of genetic engineering has and will continue to have a considerable impact on agriculture. Perhaps the advances will appear less spectacular than those in medicine but transgenic plants may affect more people and have a longer-lasting effect. Overall these changes may effect the environment in the following ways.

- The production of new subunit vaccines which should reduce the use of antibiotics in animals. The overuse of antibiotics is one of the major causes of the development of antibiotic resistance in micro-organisms.
- Increased use of genetically engineered products and monoclonal antibodies for diagnosis in both plants and animals.
- The continued introduction of transgenic plants should reduce the use of both herbicides and pesticides. However, there is public concern and resistance at the possible danger of the widespread release of transgenic plants, particularly those containing antibiotic-resistance genes. Labelling of all transgenic products may go some way to reduce these fears and market forces may limit the introduction of transgenic plants and products. For example, transgenic brewers yeast has been available for some time and has been approved for use but genetically engineered beer is not on sale to date.
- Continued introduction of transgenic animals and fish will occur but many concerns over their production may limit their production.
- Other biotechnological products such as microbial inoculants, biopesticides, and anaerobic digestion of wastes will be introduced.

9.12 Further reading

Alberts, B., Bray, D., Lewis, J., Raff, M., Roberts, K., and Watson, J.D. (2002) *Molecular Biology of the Cell*, 4th edn. Garland Publishing, New York.

Altman, A. (1998) *Agricultural Biotechnology.* Marcel Dekker, New York.

Benson, E. (1999) *Plant Conservation Biotechnology.* Taylor and Francis, Andover.

Pierpoint, W.S. and Shrewry, P.R. (1996) *Genetic Engineering of Crop Plants for Resistance to Pests and Diseases.* British Crop Protection Council, Farnham, Surrey.

Redenbaugh, K., Hiatt, W., Martineau, B., Kramer, M., Sheehy, R., Sanders, R., Houck, C., and Emlay, D. (1992) *Safety Assessment of Genetically Engineered Fruits and Vegetables.* CRC Press, Boca Raton, FL.

Stewart, C.N. (2003) *Transgenic Plants: Current Innovations and Future Trends.* Horizon Press, Norfolk.

10 Biotechnology of the marine environment

10.1	Introduction	378
10.2	Pharmaceuticals	380
10.3	Molecular biology products	384
10.4	Polymers	385
10.5	Enzymes and transgenic organisms	391
10.6	Microalgae	393
10.7	Marine pollution	393
10.8	Conclusions	400
10.9	Further reading	400

10.1 Introduction

The marine environment, which covers nearly three-quarters of the Earth's surface, is one of the most diverse, containing over 300 000 species. Marine biotechnology has been defined as the application of scientific and engineering principles to the processing of materials by marine biological agents to provide goods and services. Despite the diversity of the environment little is known about it. The marine environment is hostile, as operations underwater are difficult, access to deep oceans is limited, and marine organisms are difficult to cultivate. In addition, the difficulty of obtaining a reliable harvest of marine organisms, insufficient quantities of material for study, and the problems of culturing marine organisms have affected the development of marine biotechnology.

Marine diversity is high due to immense variations in the habitat. For example, there is a huge temperature variation; from −1.5°C in Antarctic waters to above 90°C in shallow hydrothermal systems, and 350°C in deep hydrothermal vents (Huber et al., 2000). Recent analysis of microbial diversity using rRNA sequences has shown that shallow waters contain very large populations of picoplankton with a high proportion of archaeal species, some of which have not been isolated previously. Archaeal species represent 34% of the prokaryotic biomass in the Antarctic oceans (De Long, 1997). It has been estimated that although 95% of all fish species are known less than 5% of the prokaryotic species are known.

Humans have fished the oceans for millennia and recently fishing methods have improved to such an extent that certain fisheries have failed, such as the cod stocks off Nova Scotia, Canada. Some fish species are now supplied by fish farming, such as salmon. The poisonous nature of some marine species has been recognized for at least 4000 years and marine venoms have been used for at least 2000 years. One of the first examples of a marine product was a toxin, holothurin, extracted from a sea cucumber *Actinopyga agassizi* in the 1950s. Holothurin was shown to have some anti-tumour activity in mice although it was not commercialized. In 1947 nine marine micro-organisms were shown to have antibiotic activity in a study to determine whether sea water was bacteriostatic or bactericidal. The discovery that the gorgonian *Plexusa homomalla* contained prostaglandins, important mediators in inflammatory diseases, in 1969 initiated the search among marine micro-organisms for new drugs (Proksch et al., 2002).

A limited number of marine species have been studied and thousands of chemical compounds have been isolated including pharmaceuticals, nutritional supplements, cosmetics, agrochemicals, molecular probes, enzymes, and fine chemicals.

The oceans cover some 71% of the Earth's surface and contain an immense diversity of organisms but in the past little of this diversity has been exploited. However, with developments in molecular biology, recombinant technology, and tissue culture, and wider exploration, the potential of the marine environment is beginning to be exploited. The diversity can be explained by the following:

- oceans cover 71% of the Earth's surface;
- oceans contain 97% of the world's water;
- 80% of all life is found in the oceans;
- oceans contain 95% of the habitat space on Earth;
- two-thirds of phyla are marine;
- the number of known marine species is above 275 000, including 1000 species of sea anemone (Cnidaria), under 1000 species of cephalopods (Mollusca), 1500 species of brown algae, 7000 species of echinoderms, 13 000 species of fishes, and 50 000 species of molluscs;
- global fish production exceeds terrestrial production of cattle etc;
- however, less than 5% of anti-microbial compounds are of marine origin.

This chapter describes the potential of the marine environment to provide new pharmaceuticals, polymers, enzymes, colours, adhesives, and molecular biology products. The oceans and seas are the final destination for a wide range of pollutants but the most spectacular are from oil-tanker accidents, and these and the bioremediation of the spills are discussed.

10.2 Pharmaceuticals

Over 10 000 compounds have been isolated from marine organisms and between 1969 and 1999 300 patents on marine products were issued. The majority of products under clinical trials are produced by invertebrates such as sponges (Porifera), soft corals, tunicates (sea squirts), molluscs, or bryozoans (sea mats, hard corals) (Table 10.1). These are soft bodied, sessile, and slow-moving organisms. Therefore, they have developed chemical defences, accumulating toxic or foul compounds to ward off potential predators and to ensure space. Fish and invertebrates rely heavily on their innate immune defences for protection against pathogens. Table 10.1 shows what a wide range of possible pharmaceuticals have been found in marine organisms. The compounds found are complex in structure and Fig. 10.1 shows some of the products that are undergoing clinical trials. Examples are contignasterol (IPL576,092) extracted from the sponge *Petrosia contignata* that acts as an anti-inflammatory. Bryostatin 1 is extracted from the bryozoan (sea mat) *Bugula neritina* and has been shown to be cytostatic; it inhibits the growth of cancer cells while stimulating the growth of red blood cells. Dolastatin 10 is another cancer drug, which is extracted from the mollusc *Dolabella auricularia*. Another compound in clinical trials is ecteinascidin 743, isolated from the Caribbean mangrove sea squirt *Ecteinascidia turbinata*, which is effective against lung and skin cancer. It has been shown to interfere with the interaction of minor-grove-binding transcription factors (Jin et al., 2000).

Other drugs undergoing clinical trials include didemnin B, a cyclic desipeptide from *Trididemnum solidum*. Curacin A and discodermolide are very interesting as they interact with spindle protein tubulin in a similar way to Taxol®, a plant-derived drug effective against cancers. Other commercially available marine products are given in Table 10.2.

Okadaic acid and manoalide are used as molecular probes. Okadaic acid inhibits phosphatase activity and is used to probe basic cellular phosphorylation processes. Manoalide was the first compound to inhibit phospholipase A, an enzyme involved in many inflammatory diseases. Two other unusual products are the fatty acids docosahexenoic acid and arachidonic acid, formed by the microalga *Cryptocodinium cohnii*, which are similar to those in formula milk. This microalga has been cultured in bulk for the production of formula milk (Formulaid®). Pseudopterosin C is extracted from the sea whip (soft coral) *Pseudopterogorgia elisabethae* and is an anti-inflammatory for skin treatment. The compound has been included in the Estee Lauder skin cream Resilience®. A number of marine organisms contain UV-adsorbing compounds to protect them from UV light. These compounds, mycosporine-like amino acids and scytonemin, are found in cyanobacteria, phytoplankton, and macroalgae (Nys and Steinberg, 2002).

One of the main problems with marine pharmaceuticals is the supply required for clinical trials and commercialization. Harvesting from the environment has

Table 10.1 Bioactive compounds isolated from marine organisms

Species	Compound	Use
Teleost fish		
Squalus acanthias (shark)	*Squalamine	Cancer
Pardachirus marmoaratus	Pardaxin	CAP†
Hippoglossus hippoglossus	Pleurocidin	CAP
Morone chrysops	Moronecidin	CAP
Salmo salar	Hepcidin	CAP
Oncoahynchus mykiss	Salmocidin	CAP
Tunicates		
Styela clava	Styelin	CAP
Halocynthia aurantium	Dicynthaurin	CAP
Ecteinascidia turbinata	*Ecteinascidin 743	Cancer
Aplidium albicans	*Alpidine	Cancer
Trididemnum solidum	*Didemnin B	Cancer
Chelicerates		
Tachyplesus tridentatus	Tachyplesin 1	CAP
Limulus polyphemus	Polyphemusin	CAP
Conus magnus (cone snail)	*Ziconotide	Pain
Bugula neritina	*Bryostatin 1	Cancer
Dolabella auricularia (sea hare)	*Dolastatin 10	Cancer
Amphiporus lactifloreus (marine worm)	*GTS-21	Alzheimer's disease
Pseudopterogorgia elisabethae (soft coral)	*Methopterosin	Inflammation
Crustacea		
Callinectes sapidus	Callinectin	CAP
Penaeus vannemi	Penaeidins	CAP
Molluscs		
Mytilus galioprovincialis	Mytilin	CAP
Mytilis edulis	Defensins	CAP
Sponges		
Aeglas mauritianus	*KRN7000	Cancer
Petrosia contignata	*Contignasterol (IPL576,092)	Inflammation
Luffariella variablis	*Manoalide	Inflammation
Discoderma dissoluta	Discodermolide	Spindle inhibitor
Cyanobacterium		
Lyngbya majuscule	Curacin A	Spindle inhibitor

*Undergoing clinical trials.
†CAP, marine cationic anti-microbial peptide.
Sources: Data from Pomponi (1999), Patrzykat and Douglas (2003), and Proksch et al. (2002).

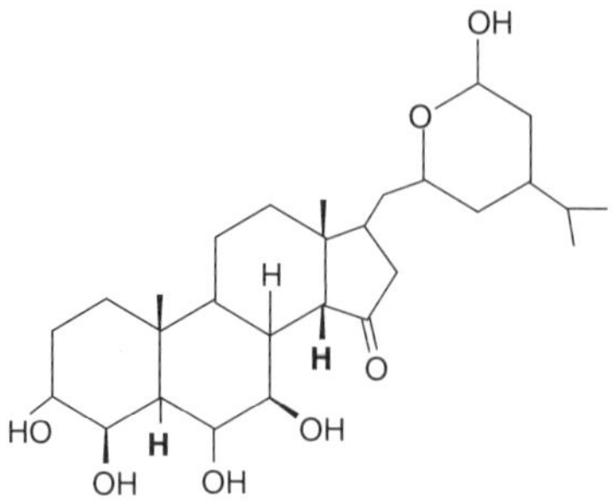

Contignasterol (IPL 576,092)
(from *Petrosia contignata*)

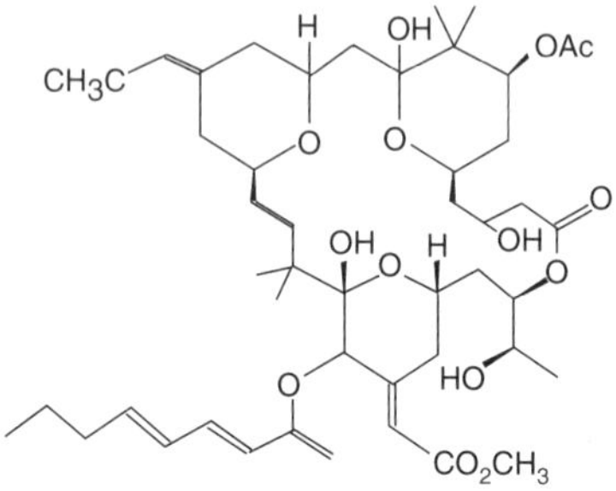

Bryostatin 1
(from *Bugula neritina*)

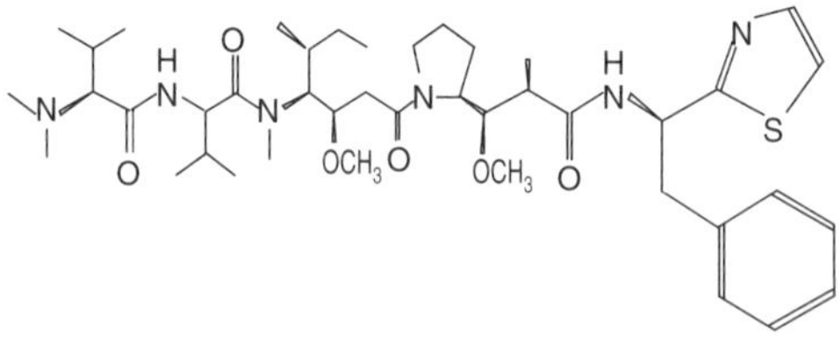

Dolastatin 10
(from *Dolabella duricularia*)

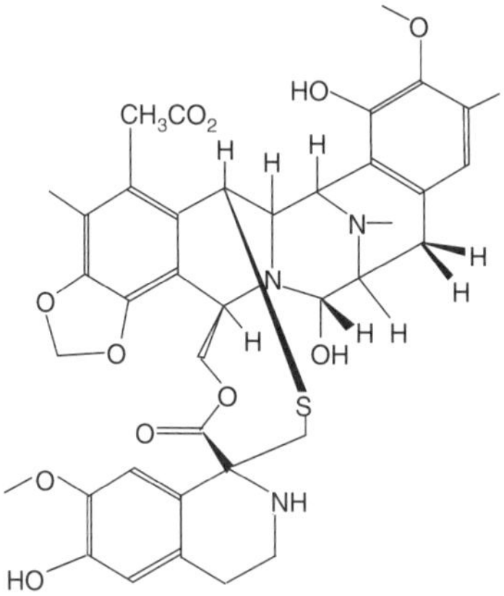

Ecteinascidin 743
(from *Ecteinascidia turbinata*)

Figure 10.1 The structures of some pharmaceutical compounds undergoing clinical trials that were isolated from marine organisms.

Table 10.2 Non-pharmaceutical marine products

Product	Source	Application
Okadaic acid	Dinoflagellate	Molecular probe, phosphatase inhibitor
Manoalide	*Luffariella variabilis* (sponge)	Molecular probe, phospholipase A2 inhibitor
Vent™ DNA polymerase	Hydrothermal-vent bacterium	PCR
Aequorin	*Aequorea victoria* (jellyfish)	Bioluminescent protein, calcium indicator
Green fluorescent protein (GFP)	*Aequorea victoria* (jellyfish)	Fluorescent protein
Phycoerythrin	Red alga	Conjugated antibodies
Pseudopterosim C	*Pseudopterogorgia elisabethae* (soft coral)	Additive to skin cream
Antifreeze protein	Flounder	Antifreeze
Adhesives	Mussels, barnacles	Bioadhesives

Sources: Data from Cowan (1997), Pomponi (1999), and Weiner (1997).

the problems of accessibility, and the product may only occur in trace amounts. Often the source organism is not sufficient to support commercial production. There are possible alternative supplies, such as the farming or mariculture of the organisms, which has been successful with some macroalgae and sponges, *in vitro* production in cultured cells, chemical synthesis, and cloning of the genes into another organism. The chemical synthesis of many marine products has been achieved but the molecules are complex and chemical synthesis is difficult (Pomponi, 1999). Therefore, the synthesis involves multiple steps, which only give low yields that are insufficient for commercial exploitation. However, discodermolide and elutherobin are being synthesized for clinical trials (Jaspars, 1999). The growth of cell cultures of organisms such as sponges and bryozoans is only in its infancy but some success has been achieved. Cultures of the sponge *Lissodendoryx* spp., *Petrosia ficiformis*, *Suberites domuncula* and *Acanthella acuta* have been prepared (Nickel et al., 2001).

10.2.1 Sea- and land-based cultivation

The shallow-water bryozoans *B. neritina* and *E. turbinata* have been cultured in the sea and in land-based systems (Pomponi, 1999). Sponge farming has proved to be successful in number of cases; indeed, 10 species have been cultivated (Osinga et al., 1998; Belarbi et al., 2003). Land-based production systems can provide better control over growth conditions but they may not be able to reproduce conditions in which the organisms grow, such as deep water. Land-based systems require the feeding of the sponges with plankton, bacteria, dissolved organic materials, and microalgae.

10.2.2 Other production systems

The application of recombinant DNA technology to the production of marine products can offer another path to their supply. Using probes based on cationic anti-microbial peptide (CAP)-encoding sequences clones encoding for pleurocidin from the winter flounder, hepcidin from bass, mytilins from mussels and penaeidins from shrimp have been obtained (Patrzykat and Douglas, 2003).

In some cases it is an associated micro-organism that is responsible for the production of the pharmaceutical rather than the host. The difficulty is to be able to culture the organism without the host, which has proved difficult. Bacteria (27 isolates) were isolated from two Mediterranean sponges, *Aplysina aerophoba* and *A. cavernicola*, and a high number of the isolates showed antimicrobial activity which indicates that marine organisms may harbor valuable microbial species (Hentschel et al., 2001).

10.3 Molecular biology products

Thermoenzymes are resistant to denaturation at temperatures of 60–120°C and a source of this type of enzymes is the hyperthermophiles, represented by both Bacteria and Archaea, that have been isolated mainly from aquatic environments. Thermoenzymes have made a significant impact on recombinant technology with the use of *Taq* DNA polymerase in PCR. DNA polymerase catalyses the addition of nucleoside monophosphates to a growing DNA chain with the other strand acting as a template. The first PCR procedure used the Klenow fragment of *E. coli* DNA polymerase I. After each cycle of DNA synthesis the mixture is heated to separate the DNA strands, which also denatures the DNA polymerase. The use of a thermostable polymerase allowed automatic thermal cycling and increased the amount of DNA produced. The enzymes also needs to have 3′–5′-exonuclease activity which removes mismatches, giving good fidelity. The first thermostable DNA polymerase, *Taq*, was isolated from non-marine *Thermus aquaticus* but had a low fidelity. Another DNA polymerase, *Tth*, (optimum, 67°C) was isolated from *Thermus thermophilus*, and this had a high fidelity. A second polymerase has been isolated from another strain of the same organism, which has an optimum of 87°C. More recently a marine PCR enzyme has been introduced, Vent^{TM}, derived from a hydrothermal-vent bacterium *Thermcoccus litoralis*, that is now produced as a recombinant protein. Green fluorescent protein (GFP)

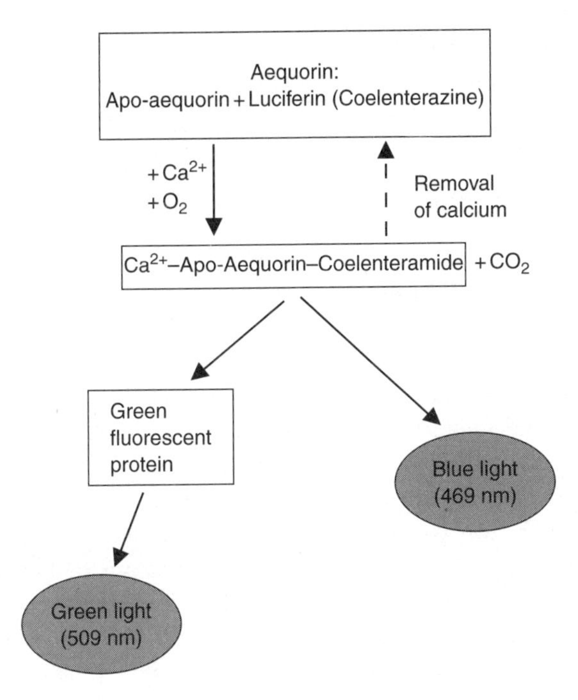

Figure 10.2 Bioluminescence of a green fluorescent protein (GFP). The green fluorescence of *Aequorea victoria* arises from a calcium-binding photoprotein, aequorin. Aequorin is an apoprotein and a luciferin, coelenterazine, bound together with oxygen. When calcium ions bind to aequorin it undergoes a conformational change to coelenteramide, which emits a blue light. However, if the GFP is present energy is transferred from aequorin and the GFP emits a green light at 509 nm (Kendall and Badminton, 1998).

extracted from the jellyfish *Aequorea victoria* was the first protein that could emit light without the need for substrates or cofactors and is ideal as a reporter gene in molecular biology (Fig. 10.2). Initially a luminescence reaction was achieved using the enzyme luciferase from firefly tails. The gene (*luc*) has been cloned and can be linked to promoters of interest so that light is produced when the gene is activated (see Chapter 3). A similar gene (*lux*) was isolated from the marine micro-organism *Photobacterium fisheri* (now known as *Vibrio fisheri*). The combination of luciferase, luciferin, and molecular oxygen produce an unstable intermediate from luciferin. When this intermediate decays it emits light.

10.4 Polymers

Marine micro- and macro-organisms are capable of accumulating intra- or extracellular polymers of many sorts that include the following:

- polysaccharides,
- emulsans,

- polyhydroxyalkanoates (PHAs),
- adhesives, and
- melanins.

10.4.1 Polysaccharides

The macroalgae, seaweeds, are a widespread and abundant source of polysaccharides that have uses in a number of industries. The polysaccharides are agar, agarose, alginates, and carrageenans. The algae that provide these polysaccharides are given in Table 10.3 and are the red and brown algae.

Agar is well known as the gelling agent used to cultivate all sorts of microorganisms on solid medium in Petri dishes. Agar is extracted commercially from red seaweeds (Rhodophyceae) including *Gelidium* spp. and *Gracilaria* spp. The main suppliers of agar are Japan, Mexico, USA, Chile, Argentina, and Brazil. Agar consists of two components, agarose (50–90%) and agaropectin. It is agarose that forms gels and can be purified by a freeze–thaw process. The structure of agarose is a disaccharide repeat formed from galactose and anhydrogalactose (Box 10.1 and Table 10.4). Agar is extracted using hot water from the dried algae followed by further purification involving bleaching and filtration. The properties and uses of agar and agarose are given in Table 10.4. Agar forms strong gels at low concentrations and melts at 60–100°C. The gels formed are strong, stable, and can be remelted and reused. Because of these properties agar is used widely as a solid microbial medium and more-purified agarose is used in biochemistry and molecular biology as a

Table 10.3 Sources of algal polymers

Polymer	Source
Agar	**Rhodophyceae (red algae)**
	Gelidiales, *Gelidium* spp., *Gracilaria* spp., *Pterocladia* spp., and *Suhria vittata*
Alginates	**Phaeophyceae (brown algae)**
	Laminariales, *Ecklonia*, *Eisenia bicyclis*, *Macrocystis pyrifera*, *Laminaria digitata*, and *Laminaria hyperborea*
	Fucales *Ascophyllium nodosum*, and *Sargassum* spp.
Carrageenans	**Rhodophyceae (red algae)**
	Gigartinales *Ahnfeltia licata*, *Chondrus crispus*, *Eucheuma* spp., *Furcellaria fastigiata*, *Gigartina* spp., *Gymnogongrus* spp., *Hypnea* spp., *Iridaea* spp., and *Phyllophora nervosa*

Source: Chapman and Gellenbech (1989).

material for column separation, gel filtration, and gel electrophoresis. In the industrial sector agar is used in a number of food products including for stabilization of confectionery, gelling in meat products, as a jelly and icing stabilizer, and in baked goods. In the medical field, as agar cannot be metabolized it is used in laxatives. The moulding properties of agar are used to make flexible moulds in dentistry and criminology.

Alginates are extracted from a number of brown algae (up to 50%; Table 10.3) and its structure and properties are given in Table 10.4. Alginate consists of a copolymer of mannuronic and guluronic acids, which occur as blocks, although alternating sequences can be found forming molecules of 80–700 units (Box 10.1). Alginates are soluble in water and will form gels at room temperature by the addition of calcium ions or other bivalent ions. The calcium ions form a link between the guluronic acid molecules and if the ions are removed the gel will break down. Alginates are extracted from milled seaweed by heating with alkali followed by precipitation by calcium chloride.

BOX 10.1 Macroalgal polysaccharides

The polysaccharides extracted from seaweeds (macroalgae) include the industrially important agar, agarose, carrageenans, and alginates. In general polysaccharides contain no more than three sugar residues or linkages and in some cases only one (i.e. glucose in starch).

Agar is a mixture of agarose and a non-gelling agaropectin. Agarose is a neutral polysaccharide consisting of a repeating structure of galactose and anhydro-α-D-galactose. The repeat structure is shown in the first figure and the three-dimensional structure in the second figure. Agar forms a strong gel at 1–2% which is transparent and can be remelted (thermoreversible). Agar is used widely to gel microbiological medium for use in Petri dishes and on slopes. It is also used in industry as a gelling agent in food, in dentistry, and in criminology. Agarose can be purified from agar and is often used as an electrophoresis gel in molecular biology.

Alternating repeats in agarose and carrageenan

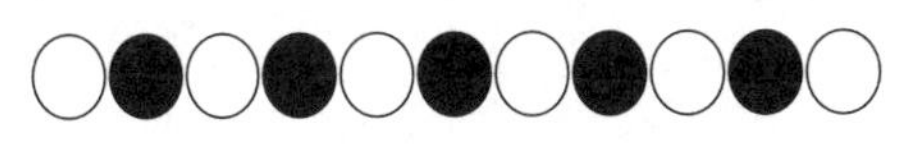

Block copolymer: alginate

Carrageenans have a similar structure to agarose with a repeating sequence of sulphated galactose and anhydro-β-D-galactose (see the second figure). The three main types of carrageenan are κ, ι, and λ. The κ and ι forms have similar repeat structures, with the anhydro-β-D-galactose sulphated in the ι version. Both these structures will form a strong gel when cooled from hot solution. λ Carrageenan is different in structure with no anhydro residues; and this stops the formation of an ordered structure and therefore this isoform does not form gels.

BOX 10.1 *(Continued)*

Agarose

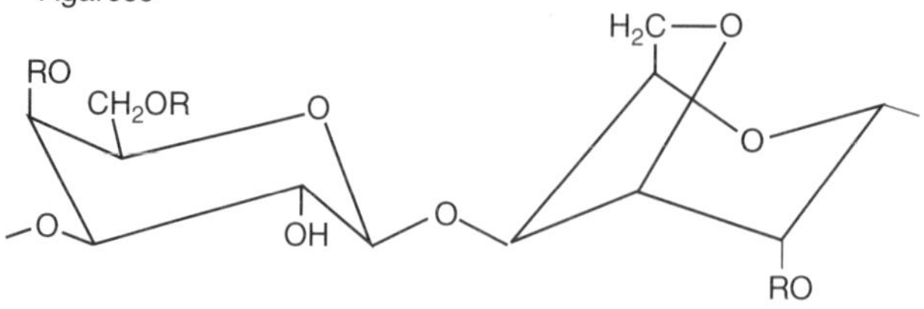

Mannuronic acid

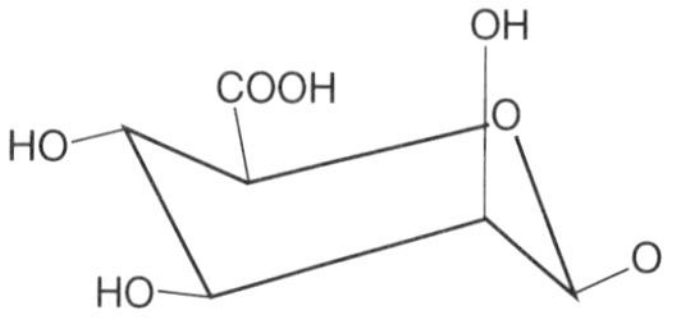

Guluronic acid

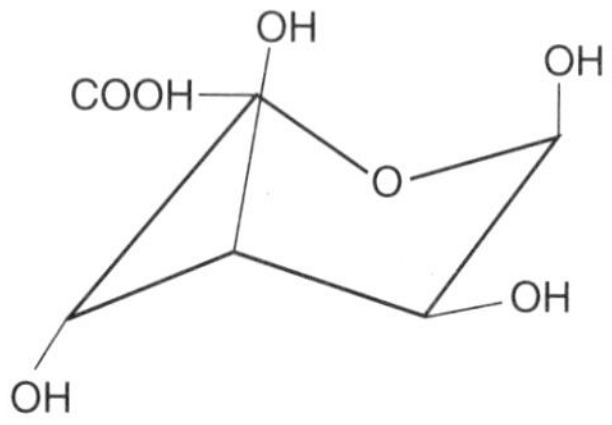

κ Carrageenan

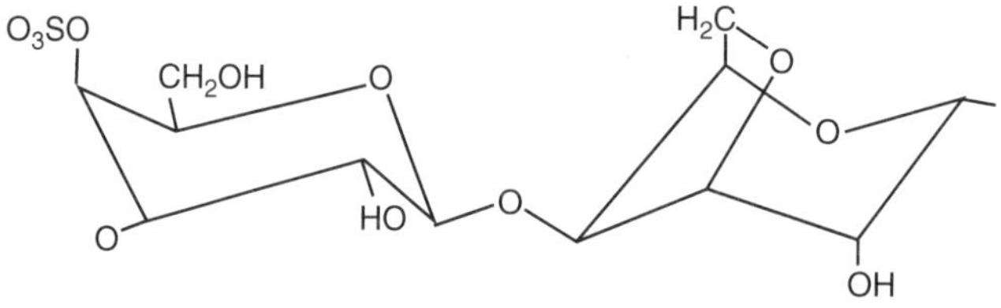

ν Carrageenan

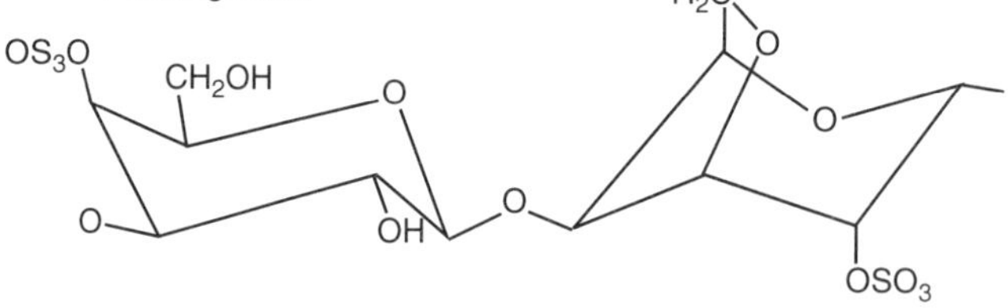

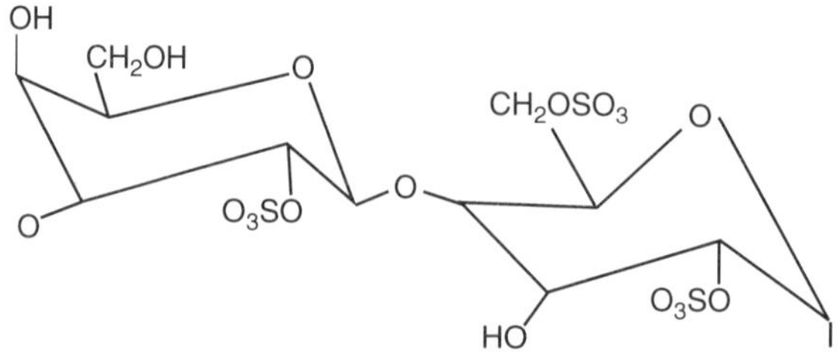

λ Carrageenan

Alginate consists of two monomers, guluronic and mannuronic acids, arranged in block copolymers. Alginate will gel in the presence of bivalent ions, usually calcium (0.5 M), which forms links between guluronic acids in adjacent polymer chains.

Alginate is used in both industry and research. The ability of alginate to be de-polymerized by removal of calcium and gelling at low temperature make it ideal for use with immobilized cells. Alginate gels cannot be reversed by heating and are not stable at low pH values. However, the high viscosity and hydration make them useful in salad dressings, frozen foods, icings, and film formation. The textile and paper industry use alginate to thicken inks, coat papers, and reduce staining. In the medical field alginate has been used to form films to cover wounds and it also has haemostatic activity.

The third polysaccharide group is the carrageenans, which are extracted from red algae (Table 10.3). Carrageenans are a repeating polymer of sulphated galactose and anhydrous galactose (Box 10.1) and occur in three main forms, κ, ι, and λ depending where the molecules are sulphated. The extraction is similar to that used for alginate where the concentration is around 28–35%. The use and properties of carrageenans are given in Table 10.4. The polymers are of a similar size to alginate and form thermoreversible gels at 0.5% and above although λ does not gel. Carrageenan is named after a coastal town in Ireland where dried seaweed had been used for years as a food gel and continues to be used in the food industry. Carrageenans are used too in milk-based products where it improves heat stability, inhibits lipase activity, stops coagulation, resists separation, and stabilizes calcium-sensitive caseins. Carrageenans are used in toothpastes, air fresheners, and pet foods.

Alginate production is about 30 000 tonnes per year and the potential supply is much greater than this. However, much of the supply is in remote regions and it is these regions that supply the best alginate. The main producers of carrageenan are the Philippines and Indonesia who produce about 30 000 tonnes per annum. Seaweed culture has been developed in the Philippines but is not yet economic so that like alginate it is extracted from wild harvested seaweed. The same is true for agar and agarose extracted from red seaweeds.

10.4.2 **Emulsans**

Emulsans are polymeric lipopolysaccharides produced by bacteria to desorb their cells from spent hydrocarbon droplets; one such species is *Acinobacter calsoaceticus*. These products are under investigation as oil-cleaning and viscosity-reducing agents.

10.4.3 **Polyhydroxyalkanoates (PHAs)**

PHAs are storage products, which are produced by a large number of bacteria (see Chapter 6) and can be used as biodegradable plastics. In marine organisms PHAs accumulate inside the cells at 25–80% of the cell's dry weight under high-carbon, low-nitrogen conditions. Until an economical high-yielding medium can be developed the marine organisms will only serve as the donors of relevant genes.

Table 10.4 Seaweed polysaccharides

Polysaccharide	Composition	Properties	Applications
Agars	Alternating 1,4-linked α-D-galactose and 3,6-anhydro-α-D-galactose substituted with methoxyl, ester sulphate, and ketyl pyruvate groups	Gels at low concentrations, thermoreversible gel, retains liquid, resists hydrolysis by microbes	Icings, jelly candies, canned meat, laxatives, microbial media, raw material for agarose
Agarose	As above	Gels at low concentrations, forms ion-dependant reversible gels, minimum protein reactivity	Electrophoresis, immunoassays, immobilized cells, microbial culture
Alginate	1,4-linked guluronic and mannuronic acids	Salts soluble in water, bind water, thickens aqueous systems, suspends solids	Frozen food, icings, salad dressing, dental impression medium, textile sizing
Carrageenans	Alternating 1,3-linked α-D-galactose and 1,4-linked 3,6-anhydro-β-D-galactose, substituted with ester sulphate	Binds moisture, stabilizes emulsions, high protein reactivity, controls flow and texture in food	Frozen deserts, chocolate milk stabilizer, low-calorie jellies, toothpaste binder, air-freshner gels, pet food

10.4.4 Adhesives

Marine organisms can attach themselves to surfaces under adverse conditions, such as temperatures of 4°C or less, under pressure, and in fast-flowing streams. The extracellular polymers involved could make marine adhesives and as they are non-immunogenic they could also be used as bioadhesives. Invertebrates such as mussels produce protein-based adhesives in order to stick to rocks etc. These proteins have a 9-amino acid repeat in a long polymer (Cowan, 1997), e.g. $(\text{Ala-Lys-Pro-Ser-Try-Hyp-Thr-Dopa-Lys})_{75}$ (where Dopa is 3,4-phenylalanine).

10.4.5 Melanins

Many marine microbes produce polymeric colours and some of the most interesting are the melanins. These are produced to protect the cells from UV damage, to enhance virulence, and to protect against protease degradation. The melanins are extracellular complex polyphenolic heteropolymers that have uses in cosmetics, dyes, and sunscreens, and as reporter genes. Only a single gene is required for accumulation of melanins as they are derived from tyrosine.

10.5 Enzymes and transgenic organisms

The marine environment contains habitats at the extremes of temperature, from less than 5°C in the Arctic and Antarctic to 350°C in hydrothermal vents. In these habitats will be found organisms adapted to heat or cold that could be a source of cold-adapted and thermotolerant proteins and enzymes.

The antifreeze proteins were first discovered in fish species living in the Antarctic, where the sea water does not freeze at 0°C because of its salt content, so that the bodies of fish are in danger of freezing. In several fish the antifreeze proteins or glycoproteins stop freezing by binding to the forming ice crystals. These antifreeze proteins may be of value to the frozen food industry, and the expression of these proteins in transgenic plants and animals has potential. The type 1 antifreeze protein gene from the flounder has been expressed in salmon, indicating its possible use (Cowan, 1997). The ice-nucleation protein from a terrestrial bacterium has already been used in artificial snow manufacture and transferred to African violet plants to reduce frosting.

Transgenic organisms are those that contain a gene(s) from another organism and transgenic plants, nematodes, fruit flies, sea urchins, frogs, mice, cows, and sheep have been produced. If the inserted gene is linked to a functional promoter the gene can be expressed when the promoter is activated. A wide range of transgenic fish species have been produced by microinjection or electroporation into fertilized or non-fertilized eggs.

Traditional fish production has been from the harvesting of the natural population and the yield has continued to increase each year as techniques have improved. However, the harvest is such that fish stocks cannot be sustained at this level and some species such as cod have seen dramatic declines in some areas. To fill this gap the aquaculture of some fish species has been developed and at present some 10 million tonnes of farmed fish are produced each year. It is in the farmed environment that transgenic fish can be commercialized. The possible improvements are an increase in growth rate, control of the growth cycle, disease resistance, and the development of new vaccines. The gene for a fish antifreeze protein has been isolated, cloned, and expressed in bacteria and plants as a method of supplying large quantities of the protein for possible commercialization. The growth hormone genes from rainbow trout and striped bass have been expressed in *E. coli* as a method for its production. The growth hormone is hydrophobic and on expression forms a precipitate that can be solubilized. Application of the growth hormone to trout at 1 μg/g per week shows an increase in growth but the expense of the hormone makes it uneconomical. However, the growth hormone has been transferred to carp, medaka, tilapia, flounder, and salmon and expression in flounder and salmon has shown significant increases in growth. A number of obstacles to the commercialization of these transgenic fish include regulation of expression levels, the avoidance of growth abnormalities, sterilization of the fish to avoid problems of escape, and public acceptance of this kind of fish.

Cold-adapted enzymes have been isolated from organisms (psychrophiles) from cold areas, mainly Antarctic waters at >1°C. These enzymes function at 0–30°C better than similar enzymes from normal (mesophilic) micro-organisms. The adaptation of these enzyme to low temperatures appears to be related to a greater flexibility of the enzyme molecule. The cold-adapted enzymes could be of use in the detergent industry as cold-wash amylases, proteases, and lipases, in the textile industry as cellulases, and in the biopolishing process. The food industry could use these enzymes in milk-based products including reduction in lactose (β-galactosidase), amylases, proteases, and xylanases. Cold-adapted enzymes could be used in the bioremediation of cold habitats and possibly in low-water conditions (Gerday et al., 2000).

10.5.1 Hyperthermophiles

A great diversity of hyperthermophilic Bacteria and Archaea have been isolated from hydrothermal, geothermal, and man-made high-temperature habitats. Deep-sea hydrothermal vents are found on the sea floor on ocean ridges and in volcanically active regions. These vents can be found at great depths. Studies of the populations on these vents have shown many more micro-organisms than were expected. The thermophilic organisms isolated belong to the domains Bacteria and Archaea with the Bacteria represented by the orders Termotogales and Aquificales, and the Archaea by the orders Archaeoglobales, Methanobacteriales, and Thermococcales (Wery et al., 2002). About 75 species have been described so far (Huber et al., 2000) and these grow above 80°C. The most extreme hyperthermophile, *Pyrolobus fumarii*, is unable to grow below 90°C and will grow at 113°C and can survive autoclaving. The cultivation of these hyperthermophiles under laboratory conditions and in bioreactors has been successful in some cases. The cultivation of the anaerobic archaean *Pyrococcus furiosus* at 100°C yielded cell densities of 3×10^8 cells/ml, corresponding to about 0.03 g of dry weight/l, which is low compared with mesophiles, at 2–10 g/l (Krahe et al., 1996). However, by using bioreactors of different designs and optimizing the conditions a stirred-tank bioreactor yielded 3×10^9 cells/ml, and a membrane bioreactor 35×10^9 cells/ml, equivalent to 3.6 g/l Thermophilic enzymes include amylases; Apu amylopullanase from *P. furiosus* was cloned and has a half-life of 44 h at 90°C and an optimum temperature of 105°C (Zeikus et al., 1998). Other enzymes are glucose isomerases from *Thermotoga neapolitana*, alcohol dehydrogenase from *Thermoanerobacterium ethanolicus*, and alkaline phosphatase from *T. ethanolicus*.

10.6 Microalgae

Historically microalgae have been seen as a problem, with algal blooms due to eutrophication, and their removal has been a priority. There have been a few traditional uses of microalgae as a food. *Spirulina* has been grown for some time as a food in alkaline lakes in Mexico and Chad.

Microalgae can be either freshwater or marine in origin and are grown for a number of products. As microalgae are often small in size (5–20 μm) and grow as single cells or in small groups they can be grown in bioreactors, small lakes, ponds, and tanks in large volumes. Freshwater microalgae have received the most attention and it was only later that marine species were investigated. Table 10.5 gives some of the products produced by both freshwater and marine microalgae. The products are pigments, unsaturated fatty acids, polysaccharides, food additives, and biofuels, and for use in wastewater treatment.

Astaxanthin, a carotinoid related to β-carotene, is found in many seafoods giving them a pink colour. Astaxanthin however is accumulated by the microalga *Haematococcus pluviatus* at 1.5–3.0% and this is used to produce commercial amounts of astaxanthin. Although *H. pluviatus* is a freshwater alga the product is used in aquaculture to add a pink colour to farmed trout, salmon, shrimp, lobsters, and crayfish.

β-Carotene, a related carotenoid, is produced by the halotolerant *Dunaliella salina* and *Dunaliella bardawil.* β-Carotene is used as a food colouring and pro-vitamin A. Fluorescent labelling in diagnostic reagents can use phycobiliproteins from cyanobacteria. Lower-value products are those used for the production of vitamins C and E and amino acids. Other possible uses include *Chlamydomonas mexicana* as a soil inoculum, green algae as single-cell protein, and *Chlorella* spp. as a biofuel.

10.7 Marine pollution

Most marine pollution occurs in coastal regions but often originates from terrestrial sources. Rivers and rain carry many agricultural products, such as fertilizers, pesticides, herbicides, and other industrial wastes—both inorganic (heavy metals) and organic (xenobiotics)—to the coastal regions. Shore-based activities such as unloading ships, repairs, washing, and use of anti-fouling compounds such tributyltin (organotin) also add material to the coastal region. In contrast, few problems occur offshore except for leakage from oil wells and ship wrecks.

Table 10.5 Products extracted from microalgae

Product	Alga	Use
Astaxanthin	*Haematococcus*	Colour in aquaculture; salmon, shrimp, crayfish
Phycobiliproteins	Red, blue-green	Food colouring
Phycocyanin	*Spirulina* spp.	Chewing gum, dairy products, jellies, cosmetics
β-Carotene	*Dunaliella* sp.	Food colouring, pro-vitamin A
Xanthophylls	Diatoms, green algae	Chicken feed
Vitamins	Green algae	Vitamins
Health foods	*Chlorella, Spirulina*	Health-food supplements
Formulaid®	*Cryptocodinium cohnii*	Formula milk, fatty acids
Polysaccharides	*Phorphyridium*	Gums
Feed for bivalves	Diatoms, *Chrysophytes*	Aquaculture
Soil inoculum	*Chlamydomonas*	Soil conditioners
Single-cell protein	Green algae	Animal feed
Fuel	*Chlorella*	Fuel supplement

10.7.1 Biosensors

Biosensors for the monitoring of marine pollution are being developed for nitrate, nitrite, organotin (an anti-fouling agent), pesticides, and endocrine disrupters. Nitrate biosensors have been produced using the enzyme nitrate reductase in an electrogenerated polymer, an enzyme from the denitrifying bacterium *Thiospaera pantotropha*, and immobilized *Agrobacterium radiobacter*. Organotin, used as an anti-fouling agent, can be detected by its toxic effect on micro-organisms linked to an oxygen electrode. Pesticides can be detected by their effect on the enzyme acetylcholinesterase.

10.7.2 Oil pollution

The widespread use of crude oil products has resulted in the discharge of oil into the environment. The marine environment has received most attention because of spectacular oil spills at sea. Land-based spills are just as important and both are dealt with in Chapter 5. Tanker spills make the headlines but millions of gallons of crude oil and oil products are released into the oceans and seas from non-accidental sources. The levels of oil released into the oceans from various sources are given in Figure 5.7 and it has been estimated that the total is between 1.7×10^6 and 8.8×10^6 tonnes.

10.7.2.1 *Natural sources*

Crude oil is a natural product formed from plant and animal materials that were incorporated into the sediments in shallow seas millions of years ago. Other strata covered these sediments and over time the organic material was converted by heat and pressure into crude oil. The crude oil can migrate to the surface or is trapped by impervious layers to form oil reservoirs. Only a proportion of the oil is trapped and therefore oil has been escaping into the environment for many millions of years. The oil that reaches the terrestrial surface will lose its volatile components, forming tar lakes and sands. The crude oil reaching the sea surface will form an oil slick where the volatiles are lost to the atmosphere. Natural seepage represents 8% of the total annual release of oil, and there are about 200 known seeps. The natural seeps occur in volcanically active areas such as offshore California and in the Arabian Gulf.

All the other sources of oil are based on industrial activity. Offshore drilling spills caused by blowouts, leaking of storage tanks, and accidental spills are responsible for 25% of the total. Atmospheric precipitation of oils as derived from vehicle exhausts is difficult to estimate but a figure of 13% has been given. By far the largest source is the municipal and industrial waste which runs-off into rivers and estuaries. Municipal wastewater can contain waste engine oil, and oil refineries and industrial operations release oil into the seas. The dumping of sewage sludge at sea has also added some oil to the seas but in Europe this process has been banned (Chapter 4). The combined run-off volume is estimated at 51% of the total.

Routine maintenance, dry docking, and other tanker operations are responsible for the release of 19% of the waste oil. A large proportion of oil is transported by sea and ballasting of empty tankers with water and the cleaning of the oil tanks at sea has seen the release considerable quantities of oil. The MARPOL protocol agreed in 1983 required that all tankers over 20 000 tonnes have a system installed that retains any tank-washing water, to be discharged when in port, which should reduce the oil released at sea.

10.7.3 **Tanker accidents**

It can be seen that the accidental spillage from tankers is far less than that from motor-oil replacement and other sources on the land which run into the oceans. However, tanker accidents do release large quantities of oil, in a small area in some cases, so that the environmental results can be significant and the accidents make national and international news. The scale of the oil spilled from some of the tanker accidents is given in Table 10.6. The frequency of spills has reduced considerably since 1980.

10.7.3.1 *Composition of crude oil*

Crude oil is a very complex mixture of organic compounds including aliphatic alkanes, monocyclic aromatics, polycyclic hydrocarbons, tars, and bitumens. The composition and structure are given in section 5.3.

10.7.3.2 *Release at sea*

Crude oil, when released at sea, will not mix with seawater and will float on the surface, allowing the escape of the volatile components; those of 12 carbons and below such as aliphatic alkanes and monocyclic aromatics such as benzene and toluene (Fig. 10.3). The fate of the oil spill will depend on a number of factors, including the weather, the distance from shore, and wind

Table 10.6 Major oil tanker spills

Tanker	Date	Location	Oil spill ($\times 10^3$ t)
Torrey Canyon	1967	Scilly Isles, UK	119
Mandoil II	1968	USA	41
World Glory	1968	South Africa	48
Texaco Denmark	1971	Belgium	107
Sea Star	1972	Gulf of Oman	123
Metula	1974	Chile	50
British Ambassador	1975	Japan	46
Urquiola	1976	Spain	100
Hawaiian Patriot	1977	Hawaii	95
Amoco Cadiz	1978	Brittany, France	223
Captain	1979	West Indies	287
Independenta	1979	Turkey	95
Irenes Serenade	1980	Greece	100
Juan Antonia Lavalleja	1980	Algeria	38
Castillo de Beliver	1983	South Africa	252
Assimi	1983	Oman	53
Pericles G.C.	1983	Persian Gulf	46
Odyssey	1988	Gulf of Oman	123
Exxon Valdez	1989	Alaska	36
Khark 5	1989	Morocco	80
New Carissa	1990	Oregon, USA	400
ABT Summer	1991	South Africa	260
Haven	1991	Italy	144
Tenyo Maru	1991	Oregon, USA	354
Aegean Sea	1992	Spain	74
Braer	1993	Shetland Islands, UK	85
Sea Empress	1996	Milford Haven, UK	72
Erika	1999	Brittany, France	10

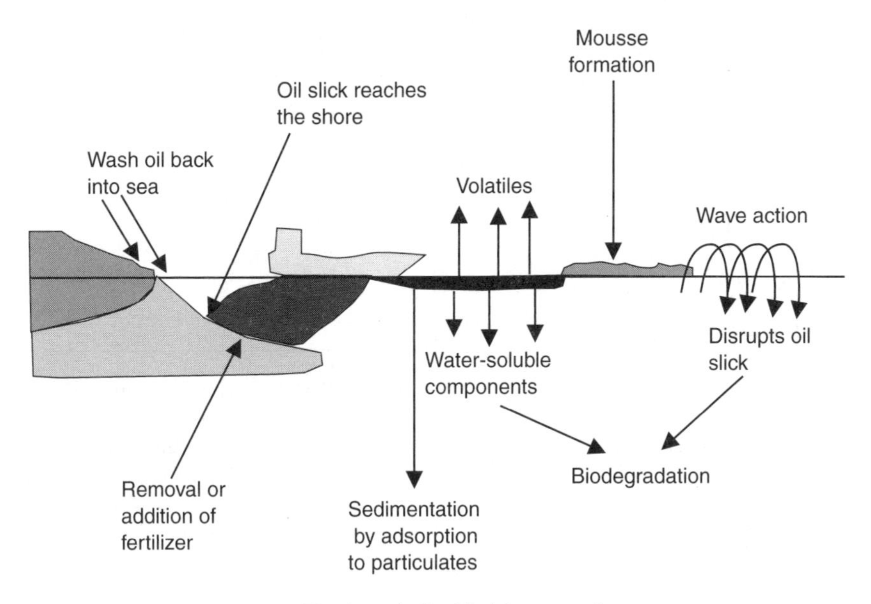

Figure 10.3 The fate of oil spilled from a tanker at sea.

direction. The floating oil can suffer a number of fates; the water-soluble components will dissolve in the water and will disperse rapidly, although this is usually around 1%. However, most of the components are not miscible with water and these can be mixed by wave action into a water and oil emulsion known as a mousse. More vigorous wave action will break the oil into small droplets (0.01–1 mm) that can dispersed in the water column. The presence of particulates may cause sedimentation of the oil as it binds to the particles. The properties of the oil spilled will also affect the rate and nature of the dispersal. An example is the oil spilled from the *Braer*, which was light crude having a higher proportion of low-molecular-weight components. The ship ran aground in a storm and 85 000 tonnes of oil were released, which gave a very high concentration of oil in the water, but because of the vigorous mixing of the light crude this was removed more rapidly than a lower concentration of oil from the *Exxon Valdez* (Fig. 10.4). In the case of the *Exxon Valdez* there was more high-molecular-weight components and a calmer sea state.

The dispersion will allow naturally occurring hydrocarbon-degrading organisms to break the oil down. Oil breakdown will occur at the interface between the oil and water and therefore the better the oil dispersion, the greater the \surface area, and the faster the degradation. Crude oil is a naturally occurring product and as such is regarded as biodegradable and it is perhaps no surprise that hydrocarbon-degrading micro-organisms are distributed widely in nature. The number of hydrocarbon-degrading micro-organisms is increased by prior exposure to hydrocarbons (Venosa and Zhu, 2003). The degradation of the oil by biological, physical, and chemical processes is often referred to as weathering. The biological degradation will become significant

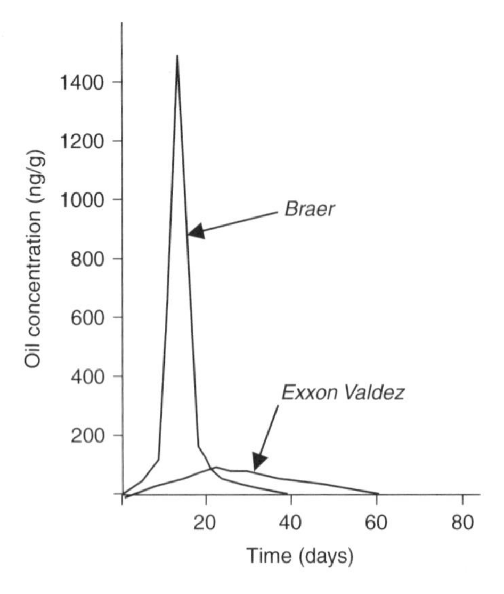

Figure 10.4 The reduction of hydrocarbons in the sea after the *Braer* and *Exxon Valdez* oil spills. The *Braer* spill involved light crude in a storm whereas the *Exxon Valdez* involved normal crude in calmer waters. *Source:* Kingston (2002).

over a number of weeks but in many cases the microbial activity is restricted by the availability of nitrogen and phosphate. In the case of the *Amoco Cadiz* microbial activity consumed 73 000 barrels of oil before it reached the shore.

The more-complex and less-soluble oil components will be degraded much more slowly than the lighter oils, and it is these high-molecular-weight components that will persist on the sand and rocks if the spill reaches shore. In many cases these high-molecular-weight components form lumps of tar which are very difficult to degrade.

10.7.4 Toxicity

Oil will affect the organisms in an environment in two ways. First, it can coat the organisms with a layer of oil, smothering the animal or plant, and it is on seabirds that this effect can be seen most dramatically. By the time the oil reaches the shore if consists of high-molecular-weight components which can affect intertidal organisms. Second, the components of oil can be toxic and it is the aromatic hydrocarbons that are the most toxic. The very high-molecular-weight components are less toxic as they are insoluble and therefore unavailable to organisms. The method used to evaluate the effect of oil spills on the subtidal area is given in Table 10.7.

10.7.5 Bioremediation

The first action in any spill is to contain any oil released with booms and barriers. This is followed by mechanical removal and sorbants (Fingas, 1995). *In situ* burning and the spraying of dispersants have also been used to remove

Table 10.7 Methods used to evaluate the effects of oil spills on the subtidal region

Method	Oil spill
Biochemistry/Physiology	
Sediment toxicity	*Exxon Valdez*, Alaska
Amphipod bioassay	*Exxon Valdez*, Alaska
Fish histopathology	*Amoco Cadiz*, France; *Exxon Valdez*, Alaska
Bile metabolites	First Gulf War, Saudi Arabia *Mobilose*, USA
Crustacean reproduction	*Exxon Valdez*, Alaska
Population changes	
Crustacean/echinoderm population structure	*Exxon Valdez*, Alaska
Increase in opportunistic species	*Amoco Cadiz*, France
Fish population	*Exxon Valdez*, Alaska
Macroalgal population	*Exxon Valdez*, Alaska

Source: Lee and Page (1997).

spills. If the spill reaches shore then physical removal can be used on sandy shores but with rock or gravel low-pressure, cold-water flushing can be used. Hot water and high pressures have been found to harm shore plants and animals. Dispersants have been used on shores but it has often been found that this harms the flora and fauna. Bioremediation of oil at sea or on shore is aimed at increasing the activity of naturally occurring oil-degrading organisms. Seeding of the polluted site with specific micro-organisms, bioaugmentations, has been used in a number of cases but the results were not very successful. Nutrient addition on the other hand has been successful both at sea and on shore (Head and Swannell, 1999). The materials were fertilizers, water soluble, slow release, and oil soluble, and the oil-soluble ones were the most effective (Fig. 10.5). Another study in Japan showed the value of adding fertilizer in the remediation of oil spills (Maki et al., 2003). Of 12 bioaugmentation products tested only three increased the rate of degradation (Aldrett et al., 1997). It has been shown that this form of bioremediation on a hydrodynamically active beach has no impact on the biota (Schratzberger et al., 2003). Recently the methyl esters of plant oils such a rapeseed oil, known as biodiesel, have been used to clean oil spills on beaches. Biodiesel acts as a non-volatile solvent dissolving many of the waxy compounds making them available for degradation, and in addition biodiesel is biodegradable (Pereira and Mudge, 2004).

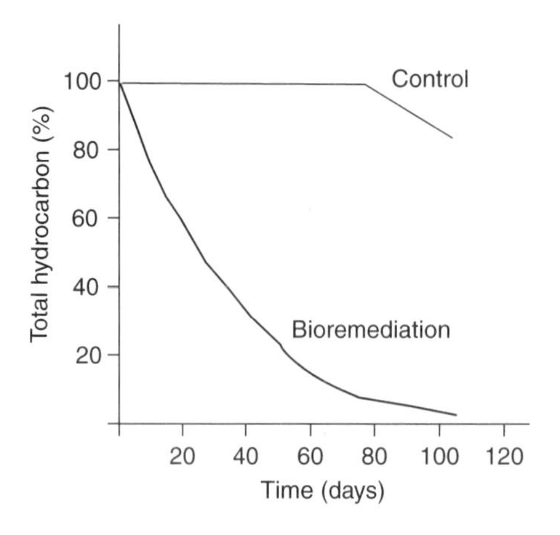

Figure 10.5 Changes in total hydrocarbons in the subsurface oil of untreated (control) and bioremediated (slow-release fertilizer) shorelines.
Source: Atlas (1995).

10.8 Conclusions

The oceans and seas are a very diverse environment and appear to contain many more species of micro-organism than was at first determined by viable plate methods. However, the application of molecular biology with the determination of rRNA sequences has shown that there are very many more Bacteria and Archaea present that are not culturable with traditional methods. Not only does the environment contain many micro-organisms but also these are found in some of the most extreme conditions, such as in deep-water hydrothermal vents (smokers) where the temperature can be a high as 350°C. Organisms in the marine environment can provide new pharmaceuticals, polymers, molecular biology products, enzymes, antifreezes, adhesives, and colourants. There are problems of supply from marine organisms as it is difficult to harvest enough from natural resources, but mariculture and aquaculture may be able to replace natural harvesting. The marine environment is the final destination for much of the pollution generated by humans, including oil. The systems for the bioremediation of oil spills have been described.

10.9 Further reading

Laws, E.A. (2000) *Aquatic Pollution*, 3rd edn. John Wiley & Sons, New York.

Tombs, M. and Harding, S.E. (1998) *An Introduction to Polysaccharide Biotechnology*. Taylor and Francis, London.

11 Specific topics

11.1	Introduction	401
11.2	*Exxon Valdez* oil spill	403
11.3	Acid mine drainage	408
11.4	Wheal Jane	415
11.5	Further reading	418

11.1 Introduction

This chapter deals with contamination or pollution that has been highlighted by a single specific accident or spill. The spectacular spill in some cases has been instrumental in assisting or initiating legislation concerning pollution.

One example of legislation being initiated by a spectacular spill is the Oil Pollution Act of 1990 (OPA90) enacted by the United States Congress in the aftermath of the *Exxon Valdez* oil spill in Alaskan waters in 1989 (Burlington, 1999). Up to that time there had been deadlock in the production of legislation. The effect on wildlife was a consistent theme in the coverage of the oil spill, with pictures of dead and dying birds and otters. The spill highlighted the conflict between the exploitation of Alaska's natural resources and its image as a wild and primitive state. The public's concern allowed the OPA90 act to be implemented and the main themes of the act are given in Table 11.1. The rules involve the construction and operation of tankers in order to reduce the number and severity of oil spills. The other rules involve the responses required should a tanker spill occur. In addition to the rules put forward to reduce the possibilities of spills, the act also introduced the 'polluter pays' principle (Ketkar, 2002). Some 10 years or more after the act the volume of oil spills in US coastal waters has fallen (Table 11.2), and although it has increased in the rest of the world, the total amount of oil spilt is down. This would indicate that there are more spills but that they are smaller.

Another very visible example of pollution is the contamination of surface waters by acid mine drainage and acid effluents from refineries. Acid mine

Table 11.1 The main rules introduced by the 1990 Oil Pollution Act (OPA90)

Rule
Double hulls
Deck spill control
Spill source control and containment
Emergency lighting and advance notice of arrival
Overfill devices
Operational measures for non-double-hulled vessels
License; certification of registration of merchant mariner's documents
Vessel-response plans
Facilities-response plans
Equipment and personnel requirements
Financial responsibility/liability

Source: Derived from Ketkar (2002).

Table 11.2 Oil spills by tankers from 1980 to 1999

Year	US tankers: number	US tankers: volume (gallons)	All other tankers: numbers	All other tankers: volume (gallons)
1980–84	348	1 740 081	7 970	11 631 497
1985–89	188	3 113 886	5 523	7 278 345
1990–94	201	1 065 379	8 864	3 246 657
1995–99	118	86 463	8 770	1 751 277

Source: Derived from Ketkar (2002).

drainage contains ferrous and ferric sulphates and other metals and on draining into less-acid waters these precipitate, giving the waters an ochre to red appearance. The pollution kills many of the plants and animals and stops the water from being used for drinking water. Acid mine drainage is produced by metal and coal mining, waste and spoil tips, and this pollution is increasing as mines become worked out and abandoned.

This chapter covers the biological processes used to clean up these pollutants using two specific cases to illustrate the processes involved, the *Exxon Valdez* oil spill and the acidic mine pollution from the Wheal Jane mine.

11.2 *Exxon Valdez* oil spill

At just past midnight on 24 March 1989 the supertanker *Exxon Valdez* ran aground on Bligh Reef in Prince William Sound, Alaska (Fig. 11.1). The tanker was coming from the port of Valdez and was full, carrying 180 000 tonnes of Alaskan crude oil. Eight of the eleven oil tanks and three out of five ballast tanks were ruptured, releasing 35 500 tonnes (41.6 million litres) of crude oil into Prince William Sound in the first 5 h. The first action in this situation is to contain the spill as soon as possible, but a barge carrying the equipment only reached the tanker after 14 h. The delay was due to equipment having to be loaded on to the barge and the cranes involved had to load a tug also bound for the tanker. The delay meant that by the time the barge reached the tanker the oil slick was too large to contain adequately, despite the good weather. In the meantime the news media had arrived and was relaying pictures of oiled shorelines, birds, and sea otters that caused a considerable public reaction. Three days later the tanker *Exxon Baton Rouge* arrived to offload the remaining oil from the stranded tanker. On 4 April offloading was completed and the next day the *Exxon Valdez* was refloated and towed to a sheltered harbour for repairs. Meanwhile the oil slick spread rapidly within the sheltered sound and the prevailing current took the oil through the Montague Strait and within 6 days the oil had begun to spread into the Gulf of Alaska. After 8 weeks the slick had travelled 750 km down the Alaskan Peninsula (Fig 11.1). Table 11.3 gives an estimate of the fate of the spilled oil. Within the first 10 days some 20% of the oil had evaporated as crude oil contains a high proportion of volatile components (toluene and benzene). About 50% was degraded by microbial activity and photolysis and only 14% was recovered. Crude oil when spilt on seawater will coalesce into a lumpy emulsion, or mousse, within a few days due to the action of wind and waves, evaporation, and dissolution. The oil slick moved southwest and broke up, as fresh oil was not being released, and formed tar balls at or near the surface.

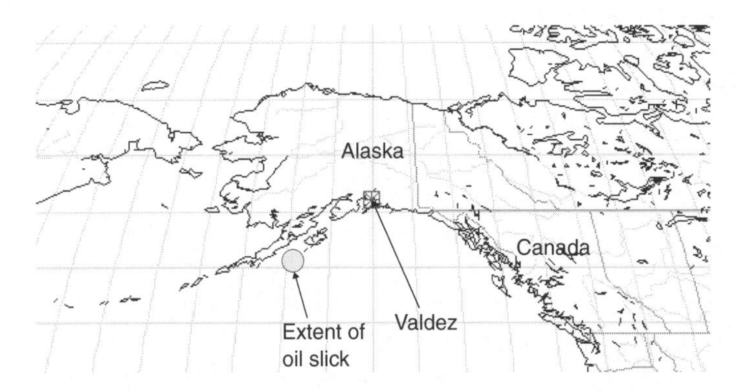

Figure 11.1 A map of Alaska showing the site of the *Exxon Valdez* oil spill.

Table 11.3 Fate of oil spilled from the *Exxon Valdez*

Fate	% of oil spill
Recovered	14
Beached	2
Subtidal sedimentation	13
Evaporated	20
Dispersed in the water	1
Biodegraded	50

Source: Laws (2000).

Overall 1700 km of shoreline was contaminated by the oil slick. Although the amount oil lost was not great in comparison with other spills its effects were important as the spill occurred in sheltered and partially enclosed water, which reduced the dispersal of the oil by wave and current activity. The waters in Alaska are cold, which would have slowed down any degradation. The setting made the spill more important as the Alaskan area was regarded as a pristine and unspoilt wilderness (Birkland and Lawrence, 2002). In truth the area of Prince William Sound was not as pristine as it was perceived to be, as the area had been used for some time by mankind for the fur trade, hunting, fishing, logging, and mining (Wooley, 2002).

11.2.1 Effect on the environment

The short-term effects on large animals and birds were the easiest to estimate as the bodies could be counted, although this can never be a complete estimate as many will sink and be lost. It was estimated that about 250 000 birds were casualties (Kingston, 2002). Some breeding colonies have recovered rapidly which suggests either an overestimate of the casualties or an influx from outside the region. In contrast, another report claims that bird populations had not recovered fully after 9 years. Other large animals killed were sea otters (2800) and harbour seals (3000). As the water is deep (100 m) in most parts of Prince William Sound the effect of the spill on fish stocks was minimal; it was the intertidal and subtidal communities that suffered. Many of the intertidal animals and plants were also harmed by the hot-water clean-up operation. An example of intertidal damage was the effect on limpet populations which had not recovered by 1991 and Fig. 11.2 shows the effect on the dominant seaweed coverage, *Fucus gardneri* (Stekoll and Deysher, 2000). *F. gardneri* (Silva) is the dominant macroalga of the intertidal area in the region of the *Exxon Valdez* spill (Fig. 11.2). The *F. gardneri* cover was reduced in all oiled areas and plants were not as reproductively competent as in unoiled areas

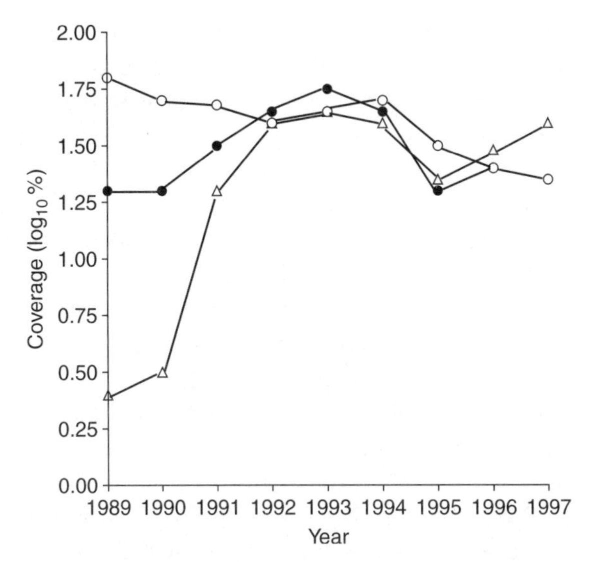

Figure 11.2 The changes in *Fucus* sp. coverage from oiled and unoiled shores following the *Exxon Valdez* oil spill. o, Control unoiled; •, oiled and unwashed; Δ, oiled and washed. From Kingston (2002).

(Stekoll and Deysher, 2000). Even in 1991 the coverage of *F. gardneri* on oiled beaches was still reduced, which suggests that recovery was slow. A number of methods, both biochemical/physiological and concerned with population levels, were used to estimate the damage and included amphipod bioassay, bile metabolites, and populations of macroalgae, crustaceans, and echinoderms. The mussel beds were not treated to remove the oil because of the potential adverse effects, as mussel beds provide food and habitat for many organisms and stabilize the intertidal area. The natural rates of oil degradation were slower in the mussel beds than expected and quantities of oil remained under the dense mussel beds 3–6 years later. Because the oil penetrated deep into the beds aeration and dispersion were poor and this reduced degradation considerably. The sediments contained more than 62 000 μg of total hydrocarbon (THC)/g of wet weight and more than 8 μg of total polynuclear aromatic hydrocarbons (TPAHs)/g of wet weight compared with the control values of 60 and 0.5 μg/g of wet weight respectively (Carls et al., 2001).

11.2.2 **Remediation**

Soon after the oil release attempts were made to burn the oil and an estimated 0.14% of the oil was burnt. After this dispersants were used but with little success as there was little wave action in the sound. More than 800 km of the polluted shoreline of Prince William Sound was subjected to a number of clean-up methods. The standard method was to try and wash the oil from the shore with cold, warm, and hot (60°C) water washes, sometimes with high-pressure hoses and dispersants. This washing only removed 20–25% of the oil and caused more damage than the oil itself. Finally slow-release fertilizer was applied to many of the beaches and this proved to be very effective for surface and near-surface oil removal. The reason for using fertilizer is that crude oil is a naturally occurring mixture of compounds and therefore there are micro-organisms in the environment that can degrade it. If the micro-organisms are well supplied with air the process of degradation can be quite rapid. In situations where conditions are regarded as 'clean' the levels of phosphate and nitrogen are very low and thus growth-limiting. It was considered that this was the case in Alaska and two fertilizers, one oleophilic (Inipol EAP 22®) and one slow release (Customblen®), were added. Both fertilizers proved successful in increasing the rate of oil degradation (Prince, 1997; Atlas, 1995); Table 11.4 shows greater microbial activity in treated gravel.

In the case of the *Exxon Valdez* the hydrocarbon levels in sediments at oiled sites reached unoiled levels after 3–4 years (Fig. 11.3). However, if a thick layer of oil is not removed from the shore by human activity or storms then anaerobic

Table 11.4 Bacterial activity in beach material from the *Exxon Valdez* oil spill

Material	Hydrocarbon concentration (mg/kg)	$^{14}CO_2$ produced from naphthalene (ng/l/h)	Cell numbers ($\times 10^6$/ml)
Clean			
Gravel	0.4	0.2	
Gravel washwater		<0.01	0.13
Sediment	7.8	0.17	
Sediment washwater		0.01	2.5
Oil-polluted			
Cobble	2 080	94.6	
Cobble washwater		30.2	5.1
Gravel	720	723.0	
Gravel washwater		3.5	0.4
Oil-polluted and treated*			
Gravel	640	600	
Gravel washwater		1.3	0.1

Source: Adapted from Button et al. (1992).
* Treated with oleophilic fertilizer.
Note: The amount of radiolabelled carbon dioxide released from radiolabelled naphthalene is an indication of the extent of naphthalene mineralization.

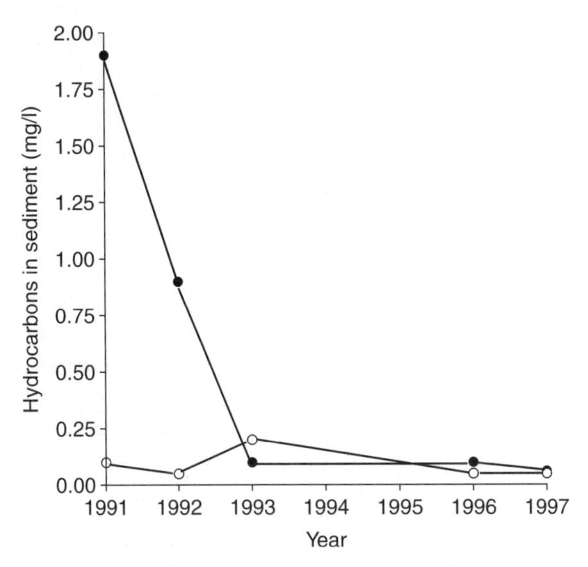

Figure 11.3 The changes in hydrocarbon levels in sediments from Prince William Sound after the *Exxon Valdez* oil spill. o, Control unoiled; •, oiled. From Kingston (2002).

conditions will develop, which slows degradation, and the oil may persist for many years. Thus the natural degradation processes were the most effective methods of oil removal and the least destructive to the indigenous population. In addition, the winter storms of 1989–90 removed about 90% of the oil from surface sediments, although the oil still remained deep in the sand and sediments (Laws, 2000).

The microbial population in terms of cell size and activity against toluene was determined in control and oil-polluted sites (Button et al., 1992; Table 11.4) Significantly large numbers of hydrocarbon-degrading micro-organisms were found at sites where the oil had polluted the shore. In clean conditions these may make up 0.1% of the population but in contaminated sites this can be up to 100%. In general no single strain is capable of degrading all the components of crude oil and biodegradation of crude oil involves a series of reactions within a microbial consortium. The effectiveness of adding of specific micro-organisms (bioaugmentation) was tried following the *Exxon Valdez* spill. Field trials failed to show enhanced degradation in the bioaugmented plots and biostimulated plots (Venosa and Zhu, 2003) although biostimulation has been shown to be successful in experimental plots.

The *Exxon Valdez* oil spill did result in a considerable impact on the plant and animal populations in the areas that the oil reached, but within 3–4 years most of the population had recovered. Only a little oil, 14%, was removed and the main biodegradation was carried out by the indigenous microbial population. The *Exxon Valdez* spill is perhaps the most-studied event of its kind and data obtained have been used to develop bioremediation strategies. In addition after the spill the US Congress passed OPA90, which was intended to reduce the probability of spills, allow containment and clean up, and establish liability for the clean-up costs (Burlington, 1999).

11.3 Acid mine drainage

Acidic metal-containing waters are formed naturally by the chemical weathering of metal sulphide-containing rocks. Acidic streams and ponds are often found as thermal springs in volcanic areas such as Yellowstone National Park in the USA. However, mining operations also expose sulphide ores in mines, mine tailings, and other wastes and thus considerably increase acid-water production. The predominant metal deposits are sulphides and the most common is iron-containing pyrite (FeS_2), although pyrite ores can also contain other metals such as gold, silver, copper, zinc, and lead. On exposure to oxygen and water the sulphide deposits are oxidized by microbial activity producing very acid water. It has been estimated that 75% of the acidic mine drainage is due to activities of micro-organisms. Acid mine drainage has a pH of below 3 and contains high concentrations of copper, iron, zinc, lead, arsenic, and cadmium and therefore constitutes an environmental problem (Table 11.5). If the drainage runs into a more-alkaline stream the metals can precipitate. The decreased pH, increased dissolved metals, and metal precipitation have a drastic effect on the benthic invertebrate and algal abundance and diversity (DeNicola and Stapleton, 2002). The metal concentration in the populations in streams is determined by a number of factors such as transport across membranes, absorption to the surface of the organisms, and intake of particles. Metal speciation, pH, the site, previous exposure, and differences between species determine the degree of toxicity to the stream population. The ferric and ferrous forms of iron are often the major components of acidic mine drainage (Table 11.5) and the ferric ion imparts a red colour in low-pH waters and an orange-yellow colour at higher pH values. In addition to metals arsenic, which can be toxic, may also be present at high concentrations, derived from the mineral arsenopyrite (FeAsS). Acidic mine drainage from a copper mine when run into the sea has been shown to reduce the coverage of rockweed (*Fucus gardneri*), a seaweed that normally dominates the seashore in British Columbia (Marsden et al., 2003).

11.3.1 The reactions involved in production of acidic mine drainage

Metal sulphides can be oxidized in a non-biotic manner by ferric ions or molecular oxygen.

$$FeS_2 + 8H_2O + 14Fe^{3+} = 15Fe^{2+} + 2SO_4^{2-} + 16H^+ \qquad (11.1)$$

Table 11.5 Characteristics of acidic mine drainage

Parameter	Rio Tinto, Spain	Iron Mountain, CA, USA	Parys mine, Wales	Wheal Jane, England
pH	2.2	0.5–1.0	2.5	3.4
Fe (total)	2 300	13 000–19 000	650	290
Fe^{2+}	1 500	13 000–19 000	650	250
Al	–	1 400–6 700	70	27
Mo	–	17–120	10	8
Cu	109	120–650	40	1.2
Zn	225	700–2 600	60	132
Sulphate	10 000	20 000–108 000	1550	400

Note: Sulphate and metal concentrations are mg/l.
Source: Derived from Johnson and Hallberg (2003).

It is the bacteria that are responsible for recycling the ferric ions from the ferrous ions in what is known as indirect leaching (Chapter 8).

$$14Fe^{2+} + 3.5O_2 + 14H^+ + \text{bacteria} = 14Fe^{3+} + 7H_2O \quad (11.2)$$

In some cases the bacteria form thiosulphate as an intermediate in the oxidation of sulphides.

$$FeS_2 + 6Fe^{3+} + 3H_2O = 7Fe^{2+} + S_2O_3^{2-} + 6H^+ \quad (11.3)$$

$$S_2O_3^{2-} + 2O_2 + H_2O = 2H^+ + 2SO_4^{2-} \quad (11.4)$$

or

$$S_2O_3^{2-} + 8Fe^{3+} + 5H_2O = 8Fe^{2+} + 2SO_4^{2-} + 10H^+ \quad (11.5)$$

Direct leaching is a situation where the micro-organism is attached to the pyrite and oxidizes it as follows.

$$FeS_2 + H_2O + 3.5O_2 = Fe^{2+} + 2SO_4^{2-} + 2H^+ \quad (11.6)$$

In some cases the pyrite can be coated with elemental suphur which can be removed by microbial activity.

$$S^0 + 1.5O_2 + H_2O = SO_4^{2-} + 2H^+ \quad (11.7)$$

In all cases the microbial oxidation of sulphides yields products which reduce the pH of the water.

11.3.2 Micro-organisms

The acidic conditions that develop in mines, mine dumps, and tailings select for acidophiles, those cells capable of growth at pH values of less than 3.

DNA-based studies of acid habitats have shown that there is a greater diversity than was at first thought from using isolation and growth studies. Within the acidophiles a number of groups can be recognized. The conditions in which acidophiles develop can vary considerably. The conditions are often dark with little organic material present so that the majority of micro-organisms found are chemoautotrophs (chemolithotrophs) that obtain their carbon from inorganic sources, generally carbon dioxide, and their energy from inorganic compounds containing iron and sulphur. On the surface of spoil heaps and in streams light encourages the growth of photoautotrophs (see Fig. 2.17). The acid conditions can also vary considerably in temperature, affecting the micro-organisms found, from mesophiles (15–45°C) in heaps and dumps, to thermophiles (40–70°C) and extreme thermophiles (70°C and above) found in hot springs and pools. In this range both divisions of prokaryotes are found; Bacteria and Archaea, with the Archaea dominating at high temperatures (Table 11.6).

Another method of dividing the acidophiles is based on the source of energy; that is, predominantly iron and sulphur oxidation and reduction. In some cases organic material is present and under these conditions heterotophic micro-organisms can be found. Table 11.7 gives the cell concentrations of various metabolic types of micro-organism found in a number of acidic mine waters. The Rio Tinto mine drainage has a pH of 2.5 and has a high level of iron and sulphur oxidizers that are extreme acidophiles and thermophiles. In contrast, Bullhouse in England, with a pH of 5.9, has a lower concentration of iron and sulphur oxidizers and in addition has heterotrophic acidophiles.

Among the bacteria found in acid conditions are those in the divisions Proteobacteria, Nitrospira, Firmicutes, and Acidobacteria. In the β-/γ-Proteobacteria contain are *Acidithiobacillus* sp. (formally *Thiobacillus ferrooxidans* and *Thiobacillus caldus*) and *Thiobacillus* sp. The Nitrospira contain the *Leptospirillum ferrooxidans* group. The Firmicutes division contains the Actinobacteria group, including *Acidimicrobium ferrooxidans*, and also *Sulfobacillus* sp. The Archaea are represented by the Thermoplasmales and Sulfolobales and are iron and iron and sulphur oxidizers. In terms of iron and sulphur oxidation the most studied are *L. ferrooxidans, Acidithiobacillus ferrooxidans*, and *Thiobacillus thiooxidans*, and *T. ferrooxidans* was the first to be described. There have been a few reports of eukaryotes in relation to acidic mine drainage which included ciliates, flagellates, fungi, yeasts, and algae (Johnson, 1995).

11.3.3 Controlling acidic mine drainage

There are two approaches to the control of acidic mine drainage:

- stop the formation, and
- treat before release and prevent the drainage reaching streams.

Table 11.6 Acidophilic metal-leaching micro-organisms

Species	Temperature optimum (°C)	Phylogenetic division
Iron oxidizers		
Leptospirillum ferrooxidans	<40	Nitrospira
L. thermoferrooxidans	40–60	Nitrospira
Ferroplasma acidiphilum	<40	Thermoplasmales*
Sulphur oxidizers		
Acidithiobacillus thioooxidans†	<40	β-/γ-Proteobacteria
Acidithiobacillus caldus†	40–60	β-/γ-Proteobacteria
Thiomonas cuprina	<40	β-Proteobacteria
Metallosphaera sp.	>60	Sulfolobales*
Sulfolobus sp.	>60	Sulfolobales*
Iron and sulphur oxidizers		
Acidithiobacillus ferrooxidans†	<40	β-/γ-Proteobacteria
Acidanus sp.	>60	Sulfolobales*
Sulfolobus metallicus	>60	Sulfolobales*
Iron reducers		
Acidiphilium sp.	<40	α-Proteobacteria
Iron oxidizers/reducers		
Acidimicrobium ferrooxidans	<40	Actinobacteria
Iron oxidizers/reducers and sulphur oxidizers		
Sulfobacillus sp.	<40 and 40–60	Firmicutes
Obligate anaerobes		
Stygiolobus azoricus	>60	Sulfolobales*
Acidilobus aceticus	>60	Sulfolobales*

Source: Data derived from Johnson and Hallberg (2003).
* Archaea.
† Formally these are *Thiobacillus* spp.

The biological methods that can be used to stop the formation of acidic mine drainage involve the inhibition of the microbial activity responsible for the acidic conditions. The micro-organisms require the presence of water and oxygen and sealing of mines and covering of heaps is possible but the size of such an undertaking does present problems. Simple inhibitors are not really available although anionic detergents have been shown to be effective in stopping growth of acidophiles in the laboratory. Grazing by protozoa has been shown to reduce the level of acidophiles.

Table 11.7 Microbiological characteristics of acidic mine drainage

Micro-organism	Rio Tinto, Spain	Parys mine, Wales	Wheal Jane, England	Bullhouse, England
Iron oxidizers, moderate acidophiles	–	1.3×10^3	3.0×10^4	–
Iron oxidizers, extreme thermophiles	1.3×10^6	3.5×10^3	1.0×10^3	$<10^2$
Sulphur oxidizers, extreme acidophiles	6.1×10^6	$<10^2$	$<10^2$	$<10^2$
Heterotrophic acidophiles	–	2.0×10^3	3.0×10^2	$<10^2$

Source: Data derived Johnson and Hallberg (2003).
Note: Data are given in cells/ml.

Conventional acid mine treatment involves its neutralization by addition of limestone, lime, and sodium hydroxide which leads to the precipitation of the metals. This requires the construction of a plant including clarifiers and precipitators, which are expensive to construct and run. The metal-containing sludge also needs to be removed and treated. Therefore, there has been considerable interest in passive biological processes. These include:

- algae,
- activated sludge,
- microbial treatment with anaerobic reducing bacteria,
- permeable, reactive barriers, and
- constructed wetlands.

11.3.3.1 *Algae*

Algae are known to adsorb and accumulate metals (Kratochvil and Voleski, 1998). Algae are effective in removing metals from acidic mine drainage over a short period but continued accumulation has proved to be toxic. However, a system has been developed which grows the algae separately before being exposed to the high-metal effluent. The algal growth bioreactors are simple, low-cost vessels (Van Hille et al., 1999) and the system is shown in Fig. 11.4. *Spirulina* sp. is cultivated using saline wastes in a high-rate algal pond of 2500 m^2 and the alkaline water from this (125 m^3) can treat 125 m^3 of mine waste daily.

11.3.3.2 *Activated sludge*

There are considerable amounts of excess activated sludge generated during normal sewage treatment and this excess has been used to adsorb metals from acid mine drainage. In one case the sludge was killed by heating at 105°C for 12 h and then used to pack a column for the removal of zinc and copper (Utgikar et al., 2000). Other non-viable bioadsorbants have included algae, peat, bacteria, and plants.

11.3.3.3 *Sulphate-reducing bacteria*

One alternative to chemical treatment is to use anaerobic bacterial sulphate reduction to convert metal sulphates to insoluble sulphides. The reactions (unbalanced) are as follows.

$$SO_4^{2-} + \text{organic acid} = H_2S + HCO_3^- \quad (11.8)$$
$$HCO_3^- = CO_2 + H_2O \quad (11.9)$$
$$7Cu^{2+} + 4H_2S = Cu_7S_4 \text{ (insoluble)} \quad (11.10)$$
$$Fe^{3+} + H_2S = FeS + 2H^+ \quad (11.11)$$

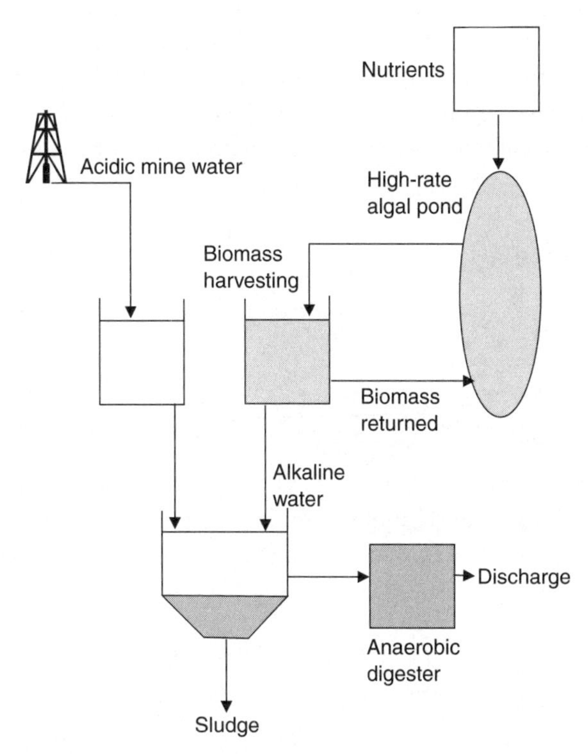

Figure 11.4 A process for using algae (*Spirulina* sp.) grown in a high-rate pond to reduce the acidity of acid mine drainage. From Van Hille et al. (1999).

The bicarbonate formation increases the pH of the medium and at pH of 6–7 many metals precipitate as hydroxides.

$$Al^{3+} + 3H_2O = Al(OH)_3 + 3H^+ \tag{11.12}$$

Bacteria were isolated from the sludge of a pyritic tailing pond in Spain and were shown to remove copper and iron provided that the pH was not less than 4.0. The pH rose over 17 days to 8.2 and in 21 days 99% of the copper and 99% of the iron was removed (Garcia et al., 2001). Figure 11.5 shows the changes in pH and removal of copper. A similar system has been used to clean up coal-mine and smelting-residue-dump leachates (Dvorak et al., 1992). In this case the sulphate system requires the following conditions:

- anaerobic conditions,
- a carbon source, generally a simple organic acid like formic acid,
- pH above 4.0, and
- some method of retaining the sulphate-reducing bacteria such as compost.

11.3.3.4 *Permeable, reactive barrier*

The principle of the barrier is to place a thick permeable layer in the groundwater flow so that the pollution has to pass through the barrier. The barrier is constructed of materials that will immobilize and feed bacteria capable of removing the pollutant (Chapter 5). These types of barrier have been used to prevent acidic mine drainage entering streams and lakes. One large example of a permeable, reactive barrier is the treatment of the acid drainage from

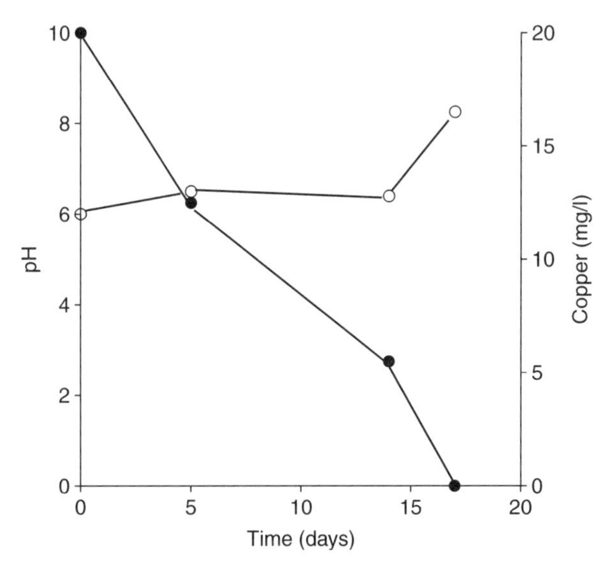

Figure 11.5 The effect of sulphate-reducing bacteria on the pH and copper content of acid mine drainage. o, pH; •, copper (mg/l). From Garcia et al. (2001).

the Shilbottle coal mine in Northumberland, England. Initially three artificial reedbeds were constructed to treat the drainage from the spoil heaps but these were not able to remove all the contamination due to channeling. The drainage has a pH of 3.5, contains 1000 mg of iron and 3000 mg of manganese/l, and where it has entered the nearest steam, the Tylaw Burn, it has turned it bright orange. Therefore, a large permeable reactive barrier has been constructed; 180 m long, 2 m wide, and 2–3 m deep. The barrier consists of 50% limestone chips, 25% green-waste compost, and 25% manure. In the anaerobic conditions of the barrier sulphate-reducing bacteria found in the manure and compost should thrive and the compost and manure should also supply the organic compounds required for growth. The reactions shown in equations 11.8–11.11 should precipitate the iron and manganese as suphides and increase the pH of the exit stream.

11.4 Wheal Jane

The closure of coal and metal mines over the years has left potential pollution as these mines flood and contaminate the surrounding area. One of the most publicized occurrences of acidic mine drainage was the release from the Wheal Jane mine in England in 1992. The Wheal Jane mine was a tin mine which was sunk some 40 years ago in what was known as a wet area, and during its operation, like most tin mines, it required continuous removal of water. Due to the competition of cheaper tin from abroad the mine was closed in February 1991, the pumps were removed, and the mine and tailing dam filled with acid water. In January 1992 the build-up was sufficient to cause the Nangiles adit (an adit is a horizontal passage leading into a mine) to burst and the acid water was released into the Carnon river, which runs into Restronguet Creek and the sea at Falmouth. The release of such a large quantity of metal-containing water caused considerable concern as the area polluted was a tourist area and contained oyster beds. The release generated an orange-brown colour in the downstream area over 6.5×10^6 m^2, reaching as far as Falmouth docks. The conspicuous nature of the release attracted worldwide attention. However, studies have concluded that there was no significant difference in heavy metals in the sediments of the Carnon river, Restronguet Creek, and Carrick Roads (a water way for ships that leads to the sea; Somerfield et al., 1994). A second study found elevated levels of arsenite (9 μg of As/l) in the saline region of the Carnon river but a year later this had dropped significantly although it remained unchanged in Restronguet Creek (Hunt and Howard, 1994). The pumps were reinstated and the effluent was treated in a tailing dam while other forms of remediation were installed.

The company Scott Wilson Piesold (formally Knight Piesold) on behalf of the Environment Agency installed both active and passive treatment systems. Passive pilot plant systems were constructed in 1994 and an active system was commissioned in 2000.

The active system was designed to treat all the mine water and was based on the precipitation of metals by raising the pH by the addition of lime. An outline of the system is given in Fig. 11.6. Water from the mine and tailings pond is pumped at 330 l/s into the first-stage tank. In this tank the water is mixed with recirculated sludge (at ratios of 25:1 and 50:1 water to sludge), which raises the pH to 8.5 from 3.9. The mixture is run into a second-stage aeration tank where 5% lime slurry is added, raising the pH to 9.25. The residence time in the first two tanks is about 30 min. A polymer flocculant (2.5 g/l) is added to the mixture and this is run into two clarifiers where the solids settle out. The sludge is removed to a holding tank or recirculated to the first stage. The sludge in the holding tank is 30–40% solids and is run into the tailings pond. The supernatant from the clarifiers is run into a holding tank and discharged into the Carnon river. The capacity of the system is 440 l/s and consent had been given for 350 l/s.

The passive treatment system was divided into three separate streams where the wastewater was treated in three different manners (Fig. 11.7). In the treatment systems chemical and biological remediation are both used. In the first system the wastewater is run into a small anoxic (anaerobic) tank followed by an anoxic limestone drain. The purpose of the initial anaerobic tank is to remove oxygen so that Fe^{3+} production is reduced so that this will not coat the limestone in the drain. The limestone drain is there to increase the pH of the wastewater before it is run into the aerobic tanks (wetlands). One consequence of the rise in pH (6.0) is the precipitation of aluminium in the

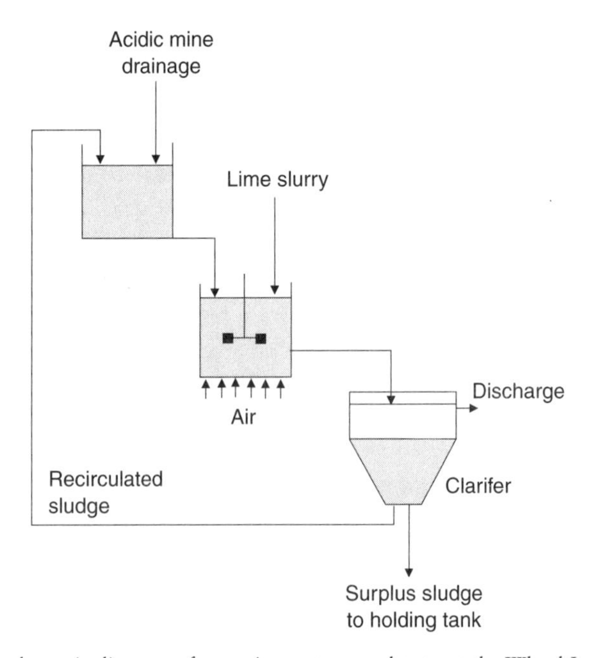

Figure 11.6 A schematic diagram of an active system used to treat the Wheal Jane mine drainage.

water as aluminium hydroxide. The aluminium hydroxide blocked the system and because of this the limestone drain was replaced with a lime-dosing tank. The water from the limestone drain was run into the biological part of the system that consisted of five artificial wetlands. These were constructed from fine mine tailings and were planted with a equal mixture of the common reed (*Phragmites* sp.) and catail (*Typha* sp.) with the addition of 100 bulrush plants (*Scirpus* sp). The first wetland had an area of 224 m^2 and this removed about 90% of the Fe^{2+} present at a removal rate of 7.6 g/m^2 per day. From the wetlands the water is passed into an anaerobic cell which is an underground chamber filled with a mixture of straw, sawdust, and manure. The straw and sawdust provide a support for the immobilization of sulphate-reducing bacteria and the manure is a source of organic carbon for the bacteria. However, these reactors did not function as expected as the effluent contained more Fe^{2+} than the feed stream. The reason for this appeared to be that solid suspended Fe^{3+} was entering the cell, and at the pH of 5.5 was reduced to Fe^{2+}, which would appear in the effluent. In another system the anaerobic cell

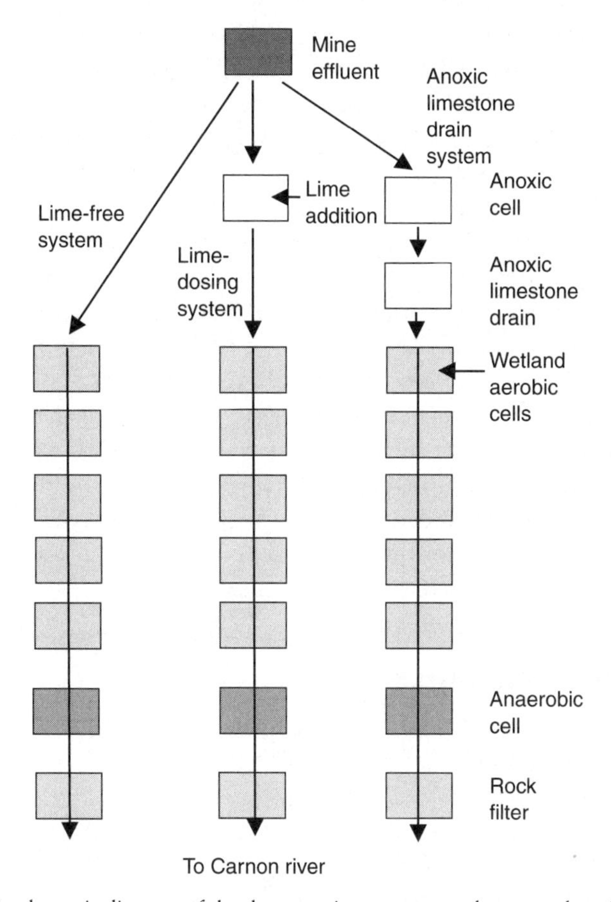

Figure 11.7 A schematic diagram of the three passive systems used to treat the mine drainage from Wheal Jane.

was sealed for 12 months for repairs but when the flow was resumed this cell functioned as expected with a pH of 6–7 and the concentrations of sulphide, Fe, and Zn were essentially zero. The shutdown clearly allowed the reactor to adapt to the conditions and the population of sulphate-reducing bacteria to increase. After the anaerobic cell the wastewater is run through a rock filter. This consists of a series of shallow pools containing small granite pebbles in order to encourage the growth of algae. The fixation of carbon dioxide by the algae would raise the pH of the water above 8, which would precipitate manganese. However, the low pH of the water from the anaerobic cell inhibited the growth of algae, which was further decreased by the oxidation of Fe^{2+}. When the anaerobic cell operated correctly algal growth was good and the pH reached 7–8 and manganese was precipitated.

In the second system the mine water was run into a tank and lime added to bring the pH up to 4.5 from 3.9. The next stage consisted of five aerobic cells (artificial wetlands) constructed in the same manner as the first system. The first cell of 187 m^2 removed some 52% of the Fe^{2+} at a rate of 4 g/m^2 per day. In the second cell of 240 m^2 the concentration of Fe^{2+} was further reduced and the two cells removed 90% of the Fe^{2+}. The effluent from the wetlands was run into an anaerobic cell and then a rock filter as for the first system. The same problems occurred with these as for the first system.

The third system had no lime dosing but the mine water was run directly into the aerobic cells (wetlands), which were somewhat larger (817.5 m^2). The first cell removed on average 77% of the iron at a rate of 5.8 g/m^2 per day. The oxidation of Fe^{2+} to Fe^{3+} also catalysed the conversion of As^{3+} to As^{5+}, causing it to precipitate. After the wetlands (aerobic cells) the effluent was run through an anaerobic cell and then a rock filter as in the other two systems. In this system the anaerobic cell was shut down and sealed for 12 months and when restarted it contained sulphate-reducing bacteria. The pH of the effluent from the anaerobic cell was between 6 and 7 and zinc and iron sulphides were below detection levels. The high pH of the water running into the rock filters gave good algal growth, probably encouraged by the organic material washed from the cell. It was concluded that the last system, which was wholly biological, was the best system for the remediation of the mine drainage water.

11.5 Further reading

Brown, M., Barley, B., and Wood, H. (2000) *Mine Water Treatment-Technology Application and Policy.* IWA Publishing, London.

Laws, E.A. (2000) *Aquatic Pollution.* John Wiley & Sons, New York.

Younger, P.L., Banwart, S.A., and Hedin, R.S. (2002) *Mine Water: Hydrology, Pollution, Remediation.* Kluwer Academic Publishers, Dordrecht.

Exxon Valdez web site: www.noaa.gov

Wheal Jane web site: www.rdg.ac.uk/whealjane

References

Agate, A.D. (1996) Recent advances in microbial mining. *World J. Micro. Biotechnol.* **12**, 487–495.

Akcil, A. and Mudder, T. (2003) Microbial destruction of cyanide wastes in gold mining: process review. *Biotechnol. Letters* **25**, 445–450.

Aldrett, S., Bonner, J.S., Mills, M.A., Autenrieth, R.L., and Stephens, F.L. (1997) Microbial degradation of crude oil in marine environments tested in a flask experiment. *Water Research* **31**, 2840–2848.

Ali, Y., Hanna, M.A., and Leviticus, L.I. (1995) Emissions and power characteristics of diesel engines on methyl soyate and diesel fuel blends. *Bioresource Technology* **52**, 185–195.

Allard, A.-S. and Neilson, A.H. (1997) Bioremediation of organic waste sites: a critical review of microbiological aspects. *Int. Biodeterioration Biodegradation* **39**, 253–285.

Amann, R. and Ludwig, W. (2000) Ribosomal RNA-targeted nucleic acid probes for studies in microbial ecology. *FEMS Microbiology Reviews* **24**, 555–565.

Anderson, J.G. and Smith (1987) Composting. In Sidwick, J.M. and Holdom, R.S. (eds) *Biotechnology of Waste Treatment and Exploitation*. Ellis Horwood, Chichester, UK.

Andrade, L., Gonzalez, A.M., Araujo, F.V., and Paranthos, R. (2003) Flow cytometry assessment of bacterioplankton in tropical marine environments. *J. Microbiological Methods* **55**, 841–850.

Anonymous (1997) World oil supply and demand. *Euro Energy Information Newsletter*, **15**.

Ansola, G., Gonzalez, J.M., Cortijo, R., and de Luis, E. (2003) Experimental and full-scale pilot plant constructed wetlands for municipal wastewaters treatment. *Ecological Engineering* **21**, 43–52.

Antolin, G., Tinaut, F.V., Briceno, Catano, V., Perez, C., and Ramerirez, A.I. (2002) Optimization of biodiesel production by sunflower oil transesterification. *Bioresource Technology* **83**, 11–114.

Arienzo, M., Adamo, P., and Cozzolino, V. (2004) The potential of *Lolium perenne* for revegetation of contaminated soil from a metallurgical site. *The Science of the Total Environment* **319**, 13–25.

Atagana, H.I. (2003) Bioremediation of creosote-contaminated soil: a pilot-scale landfarming evaluation. *World J. Micro. & Biotechnol.* **19**, 571–581.

Atlas, R.M. (1995) Bioremediation of petroleum pollutants. *International Biodeterioration Biodegradation* **30**, 317–327.

Attewell, P. (1993) *Ground Pollution*. E & FN Spon, London.

Babich, I.V. and Moulijn, J.A. (2003) Science and technology of novel processes for deep desulfurization of oil refinery streams: a review. *Fuel* **82**, 607–631.

Baker, B.J. and Banfield, J.F. (2003) Microbial communities in acid mine drainage. *FEMS Microbiology Ecology* **44**, 139–152.

Baker, J.A., Entsch, B., Neilan, B.A., and McKay, D.B. (2002) Monitoring changing toxigenicity of a cyanobacterial bloom by molecular methods. *Appl. Environ. Microbiol.* **68**, 6070–6076.

Bakken, L.R. and Lindahl, V. (1995) Recovery of bacterial cells from soil. In *Nucleic Acids in the Environment* (eds Trevors, J.T., van Elsas, J.D.) pp 9–27, Springer-Verlag, Heidelberg.

Ban, K., Kaieda, M., Matsumoto, T., Kondo, A., and Fukuda, H. (2001) Whole cell biocatalyst for biodiesel production utilizing *Rhizopus oryzae* cells immobilized within biomass support particles. *Biochemical Engineering J.* **8**, 39–43.

Belarbi, E.H., Gomez, A.C., Chisti, Y., Camacho, F.G., and Grima, E.M. (2003) Producing drugs from marine sponges. *Biotechnology Advances* **21**, 585–598.

Belghith, H., Ellouz-Chaabouni, S., and Gargouri, A. (2001) Biostoning of denims by *Penicillium occitanis* (Pol6) cellulases. *J. Biotechnology* **89**, 257–262.

Benedik, M.J., Gibbs, P.R., Riddle, R.R., and Willson, R.C. (1998) Microbial denitrogenation of fossil fuels. *TIBTECH* **16**, 390–395.

Berg, C. (2001) World ethanol production 2001. **www.distill.com**

Biddlestone, A.J., Gray, K.R., and Job, G.D. (1991) Treatment of dairy farm wastewater in engineered reed bed systems. *Process Biochem.* **26**, 265–268.

Birkland, T.A. and Lawrence, R.G. (2002) The social and political meaning of the *Exxon Valdez* oil spill. *Spill Science Technology Bulletin* **7**, 17–22.

Bizily, S.P., Rugh, C.L., and Meagher, R.B. (2000) Phytodetoxification of hazardous organomercurials by genetically engineered plants. *Nature Biotechnology* **18**, 213–217.

Black, J.G. (2002) *Microbiology Principles and Explorations.* 5th Edition, John Wiley & Sons, Inc., New York.

Bossert, I.D. and Compeau, G.C. (1995) Cleanup of petroleum hydrocarbon contamination in soil, in Young, L.Y. and Cerniglia, C.E. eds. *Microbial Transformation and Degradation of Toxic Organic Chemicals*, pp 77–125, Wiley-Liss, New York.

Bottger, E.C. (1996) Approaches for identification of microorganisms, *ASM News*, **62**, 247–250.

Boyajian, G.E. and Carreira, L.H. (1997) Phytoremediation: A clean transition from laboratory to marketplace? *Nature Biotechnology*, **15**, 127–128.

Boyle, G. (1996) *Renewable Energy: Power for a sustainable future.* Oxford University Press, Oxford.

Braun, R. (2002) People's concerns about biotechnology: some problems and some solutions. *J. Biotechnology* **98**, 3–8.

Brindle, K. and Stephenson, T. (1996) The application of membrane biological reactors for the treatment of wastewaters. *Biotechnol. Bioeng.* **49**, 601–610.

British Standards Institute (1988) *Code of practice for the identification of contaminated land and its investigation* DD175, HMSO, London.

Bruno, E. and Eklund, B. (2003) Two new growth inhibition tests with filiamentous algae *Ceramium strictum* and *C. tenuicorne* (Rhodophyta). *Environmental Pollution* **125**, 287–293.

Bryner, L.C., Beck, J.F., Davis, B.D., and Wilson, D.G. (1954) Microorganisms in leaching sulfide minerals. *Ind. Eng. Chem.* **46**, 2587–2592.

Bull, A.T., Bunch, A.W., and Robinson, G.K. (1999) Biocatalysts for clean industrial products and processes. *Current Opinion Microbiology* **2**, 246–251.

Burgess, J.E., Parson, S.A., and Stuetz, R.M. (2003) Developments in odour control and waste gas treatment biotechnology; a review. *Biotechnology Advances* **19**, 35–63.

Burlington, L.B. (1999) Ten year historical perspective of the NOAA damage assessment and restoration program. *Spill Science Technology Bulletin* **5**, 109–116.

Button, D.K., Robertson, B.R., McIntosh, D., and Juttner, F. (1992) Interactions between marine bacteria and dissolved-phase and beached hydrocarbons after the *Exxon Valdez* oil spill. *Appl. Environmental Microbiology* **58**, 243–251.

Cainey, T. (1987) *Reclaiming Contaminated Land.* Blackie, London.

Calvin, M. (1985) Fuel oils from plants. In Fuller, K.W. and Gallon J.R. (eds) *Plant products and the New Technology*, Oxford, Clarendon Press, pp 147–160.

Candido, E.P.M. and Jones, D. (1996) Transgenic *Caenorhabditis elegans* strains as biosensors. *Tibtech.* **14**, 125–129.

Cannell, M.G.R. (2003) Carbon sequestration and biomass energy offset: theoretical, potential and achievable capacities globally, in Europe and the UK. *Biomass & Bioenergy* **24**, 97–116.

Carls, M.G., Babcock, M.M., Harris, P.M., Irvine, G.V., Cusick, J.A., and Rice, S.D. (2001) Persistence of oiling in mussel beds after the *Exxon Valdez* oil spill. *Marine Environmental Research* **51**, 167–190.

Castanier, L.M. and Brigham, W.E. (2003) Upgrading of crude oil via in situ combustion. *J. Petroleum Science & Engineering* **39**, 125–136.

Cerniglia, C.E. (1993) Biodegradation of polycyclic aromatic hydrocarbons. *Current Opinion in Biotechnology* **4**, 331–338.

Chaney, R.L., Malik, M., Li, Y.M., Brown, S.L., Brewer, E.P., Angel, J.S., and Baker, A.J.M. (1997) Phytoremediation of soil metals. *Current Opinion in Biotechnology* **8**, 279–284.

Chang, I.S., Jang, J.K., Gil, G.C., Kim, M., Kim, H.J., Cho, B.W., and Kim, B.H. (2004) Continuous determination of biochemical oxygen demand using microbial fuel cell type biosensor. *Biosensors Bioelectronics* **19**, 607–613.

Chapman, D.T. and Gellenbach, K.W. (1989) Products from algae. In: *Algal and Cyanobacterial Biotechnology* (eds Cresswell, RC., Rees, T.A.V. and Shah, N.). Longman Science and Technology, Harlow, UK.

Chung, C.-H., Kwon, O.-C., Yi, Y.-B., and Lee, S.-Y. (1998) Isolation of quality genomic DNA from tenacious seeds of sesame. *Plant Tissue Cult. & Biotechnol.*, **4**, 42–48.

Churchill, P.F., Dudley, R.J., and Churchill, S.A. (1995) Surfactant-enhanced bioremediation. *Waste Management* **15**, 371–377.

Clausen, G.B., Larsen, L., Johnsen, K., de Lipthay, J.R., and Aamand, J. (2002) Quantification of the atrazine-degrading Pseudomonas sp. Strain ADP in aquifer sediment by quantitative competitive polymerase chain reaction. *FEMS Microbiology Ecology* **41**, 221–229.

Cleuvers, M. and Ratte, H.-T. (2002) Phytotoxicity of coloured substances: is *Lemna* Duckweed an alternative to the algal growth inhibition test? *Chemosphere* **49**, 9–15.

Colls, J. (1997) *Air Pollution An Introduction.* E & FN Spon, London.

Comber, S.D.W., Gardner, M.J., and Gunn, A.M. (1996) Measurement of absorbance and fluorescence as potential alternatives to BOD. *Environ. Tech.* **17**, 771–776.

Conti, M.E. and Botre, F. (2001) Honeybees and their products as potential bioindicators of heavy metals contamination. *Environmental Monitoring & Assessment* **69**, 267–282.

Conti, M.E. and Cecchetti, G. (2001) Biological monitoring: lichens as bioindicators of air pollution assessment – a review. *Environmental Pollution* **114**, 471–492.

Cook, J. and Beyea, J. (2000) Bioenergy in the United States: progress and possibilities. *Biomass Bioenergy* **18**, 441–455.

Cort, T.L., Song, M.-S., and Bielefeldt, A.R. (2002) Nonionic surfactant effects on pentachlorophenol biodegradation. *Water Research* **36**, 1253–1263.

Costerton, J.W., Lewandowski, Z., Caldwell, D.E., Korber, D.R., and Lappin-Scott, H.M. (1995) Microbial biofilms *Annu. Rev. Microbiol.* **49**, 711–745.

Cowan, D.A. (1997) The marine biosphere: a global resource for biotechnology. *TIBTECH* **15**, 129–131.

Cox, H.H.J., Moerman, R.E., VanBaalen, S., VanHeiningen, W.N.M., Doddema, H.J., and Harder, W. (1997) Performance of a styrene-degrading biofilter containing the yeast *Exophiala jeanselmei. Biotechnol. Bioeng.* **53**, 259–266.

Craggs, R.J., McAuley, P.J., and Smith, V.J. (1997) Wastewater nutrient removal by marine microalgae grown on a corrugated raceway. *Water Research* **31**,1701–1707.

Crueger, W. and Crueger, A. (1990) *Biotechnology: A Textbook of Industrial Microbiology*. Sinauer Associates Inc., Sunderland, MA, USA.

Crundwell, F.K. (2003) How do bacteria interact with minerals? *Hydrometallurgy* **71**, 75–81.

Curtin, M.E. (1983) Microbial mining and metal recovery, *Biotechnology* **1**, 228–235.

Daniell, H., Datta, R., Varma, S., Gray, S., and Lee, S.-B. (1998) Containment of herbicide through genetic engineering of the chloroplast genome. *Nature Biotechnol.* **16**, 345–348.

Daugulis, A.J. and Boudreau, N.G. (2003) Removal and destruction of high concentrations of gaseous toluene in a two-phase partitioning bioreactor by *Alcaligenes xylosoxidans. Biotechnology Letters* **25**, 1421–1424.

Davey, H.M. and Kell, D.B. (1996) Flow cytometry and cell sorting of heterogenous microbial populations: the importance of single cell analyses. *Microbiology Reviews* **60**, 641–696.

De Long, E. (1997) Marine microbial diversity: the tip of the iceberg. *TIBTECH* **15**, 203–207.

De Paula, G.O. and Cavalcanti, R.N. (2000) Ethics: essence for sustainability. *J. Cleaner Production* **8**, 109–117.

De Rore, H., Demoulder, K., DeWilde, K., Top, E.M., Houven, F., and Venstraete, W. (1994) Transfer of the catabolic plasmid RP4:Tn371 to indigenous soil bacteria and its effect on respiration and biphenyl breakdown. *FEMS Micobiology Ecology* **15**, 71–81.

De Wulf-Durand, P., Bryany, L.J., and Sly, L.I. (1997) PCR-mediated detection of acidophilic bioleaching-associated bacteria. *Appl. Environ. Micro.* **63**, 2944–2948.

Dempsey, D.A., Silva, H., and Klessig, D.F. (1998) Engineering disease and pest resistance in plants. *Trends in Microbiology* **6**, 54–60.

DeNicola, D.M. and Stapleton, M.G. (2002) Impact of acid mine drainage on benthic communities in streams: the relative roles of substratum vs aqueous effects. *Environmental Pollution* **119**, 303–315.

Denizen, M.J. and Turner, A.P.F. (1995) Biosensors for environmental monitoring *Biotec. Advances* **13**, 1–12.

Department of Trade and Industry (2003) *Our energy future – creating a low carbon economy*. The Stationery Office, London.

DeSouza, M.L., Newcombe, D., Alvey, S., Crowley, D.E., Hay, A., Sadowsky, M.J., and Wackett, L.P. (1998) Molecular basis of a bacterial consortium: Interspecies catabolism of atrazine. *Appl. Environ. Micro.* **64**, 178–184.

Diels, L., Spaans, P.H., VanRoy, S., Hooyberghs, L., Ryngaert, A., Wouters, H., Walter, E., Winters, J., Macaskie, L., Finlay, J., Perfuss, B., Woebking, H., Pumpel, T., and Tsezos, M. (2003) Heavy metals removal by sand filters inoculated with metal sorbing and precipitating bacteria. *Hydrometallurgy* **71**, 235–241.

Doty, S.L., Shang, T.Q., Wilson, A.M., Tabgen, J., Westergreen, A.D., Newman, L.A., Strand, S.E., and Gordon, M.P. (2000) Enhanced metabolism of halogenated hydrocarbons in transgenic plants containing mammalian cytochrome P450 2E1. *Proc. Natl. Acad. Sci.* **97**, 6287–6291.

Durre, P. (1998) New insights and novel developments in clostridial acetone/butanol/isopropanol fermentation. *Appl. Microbiol. Biotechnol.* **49**, 639–648.

Durre, P. and Bahl, H. (1995) Microbial production of acetone/butanol/isopropanol. In M.Roehr (ed) *Biotechnology*, 2nd edn. VCH, Weinheim. pp 229–268.

Durre, P., Fischer, R.-J., Kuhn, A., Lorenz, K., Schreiber, W., Sturzenhofecker, B., Ullmann, S., Winzer, K., and Sauer, U. (1995) Solventogenic enzymes of *Clostridium acetobutylicum*: catalytic properties, genetic organisation and transcriptional regulation. *FEMS Microbiology Reviews* **17**, 251–262.

Dvorak, D.H., Hedin, R.S., Edenborn, H.M., and McIntire, P.E. (1992) Treatment of metal-contaminated water using bacterial sulfate reduction: Results from pilot-scale reactors. *Biotechnology Bioengineering* **40**, 609–616.

Eichler, J. (2001) Biotechnological uses of archaeal extremozymes. *Biotechnology Advances* **19**, 261–278.

Eklund, B.T. and Kautsky, L. (2003) Review on toxicity testing with marine macroalgae and the need for method standardization – exemplified with copper and phenol. *Marine Pollution Bulletin* **46**, 172–181.

Energy Information Administration (2003) Annual energy outlook 2003. US Dept. Energy, Washington, DC. (**www.eia.doe.gov**)

Energy Information Administration (EIA) (2003) *Annual Energy Outlook 2003*.

Environmental Protection Agency (EPA) (1971) Algal assay procedure bottle test. *National Eutrophication Research Program*, Corvallis, Oregon, USA.

Erb, R.W., Eichner, C.A., WagnerDobler, I., and Timmis, K.N. (1997) Bioprotection of microbial communities from toxic phenol mixtures by genetically designed pseudomonad. *Nature Biotechnology* **15**, 372–382.

Escobar, B., Huenupi, E., and Wiertz, J.V. (1997) Chemical and biological leaching of enargite. *Biotechnol. Letters* **19**, 719–722.

European Federation of Biotechnology (1982) Definition of Biotechnology, *EFB Newsletter*, September, Nr. **5**, p 2.

European Foundation for the Improvement of Living & Working Conditions (2000) Design for sustainable enterprise crops for sustainable enterprise. Dublin, Ireland. (**www.eurofound.ie**)

Evans, J. (1999) Call out the reserves. *Chemistry in Britain*, Aug. 38–41.

Evans, J. (2000) Power to the people. *Chemistry in Britain*, August, 30–33.

Farouq Ali, S.M. and Thomas, S. (1996) The promise and problems of enhanced oil recovery methods. *J. Canadian Petroleum Tech.* **35**,57–63.

Ferrer, M., Golyshin, P., and Timmis, K.N. (2003) Novel maltotriose esters enhance biodegradation of Aroclor 1242 by *Burkholderia cepacia* LB400. *World J. Micro. & Biotech.* **19**, 637–643.

Fingas, M. (1995) Oil spills and their cleanup. *Chemistry and Industry*, December, 1005–1008.

Fox, J.L. (1997) Farmers say Monsanto's engineered cotton drops bolls. *Nature Biotechnol.* **15**, 1233.

French, C.E., Rosser, S.J., Davies, G.J., Nicklin, S., and Bruce, N.C. (1999) Biodegradation of explosives by transgenic plants expressing pentaerythritol tetranitrate reductase. *Nature Biotechnology* **17**, 491–494.

Frische, T. (2002) Screening for soil toxicity and mutagenicity using luminescent bacteria – a case study of the explosive 2,4,6-trinitrotoluene (TNT). *Ecotoxicity Environmental Safety* **51**, 133–144.

Frohlich, J. and Konig, H. (2000) New techniques for isolation of single prokaryotic cells. *FEMS Microbology Reviews* **24**, 567–572.

Fulkerson, W., Judkins, R.R., and Sanghvi, M.K. (1990) Energy from fossil fuels. *Scientific American*, Sept., 83–89.

Gadd, G.M. and White, C. (1993) Microbial treatment of metal pollution – a working biotechnology. *Trends in Biotechnology* **11**, 353–359.

Gadd, G.M. (2000) Bioremedial potential of microbial mechanisms of metal mobilization and immobilization. *Current Opinion in Biotechnology* **11**, 271–279.

Garbisu, C. and Alkarta, I. (2001) Phytoextraction: a cost-effective plant-based technology for the removal of metals from the environment. *Bioresource Technology* **77**, 229–236.

Garcia, C., Moreno, D.A., Ballester, A., Blazquez, M.L., and Gonzalez, F. (2001) Bioremediation of an industrial acid mine water by metal-tolerant sulphate-reducing bacteria. *Minerals Engineering* **14**, 997–1008.

Gentry, T.J., Josephson, K.L., and Pepper, I.L. (2004) Functional establishment of introduced chlorobenzoate degraders following bioaugmentation with newly activated soil. *Biodegradation* **15**, 67–75.

Gerday, C., Aittaleb, M., Bentahir, M., Chessa, J.-P., Claverie, P., Collins, T., D'Arnico, S., Durnont, J., Garsoux, G., Georiette, D., Hoyoux, A., Lonhienne, T., Meuwis, M.-A., and Feller, G. (2000) Cold-adapted enzymes: from fundamentals to biotechnology. *TIBTECH* **18**, 103–107.

Gerhardson, B. (2002) Biological substitutes for pesticides. *Trends in Biotechnology* **20**, 338–343.

Gestel, K.V., Mergert, J., Swings, J., Coosemans, J., and Ryckeboer, J. (2003) Bioremediation of diesel oil-contaminated soil by composting with biowaste. *Environmental Pollution* **125**, 361–368.

Ghirardi, M.L., Zhang, L., Lee, J.W., Fynn, T., Seibert, M., Greenbaum, E., and Melis, A. (2000) Microalgae: a green source of renewable H_2. *TIBTECH* **18**, 506–511.

Giesy, J.P., Hilscherova, K., Jones, P.D., Kannan, K., and Machala, M. (2002) Cell bioassays for detection of aryl hydrocarbon (AhR) and estrogen receptor (ER) mediated activity in environmental samples. *Marine Pollution Bulletin* **45**, 3–16.

Gisbert, C., Ros, R., De Haro, A., Walker, D.J., Bernal, M.P., Serrano, R., and Navarro-Avino, J. (2003) A plant genetically modified that accumulates Pb is especially promising for phytoremediation. *Biochem. Biophys. Res. Comm.* **203**, 440–445.

Glazer, A.N. and Nikaido, H. (1994) *Microbial Biotechnology*, Freeman & Co, New York.

Glick, B.R. (2003) Phytoremediation: synergistic use of plants and bacteria to clean up the environment. *Biotechnol. Advances* **21**, 383–393.

Goddijn, O.J.M. and Pen, J. (1995) Plants as bioreactors. *TIBTECH* **13**, 379–387.

Golub, E.S. (1997) Genetically enhanced food for thought. *Nature Biotechnology* **15**, 112.

Gomez, E., Ballester, A., Gonzalez, F., and Blazquez, M.L. (1999) Leaching capacity of a new extremely thermophilic microorganism, *Sulfolobus rivotincti. Hydrometallurgy* **52**, 349–366.

Graham, A. (1994) A haystack of needles: applying the polymerase chain reaction *Chemistry & Industry* Sept., 718–721.

Gray, D.J, Hiebert, E., Lin, C.M., Compton, M.E., McColley, D.W., Harrison, M., and Gaba, V.P. (1994) Simplified construction and performance of a device for particle bombardment. *Plant Cell Tissue Org.Cult.* **37**, 179–184.

Gray, K.A., Pogrebinsky, O.S., Mrachko, G.T., Xi, L., Monticello, D.J., and Squires, C.H. (1996) Molecular mechanisms of biocatalytic desulphurisation of fossil fuels. *Nature Biotechnology* **14**, 1705–1709.

Gray, N.F. (1989) *Biology of Waste Water Treatment*, Oxford University Press, Oxford.

Gunzel, B., Yonsel, S., and Deckwer, W.-D. (1991) Fermentative production of 1,3-propansdiol from glycerol by *Clostridium butyricum* up to a scale of 2 m^2. *Appl. Microbiol. Biotechnol.* **36**, 289–294.

Gupta, R., Beg, Q.K., and Lorenz, P. (2002) Bacterial alkaline proteases: molecular approaches and industrial applications. *Appl. Microbiol. Biotechnology* **59**, 15–32.

Gupta, R., Gigras, P., Mohapatra, H., Goswami, V.K., and Chauhan, B. (2003) Microbial α-amylase: a biotechnological perpective. *Process Biochem.* **38**, 1599–1616.

Hahn, M.E. (2002) Biomarkers and bioassays for detecting dioxin-like compounds in the marine environment. *The Science of the Total Environment* **289**, 49–69.

Haki, G.D. and Rakshit, S.K. (2003) Developments in industrially important thermostable enzymes: a review. *Bioresource Technology* **89**, 17–34.

Hall, J. and Crowther, S. (1998) Biotechnology: the ultimate cleaner production technology for agriculture? *J. Cleaner Production* **6**, 313–322.

Halliwell, B. and Aruoma, O.I. (1993) *DNA and Free Radicles.* Ellis Horwood Ltd., Chichester.

Hansen, G. and Wright, M.S. (1999) Recent advances in the transformation of plants. *Trends in Plant Science* **4**, 226–231.

Harmsen, H.J.M., Prieur, D., and Jeanthon, C. (1997) Distribution of microorganisms in deep-sea hydrothermal vent chimneys investigated by whole-cell hybridization and enrichment culture of thermophilic subpopulations. *Appl. Environ. Microbiol.* **63**, 2876–2883.

Harris, B. (1999) Exploiting antibody-based technologies to manage environmental pollution. *TIBTECH* **17**, 290–296.

Harvey, D. (1995) *Global Warming: The Hard Science.* Addison Wesley & Longmann, Harlow, UK.

Head, I.M. and Swannell, R.P.J. (1999) Bioremediation of petroleum hydrocarbon contaminants in marine habitats. *Current Opinion in Biotechnology* **10**, 234–239.

Hentschel, U., Schmid, M., Wagner, M., Fieseler, L., Gernert, C., and Hacker, J. (2003) Isolation and phylogenetic analysis of bacteria with antimicrobial activities from the Mediterranean sponges *Aplysina aerophoba* and *Aplysina cavernicola. FEMS Mirobiol. Ecology* **35**, 305–312.

Hoefel, D., Groogy, W.L., Monis, P.T., Andrews, S., and Saint, C.P. (2003) Enumeration of water-borne bacteria using viability assays and flow cytometry: a comparison to culture-based techniques. *J. Microbiological Methods* **55**, 585–597.

Holdgate, M.W. (1979) *A Perspective of Environmental Pollution*, Cambridge University Press, Cambridge.

Holliger, C. and Zehnder, A.J.B. (1996) Anaerobic biodegradation of hydrocarbons. *Current Opinion in Biotechnology* **7**, 326–330.

Holmberg, N. and Bulow, L. (1998) Improved stress tolerance in plants by gene transfer. *Trends in Plant Science* **3**, 61–66.

Hood, E.E. (2002) From green plants to industrial enzymes. *Enzyme Microbial Technology* **30**, 279–283.

Hood, E.E. and Jilka, J.M. (1999) Plant-based production of xenogenic proteins. *Current Opinion in Biotechnology* **10**, 382–386.

Houghton, J. (1996) Climate change calls for action now. *Chemistry & Industry*, March, 232.

Houghton, J.T., Jenkins, G.J., and Ephraums, J.J. (1990) *Climate Change*. Cambridge University Press, Cambridge.

Houghton, J.T., Ding, Y., Griggs, D.J., Noquer, M., van der Linden, P.J., and Xiaosu, D. (2001) Intergovernmental Panel on Climate Change (IPCC), *Climate Change 2001*. Cambridge University Press, Cambridge, **www.ipcc/ch/pub/reports**

Houwink, E.H. (1989) *Biotechnology: Controlled Use of Biological Information*, Dordrecht: Kluwer Academic Publishers.

Huber, H. and Stetter, K.O. (1998) Hyperthermophiles and their possible potential in biotechnology. *J. Biotechnology* **64**, 39–52.

Huber, R., Huber, H., and Stetter, K.O. (2000) Towards the ecology of hyperthermophiles: biotopes, new isolation strategies and novel metabolic properties. *FEMS Microbiol. Reviews* **24**, 615–623.

Hugenholtz, P. and Pace, N.R. (1996) Identifying microbial diversity in the natural environment: a molecular phylogenetic approach. *TIBTECH* **14**, 190–197.

Hunt, L.E. and Howard, A.G. (1994) Arsenic speciation and distribution in the Carnon estuary following the acute discharge of contaminated water from a disused mine. *Marine Pollution Bulletin* **28**, 33–38.

International Energy Agency (IEA) (2002) Renewables in global energy supply. Paris, France. (**www.iea.org**)

Jahren, S.J., Rintala, J.A., and Odegaard, H. (2002) Aerobic moving bed biofilm reactor treating thermomechanical pulping whitewater under thermophilic conditions. *Water Research* **36**, 1067–1075.

Jain, S.M. and DeKlerk, G.-J. (1998) Somaclonal variation in breeding and propagation of ornamental crops. *Plant Tissue & Biotechnol.* **4**, 63–75.

Jansson, J.K. (1995) Tracking genetically engineered microorganisms in nature. *Current Opinion in Biotechnology* **6**, 275–283.

Jaspars, M. (1999) Testing the water. *Chemistry & Industry* 18 Jan., 51–55.

Jetten, M.S.M., Wagner, M., Fuerst, J., van Loosdrecht, M., Kuenen, G., and Strous, M. (2002) Microbiology and application of the anaerobic ammonium oxidation (anammox) process. *Current Opinion in Biotechnology* **12**, 283–288.

Jin, S., Gorfajn, B., Faircloth, G., and Scotto, K.W. (2000) Ecteinascidin 743, a transcription-targeted chemotherapeutic that inhibits MDRI activation. *Proc. Natl. Acad. Sci.* **97**, 6775–6779.

Johnson, D.B. (1995) Acidophilic microbial communities: Candidates for bioremediation of acidic mine effluents. *International Biodeterioration Biodegradation* 41–58.

Johnson, D.B. (1998) Biodiversity and ecology of acidophilic microorganisms. *FEMS Microbiology Ecology* **27**, 307–317.

Johnson, D.B. and Hallberg, K.B. (2003) The microbiology of acid mine waters. *Research in Microbiology* **154**, 466–473.

Jones, D.T. and Keis, S. (1995) Origins and relationships of industrial solvent-producing clostridial strains. *FEMS Microbiology Reviews* **17**, 223–232.

Kamal, M., Ghaly, A.E., Mahmoud, N., and Cote, R. (2004) Phytoaccumulation of heavy metals by aquatic plants. *Environment International* **29**, 1029–1039.

Kareiva, P. and Stark, J. (1994) Environmental risks in agricultural biotechnology. *Chemistry & Industry*, January, 52–55.

Karp, A. (1995) Somaclonal variation as a tool for crop improvement. *Euphytica* **85**, 295–302.

Karp, A., Edwards, K.J., Bruford, M., Funk, S., Vosman, B., Morgante, M., Seberg, O., Kremer, A., Boursot, P., Arctander, P., Tautz, D., and Hewitt, G.M. (1997) Molecular technologies for biodiversity evaluation: Opportunities and challenges. *Nature Biotechnology* **15**, 625–628.

Keith, L.H. (1988) (ed) *Principles of Environmental Sampling*, American Chemical Society, Washington.

Kemp, D.L. and Quickenden. J. (1989) Whey processing for profit – a worthy alternative. In R.Greenshields (ed) *Resources and Applications of Biotechnology: The New Wave.* Macmillan, London, UK.

Kendall, J.M. and Badminton, M.N. (1998) *Aequorea victoria* bioluminescence moves into an exciting new era. *TIBTECH*, **16**, 216–224.

Kennedy, M.J. (1991) The Evolution of the Word 'Biotechnology'. *Trends in Biotechnology* **9**, 218–220.

Kerk, N.M., Ceserani, T., Tausta, S.L., Sussex, I.M., and Nelson, T.M. (2003) Laser capture microdissection of cells from plant tissues. *Plant Physiol.* **132**, 27–35.

Ketkar, K.W. (2002) The oil pollution act of 1990: A decade later. *Spill Science Technology Bulletin* 7,45–52.

Kikkert, J.R. (1993) The biolistic PDS-1000/He device. *Plant Cell Tissue Org. Cult.* **33**, 221–226.

Kilbane, J.J. (1989) Desulphurisation of coal: the microbial solution. *TIBTECH* 7, 97–101.

Kim, B.S. (2000) Production of poly(3-hydroxybutyrate) from inexpensive substrates. *Enzyme Microb. Technol.* **27**, 774–777.

Kingston, P.F. (2002) Long-term environmental impact of oil spills. *Spill Science & Technol. Bulletin* 7, 53–60.

Kirk, O., Borchert, T.V., and Fugisang, C.C. (2002) Industrial enzyme applications. *Current Opinion Biotechnology* **13**, 345–351.

Kirschbaum, M.U.F. (2003) To sink or burn? A discussion of the potential contributions of forests to greenhouse gas balances through storing carbon or providing biofuels. *Biomass & Bioenergy* **24**, 297–330.

Kobayashi, N., Horikoshi, T., Katsuyama, H., Handa, T., and Takayanagi, K. (1998) A simple and efficient DNA extraction method for plants especially woody plants. *Plant Tissue Cult. & Biotechnol.* **4**, 76–80.

Kok, E.J. and Kuiper, H.A. (2003) Comparative safety assessment for biotech crops. *Trends in Biotechnology* **21**, 439–444.

Kolmart, A., Henrysson, T., Hallberg, R., and Mattiasson, B. (1997) Optimization of sulphide production in an anaerobic continuous biofilm process with sulphate reducing bacteria. *Biotechnology Letters* **19**, 971–975.

Kononov, M.E., Bassuner, B., and Gelvin, S.B. (1997) Integration of T-DNA binary vector "backbone" sequences into the tobacco genome: Evidence for multiple complex patterns of integration. *Plant Journal* **11**, 945–957.

Kosaric, N. and Velikonja, J. (1995) Liquid and gaseous fuels from biotechnology: challenge and opportunities. *FEMS Microbiol. Rev.* **16**, 111–142.

Koutinas, A.A., Wang, R., and Webb, C. (2003) Estimation of fungal growth in complex, heterogeneous culture. *Biochem. Eng. J.* **14**, 93–100.

Krahe, M., Antranikian, G., and Markl, H. (1996) Fermentation of extremophile microorganisms. *FEMS Microbiol. Reviews* **18**, 271–285.

Kratochvil, D. and Volesky, B. (1998) Advances in the biosorption of heavy metals. *TIBTECH* **16**, 291–300.

Krebs, W., Brombacher, C., Bosshard, P.P., Bachofen, R., and Brandl, H. (1997) Microbial recovery of metals from solids. *FEMS Microbiol. Reviews* **20**, 605–617.

Kyambadde, J., Kansiime, F., Gumaelius, L., and Dalhammar, G. (2004) A comparative study of *Cyperus papyrus* and *Miscanthidium violaceum*-based constructed wetlands for wastewater treatment in a tropical climate. *Water Research* **38**, 475–485.

Laforgia, D. and Ardito, V. (1995) Biodiesel fueled IDI engines: performances, emissions and heat release investigation. *Bioresource Technology* **51**, 53–59.

Langwaldt, J.H. and Puhakka, J.A. (2000) On-site biological remediation of contaminated groundwater: a review. *Environmental Pollution* **107**, 187–197.

Larson, E.D., Worrell, E., and Chen, J.S. (1996) Clean fuels from municipal solid waste for fuel cell buses in metropolitan areas. *Resources Conservation & Recycling* **17**, 273–298.

Lassen, J., Madsen, K.H., and Sandee, P. (2002) Ethics and genetic engineering – lessons to be learned from GM foods. *Bioprocess Biosystems Engineering* **24**, 263–271.

Laws, E.A. (2000) *Aquatic Pollution: An Introductory Text.* John Wiley & Sons, Inc., New York, USA.

Le Borgne, S. and Quintero, R. (2003) Biotechnological processes for the refining of petroleum. *Fuel Processing Technology* **81**, 155–169.

Ledin, M. and Pedersen, K. (1996) The environmental impact of mine wastes – role of microorganisms and their significance in treatment of mine wastes. *Earth-Science Rev.* **41**: 67–108.

Lee, C.M. and Gongaware, D.F. (1997) Optimization of SFE conditions for the removal of diesel fuel. *Environmental Technology* **18**, 1157–1161.

Lee, E.Y., Ye, B.D., and Park, S. (2003) Development and operation of a trickling biofilter system for continuous treatment of gas-phase trichloroethylene. *Biotechnology Letters* **25**, 1757–1761.

Lee, R.F. and Page, D.S. (1997) Petroleum hydrocarbons and their effects in subtidal regions after major oil spills. *Marine Pollution Bulletin* **34**, 928–940.

Lee, S.Y. (1996a) Plastic bacteria? Progress and prospects for polyhydroxyalkanoate production in bacteria. *TIBTECH* **14**, 431–438.

Lee, S.Y. (1996b) Bacterial polyhydroxyalkanoates. *Biotechnol. Bioeng.* **49**, 1–14.

Lee, Y.D., Shin, E.B., Choi, Y.S., Yoon, H.S., Lee, H.S., Chung, L.J., and Na, J.S. (1997) Biological removal of nitrogen and phosphorus from wastewater by a single sludge reactor. *Environ. Tech.* **18**, 975–986.

Lei, X.G. and Stahl, C.H. (2001) Biotechnological development of effective phytases for mineral nutrition and environmental protection. *Appl. Microbiol. Biotechnol.* **57**, 474–481.

Lestan, D. and Lamar, R.T. (1996) Development of fungal inocula for bioaugmentation of contaminated soils. *Appl. Environ. Micro.* **62**, 2045–2052.

Lewis, M.A. (1995) Use of freshwater plants for phytotoxicity testing: a review. *Environmental Pollution* **87**, 319–336.

Liebier, D.C. (2002) Proteomic approaches to characterize protein modifications: new tools to study the effects of environmental exposures. *Environmental Health Perspectives* **110**, 3–9.

Liu, Y. and Tay, J.-H. (2001) Strategy for minimization of excess sludge production from the activated sludge process. *Biotechnology Advances* **19**, 97–107.

Louwrier, A. (1998) Biodiesel: tomorrow's gold. *Biologist*, **45**, 1721.

Ma, F. and Hanna, M.A. (1999) Biodiesel production: a review. *Bioresource Technology* **70**, 1–15.

Machacon, H.T.C., Shiga, S., Karasawa, T., and Nakamura, H. (2001) Performance and emission characteristics of a diesel engine fueled with coconut oil-diesel fuel blend. *Biomass & Bioenergy* **20**, 63–69.

Macnaughton, S.J., Stephen, J.R., Venosa, A.D., Davis, G.A., Chang, Y.-J., and White, D.C. (1999) Microbial population changes during bioremediation of an experimental oil spill. *Appl. Environ. Microbiol.* **65**, 3566–3574.

Madison, L.L. and Huisman, G.W. (1999) Metabolic engineering of poly (3-hydroxyalkanoates): from DNA to plastic. *Microbiology Molecular Biology Rev.* **63**, 21–53.

Maki, H., Hirayama, N., Hiwatari, T., Kohata, K., Uchiyama, H., Watanabe, M., Yamasaki, F., and Furuki, M. (2003) Crude oil bioremedition field experiment in the Sea of Japan. *Marine Pollution Bulletin* **47**, 74–77.

Makova, M., Macek, T., Kucerova, P., Burkhard, J., Pazlarova, J., and Demnerova, K. (1997) Degradation of polychlorinated biphenyls by hairy root culture of *Solanun nigrum*. *Biotechnology Letters*, **19**, 787–790.

Margesin, R. and Schinner, F. (1997) Bioremediation of diesel-oil-contaminated alpine soils at low temperatures. *Appl. Microbiol. Biotechnol.* **47**, 462–468.

Markx, G.H. and Davey, C.L. (1999) The dielectric properties of biological cells at radiofrequencies: Applications in biotechnology. *Enzyme Microb.Technol.* **25**, 161–171.

Marschner, H. (1995) *Mineral Nutrition of Higher Plants*. Academic Press, London.

Marsden, A.D., DeWreede, R.E., and Levings, C.D. (2003) Survivorship and growth of *Fucus gardneri* after transplant to an acid mine drainage-polluted area. *Marine Pollution Bulletin* **46**, 65–73.

Mason, C.F. (1996) *Biology of Freshwater Pollution,* 3rd edition, Longman Scientific and Technical, Harlow.

McCabe, D. and Christou, P. (1993) Direct DNA transfer using electric discharge particle acceleration (ACCELL™ technology). *Plant Cell Tissue Org.Cult.* **33**, 227–236.

McCann, J.H., Greenberg, B.M., and Solomon, K.R. (2000) The effect of creosote on the growth of an axenic culture of *Myriophyllum spicatum* L. *Aquatic Toxicology* **50**, 265–274.

McEldowney, J.F. and McEldowney, S. (1996) *Environment and the Law.* Addison Wesley Longman, Harlow, UK.

McEldowney, S., Hardman, D.J., and Waite, S. (1993) *Pollution: Ecology and Biotreatment.* Longman Scientific & Technical, Harlow, UK.

McFarland, B.L. (1999) Biodesulfurization. *Current Opinion Biotechnology* **2**, 257–264.

McFarland, G.R., Booth, D.J., and Brown, K.R. (2000) The semaphore crab, *Heleocius cardiformis*: bio-indication potential for heavy metals in esturine systems. *Aquatic Toxicology* **50**, 153–166.

McGrath, S.P. and Zhao, F.-J. (2002) Phytoextraction of metals and metalloids from contaminated soils. *Current Opinion in Biotechnology* **14**, 277–282.

McNeil, B. and Harvey, L.M. (1993) Viscous fermentation products. *Critical Reviews in Biotechnology* **13**, 275–303.

McNevin, D. and Barford, J. (2000) Biofiltration as an odour abatement strategy. *Biochemical Engineering J.* **5**, 231–242.

Meagher, R.B. (2000) Phytoremediation of toxic elemental and organic pollutants. *Current Opinion in Plant Biology* **3**, 153–162.

Mesarch, M.B. and Nies, L. (1997) Modification of heterotrophic plate counts for assessing the bioremediation potential of petroleum-contaminated soils. *Environ. Tech.* **18**, 639–646.

Mihelcic, J.R., Lueking, D.R., Mitzell, R.J., and Stapleton, J.M. (1993) Bioavailibility of sorbed- and separate-phase chemicals, *Biodegradation*, **4**, 141–153.

Miki, B. and McHugh, S. (2004) Selectable marker genes in transgenic plants: applications, alternatives and biosafety. *J. Biotechnology* **107**, 193–232.

Milcic-Terzic, J., Lopez-Vidal, Y., Vrvic, M.M., and Saval, S. (2001) Detection of catabolic genes in indigenous microbial consortia isolated from a diesel-contaminated soil. *Bioresource Technol.* **78**, 47–54.

Millington, A., Price, D., and Hughes, R. (1994) In situ combustion for oil recovery. *Chemistry & Industry*, August, 632–635.

Misteli, T. and Spector, D.L. (1997) Applications of the green fluorescent protein in cell biology and biotechnology. *Nature Biotechnology* **15**, 961–964.

Monticello, D.J. (2000) Biosulfurization and the upgrading of petroleum distillates. *Current Opinion Biotechnology* **11**, 540–546.

Morikawa, H. and Erkin, O.C. (2003) Basic processes in phytoremediation and some applications to air pollution control. *Chemosphere* **52**, 1553–1558.

Moses, V. (1991) Oil production and processing. In *Biotechnology The Science and Business*, (ed) Moses, V. and Cape, R.E. pp 537–565, Harwood Academic Press, UK.

Moses, V. (2002) Agricultural biotechnology and the UK public. *Trends in Biotechnology* **20**, 402–404.

Moshiri, G.A. (1993) *Constructed Wetlands for Water Quality Improvement.* CRC Press, Boca Raton, Fl., USA.

Mourges, F., Brisset, M.N., and Cheveau, E. (1998) Strategies to improve plant resistance to bacterial diseases through genetic engineering. *TIBTECH* **16**, 203–209.

Murphy, D.J. (1996) Engineering oil production in rapeseed and other oil crops. *TIBTECH* **14**, 206–213.

Namkoong, W., Hwang, E.-Y., Park, J.-S., and Choi, J.-Y. (2002) Bioremediation of diesel-contaminated soil with composting. *Environmental Pollution* **119**, 23–33.

Ndon, U.J. and Dague, R.R. (1997) Ambient temperature treatment of low strength wastewater using anaerobic sequencing batch reactor. *Biotechnol. Letters* **19**, 319–323.

Neves, A.A., Pereira, D.A., Vieira, L.M., and Menezes, J.C. (2000) Real time monitoring biomass concentration in *Streptomyces clavuligerus* cultivations with industrial media using a capacitance probe. *J. Biotechnology* **84**, 45–52.

Nickel, M., Leininger, S., Proll, G., and Brummer, F. (2001) Comparative studies on two potential methods for the biotechnological production of sponge biomass. *J. Biotechnology* **92**, 109–178.

Nys, R. and Steinberg, P.D. (2002) Linking marine biology and biotechnology. *Current Opinion in Biotechnology* **13**, 244–248.

Oard, J. (1993) Development of an airgun device for particle bombardment. *Plant Cell Tissue Org. Cult.* **33**, 247–250.

Ogram, A., Sayler, G.S., and Barkay, T. (1987) The extraction and purification of microbial DNA from sediments. *J. Microbiology Methods* **7**, 57–66.

Olguin, E.J. (2003) Phycoremediation: key issues for cost-effective nutrient removal processes. *Biotechnology Advances* **22**, 83–93.

Onianwa, P.C. (2001) Monitoring atmospheric metal pollution: a review of the use of mosses as indicators. *Environmental Monitoring & Assessment* **71**, 13–50.

Osinga, R., Tramper, J., and Wijffels, R.H., (1998) Cultivation of marine sponges for metabolite production applications for biotechnology? *TIBTECH* **16**, 130–134.

Paitan, Y., Biran, D., Biran, I., Shechter, N., Babai, R., Rishpon, J., and Ron, E.Z. (2003) On-line and in situ biosensors for monitoring environemnetal pollution. *Biotechnology Advances* **22**, 27–33.

Papanikolaou, S., Ruiz-Sanchez, P., Pariset, B., Blanchard, F., and Fick, M. (2000) High production of 1,3-propanediol from industrial glycerol by newly isolated *Clostridium butyricum* strain. *J. Biotechnology* **77**, 191–208.

Patrzykat, A. and Douglas, S.E. (2003) Gone gene fishing: how to catch novel marine antimicrobials. *Trends in Biotechnology* **21**, 362–369.

Pawlik-Skowronska, B. (2001) Phytochelatin production in freshwater algae *Stigeoclonium* in response to heavy metals contained in mining water; effects of some environmental factors. *Aquatic Toxicology* **52**, 241–249.

Pehr, K., Sauermann, P., Traeger, O., and Bracha, M. (2001) Liquid hydrogen for motor vehicles – the world's first public LH_2 filling station. *Hydrogen Energy* **26**, 777–782.

Pereira, M.G. and Mudge, S.M. (2004) Cleaning oiled shores: laboratory experiments testing the potential use of vegetable oil biodiesels. *Chemosphere* **54**, 297–304.

Peterson, C.L., Reece, D.L., Thompson, J.C., Beck, S.M., and Chase, C. (1996) Ethyl ester of rapeseed used as a biodiesel fuel – a case study. *Biomass & Bioenergy* **10**, 331–336.

Phillips, P.W.B. (2002) Biotechnology in the global agri-food system. *Trends in Biotechnology* **20**, 376–381.

Piel, W.J. (2001) Transportation fuels of the future. *Fuel Processing Technology* **71**, 167–179.

Plat, J.Y., Sayag, D., and Andre, L. (1984) High-rate composting of wool industry waste. *Biocycle* **25**, 39–42.

Podila, G.K. and Karnosky, D.F. (1996) Fibre farms of the future: genetically engineered trees. *Chemistry & Industry* Dec. 976–981.

Poirier, Y., Nawrath, C., and Somerville, C. (1995) Production of polyhydroxyalkanoates, a family of biodegradable plastics and elastomers, in bacteria and plants. *Biotechnology* **13**, 142–150.

Polz, M.F. and Cavanaugh, C.M. (1997) A simple method for quantification of uncultured microorganisms in the environment based on in vitro transcription of the 16S rRNA. *Appl. Environ. Micro.* **63**, 1028–1033.

Pomponi, S.A. (1999) The bioprocess-technological potential of the sea. *J. Biotechnology* **70**, 5–13.

Potter, M. (1994) BST: science's own goal. *Chemistry & Industry*, October, 836.

Prenafeta-Boldu, F.X., Ballerstedt, H., Gerritse, J., and Grotenhuis, J.T.C. (2004) Bioremediation of BTEX hydrocarbons: Effect of soil inoculation with the toluene-growing fungus *Cladophialophora* sp. Strain T1. *Biodegradation* **15**, 59–65.

Prince, R.C. (1997) Bioremediation of marine oil spills. *TIBTECH* **19**, 158–159.

Proksch, P., Edrada, R.A., and Ebel, R. (2002) Drugs from the seas-current status and microbiological implications. *Appl. Microbiol. Biotechnol.* **59**, 125–134.

Pulford, I.D. and Watson, C. (2003) Phytoremediation of heavy metal contaminated land by trees – a review. *Environment International* **29**, 529–540.

Punja, Z.K. and Utkhede, R.S. (2003) Using fungi and yeasts to manage vegetable crop diseases. *Trends in Biotechnology* **21**, 400–407.

Quan, X., Shi, H., Liu, H., Lv, P., and Qian, Y. (2004) Enhancement of 2,4-dichlorophenol degradation in conventional activated sludge systems bioaugmented with mixed special culture. *Water Research* **38**, 245–253.

Ra, C.S., Lo, K.V., and Mavinic, D.S. (1998) Real-time control of two-stage sequencing batch reactor system for the treatment of animal wastewater. *Environ. Technol.* **19**, 343–356.

Rawlings, D.E. and Silver, S. (1995) Mining with microbes. *Biotechnology* **13**, 773–778.

Rawlings, D.E., Dew, D., and du Plessis, C. (2003) Biomineralization of metal-containing ores and concentrates. *Trends in Biotechnology* **21**, 38–44.

Reeve, R.N. (1994) *Environmental Analysis*. John Wiley & Sons, Chichester, UK.

Regioli, F. and Principato, G. (1995) Glutathione, glutathione-dependant and antioxidant enzymes in mussel, *Mytilus galloprovincialis*, exposed to metals under field and laboratory conditions: implications for the use of biochemical biomarkers. *Aquatic Toxicology* **31**, 143–164.

Ren, S. and Frymier, P.D. (2003) Kinetics of the toxicity of metals to luminescent bacteria. *Advances in Environmental Res.* **7**, 537–547.

Ren, X. (2003) Biodegradable plastics: a solution or a challenge? *J. Cleaner Production* **11**, 27–40.

Repetto, G., Jos, A., Hazen, M.J., Molero, M.L., Peso, A.D., Salguero, M., Castillo, P.D., Rodriguez-Vincente, M.C., and Repetto, M. (2001) A test battery for the ecotoxicological evaluation of pentachlorophenol. *Toxicology in Vitro* **15**, 503–509.

Rhee, S.K, Lee, J.J., and Lee, S.T (1997) Nitrite accumulation in a sequencing batch reactor during the aerobic phase of biological nitrogen removal. *Biotechnol. Letters*, **19**, 195–198.

Roberts, L.E.J., Liss, P.S., and Saunders, P.A.H. (1990) *Power Generation and the Environment*. Oxford University Press, Oxford.

Rodriguez, Y., Ballester, A., Blazquez, M.L., Gonzalez, F., and Munoz, J.A. (2003) New information on the pyrite bioleaching mechanism at low and high temperatures. *Hydrometallurgy* **71**, 37–46.

Romig, C. and Spataru, A. (1996) Emissions and engine performance from blends of soya and canola methyl esters with ARB\2 diesel in a DDC 6V92TA MUI engine. *Bioresource Technology* **56**, 25–34.

Rowley, A.G. (1993) Time to clean up the act? *Chemistry in Britain*, Nov., 959–963.

Rugh, C.L., Senecoff, J.F., Meagher, R.B., and Merkle, S.A. (1998) Development of transgenic yellow poplar for mercury phytoremediation. *Nature Biotechnology* **16**, 325–328.

Saano, A., Piipola, S., Linstrom, K., and Van Elsas, J.D. (1995) Extraction and analysis of microbial DNA from soil, in Trevors, J.T. and Van Elsas, J.D. (eds) *Nucleic Acids in the Environment: Methods and Applications*, pp 49–67, Springer-Verlag, Heidelberg.

Saida, H., Maekawa, T., Satake, T., Higashi, Y., and Seki, H. (2000) Gram stain index of a natural bacterial community at a nutrient gradient in the freshwater environment. *Environ.Pollution* **109**, 293–301.

Salem, S., Berends, D.H.J.G., Heijnen, J.J., and van Loosdrecht, M.C.M. (2003) Bioaugmentation by nitrification with return sludge. *Water Research* **37**, 1794–1804.

Sami, M., Annamalai, K., and Wooldridge, M. (2001) Co-firing of coal and biomass fuels blends. *Progress in Energy & Combustion Sci.* **27**, 171–214.

Sanford, J.C. (1988) The biolistic process. *Trends in Biotechnology* **6**, 299–302.

Sano, T., Nagayama, A., Ogawa, T., Ishida, I., and Okada, Y. (1997) Transgenic potato expressing a double-stranded RNA-specific ribonuclease is resistant to potato spindle tuber viroid. *Nature Biotechnol.* **15**, 1290–1294.

Schaefer, M. (2004) Assessing 2,4,6-trinitrotoluene (TNT) contaminated soil using three different earthworm test methods. *Ecotoxicology Environmental Safety* **57**, 74–80.

Schratzberger, M., Daniel, F., Wall, C.M., Kilbride, R., Macnaughton, S.J., Boyd, S.E., Rees, H.L., Lee, K., and Swannell, R.P.J. (2003) Response of estuarine meio- and macrofauna to in situ bioremediation of oil-contaminated sediment. *Marine Pollution Bulletin* **46**, 430–443.

Schroeder, E.D. (2002) Trends in application of gas-phase bioreactors. *Reviews in Environmental Science & Biotechnology* **1**, 65–74.

Scott, D.L., Ramanathan, S., Shi, W., Rosen, B.P., and Daunert, S. (1997) Genetically engineered bacteria: electrochemical sensing systems of antimonite and arsenite. *Anal Chem.*, **69**, 16–20.

Scott, S.E. and Wilkinson, M.J. (1999) Low probability of chloroplast movement from oilseed rape (*Brassica napus*) into wild *Brassica rapa*. *Nature Biotechnology* **17**, 390–393.

Scragg, A.H., Illman, A.M., Carden, A., and Shales, S.W. (2002) Growth of microalgae with increased calorific values in a tubular bioreactor. *Biomass Bioenergy* **23**, 67–73.

Scurlock, J.M.O., Hall, D.O., House, J.I., and Howes, R. (1993) Utilising biomass crops as an energy source: a European perspective. *Water, Air & Soil Pollution*, **70**, 499–518.

Semprini, L. (1997) Strategies for the aerobic co-metabolism of chlorinated solvents. *Current Opinion in Biotechnology* **8**, 296–308.

Seviour, R.J., Mino, T., and Onuki, M. (2003) The microbiology of biological phosphorus removal in activated sludge systems. *FEMS Microbiology Reviews* **27**, 99–127.

Shan, X., Wang, H., Zhang, S., Zhou, H., Zheng, Y., Yu, H., and Wen, B. (2003) Accumulation and uptake of light rare earth elements in a hyperaccumulator *Dicropteris dichotoma. Plant Science* **165**, 1343–1353.

Sharma, H.C., Crouch, J.H., Sharma, K.K., Seetharama, N., and Hash, C.T. (2001) Applications of biotechnology for crop improvement: prospects and constraints. *Plant Science* **163**, 381–395.

Shaw, I.C. and Chadwick, J. (1998) *Principles of Environmental Toxicology*. Taylor and Francis, London, UK.

Shaw, J.J., Dane, F., Geiger, D., and Kloepper, J.W. (1992) Use of bioluminescence for detection of genetically engineered microorganisms released into the environment. *Appl. Environ. Technol.* **58**, 267–273.

Shay, E.G. (1993) Diesel fuel from vegetable oils: status and opportunities. *Biomass & Bioenergy* **4**, 227–242.

Shim, J.K., Yoo, I.-K., and Lee, Y.M. (2002) Design and operation considerations for wastewater treatment using a flat submerged membrane bioreactor. *Process Biochemistry* **38**, 279–285.

Sidwick, J.M. and Holdom, R.S. (1987) *Biotechnology of Waste Treatment and Exploitation*, Ellis Horwood Ltd., Chichester.

Siemens, J. and Schieder, O. (1996) Transgenic plants: genetic transformation – recent developments and state of the art. *Plant Tissue Cult. & Biotechnol.* **2**, 66–75.

Simpson, C.L. and Stern, D.B. (2002) The treasure trove of algal chloroplast genomes. Surprises in architecture and gene content, and their functional implications. *Plant Physiol.* **129**, 957–966.

Singleton, P. (1999) *Bacteria in Biology, Biotechnology and Medicine*. John Wiley & Sons, New York.

Skulberg, O.M. (1964) Algal problems related to the eutrophication of European water supplies, and a bioassay method to assess fertilizing influences of pollution of inland water, in *Algae and Man*. Plenum Press, New York.

Slater, S., Mitsky, T.A., Houmiel, K.L., Hao, M., Reiser, S.E., Taylor, N.B., Tran, M., Valentin, H.E., Rodriguez, D.J., Stone, D.A., Padgette, S.R, Kishore, G., and Gruys, K.J. (1999) Metabolic engineering of *Arabidopsis* and *Brassica* for poly(3-hydroxybutyrate-co-3-hydroxyvalerate) coploymer production. *Nature Biotechnology* **17**, 1011–1016.

Smalla, K., Cresswell, N., Mendonca-Hagler, L.C., Wolters, A.C., and Van Elsas, J.D. (1993) Rapid DNA extraction protocol from soil for polymerase chain reaction-mediated amplification. *J. Appl. Bacteriol.* **74**, 78–85.

Snaidr, J., Amann, R., Huber, I., Ludwig, W., and Schleifer, K.-H. (1997) Phylogenetic analysis and in situ identification of bacteria in activated sludge. *Appl. Environ. Microbiol.*, **63**, 2884–2896.

Snell, K.D. and Peoples, O.P. (2002) Polyhydroxyalkanoate polymers and their production in transgenic plants. *Metabolic Engineering* **4**, 29–40.

Somerfield, P.J., Gee, J.M., and Warwick, R.M. (1994) Benthic community structure in relation to a instantaneous discharge of waste water from a tin mine. *Marine Pollution Bulletin* **28**, 363–369.

Sorensen, A.H. and Ahring, B.K. (1997) An improved enzyme-linked immunosorbent assay for whole-cell determination of methanogens in samples from anaerobic reactors. *Appl. Environ. Microbiol.* **63**, 2001–2006.

Speece, R.E. (1983) Anaerobic biotechnology for industrial wastewater treatment. *Environ. Sci. Technol.* **17**,416–427.

Spinks, A. (1980) *Biotechnology*, Report of a Joint Working Party, London, HMSO.

Srivastava, A. and Prasad, R. (2000) Triglyceride-based diesel fuels. *Renewable & Sustainable Energy Reviews* **4**, 111–133.

Stafford, D.A. (1989) The anaerobic digestion of food processing waste, in Greenshields, R. (ed) *Resources and Applications of Biotechnology: The New Wave*, MacMillan Press, Basingstoke, pp 305–32.

Stams, A.J.M. and Oude Elferink, S.J.W.H. (1997) Understanding and advancing wastewater treatment. *Current Opinion in Biotechnology* **8**, 328–334.

Steen, H. and Pandley, A. (2002) Proteomics goes quantitative: measuring protein abundance. *Trends in Biotechnology* **20**, 361–364.

Stekoll, M.S. and Deysher, L. (2000) Response of the dominant alga *Fucus gardneri* (Silva) (Phaeophyceae) to the *Exxon Valdez* oil spill and clean-up. *Marine Pollution Bulletin* **40**, 1028–1041.

Stewart, C.N. (2001) The utility of green fluorescent protein in transgenic plants. *Plant Cell Reports* **20**, 376–382.

Sticher, P., Jaspers, M.C.M., Stemmler, K., Harms, H., Zehnder, A.J.B., and van de Meer, J.R. (1997) Development and characterization of a whole-cell bioluminescent sensor for bioavailable middle-chain alkanes in contaminated groundwater samples. *Appl. Environ. Microbiol.* **63**, 4053–4060.

Sul, I.-W. and Korban, S.S. (1996) A highly efficient method for isolating genomic DNA from plant tissues. *Plant Tissue Cult. & Biotechnol.* **2**, 114–117.

Sung, K., Munster, C.L., Rhyerd, R., Drew, M.C., and Corapcioglu, M.Y. (2003) The use of vegetation to remediate soil freshly contaminated by recalcitrant contaminants. *Water Research* **37**, 2408–2418.

Sutherland, I.W. (1998) Novel and established applications of microbial polysaccharides. *TIBTECH* **16**, 41–46.

Suzuki, I. (2001) Microbial leaching of metals from sulfide minerals. *Biotechnology Advances* **19**, 119–132.

Taherzadeh, M.J., Adler, L., and Liden, G. (2002) Strategies for enhanced fermentative production of glycerol – a review. *Enzyme Microb. Technol.* **31**, 53–66.

Tam, N.F.Y., Chong, A.M.Y., and Wong, Y.S. (2002) Removal of tributyltin by live and dead microalgal cells. *Marine Pollution Bulletin* **45**, 362–371.

Taylor, L.T. and Jones, D.M. (2001) Bioremediation of coal tar PAH in soils using biodiesel. *Chemosphere* **44**, 1131–1136.

Tchobanoglous, G. and Burton, H. (1991) *Wastewater Engineering*. 3rd edition. McGraw-Hill, New York.

Tebbutt, T.H.Y. (1998) *Principles of Water Quality Control*, 5th edition, Butterworth-Heineman, Oxford, UK.

Tellez, G.T., Nirmalakhandan, N., and Gardea-Torresdey, J.L. (2002) Performance evaluation of an activated sludge system for removing petroleum hydrocarbons from oilfield produced waste. *Advances in Environmental Research* **6**, 455–470.

Thurnbeer, T., Gmur, R., and Guggenheim, B. (2004) Multiplex FISH analysis of a six-species bacterial film. *J. Microbiology Methods* **56**, 37–47.

Tickell, J. (2000) *From the Fryer to the Fuel Tank*. Joshua Tickell Media Productions, Los Angeles, USA.

Top, E.M. and Springael, D. (2003) The role of mobile genetic elements in bacterial adaptation to xenobiotic organic compounds. *Current Opinion in Biotechnology* **14**, 262–269.

Top, E.M., Springael, D., and Boon, N. (2002) Catabolic mobile genetic elements and their potential use in bioaugmentation of polluted soils and waters. *FEMS Microbiology Ecology* **42**, 199–208.

Torsvik, V., Daae, F.L., and Goksoyr, J. (1995) Extraction, purification and analysis of DNA from soil bacteria in Trevors, J.T., and Van Elasa, J.D. (eds) *Nucleic Acid in the Environment: Methods and Applications*, pp 29–48, Springer-Verlag, Heidelberg.

Trick, H.N., Dinkins, R.D., Santarem, E.R., Di, R., Samoyolov, V., Meurer, C.A., Walker, D.R., Parrott, W.A., Finer, J.J., and Collins, G.B. (1997) Recent advances in soybean transformation. *Plant Tissue Cult. & Biotechnol.* 3, 9–26.

Troquet, J., Larroche, C., and Dussap, C.-G. (2003) Evidence for the occurrence of an oxygen limitation during soil bioremediation by solid state fermentation. *Biochemical Engineering J.* **13**, 103–112.

Tsai, Y.L. and Olson, B.H. (1991) Rapid method for direct extraction of DNA from soil and sediments. *Appl. Environ. Microbiol.* **57**, 1070–1074.

Tzanov, T., Calafell, M., Guebitz, G.M., and Cavaco-Paulo, A. (2001) Bio-preparation of cotton fabrics. *Enzyme Microb. Technol.* **29**, 357–362.

Utgikar, V., Chen, B.-Y., Tabak, H.H., Bishop, D.F., and Govind, R. (2000) Treatment of acid mine drainage: I. Equilibrium biosorption of zinc and copper on non-viable activated sludge. *International Biodeterioration Biodegradation* **46**, 19–28.

Van Hille, R.P., Boshoff, G.A., Rose, P.D., and Duncan, J.R. (1999) A continuous process for the biological treatment of heavy metal contaminated acid mine water. *Resources Conservation Recycling* **27**, 157–167.

Venosa, A.D. and Zhu, X. (2003) Biodegradation of crude oil contaminating marine shorelines and freshwater wetlands. *Spill Science & Technology Bulletin* **8**, 163–178.

Vilchez, C., Garbayo, I., Lobato, M.V., and Vega, J.M. (1997) Microalgae-mediated chemicals production and wastes removal. *Enzyme Microb. Technol.* **20**, 562–572.

Vinyard, G.L. (1996) A chemical and biological assessment of water quality impacts from acid mine drainage in a first order mountain stream, and a comparison of two bioassay techniques. *Environmental Toxicology* **17**, 273–283.

Vives-Rego, J., Lebaron, P., and Caron, G.N.-V. (2000) Current and future applications of flow cytometry in aquatic microbiology. *FEMS Microbiology Reviews* **24**, 429–448.

Volesky, B. and Holan, Z.R. (1995) Biosorption of heavy metals. *Biotechnol. Progress*, **11**, 235–250.

von Westernhagen, H., Dethlefsen, V., and Haarich, M. (2001) Can a pollution event be detected using a single biological effects monitoring method? *Marine Pollution Bulletin* **42**, 294–297.

Vymazal, J. (2002) The use of sub-surface constructed wetlands for wastewater treatment in the Czech Republic: 10 years experience. *Ecological Engineering* **18**, 633–646.

Wackett, L.P. (1996) Co-metabolism: Is the emperor wearing any clothes? *Current Opinion in Biotechnology* 7, 321–325.

Wang, C.C., Chang, C.W., Chu, C.P., Lee, D.J., Chang, B.-V., and Liao, C.S. (2003) Producing hydrogen from wastewater sludge by *Clostridium difermentans*. *J. Biotechnology*, **102**, 83–92.

Webster, G., Newberry, C.J., Fry, J.C., and Weightman, A.J. (2003) Assessment of bacterial community structure in the deep sub-seafloor biosphere by 16S rDNA-based techniques: a cautionary tale. *J. Microbiol. Methods* **55**, 155–164.

Weinbauer, M.G., Beckmann, C., and Hofle, M.G. (1998) Utility of green fluorescent nucleic acid dyes and aluminum oxide membrane filters for rapid epifluorescence enumeration of soil and sediment bacteria. *Appl. Environ. Microbiol.* **64**, 5000–5003.

Weiner, R.M. (1997) Biopolymers from marine prokaryotes. *TIBTECH* **15**, 390–394.

Wery, N., Cambon-Bonavita, M.-A., Lesongeur, F., and Barbier, G. (2002) Diversity of anaerobic heterotrophic thermophiles isolated from deep-sea hydrothermal vents of the Mid-Atlantic Ridge. *FEMS Microbiol. Ecology* **41**, 105–114.

Wheals, A.E., Basso, L.C., Alves, D.M.G., and Amorim, H.V. (1999) Fuel ethanol after 25 years. *TIBTECH* **17**, 482–487.

Williamson, A.M. and Badr, O. (1998) Assessing the viability of using rape methyl ester (RME) as an alternative to mineral diesel fuel for powering road vehicles in the UK. *Applied Energy* **59**, 187–214.

Willmut, I., Schnieke, A.E., McWhir, J., Kind, A.J., and Campbell, K.H.S. (1997) Viable offspring derived from fetal and adult mammalian cells. *Nature* **385**, 810–813.

Winpenny, J., Marrz, W., and Szewzyk, U. (2000) Heterogenicity in biofilms. *FEMS Microbiology Reviews* **24**, 661–671.

Woese, C., Kandler, O., and Wheelis, M.L. (1990) Towards a natural system of organisms: Proposal for the domains Archaea, Bacteria, and Eucarya. *Proc. Natl. Acad. Sci.* **87**, 4576–4579.

Wooley, C. (2002) The myth of the 'pristine environment': past human impacts in Prince William Sound and the northern gulf of Alaska. *Spill Science Technology Bulletin* 7, 89–104.

World Commission on Environment and Development (1987) *Our Common Future*. Oxford University.

World Energy Council (1994) *New renewable energy resources*, Kogan Press, London.

Wrobel, M., Zebrowski, J., and Szopa, J. (2004) Polyhydroxybutyrate synthesis in transgenic flax. *J. Biotechnology* **107**, 41–54.

Wuebbles, D.J., Jain, A., Edmonds, J., Harvey, D., and Hayhoe, K. (1999) Global change: state of the science. *Environmental Pollution* **100**, 57–86.

Xia, H.P. (2004) Ecological rehabilitation and phytoremediation with four grasses in shale mined land. *Chemosphere* **54**, 345–353.

Yakimov, M.M., Amro, M.M., Bock, M., Boseker, K., Fredrickson, H.L., Kessel, D.G., and Timmis, K.N. (1997) The potential of *Bacillus licheniformis* strains for *in situ* enhanced oil recovery. *J.Petroleum Science and Engineering* **18**, 147–160.

Zeikus, J.G., Vieille, C., and Savchenko, A. (1998) Thermozymes: biotechnology and structure-function relationships. *Appl. Microbiology Biotechnology* **57**, 179–183.

Zhang, W. (2003) Nanoscale iron particles for environmental remediation: an overview. *J.Nanoparticle Research* 5, 323–332.

Zimmerley S.R, Wilson D.G., and Pratter J.F. (1958) Cyclic leaching process employing iron oxidising bacteria. *US Patent No.* 2,829,964.

Index

A

Absidia cylindrospora 193
Acanthella acuta 383
ACC (1-aminocyclopropane-1carboxylic acid synthase) 215
acetic acid (vinegar) 244
acetone/butanol/ethanol fermentation 239–43
acetylcholine esterase 108–9
Acetobacter spp 244
acetylCoA 55
Acidithiobacillus spp (was *Thiobacillus ferroxidans*) 410
Acidimicrobium ferroxidans 410
acid mine drainage 401–2, 408–15
  treatment 410–15
acid rain 267, 286–89
acidophile 410
Acinetobacter calsoaceticus 389
Acinetobacter radioresistens 191
Acinetobacter spp 328
Actinopyga agassizi 379
acrylamide synthesis 244
activated sludge 124–5, 135–7, 203
  aeration 134–9
  aeration Kessener 134
  aeration Pasverr ditch 139
  aeration Simcar 134
  disposal composting 161–4
  disposal incineration 160–1
  disposal landfill 155–60
  operation 138–9
adhesives biological 390
Aequorea Victoria 99, 357, 385
agar agarose 386–8
Agent Orange 178
Agrobacterium spp 247, 255, 330, 360–2
  border sequences 361
  co-cultivation 360
  vacuum infiltration
Agrobacterium radiobacter 394
airlift bioreactor (see bioreactors)
Alcaligenes eutrophus (now known as *Ralstonia eutropha*) 249, 251–3
Alcaligenes latus 251
Alcaligenes xylosoxidans 224
Alcohol 310–11
alder (*Alnus)* 210
algae (see macro and microalgae)
AlgaSORB 218
alginate 386–9
alkaliphile 54
Alyssum bertoloni 206
Alyssum lesbiacum 206
Alyssum sp 211
Ames test (see toxicity testing)
ammonia removal
  Anammox 153
  CANNON 153
  SHARON 153
Amoco Cadiz 2, 74, 398
Anabena cyindrica 318
anaerobic digestion
  fixed-bed reactor 169
  fluidised-bed 169
  process 165–72
  upflow anaerobic sludge blanket (USAB) 168
Analytical Profile Index (API) 71
anther culture 351–2
antifreeze protein 391
antisense RNA 370
Aplysina aerophoba 384
Aplysina cavernicola 384
Arabdopsis halleri 208
Arabdopsis spp 216
Arabidopsis thaliana 216, 252
Archaea (Archaebacteria) 18, 21, 262, 328
Aroclor 1242 188
Artemia salina 101
Arthrobacter oxydans 224
Arthrobacter spp 227
artificial wetlands 212–15, 416–18
aryl hydrocarbon hydroxylase (AHH) 97
Aspergillus niger 258, 313
Aspergillus spp 244
astaxanthin 393
atomic adsorption spectrophotometry 86
atrazine 42, 167–7, 227
Azolla pinnata 209
Azotobacter vinelandii 251

B

Bacillus licheniformis 328
Bacillus megaterium 248
Bacillus spp 328
Bacillus subtilis 238, 255
Bacillus thuringiensis 367
  Bacillus thuringiensis toxin (BT toxin) 367
BacLight™ (see viable count)
bacteriophage (see vector)
Bacteroides spp 166
bagasse 311–12, 314
Betula (birch) 210
benzene, toluene, ethylbenzene, xylene (BTEX) 179–82, 224
Berkheya coddii 206
best available technology not entailing excessive cost (BATNEEC) 9
binary fission 26, 61
bioaugmentation 148, 192–93, 198, 407
bioconcentration factor 6, 176, 206
biodegradable plastics 9, 248–53
biodiesel 192, 301, 306–7, 399
biodiversity 376
biofilm 38, 66
biofilter
  using activated sludge 223–24
  bioscrubber 220–23
  membrane 223
  trickling 223
biofuel
  biodiesel (see biodiesel)
  biogas 298–301
  biomass 293–8
  co-firing 295
  ethanol (see ethanol)
  gasification 295
  hydrogen (see hydrogen)
  pyrolysis 295
bioindicators 5, 94–6
bioleaching
  bioreactor 336, 340–2
  Biox® 340
  copper 341
  direct 331, 333–34, 409
  dump system 336–40
  gold 336, 342–43
  indirect 331–33, 409
  in situ 336
  uranium 341–42
biolistics 358–9
biological oxygen demand (BOD_5) 91–5, 106, 115–16, 120–21, 132–39, 166, 170–71
  BOD sag 113
  BOD sensor 106–10
biomagnification 174, 178
biomarker 5, 94, 96–7
bio-oil 296
biopharming 10, 371, 373
biopile 199–200
bioplastics (see microbial plastics)
Biopol™ (see microbial plastics)
biopolymers (see microbial polymers)
bioreactor
  activated sludge 124–5
  airlift 142–44
  biofilter 220–24
  bioscrubber 220–23
  cell recycle 125–30
  corrugated raceway 215
  deep shaft 143–44
  fluidized bed 141–42, 202, 218
  fixed-bed 202, 218
  membrane 146–47
  moving-bed biofilm 143
  trickling filter 121–24
  stirred tank 50
  rotating disk 218
  rotating drum 202
  upflow anaerobic sludge blanket (USAB) 168, 219–20
  upflow fixed film 169
Bioreactor operation 49–51
  continuous 127–30
Bioremediation
  bioaugmentation 148, 192–93, 198, 407
  Ex situ 195–98
  gaseous 220–24
  genetic manipulated organisms (GMOs) 194–95
  In situ 195–98
  marine oil spills 396–400
  metals 216–20
biosensor 102–10
  BOD sensor 106–10
biosorption 216–18
biosparging 197–8
biosphere 271–2
biostimulation 199
biosurfactants 190–1, 245–46
biotower 140
Biotox™ (see toxicity testing)
bioventing 197
birch (*Betula* spp)
blooms (see eutrophication)
Bophal 2
Botyrococcus braunii 303
bovine somatotrophin (BST) 373–74
bovine spongiform encephalopathy (BSE) 11
Braer 397–98
Brassica campestris 215
Brassica napus (rapeseed) 252, 363
Brassica rapa 363
Brevibacterium spp 258
British Thermal Unit (BTU) 266

Brocadia anammoxidans 153
Brundtland Commission 230
budding (yeast) 25–6
Bugula neritina 380, 383
bulrush (*Scirpus* spp) 417
Burkholderia cepacia 224

C
Caenorhabditis elegans 102
callus culture (see plant cell culture)
Calotropis procera 302
Candida spp 244
CANNON (see ammonia removal)
capacitance 51
capillary number 323–24
capsule 23, 245
Captor process 145–46
carbon cycle 273, 291
Cardaminopsis halleri 206
carrageenan 385–89
catail (*Typha* spp) 212, 417
Catharanthus roseus 376
Caulobacter spp 22, 67
Ceramium strictum 101
Ceramium tenuicorne 101
cetane number 301
Champia parvula 101
chemical floods (see enhanced oil recovery)
chemical oxygen demand (COD) 86–8, 92, 166
chemoheterotrophs 55, 57–8
chemolithotrophs (chemoautotrophs) 55, 57–8, 257, 331, 334, 336, 410
Chernobyl 2, 292
Chlamydomonas mexicana 393
Chlorella minuta 215
Chlorella sorokiniana 215
Chlorella vulgaris 101, 215
chloroflurocarbons (CFCs) 12, 275, 280–1, 284
chloroplast 21–2
citric acid production 244
Cladophialophora spp 193
Clark oxygen electrode 106
Clavibacter spp 227
Clean Air Act (UK) 289
clean technology 4, 9, 253–54
Clostridium acetobutylicum 239, 241
Clostridium bifermentans 319
Clostridium butylicum 243, 319
Clostridium saccharobutylicum 241
Clostridium spp 166, 243, 327–8
climate forcing 271
coenocytic 24
combined heat and power (CHP) 262, 264
co-metabolism 188, 190
common reed (*Phragmites* spp) 212, 417
complementary DNA (cDNA) 88–90
composting 161–64, 199
confocal scanning laser microscope (CSLM) 38, 66
coning (in oil extraction) 322, 329
continuous culture 127–29
Copaifera multijuga (coconut) 302–3
corrugated raceway 215
Coulter Counter 29–32, 88
crude oil composition 179–81, 395
cryosphere 271–2
Cryptocodinium cohmii 380
curdlan 247, 325–27, 33
Cyperus papyrus (papyrus) 214
cytochrome P-450 97, 100, 211, 216, 224

D
DAPI (4,6-diamidino-2-phenylindole) 29, 37
decline phase 62
deep shaft (see bioreactors)
dehalogenation 225–27
denaturing gradient gel electophoresis (DGGE) 46–7, 89, 192
denitrification 149–54, 276–78
denitrification of oil 258–59
Desulfotomaculum spp 218
Desulphovibrio spp 166, 218, 328, 338
desulphurisation of coal and oil 254–58
detergent biological 260–61
dextran 246–47
dibenzothiophene (DBT) 255
dichlorodiphenyltrichloroethane (DDT) 95
2,4-dichlorophenyloxyacetic acid (2,4-D) 176, 193, 201
Dicropteris dichotoma 209
diesel 301, 304–7
direct leaching (see bioleaching)
DNA
  extraction 41–2
  microarrays 91
  polymerase Taq 43
Dolabella auricularia 380
doubling time 62–3
drilling muds 245–46, 330
dump system (see bioleaching)
Dunaliella bardawil 393
Dunaliella salina 393
Dunaliella tertiolecta 238

E
Eckonia radiate 218
Ecteinascidia turbinate 380, 383
Eichhornia crassipes (water hyacinth) 210
electroporation 357, 360
Eloidia canadensis 101
Embden-Meyerhof pathway (glycolysis) 57, 308

embryogenesis (see plant cell culture)
Energy Information Administration (EIA) 266–67
enhanced oil recovery (EOR) 323–27
  chemical floods 324–26
  thermal 326–27
Enterobacter aerogenes 319
Enterobacter agglomerans 193
Enterotube Multitest System 71
Entner-Doudoroff pathway 309
Environment Protection Act 1990 183, 395
Environmental Protection Agency (EPA) 6, 75
enzyme-linked immunoadsorbent assay (ELISA) 38–9
epifluorescent microscope 29, 36
EPSP (3-enolpyruvylshikimic acid-5-phosphate) 366
Epulopiscium fishelsoni 22
Erwinia carotovora atroseptica 368
Eschrichia coli (*E.coli*) 10, 17, 55, 102, 191, 193, 218, 252, 362, 374–75, 384, 391
ethanol 237, 307–16
  in Brazil 311–12
  from cellulose 314–16
  improvements 314
  simultaneous saccharification and fermentation (SSF) 316
  in the USA 312–13
ethoxyresorufin-o-dethylase (EROD) 97
Eubacterium spp 166
Eucalptus globus 302
Euphorbia lathyris 302
eutrophication (blooms) 66, 153
Evans blue 36
explant (see plant cell culture)
exponential phase (log) 62–5, 130
extracellular polysaccharides (see microbial polymers)
extreme thermophiles 53–4, 392
extremophiles 53–4, 259, 262–63
extremozymes 262
Exxon Baton Rouge 403
Exxon Valdez 2, 183, 397–98, 401–8

F
faculative pond 119–21
fermentation 58, 308, 313–14
fermenter (see bioreactor)
fimbriae 24
fingering in oil extraction 323
fixed bed reactor (see bioreactor)
flagella 24, 26, 71
Flavobacterium spp 227
Flavr Savr™ 10, 352, 369
flax (*Linum usitatissimum*) 232
flow cell technology 66
flow cytometer 29–32, 88
fluidised bed (see bioreactor)
fluorescein diacetate (FDA) 36, 88
fluorescein isothiocyanate (FITC) 32
fluorescent *in situ* hybridization (FISH) 46–7, 89
flurochrome (fluorescent dyes) 29
Food and Drugs Administration (FDA) 362
fumaric acid production 244
Fucus gardneri 404–5, 408

G
gas liquid chromatography (GLC) 86
gasification of biomass (see biomass)
gasohol 307
  E10 310
  E85 310
gene isolation plants 353
gene selection 356
genetic engineered microorganisms (GEMS) (see GMOs)
genetic engineered plants (see transformed plants)
genetic engineering (see recombinant DNA technology)
genetic manipulated organism (GMO) 9–11, 194–95, 352, 362–63
genomics 354–55
geothermal power 293
Gluconobacter spp 244
global warming 8, 231, 267
germplasm 376
glutathione 97
  conjugate 211, 215
glutathione-S-transferase 97
glycerol production 238–39, 307
glycolysis 57, 308
glyphosate 357, 366
gold 336, 342–43
Gordona spp 255
Gram stain 69–71
Granulobacter butylicus 239
Granulobacter saccharobutyricum 239
gratuitous metabolism 188
green fluorescent protein (Gfp) 357, 385
greenhouse gases 8, 273–75
growth curve 61–5
  hormone 391–92
  kinetics 62–5
  rate 62–5
gus (β-glucuronidase) 357

H
haemocytometer (Petroff-hauser chamber) 28
Haemotococcus pleuviatus 393
half-life enzyme 10

halophiles 52
haploid plants (see anther culture)
Heloecius cordiformis (semaphore crab) 95
Hemmthria sp (limpograss) 298
hemp (*Cannabis sativa*) 232
Hevea brasiliensis 302
hexidium iodide 36
high performance liquid chromatography (HPLC) 86
honeybees 95
hydraulic retention time (HRT) 132
hydroelectric power 292
hydrogen production for power 316–19
  biological production 317–19
hydrosphere 271–72
hydrothermal vents 44, 90, 378, 391–92
hyperaccumulator 206
hyperthermophiles (see extreme thermophiles)
hyphae 22, 24
Hyphomicrobium spp 224

I
immobilization (enzymes and cells) 103–5, 259, 307
in situ combustion (see enhanced oil extraction)
incineration 160
indirect leaching (see bioleaching)
International Convention for the Protection of Pollution from Ships (see MARPOL)
International Energy Agency (IEA) 266, 269

K
Kessener aerator (see aeration of activated sludge)
Klebsiella spp 255
Kluyveromyces fragilis 316
Kluyveromyces marxianus 316
Krebs cycle (citric acid cycle) 224, 227, 249
Kodama pathway 256, 259
Kuenenia stuttgartiensis 153
Kyoto protocol 12, 231

L
lactic acid 244
Lactobacillus delbrucckii 244
lag phase 61–2
lagoons 119
land farming 195
landfill 155–60, 248, 300
  gas 157–60
  leachates 158–59
Lemna minor 209
Lemna spp 101
Leptospirillum ferroxidans 336, 410
lichens 24, 94
Limanda limanda (dab) 95
limpograss (*Hemmthria)* 298
liposomes 357
Liriodendron tulipifera 216
Lissodendoryx spp 383
logarithmic phase 62–4
Lolium perenne 212
Ludwiigina palustris 209

M
macroalgae (seaweeds) 386–89
maize (corn, *Zea mays*) 233
marine oil spills 395–98, 401–2
  remediation 398, 405–8
marine products
  pharmaceuticals 380–3
  pollution 393
  polymers 385–90
  molecular biology 384–85
MARPOL 183, 248, 395
matrix acidising 329
melanin 390–1
membrane reactors (see bioreactor)
Mendel 345
Mentha aquatica 306
mesophiles 335–36
meta/ortho cleavage 224
metal recovery 330–43
metallothioneins 97, 206, 216–18
methane (biogas) 278, 281
Methanobacillus spp 166
Methanococcus spp 166
Methanosarcina spp 166
Methnobacterium spp 166, 328
Methylosinus trichosporium 193
methyl tertiary-butyl ether (MTBE) 211, 310
microaerophiles 54–5
microalgae 393
  oils 303
microbial enhanced oil recovery (MEOR) 327–30
  matrix acidizing 329
  profile improvement 329
microbial plastics 248–53
  Biopol™ 249
microbial polymers (polysaccharides) 245–47, 330
Micrococcus spp 328
microinjection 360
micromanipulation 18–19
micropropagation 347
Microtox™ (see toxicity testing)
Miles and Misra (plate method) 43, 88
Minamata 184
Ministry of Agriculture Food and Fisheries (MAFF) 294

Miscanthus spp 298
Miscanthidium violaceum 214
mitochondria 21–2
mixed liquor suspended solids (MLSS) 131–34, 203
mobility ratio 323–24
Monera 17
Monod equation 125–26, 129
Monsanto 249
Montreal Protocol 12, 284
Most Probable Number (MPN) 34–5
municipal solid waste (MSW) 299
mycelium 22, 24
Mycobacterium leprae (leprosy) 53
Mycobacterium spp 256
mycorrhizae (fungi) 26
Myriophyum aquaticum 209
Myriophyum spicatum 210
Mytilis edulis (mussels) 64–5
Mytilis galloprovincialis 95

N
napiergrass (*Pennisetum* spp) 2, 53
nanoscale iron particles 204
nettle (*Urtica dioica*) 233
Nicotiana glauca 215
nitrification 148–50
Nitrobacter spp 148–49
nitrogen dioxide (NO_2) 284, 289
Nitrosococcus spp 148
Nitrosocystis spp 148
Nitrosogloea spp 148
Nitrosomonas spp 148–49
Nitrosomonas eutropha 153
Nitrospira spp 148
nitrous oxides (NOx) 278, 280, 289, 290, 300
nitrous oxide (N_2O) 276–78
Nocardia globelula 255
non-fossil fuel obligation (NFFO) 294, 300
nuclear power 292

O
octane rating 309–10
Oil pollution Act (1990) 2, 401
oil pollution 294
oil recovery
  enhanced oil recovery (EOR) 323–24
  primary 321–22
  secondary 322–23
oil resevoir 321
o-nitrophenyl-β-galactopyranoside (ONPG) 102
optical tweezers 19
Organisation for Economic Co-operation and Development (OECD) 2, 253
organogenesis (see plant cell culture)
organotin 394
ortho cleavage (see meta cleavage)
oxygen electrode 86
ozone layer 274, 282–85

P
packed cell volume (PCV) 37, 88
Paenibacillus spp 256
Paspalum notatum 209
Pasverr ditch (aeration of activated sludge)
Pennisetum glaucum (napier grass) 209
pentachlorophenol (PCP) 193, 227
permeable reactive barrier 204, 414–15
petrochemical spill 182–83
Petrosia contignata 380
Petrosia ficiformis 383
Petunia hybrida 366
Phaedactylum tricornutum 101
phagotrophic 26
Phanerochaete chryosoporium 227
Phospholipids ester-linked fatty acids (PLFAME) 39
phospholipid fatty acids (PLFA) 39–40
photoautotrophic (photolithotrophic, photosynthetic) 55, 58–9
Photobacterium fisheri (was *Vibrio fisheri*) 101, 385
Photobacterium phosphoreum 101, 166
photoheterotroph 55, 59–61
Phragmites spp (common reed) 212, 417
phycoremediation 215
phylotype 18
phytoalexins 368
phytochelatin 97, 206
phytodegradation 211
phytoextraction (phytoaccumulation) 208–9
phytoremediation 204–15
phytostabilization 211
phytovolatilization 211
pili 24
Pittosporum resiniferum 302
Pittosporum tabira 216
plant cell culture
  callus 347–49
  embryogenesis 348–49
  explant 343
  organogenesis 347–49
  suspension 347–49
plant oils 301–3
  blending 304–5
  pyrolysis 306
  transesterification (see biodiesel)
plant transformation 357–62
  Agrobacterium 360–62
  biolistics 359
  electroporation 360

microinjection 360
silicone whiskers 360
plasmid (see vector)
plate count (see viable count)
Plexusa homomalla 379
pollution
chemical 114
nutritional 114
physical 114
ponds 119–21
Populus (poplar) 210, 298
polyalkanoates (PHA) 248–53, 389
polyaromatic hydrocarbons (PAH) 179–80, 211
polychlorobiphenyls (PCB, PCDD, PCDF) 95, 176
polyhydroxy butyrate (PHB) 248–53, 372
polymers (see microbial polymers)
polymerase chain reaction (PCR) 18, 42–6, 262, 343, 346, 353, 384
primary products 234–37
propidium iodide 29, 32, 36
proteomics 47–8, 91
Protomonas extorquens 261
Protista (protists) 17
protoplast 347, 349, 357
fusion 350–51
Pseudomonas aeruginosa 259
Pseudomonas oleovorans 99, 251
Pseudomonas putida 194, 224
Pseudomonas spp 194, 227, 244, 255–56, 328
Pseudomonas TG 232
pseudopodia 26
Pseudopterogorgia elisabethae (soft coral) 380
psychrophiles 53
Pteris vittata 206
pyrites (iron) 257, 290, 331, 341–42, 408
Pyrolobus fumarii 392
Pyrococcus furiosus 392
pyrolysis of biomass (see biomass)

Q
Quorn 26

R
Rabora® 374
Ralstonia eutrophia (was *Acaligenes eutrophus*) 249, 251–52
Raphiolepis umbellata 216
recombinant DNA (gene) technology 9–11, 40, 88–91, 352, 362–63
reed bed (see artificial wetlands)
restriction fragment polymorphism (RFLP) 46
Rhizobium spp 247, 330
rhizofiltration 212–14
Rhizopus spp 244
rhizosphere 212
rhizostimulation 214–15
Rhodobacter sphaeroides 319
Rhodococcus spp 227, 256, 258
Rhodococcus ECRD-1 255
Rhodocossus erythropolis 255
Rhodospirillum rubrum 319
ribosomal RNA (see small subunit)
ribosome 22
Rio de Janeiro 1990 230, 376
rotating biological reactor (RBC) (see bioreactor)
rotating drum bioreactor (see bioreactor)
RNA extraction 41–2, 89

S
simultaneous saccharification and fermentation (SSF) (see ethanol)
Saccharomyces cerevisiae 10, 17, 55, 238, 308, 313, 316
Saccharum offinarum (sugarcane) 311
Salix atrocinerea 212
Salix (willow) 211
Salmonella spp 88
Salmonella typhimurium 210
sand filter 220
scheroglucan 325–27, 330
Scirpus spp 417
Sclerotium glucanium 247, 330
Sea Empress 183
secondary products 234–37
*Selenastrum capriconutum (*now *Pseudokirchneriella subcapitata)* 100–1
Senedesmus dimorphus 215
Senedesmus platydiscus 215
Senedesmus subspicatus 101
sequencing batch reactor (SBR) 150–52
Seveso 74
Sewage composition 115
sewage treatment
Captor 145
oxygen addition 145
ponds 119
preliminary 117–19
primary 117–19
secondary 117–19
tertiary 117–19
Shilbottle coal mine 415
Silent Spring 2
Simcar aerator (see aeration of activated sludge)
SimPlate™ (see viable count)
sludge
age 131
disposal 154
loading rate (SLR) 132–34

sludge (*cont.*)
recycle 133
residence time (SRT) 131–34
settling index (SDI) 133
volume index (SVI) 133
small subunit ribosomal RNA (ssrRNA) 18, 42–6, 346
smog 288–89
solar energy 293
solids retention time (SRT) 203
somaclonal variation 247, 350
Southern blotting 44–5
Spinks Report 2
Spirulina spp 393, 412
spore bacterial 67–9, 71
stationary phase (see growth kinetics)
steam assisted gravity distillation (SAGD) 326
Stenotaphrum scundatium 209
stirred tank bioreactor (see bioreactor)
streptavidin 44
Streptomyces clavuligerus 51
Streptomyces spp 69
sugarcane (*Saccharum officinarum*) 311
Sulphobacillus spp 410
Sulpholobus acidocaldarius 257
Sulfolobus brierleyi 336
Sulfolobus spp 258
sulphate-reducing bacteria 336, 413–17
sulphur dioxide 285–88
surfactant (see biosurfactant)
suspended solids sewage 115
suspension cultures (see plant cell culture)
Svedberg units (S) 18
SYBR Green 1 37
sycamore (*Acer* spp)
syngas 296, 317
Syntrophobacter spp 166
Syntrophomonas spp 166

T
Taq polymerase 262, 284
Tar sands 327
tetrazolium 36
tidal power 292
temperature gradient gel electrophoresis (TGGE) 46, 89, 192
Thaspi caerulescens 206, 208
Thermococcus litoralis 384
thermophiles 53–4, 328, 336, 410
extreme 54, 328, 336, 410
Thermus aquaticus 262, 384
Thermus thermophilus 384
thief zones 323, 329
Thiobacillus ferrooxidans 257, 333, 336, 341
Thiobacillus spp 224, 330, 338, 348, 410
Thiobacillus TH-1 336
Thiobacillus thiooxidans 257, 336
thiophenes 255–58, 290
Thiospaera pantotropha 394
Three Mile Island 292
Ti plasmid 360–62
tidal power 292
tobacco mosaic virus (TMV) 368
total organic carbon (TOC) 86–8, 92
toxicity testing
Ames test 102
Biotox™ 101
Lumistox® 101
Microtox™ 101
Seed germination 101
transesterification (see biodiesel)
transformation of plants (see plant transformation)
transgenic animals 373–74
transgenic plants 352, 358, 364–66
biopharming 371
nutritional 368–70
pathogens 368
resistance to insects 366
safety 362–64
salinity, drought 370–71
tolerance to biocides 366
Treaty of Maastrict 12
trichloroethylene (TCE) 193, 211, 216
trickle-bed bioreactor (see bioreactor)
trickling filter (see bioreactors)
Trichoderma horzianum 343
Trichoderma reesei 315–16
Tridiemnum solidum 380
trinitrotoluene (TNT) 99, 190, 210–11
Trypan blue 36
Typha latifolia 212

U
United Nations Conference on Environment and Development (UNCED) 12
UNOX system
upflow anaerobic sludge blanket bioreactor (UASB) (see bioreactor)
upflow fixed film bioreactor (see bioreactor)
uranium extraction 341–42

V
vaccine 347, 374
subunit 374
vacuum infiltration (see plant transformation)
vector 193, 353, 355–56
Vetiveria zizaniodes 209
viable count 32, 88
BacLight™ 36, 88
Fluorescein diacetate (FDA) 36, 88
SimPlate™ 36, 88
Vibrio fisheri 99, 385

Vibrio harvegi 99
vinegar (see acetic acid)
volatile organic chemical (VOC) 8, 272

W
water activity 52
water flooding 325
wave power 292
Weizmann 239–40
wetlands (see artificial wetlands)
Wheal Jane 212, 402, 415–18
Whittaker taxonony 17
willow (*Salix* sp) 298
wind power 293
windrow composting (see compost)

X
xanthan 23, 245–47, 325–27, 330
Xanthomonas campestris 23, 99, 245, 330
Xanthomonas orzyae 368
Xanthomonas spp 224
X-gal (5-bromo-4-chloro-3-indolyl β-D-galactoside) 98

Y
yield coefficient 129–30

Z
Zeldovitch mechanism 278–79
Zeneca 249, 369–70
Zymomonas mobilis 309